W0018477

Mathematical Reasoning

Mathematical Reasoning

Patterns, Problems, Conjectures, and Proofs

Raymond S. Nickerson

Routledge
Taylor & Francis Group

LONDON AND NEW YORK

First published in 2010 by Psychology Press

Published 2016 by Routledge
2 Park Square, Milton Park, Abingdon, Oxfordshire OX14 4RN
711 Third Avenue, New York, NY 10017

First issued in paperback 2015

Routledge is an imprint of the Taylor and Francis Group, an informa business

© 2010 by Taylor and Francis Group, LLC

All rights reserved. No part of this book may be reprinted or reproduced or utilised in any form or by any electronic, mechanical, or other means, now known or hereafter invented, including photocopying and recording, or in any information storage or retrieval system, without permission in writing from the publishers.

Notice:
Product or corporate names may be trademarks or registered trademarks, and are used only for identification and explanation without intent to infringe.

ISBN 13: 978-1-138-98058-7 (pbk)
ISBN 13: 978-1-84872-827-1 (hbk)

Library of Congress Cataloging-in-Publication Data

Nickerson, Raymond S.
  Mathematical reasoning : patterns, problems, conjectures, and proofs/
Raymond Nickerson.
    p. cm.
  Includes bibliographical references and index.
  ISBN 978-1-84872-827-1
  1. Mathematical analysis. 2. Reasoning. 3. Logic, Symbolic and mathematical.
4. Problem solving. I. Title.

QA300.N468 2010
510.1'9--dc22   2009021567

Visit the Taylor & Francis Web site at
http://www.taylorandfrancis.com

CONTENTS

THE AUTHOR

Raymond S. Nickerson is a research professor at Tufts University, from which he received a PhD in experimental psychology, and is retired from Bolt Beranek and Newman Inc. (BBN), where he was a senior vice president. He is a fellow of the American Association for the Advancement of Science, the American Psychological Association, the Association for Psychological Science, the Human Factors and Ergonomics Society, and the Society of Experimental Psychologists.

Dr. Nickerson was the founding editor of the *Journal of Experimental Psychology: Applied* (American Psychological Association), the founding and first series editor of *Reviews of Human Factors and Ergonomics* (Human Factors and Ergonomics Society), and is the author of several books. Published titles include:

The Teaching of Thinking (with David N. Perkins and Edward E. Smith)
Using Computers: Human Factors in Information Systems
Reflections on Reasoning
Looking Ahead: Human Factors Challenges in a Changing World
Psychology and Environmental Change
Cognition and Chance: The Psychology of Probabilistic Reasoning
Aspects of Rationality: Reflections on What It Means to Be Rational and Whether We Are

PREFACE

What does it means to reason well? Do the characteristics of good reasoning differ from one context to another? Do engineers reason differently, when they reason well, than do lawyers when they reason well? Do physicians use qualitatively different reasoning principles and skills when attempting to diagnose a medical problem than do auto mechanics when attempting to figure out an automotive malfunction? Is there anything about mathematics that makes mathematical reasoning unique, or at least different in principle from the reasoning that, say, nonmathematical biologists do?

As a psychologist, I find such questions intriguing. I do not address them directly in this book, but mention them to note the context from which my interest in reasoning in mathematics stems. Inasmuch as I am not a mathematician, attempting to write a book on this subject might appear presumptuous—and undoubtedly it is. My excuse is that I wished to learn something about mathematical reasoning and I believe that one way to learn about anything—an especially good one in my view—is to attempt to explain to others, in writing, what one thinks one is learning. But can a nonmathematician hope to understand mathematical reasoning in a more than superficial way? This is a good question, and the writing of this book represents an attempt to find out if one can. I beg the indulgence of my mathematically sophisticated friends and colleagues if specific attempts at exposition reveal only mathematical naiveté. I count on their kindness to give me some credit for trying.

Although much has been written about the importance of the teaching and learning of mathematics at all levels of formal education, and much angst has been expressed about the relatively poor job that is being done in American schools in this regard, especially at the primary and secondary levels, the fact is that mathematics is not of great interest to most people. Hammond (1978) refers to mathematics as an "invisible culture" and raises the question as to what it is in the nature of "this unique human activity that renders it so remote and its practitioners so isolated from popular culture" (p. 15).

One conceivable answer is that the fundamental ideas of mathematics are inherently difficult to grasp. Certainly the world of mathematics is populated with objects that are not part of everyday parlance: vectors, tensors, twisters, manifolds, geodesics, and so forth. Even concepts that are relatively familiar can quickly become complex when one begins to explore them; geometry, for example, which we know from high school math deals with properties of such mundane objects as points, lines, and angles, encompasses a host of more esoteric subdisciplines: differential geometry, projective geometry, complex geometry, Lobachevskian geometry, Riemannian geometry, and Minkowskian geometry, to name a few. But, even allowing that there are areas of mathematics that are arcane, and will remain so, to most of us, mathematics also offers countless delights and uses for people with only modest mathematical training. I hope I am able to convey in this book some sense of the fascination and pleasure that is to be found even in the exploration of only limited parts of the mathematical domain.

Although there are a few significant exceptions, mathematicians generally are notoriously bad about communicating their subject matter to the general public. Why is that the case? Undoubtedly, many mathematicians are sufficiently busy doing mathematics that they would find an attempt to explain to nonmathematicians what they are doing to be an unwelcome, time-consuming distraction. There is also the possibility that the abstract nature of much of mathematics is extraordinarily difficult to communicate in lay terms. Steen (1978) makes this point and contrasts the "otherworldly vocabulary" of mathematics with the somewhat more concrete terms ("molecules, DNA, and even black holes") that provide chemists, biologists, and physicists with links to material reality with which they can communicate their interests. "In contrast, not even analogy and metaphor are capable of bringing the remote vocabulary of mathematics into the range of normal human experience" (p. 2).

Whatever the cause of the paucity of books about *the doing of mathematics* or about the *nature of mathematical reasoning* written by mathematicians, we should be especially grateful to those mathematicians who have proved to be exceptions to the rule: G. H. Hardy, Mark Kac, Imre Lakatos, George Polya, and Stanislav Ulam, along with a few contemporary writers, come quickly to mind. (Hardy wrote *about* the doing of mathematics only when he considered himself too old to be able to *do* mathematics effectively, and expressed his disdain for the former activity; nevertheless, his *Apology* provides many insights into the latter.) I have found the writings of these expositors of mathematical reasoning to be not only especially illuminating, but easy and pleasurable to read.

I hope in this book to convey a sense of the enriching experience that reflection on mathematics can provide even to those whose mathematical knowledge is not great.

I owe thanks to several people who generously read drafts of sections of the book and gave me the benefit of much insightful and helpful feedback. These include Jeffrey Birk, Susan Chipman, Russell Church, Carol DeBold, Francis Durso, Ruma Falk, Carl Feehrer, Samuel Glucksberg, Earl Hunt, Peter Killeen, Thomas Landauer, Duncan Luce, Joseph Psotka, Judah Schwartz, Thomas Sheridan, Richard Shiffrin, Robert Siegler, and William Uttal. I am especially grateful to Neville Moray, who read and commented on the entire manuscript. Stimulating and enlightening conversations on matters of psychology and math with son, Nathan Nickerson, and colleagues at Tufts, especially Susan Butler, Richard Chechile, and Robert Cook, have been most helpful and enjoyable. Special thanks go also to granddaughters Amara Nickerson for critically reading the chapters on learning math and problem solving from the perspective of a first-year teacher with Teach America, and Laura Traverse for pulling the cited references out of a cumbersome master reference file and catching various grammatical blunders in the manuscript in the process. Thanks also to Paul Dukes and Marsha Hecht of Psychology Press for their skillful and amicable guidance of the manuscript to the point of publication. As always, I am profoundly grateful to my wife, Doris, whose constant love and support are gifts beyond measure.

It is a great pleasure to dedicate this book to our youngest grandson, Landon Traverse, whose progress in learning to count and reckon is fascinating and wonder-evoking to observe.

CHAPTER 1

What Is Mathematics?

> Mathematics is not primarily a matter of plugging numbers into formulas and performing rote computations. It's a way of thinking and questioning that may be unfamiliar to many of us, but is available to almost all of us. (Paulos, 1995, p. 3)

> Mathematics is permanent revolution. (Kaplan & Kaplan, 2003, p. 262)

> Many have tried, but nobody has really succeeded in defining mathematics; it is always something else. (Ulam, 1976, p. 273)

What is mathematics? Is it the "queen of the sciences," as, thanks to Carl Frederich Gauss,* it is often called? Or the "most original creation of the human spirit," as Alfred North Whitehead† suggests (1956, p. 402)? Or, more prosaically, is it, as George Polya claims it appears to be to many students, "a set of rigid rules, some of which you should learn by heart before the final examinations, and all of which you should forget afterwards" (1954b, p. 157)? Is it the one area of knowledge in which absolute truth is possible? Or is it "fundamentally a human enterprise arising from human activities" (Lakoff & Núñez, 2000, p. 351), and therefore "a necessarily imperfect and revisable endeavor" (Dehaene, 1997, p. 247)? Do the truths of mathematics exist independently of the minds that discover them, or are they human inventions?

---

* Birth and death dates of (deceased) mathematicians, logicians, and philosophers mentioned in this book are given in the Appendix.
† Whitehead noted that music might also make this claim, and he did not attempt to settle the matter.

Are they timeless and culture free? Or do they rest on assumptions that can differ over time and place? Are mathematics and logic one and the same? Does mathematics spring from logic, or logic from mathematics? Or is it the case, as Polkinghorne argues, that "mathematical truth is found to exceed the proving of theorems and to elude total capture in the confining meshes of any logical net"? (1998, p. 127). Is mathematics, as Bertrand Russell famously said, "the subject in which we never know what we are talking about, nor whether what we are saying is true" (1901/1956a, p. 1576)?

Not surprisingly, a diversity of opinions can be found regarding the answers to these and many related questions. In particular, as Hammond (1978) reminds us, "Mathematicians do not agree among themselves whether mathematics is invented or discovered, whether such a thing as mathematical reality exists or is illusory" (p. 16). In this book, I shall use the terms *invention* and *discovery* more or less interchangeably in reference to mathematical advances, having not found an entirely convincing argument to prefer one over the other. There are numerous views as to what constitutes the essence of mathematics, especially among mathematicians. A major purpose of this book is to explore some of those views and to catch a glimpse of what it means to reason mathematically.

I suspect that for many people, mathematics is synonymous with computation or calculation. Doing mathematics, according to this conception, amounts to executing certain operations on numbers—addition, subtraction, multiplication, division. More complex mathematics might involve still other operations—raising a number to a specified power, finding the nth root of a number, finding a number's prime factors, integrating a function.

Computation is certainly an important aspect of mathematics and, for most of us, perhaps the aspect that has the greatest practical significance. Knowledge of how to perform the operations of basic arithmetic is what one needs in order to be able to make change, balance a checkbook, calculate the amount of a tip for service, make a budget, play cribbage, and so on. Moreover, the history of the development of computational techniques is an essential component of the story of how mathematics got to be what it is today. But, important as computation is, it plays a minor role, if any, in much of what serious mathematicians do when they are engaged in what they consider to be mathematical reasoning.

Whatever else may be said about mathematics, even the casual observer will be struck by the rich diversity of the subject matter it subsumes. Ogilvy (1956/1984) suggests that mathematics can be roughly divided into four main branches—number theory, algebra, geometry, and analysis—but each of these major branches subsumes many subspecialties, each of which can be portioned into narrower subsubspecialties. What is it that the myriad forms of mathematical activity have in common that justifies referring to them all with the same name?

Sternberg (1996) begins a commentary on the chapters of a book on the nature of mathematical thinking that he edited with Ben-Zeev (Sternberg & Ben-Zeev, 1996), with the observation that the chapters make it clear that "there is no consensus on what mathematical thinking is, nor even on the abilities or predispositions that underlie it" (p. 303). He cautions the futility of the hope of understanding mathematical thinking in terms of a set of features that are individually necessary and jointly sufficient to define it, and expresses doubt even of the possibility of characterizing it in terms of a prototype, as has proved to be effective with other complex concepts.

This seems right to me. There are many varieties of mathematical thinking. And mathematicians are a diverse lot of people, reflecting an unbounded assortment of interests, abilities, attitudes, and working styles. Nevertheless, there are, I believe, certain ideas that are especially descriptive of the doing of mathematics and that hold it together as a unified discipline. Among these are the ideas of pattern, problem solving, conjecture, and proof.

☐ Mathematics as the Study of Pattern

> Where there is life, there is pattern, and where there is pattern there is mathematics. (Barrow, 1995a, p. 230)

At the heart of mathematics is the search for regularity, for structure, for pattern. As Steen (1990) puts it, "Mathematics is an exploratory science that seeks to understand every kind of pattern—patterns that occur in nature, patterns invented by the human mind, and even patterns created by other patterns" (p. 8). And, as Whitehead (1911) and Hammond (1978) note, it is the most powerful technique for analyzing relations among patterns. Sometimes mathematics is referred to simply as "the science of patterns" (Devlin, 2000a, p. 3; Steen, 1988, p. 611).

The types of patterns that mathematicians look for and study include patterns of shapes, patterns of numbers, patterns in time, and patterns of patterns. What is the pattern of constant-radius spheres when they are packed in the most efficient way possible? What is the pattern that describes the distribution of prime numbers? How does one tell whether a pattern that describes a relationship between or among mathematical entities in all known instances is descriptive of the relationship generally? (For all cases checked, every even number greater than 2 is the sum of two primes; how does one know whether it is true of all numbers?) Is it possible to find a single pattern of relationships in the falling of a feather and a stone, and the motion of the moon around the Earth and that of the planets around the sun?

The detection of patterns has been of great interest to scientists as well as to mathematicians. At least one account of human cognition makes pattern recognition the fundamental basis of all thought (Margolis, 1987). That the same patterns are observed again and again in numerous contexts in nature—in crystals, in biological tissues, in structures built by organisms, in effects of physical forces on inanimate matter (Stevens, 1974)—demands an explanation. The importance of pattern in art is obvious; it is expressed in numerous ways, and one need not be a mathematician to appreciate or to produce it. Maor (1987) says of Johann Sebastian Bach and Maurits C. Escher: "Both had an acute sense for pattern, rhythm, and regularity—temporal regularity in Bach's case, spatial in Escher's. Though neither would admit it (or even be aware of it), both were experimental mathematicians of the highest rank" (p. 176). Hofstadter (1979) gives numerous examples of the role of pattern in the work of both men.

The detection of a local pattern, some regularity among a limited set of mathematical entities (numbers, shapes, functions), can be the stimulus to a search for a general principle from which the observed pattern would follow. Often such observations have prompted conjectures, and these conjectures sometimes have stood as challenges to generations of mathematicians who have sought to prove them and elevate them to theorem status. Cases in point include Gauss's conjecture, made when he was 15, that the number of prime numbers between 1 and n is approximately $n/\log_e n$; Christian Goldbach's conjecture, dating from 1742, that every even number greater than 2 is the sum of two primes; Pierre de Fermat's "last theorem," that $x^n + y^n = z^n$ is not solvable for $n > 2$; and Georg Riemann's zeta conjecture. These conjectures and others will be noted again in subsequent chapters.

Unlike the conjectures of Gauss, Goldbach, and Fermat, that of Riemann—usually referred to as the Riemann *hypothesis*, is likely to appear arcane to nonmathematicians, but the—so far unsuccessful—search for a proof of it has been a quest of many first-rate mathematicians. For present purposes it suffices to note that the Riemann hypothesis has to do with the distribution of prime numbers and involves the *zeta* function,

$$\zeta(x) = \sum_{n=1}^{\infty} \frac{1}{n^x}$$

first noted by Leonard Euler, and about which more is given in Chapter 3. To all appearances the distribution of primes is chaotic and completely unpredictable, but proof of the Riemann hypothesis would make it possible to locate easily a prime of any specified number of digits, and would have other implications for number theory as well. The interested reader is referred to a beautiful book-length treatment of the Riemann hypothesis by Marcus du Sautoy (2004), who, in the subtitle of his book, refers to the question of the distribution of primes as "the greatest mystery of mathematics."

Local patterns do not always signal more general relationships, so one cannot simply extrapolate beyond what one has actually observed, unless one has the authority of a proof on which to base such an extrapolation. We will return to this point in Chapter 5, but will illustrate it now with an example from solid geometry. A convex deltahedron is a solid, all the faces of which are equilateral triangles of the same size. (The term *deltahedron* comes from the solid's triangular face resembling the Greek letter delta.) The smallest possible deltahedron is a tetrahedron, and the largest is an icosahedron. These are the smallest (4-faced) and largest (20-faced) of the five Platonic solids. That it is impossible to make a solid with less than four equilateral triangles should be obvious. That it is impossible to make one with more than 20 such triangles follows from the necessity, in order to do so, to have six triangles meet at a corner (five meet at a corner in the icosahedron), and the only way to do this is to have all six be in the same plane. How many deltahedra should it be possible to make counting the smallest and largest possibilities?

Because every triangle has three sides and every edge of a deltahedron must be shared by two triangles, the finished shape with n triangles will have $3n/2$ edges, and because the shape must have an integral number of edges, the number of triangles used, n, must be a multiple of 2. If we tried making a six-faced deltahedron, we would find it is possible to do so. We would also discover, if we tried, that we could make deltahedra with 8 faces, with 10, with 12, and so forth. At some point in this experiment, we might begin to suspect, perhaps even to convince ourselves, that deltahedra can be made with any even number of triangles between 4 and 20 inclusive. If we continued making these things, our confidence in this conjecture would probably be quite high by the time we succeeded in making 14- and 16-faced forms. But surprise! Try as we might, we would not succeed in making one with 18 faces.

In short, local regularities can be misleading. Inferring a general pattern from a local pattern is risky. If there is a general pattern, a local pattern will be observed, but the converse is not necessarily true, which is to say that if a local pattern is observed, a general pattern of which it is an instance may or may not exist. Not surprisingly, mathematicians are especially interested in finding patterns that are general; they take it as a challenge to distinguish between those that are and those that are not.

☐ Mathematics as Problem Solving

As well as regarding mathematics as the study of patterns, mathematics can be viewed, pragmatically, as a vast collection of problems of certain types and of approaches that have proved to be effective in solving them. Pure mathematicians may not appreciate this view, but it is a viable one nevertheless. Some mathematicians see the essence of mathematics to be problem solving (Halmos, 1980; Polya, 1957, 1965). Casti (2001) puts it this way: "The real raison d'etre for the mathematician's existence is simply to solve problems. So what mathematics really consists of is problems and solutions" (p. 3). The really "good" problems, in Casti's view—those that become recognized as "mathematical mountaintops"—are those that challenge the best mathematical minds for centuries. Really good problems are really good, not only by virtue of being difficult, but because attempts to solve them commonly contribute to the development of whole new fields of mathematics. Casti's selection of the five most famous mathematical problems of all time is shown in Table 1.1. More will be said about some of these problems in subsequent chapters.

A distinction is sometimes made between knowing mathematics and doing mathematics. Romberg (1994b) illustrates the distinction by drawing an analogy with the difference between having knowledge about other activities (flying an airplane, playing a musical instrument) and actively engaging in them. Schoenfeld (1994b) likens the doing of mathematics with the doing of science—"a 'hands-on,' data-based enterprise" with "a significant empirical component, one of data and discovery. What makes it mathematics rather than chemistry or physics or biology is the unique character of the objects being studied and the tools of the trade" (p. 58).

The nature of the problems that present themselves in pure and applied math may differ, but the need for problem solving spans the range from the purest to the most highly applied. Among other benefits, mathematical training gives one the ability to solve many complex problems in a stepwise fashion, by formulating the problems in such a way that the successive application of well-defined symbol transformation rules will take one from the problem statements to the expressions of the desired solutions.

There is the view too that one of the purposes that mathematics serves is to make problem solving—or reasoning more generally—easier or, in some cases, unnecessary. Austrian physicist-philosopher Ernst Mach (1906/1974) attributes to Joseph Louis Lagrange the objective, in his *Méchanique analytique*, "to dispose once and for all of the reasoning necessary to resolve mechanical problems, by embodying as much as possible of it in a single formula" (p. 561). Mach believed that Lagrange

---

TABLE 1.1. The Five Most Famous Mathematical Problems of All Time, According to Casti (2001)

---

1. Determination of the solvability of a Diophantine equation. "To devise a process according to which it can be determined by a finite number of operations whether a Diophantine equation is solvable in rational integers" (p. 12). A Diophantine equation is a polynomial equation, all the constants and variables of which, as well as the solutions of interest, are integers.

2. Proof of the four-color conjecture: The conjecture that four colors suffice to color a two-dimensional map so that no two contiguous regions have the same color.

3. Proof of Cantor's continuum hypothesis: There is no infinity between that of the natural numbers and that of the reals.

4. Proof of Kepler's conjecture: Face-centered cubic packing of spheres of the same radius is the optimal packing arrangement (yields the smallest ratio of unfilled-to-filled space).

5. Proof of Fermat's "last theorem": The equation $x^n + y^n = z^n$ has no solutions for $n > 2$.

---

was successful in attaining this objective, that his method made it possible to deal with a class of problems mechanically and unthinkingly. "The mechanics of Lagrange," Mach contends, "is a stupendous contribution to the economy of thought" (p. 562). It seems clear that discoveries of mathematical equations that are descriptive of relationships in the physical world have made reasoning about those relationships easier, or even perhaps unnecessary in some instances, but it seems equally clear that the same discoveries have had the effect of extending the range of reasoning, unburdening it in some respects and thereby enabling it to function at higher levels.

☐ Mathematics as Making Conjectures

Quibbles aside, every theorem is born as a conjecture. Like the rest of us, the mathematician's seeing-that precedes reasoning-why. But it is an important point (as many mathematicians will confirm) that the mathematician comes to see his conjecture as a theorem at some stage prior to, and indeed usually a good deal prior to, being able to prove it. (Margolis, 1987, p. 84)

The idea that theorems typically begin life as conjectures is easy to accept, but what evokes conjectures that eventually become theorems is not known very precisely. Inductive reasoning, searching for patterns and regularities, and playing with ideas often appear to be involved. Conjectures in this context are not wild guesses; mathematicians usually attempt to prove only conjectures they believe likely to be true (vis-à-vis the axioms of some mathematical system). They often have claimed to have been convinced of the truth of a conjecture long before being able to construct a rigorous proof of it. And sometimes they have described "seeing" a proof in its entirety before laying out the sequential argument explicitly, which is not unlike the reported experience of some composers of having had a conception of a complex composition as a whole before writing down any music. Penrose (1989) gives a hint of this ability to see arguments, in some sense, in their entirety before being aware of their parts. "People might suppose that a mathematical proof is conceived as a logical progression, where each step follows upon the ones that have preceded it. Yet the conception of a new argument is hardly likely actually to proceed in this way. There is a globality and seemingly vague conceptual content that is necessary in the construction of a mathematical argument; and this can bear little relation to the time that it would seem to take in order fully to appreciate a serially presented proof" (p. 445).

In her biography of John Nash, Sylvia Nasar (1998) says, "Nash always worked backward in his head. He would mull over a problem and, at some point, have a flash of insight, an intuition, a vision of the solution he was seeking. These insights typically came early on, as was the case, for example, with the bargaining problem, sometimes years before he was able, through prolonged effort, to work out a series of logical steps that would lead one to his conclusion" (p. 129). Nasar notes that other great mathematicians, including Riemann, Poincaré, and Wiener, worked in a similar way.

It is not necessary to assume that all such "flashes of insight" have turned out to be correct in order to appreciate the importance of the kind of reasoning that has led to them and the difference between it and the "series of logical steps" that could be used to convince others of the truth—in the mathematical sense—of those that have proved to be true. Any adequate theory of mathematical reasoning must be able to account both for the exploratory ruminations that produce the insights—or productive conjectures—and for the process of constructing logically tight arguments that justify the conjectured conclusions.

Just as individual mathematicians make use of hunches and conjectures before they have been proved, the history of mathematics has many examples of concepts being used long before they have been justified or defined in any formal way. Moreover, concepts that we are likely to consider to have intuitively obvious meanings may have become "intuitively obvious" only as a consequence of familiarity through common usage over a long time; many of these concepts were used much more tentatively, if at all, by mathematicians of previous centuries.

There are few concepts, for example, that are more useful in mathematics today than those of variable and function. A variable is an entity whose value is not fixed. A function expresses the relationship between two or more variables; in particular, it shows how the value of one variable (the dependent variable) depends on the value or values of one or more other variables (the independent variables). A more general definition of a function could be "any rule that takes objects of one kind and produces new objects from them" (Devlin, 2002, p. 26). The idea of dependence among variables is basic not only to mathematics, but to science as well, and it may come as something of a surprise that neither the idea of variable nor that of function was prominent before the time of René Descartes.

Though Isaac Newton and Gottfried Leibniz both succeeded in developing the differential calculus—the study of continuous change—to the point at which it could be applied effectively to problems of interest, neither was able to provide an adequate logical foundation for the subject, nor was anyone else, in spite of concerted efforts to do so, for about 150 years. Both Newton and Leibniz were guided more by intuition than by logic. American mathematician-historian of mathematics Morris Kline (1953a) refers to the recognition of the relationship between the general concept of rate of change and the determination of lengths, areas, and volumes as "the greatest single discovery made by Newton and Leibniz in the Calculus" (p. 224), and to the calculus itself as the richest of all the veins of thought explored by geniuses of the 17th century. He also points to the roles of Newton and Leibniz in the development of the calculus as compelling refutations to the popular conception of mathematicians reasoning perfectly and directly to conclusions.

Probably no one has done more to call attention to the importance of conjectural and inductive thinking in mathematics than Hungarian-American mathematician George Polya, who distinguishes "finished mathematics" from "mathematics in the making." The former consists of the demonstrative reasoning of deductive proofs; the latter resembles other human knowledge in the making: "You have to guess the mathematical theorem before you prove it; you have to guess the idea of the proof before you carry through the details. You have to combine observations and follow analogies: you have to try and try again. The result of the mathematician's creative work is demonstrative reasoning, a proof; but the proof is discovered by plausible reasoning, by guessing" (Polya, 1954a, p. vi). It is important, Polya held, for the student to learn not only to distinguish a proof from a guess, but also to tell the difference between more and less reasonable guesses. "To be a good mathematician, or a good gambler, or good at anything, you must be a good guesser" (p. 111).

At the beginning of this section, I quoted Margolis's observation that "every theorem is born as a conjecture." It is not the case, however, that every conjecture ends up being proved; some do and others do not. There are many famous conjectures that have been conjectures for a very long time and remain so despite the countless hours that first-rate mathematicians have spent trying to prove them. Others have been shown to be false. No one knows what percentage of the conjectures that mathematicians make eventually prove to be true. It could be that most conjectures are wrong. If that is the case, it does not follow that the making of conjectures is a waste of time. The exploration of conjectures that turned out to be wrong has often led to important discoveries and the development of new areas of mathematical inquiry. Arguably, this is because the conjectures that mathematicians make are generally made in a context that is rich in mathematical knowledge. Seldom are inspired mathematical conjectures made by people who have little understanding of mathematics.

The reasoning of the mathematician and that of the scientist are similar to a point. Both make conjectures often prompted by particular observations. Both advance tentative generalizations and look for supporting evidence of their validity. Both consider specific implications of their generalizations and put those implications to the test. Both attempt to understand their generalizations in the sense of finding explanations for them in terms of concepts with which they are already familiar. Both notice fragmentary regularities and—through a process that may include false starts and blind alleys—attempt to put the scattered details together into what appears to be a meaningful whole. At some point, however, the mathematician's quest and that of the scientist diverge. For scientists, observation is the highest authority, whereas what mathematicians seek ultimately for their conjectures is deductive proof.

☐ Mathematics as Proof Making

Mathematics seems to be a totally coherent unity with complete agreement on all important questions; especially with the notion of *proof*, a procedure by which a proposition about the unseen reality can be established with finality and accepted by all adherents. It can be observed that if a mathematical question has a definite answer, then different mathematicians, using different methods, working in different centuries, will find the same answer. (Davis & Hersh, 1981, p. 112)

If conjecture is the engine that powers the mathematical train, proof is the intended destination. The centrality of the idea of proof in mathematics is widely acknowledged. Devlin (2000a) refers to proof as the only game in town when it comes to establishing mathematical truth. Romberg (1994) expresses the same idea in noting that for a proposition to be considered a mathematical product it must be rigorously proved by a logical argument. Aczel (2000) says that statements that lack proofs carry little weight in mathematics. Noted American historian of mathematics Eric Temple Bell (1945/1992) sees proof as the *sine qua non* of mathematics: "Without the strictest deductive proof from admitted assumptions, explicitly stated as such, mathematics does not exist" (p. 4). Whether the proofs that mathematicians achieve are ever fully deductive is debatable; more will be said on this issue in subsequent chapters, especially Chapter 5.

Proof in mathematics is sometimes characterized as the counterpoint to intuition. The need for proofs is seen in an observation by Ogilvy (1956/1984) that "in mathematics, so many of the things that are 'obviously' true aren't true" (p. 19). Kaplan (1999) speaks of "the ever present tension between intuition and proof," which he describes lyrically as follows.

These are the two poles of all mathematical thought. The first centers the free play of mind, which browses on the pastures of phenomena and from its rumination invents objects so beautiful in themselves, relations that work so elegantly, both fitting in so well with our other inventions and clarifying their surroundings, that world and mind stand revealed each as the other's invention, conformably with the unique way that Things Are.

After invention the second activity begins, passing from admiring to justifying the works of mind. Its pole is centered in the careful, artful deliberations which legalize those insights by deriving them, through a few deductive rules, from the Spartan core of axioms (a legal fiction or two may be invented along the way, but these will dwindle to zero once their facilitating is over). What emerges, safe from error and ambiguity, others in remote places and times may follow and fully understand. (p. 159)

Without denying the usefulness of the distinction between intuition and proof, I believe it can be drawn too sharply; intuition plays an essential role in the making and evaluating of proofs and is sometimes changed as a consequence of these processes. In this respect, the distinction is like that between creative and critical thinking; while this distinction too is a useful one, it is not possible to have either in any very satisfactory sense without the other.

Greeno (1994) sees an analogy between the role that proofs play in mathematics and the one that observations and experiments play in science. Just as observations and experiments provide evidence of the tenability of theoretical assertions in science, so proofs provide evidence of the truth of theorems in mathematics. The evidence provided by a proof in mathematics is qualitatively different from that provided by empirical observations in science, but the analogy holds in the sense that learning mathematics without an appreciation of the role of proofs would be as disabling as learning science without appreciation of the role of empirical evidence.

Why Is Mathematics Important?

I would not wish to have you possessed by the notion that the pursuit of mathematics by human thought must be justified by its practical uses in life. (Forsyth, 1928/1963, p. 45)

The presumed practical importance of numeracy in modern society is reflected in the stress placed on its acquisition by formal education. Much of the curriculum during the first few years of school is devoted to learning the natural numbers and methods of manipulating them. This emphasis is often justified by assertions of the importance of mathematical competence for the demands of modern life. The editors of the report of the National Research Council's Mathematics Learning Study Committee contend that "the growing technological sophistication of everyday life calls for universal facility with mathematics" (Kilpatrick, Swafford, & Findell, 2001, p. 16). The more recent final report of the National Mathematics Advisory Panel (2008) states that while there is reason enough for the often expressed concern about the implications of mathematics and science for "national economic competitiveness and the economic well-being of citizens and enterprises," the more fundamental recognition is "that the safety of the nation and the quality of life—not just the prosperity of the nation—are at issue" (p. xi).

Somewhat ironically, despite the great emphasis that is put on the practical importance of mathematics as a justification for requiring mathematics beyond arithmetic in secondary education, even teachers of math courses may be hard-pressed to make a convincing case. In a small survey of middle school teachers, Zech et al. (1994) found that only 3 of 25 were able to identify uses of geometry other than the calculation of area and volume, and 5 could give no ideas of geometry's practical usefulness. Before condemning the teachers for their inability to answer this question, one might do well to try to imagine how many people of one's acquaintance actually use in their daily lives geometry, algebra, or any other area of mathematics—as formally taught in school—beyond arithmetic.

Unquestionably, today mathematics is essential to certain professions—engineering, architecture, accounting, scientific research. But what about people who are not working in these professions, which is to say most people? From the considerable emphasis that is placed on the importance of having strong math courses in the public secondary schools, one would judge that the prevailing assumption is that for most of us the acquisition of nontrivial competence in mathematics is a pressing need. Most students are encouraged, if not required, to have courses in several areas of math beyond elementary arithmetic in high school (algebra, geometry, trigonometry, calculus).

I support this emphasis and do not mean to suggest that it is misguided, but I think it important to consider the possibility that for the majority of students it is not clear that this emphasis on mathematics beyond basic arithmetic serves any practical purpose. I venture the guess that the large majority of them will seldom, if ever, solve a problem with algebra, or trigonometry, or calculus when they no longer have to do so to satisfy a course requirement. And in most cases, this will not be a major handicap. One need not know how to solve problems in algebra or other nonelementary areas of mathematics in order to balance a checkbook, shop for groceries, drive a car, cook, maintain a home, and do the countless other things that daily life demands. Nor is mathematical expertise essential to the performance of most jobs, even those that are intellectually demanding in other respects. (Whether most jobs of the future will be more or less demanding of mathematical expertise than most jobs of today is an open question; one can find predictions both ways.) Unless one is among the minority of people whose jobs require the use of higher math, one generally can get along without it quite well. Undoubtedly, poor mathematical skills can limit one's comprehension of much of the news and commentary that appears in daily newspapers and other media (Paulos, 1995), and can make one vulnerable to ill-advised financial decisions (Taleb, 2004) and predatory scams as well as to the reporting of true but misleading statistics for political or other purposes (Huff, 1973), but my sense is that poor mathematical skills in these contexts generally means the inability to do basic arithmetic or, in some cases, to think carefully and logically.

If I believe this to be so, why do I believe it to be good for high schools to put the emphasis on mathematics that they do? Without denying the practical importance of math for those students who will eventually use it in their work, I want to argue that some acquaintance with higher math is beneficial to everyone for at least four reasons. The first reason is that if mathematics is to continue to advance as a discipline, there is a perennial need for the training of individuals with the potential and desire to become professional mathematicians. And how better to discover such individuals than by means of their mathematical performance in their early and intermediate school years. From the point of view of the community of professional mathematicians, this is arguably the preeminent consideration. Dubinsky (1994b) puts it this way: "The issue of concern for most professional mathematicians is the continuation and preservation of the species [of professional mathematicians]: the education and production of first-rate research mathematicians, people who are from the beginning very talented in mathematics" (p. 47). Dubinsky distinguishes this concern from what would likely be considered more fundamental by mathematical educators, namely, mathematical literacy—"the raising of the level of mathematical understanding among the general population."

A second reason that I believe it to be good for elementary and secondary schools to put considerable emphasis on mathematics is the possibility, expressed clearly by Judah Schwartz (1994), that successful mathematical education means "changing habits of mind so that students become critical, in a constructive way, of all that is served up on their intellectual plates" (p. 2). "Students who have been successfully educated mathematically," Schwartz contends, "are skeptical students, who look for evidence, example, counterexample, and proof, not simply because school exercises demand it, but because of an internalized compulsion to know and to understand" (p. 2). I wish I could point to compelling empirical evidence that this claim is true. Unhappily—and ironically perhaps in view of the nature of the claim—I cannot. But if it is true, it establishes the importance of a solid grounding in mathematics beyond doubt in my view. And determining whether it is true strikes me as a more important question for research than many on which much greater effort is expended. In response to a query about the claim, Schwartz (personal communication) expressed the belief that it is more likely to be true with people whose mathematical education is put to use in scientific and engineering contexts than in purely mathematical contexts.

Third, nontrivial knowledge of mathematics is important in the sense in which some acquaintance with great literature and art, influential philosophies, cultures other than one's own, and the history of science is important. Some acquaintance with these topics is part of what it means to be well educated. More importantly, familiarity with such areas of human knowledge and creative work enriches one's intellectual life immeasurably. It is a disservice to students, I think, to teach mathematics as though the only, or even the primary, reason for learning it is the practical use one might make of it. One would not think of trying to justify the teaching of music, literature, history, or philosophy solely on the basis of the practical utility of exposure to these subjects (not to deny there undoubtedly is some); it makes no more sense to me to do so in the case of mathematics.

Fourth, I believe that, as Friend (1954), Court (1961), Devi (1977), and Pappas (1989, 1993), among others, remind us, there is much pleasure to be had in the pursuit of mathematics, even at modest levels of expertise. My fuzzy memory of high school math (geometry, algebra, trigonometry) is sadly devoid of any effort by any teacher to convince me, or to demonstrate with personal enthusiasm, how exciting and wonderful (full of wonder) mathematics can be. I do not mean to demean my teachers; they were good and conscientious people, dedicated to their jobs and the progress of their students. My guess is that they themselves thought of high school math as something one ought to learn, much as one ought to eat one's spinach, and that the idea that it could, if presented properly, be fun would have struck them as more than slightly odd. I have no way of knowing whether my experience is representative of that of others, but would be surprised to learn that it was unique or even highly unusual. And that, I believe, is a sad reality.

There are undoubtedly compelling reasons other than those I have mentioned to teach mathematics in school, beginning at an early age, but even if that were not the case, these suffice in my view. My hope for this book is that it will help make not only the usefulness of mathematics clear, but also its charm.

☐ Mathematics and Psychology

To the psychologist, the development of mathematical competence, both by the species over millennia and by individuals over their lifetimes, is a fascinating aspect of human cognition. Many questions of psychological interest arise. When and why did the rudiments of mathematical capability first appear among human beings? How and why has mathematics grown into the richly branching complex of specialties that it is today?

From where comes our fascination with and propensity for abstract reasoning? What prompts the emergence of concepts—like the infinite and the infinitesimal—that appear to be descriptive of nothing in human experience? Why does abstract mathematics, developed with no thought of how it might be applied to the solution of practical problems, often turn out to be useful in unanticipated ways?

What are the fundamental concepts of mathematics? What is a number? Do species other than human beings have a sense of number? Are they capable of counting, or of doing elementary mathematics in any meaningful sense?

Why do essentially all modern cultures use the same system for representing numbers and counting, despite that they do not all speak the same language? How did the system that is now used nearly universally come to be what it is?

What is the basis of mathematicians' compulsion to prove assertions? What makes a proof a proof, that is, cognitively compelling? How can one be sure that proofs (e.g., about infinite sets, infinitesimals, and other unobservables) that cannot be verified empirically are correct? What is it about certain mathematical problems that motivate people to work obsessively on them, sometimes with little to show for their efforts, for years?

Are the truths of mathematics discoveries or inventions? What are mathematicians seeing when they describe a mathematical entity (proof, theorem, equation) as beautiful? How is it that mathematical ideas that seem absurd to one generation can be accepted with equanimity by another?

What types of mathematical awareness do children acquire spontaneously? What do children need to know, what concepts and skills must they have, in order to be able to do well when first introduced to elementary mathematics? How is the potential for mathematical reasoning best developed through instruction?

To what extent are the considerable differences that people show in their interest in mathematics and in the level of mathematical competence they attain attributable to genetics, or to experience? Is there such a thing as mathematical potential that is distinct from general intelligence? Or a specifically mathematical disability that is distinguishable from a general cognitive deficit? Do mathematicians share a set of characteristics that distinguish them, as a group, from nonmathematicians? Why have the vast majority of notable mathematicians been men?

It is questions of these sorts that motivate this book. I do not pretend to answer them, but I do hope to provide some food for thought that is relevant to them, and to many others of a similar ilk.

Counting

I regard the whole of arithmetic as a necessary, or at least natural consequence of the simplest arithmetic act, that of counting. (Dedekind, 1872/1901, p. 4)

Ernst Mach (1883/1956) once defined mathematics as "the economy of counting" (p. 1790). The dependence of mathematics and the quantitative sciences on this ability is obvious; perhaps less immediately apparent but no less real is the importance of counting to the very existence of any technology or organized society beyond the most primitive. How and when counting was invented, or discovered, no one knows; we can only speculate. Our debt to the fact that it was is immeasurable.

Aristotle believed that the ability to count is among the things that make human beings unique. Is counting a uniquely human capability? Asking that question presumes agreement on what constitutes counting. Is there such agreement? What does constitute counting? These and related questions motivate this chapter. Inasmuch as the focus of this book is mathematical reasoning as it is done by human beings, discussion of whether animals count may seem out of place, but a brief digression on the topic is useful in rounding out the discussion of what it means to count, and in providing a broader context in which to understand what may be unique about human capabilities in this regard.

☐ What Counts as Counting?

In English, the word *count*, when used as a verb, can be either transitive or intransitive. When used as a transitive verb, it refers to the enumeration of a set of entities—the determination of the number of items in the set by a process that focuses successively on each item in the set. To count the words in this sentence is to determine, by enumeration, that there are 18 of them. When used as an intransitive verb, it refers to the recitation, in order, of the natural numbers. When a proud parent notes that two-year-old Johnny can count, she is probably claiming that, when asked to display his precocity in this regard, Johnny will recite "one, two, three, ..." Von Glasersfeld (1993) refers to such recitation of number words, apart from coordination with countable items, as "vacuous counting."

In the study of counting the focus has generally been on counting in the first sense, that of enumeration, the determination of quantity. The process of counting in this sense probably appears to most of us to be straightforward and simple to grasp conceptually. But is it really? The simplicity becomes somewhat less obvious when we consider the question of what should be taken as evidence that someone, or something, is counting.

Suppose we ask a child, "How many fingers do you have?" and he holds up both hands and says, "That many." Should we take this as evidence that he can count? Or imagine that we ask, "How many sisters do you have?" and the child holds up three fingers. Assuming that he really has three sisters, should we take that as evidence that he can count? The second gesture seems to be better evidence of counting ability than the first. In the first case, the child could be simply showing us the things we asked about (his fingers) and saying in effect, "You count them." In the second case, there is at least an implicit mapping operation involved. The child is not holding up his sisters for enumeration, but rather a set of fingers that has in common with his sisters only their number. He is implying, knowingly or not, that if we count his fingers, we will get the same number that we would get if we counted his sisters. He is using one set to represent another with respect to the property of number.

In order to represent the number of his sisters by holding up three fingers, the child need not be able to attach a name (three, trois, drei) to the number he is representing. Nor is it necessary that he realize that a quantity of three is greater than one of two and less than one of four. That is, he need not understand numbers in either their cardinal (how many) or ordinal (position in an ordered set) sense. He needs only to know that the number of fingers that he is holding up is the same as the number of sisters he has. (Of course, if he has been taught to hold up his hand in a certain way when asked how many sisters he has, he need not know even that.) This is not to suggest that when children use fingers to represent quantities that that is all they know, but simply to say that they need not know more than that.

Being able to verbalize a quantity also is not compelling evidence of an ability to count. The ability to say, for example, that one has three sisters, or that one is three years old, does not require the knowledge that three is greater than two and less than four. It does not imply an understanding of greater than and less than, or any concepts of quantity at all. One may learn that one has three sisters, or that one is three years old, in much the same what that one learns that one's name is Sam, or that one's address is 17 Oak Street. Similarly, one may learn to repeat a sequence of words—one, two, three, four, ...—without understanding how to apply these words to the task of counting.

Suppose that an individual is able to reproduce a sequence of, say, four taps accurately but is unable to say how many taps there are in the sequence and is unable to produce four taps on request without having a model to copy. It is easy to imagine being able to reproduce a pattern of a sequence of a small number of events (four taps) without having a concept of number. One need only store a rhythmic pattern (tap-tap-tap-tap) and reproduce it as a whole. One suspects that moderately long sequences can be stored and reproduced without counting, by the use of rhythmic groupings (di-dah-di-dah—di-dah-di-dah—di-dah-di-dah). It is not clear what the upper limit of one's ability to retain and reproduce such sequences without resorting to counting is, but it is conceivable that one might be able to reproduce such sequences of modest length accurately—as, for example, in scat singing of jazz—even if one had no concept of number and was unable to specify which of two sequences had the greater number of elements.

The ability to distinguish many from few is not compelling evidence that one can count, inasmuch as such distinctions can be made on the basis of differences in gross perceptual features of collections. The same may be said regarding the ability to distinguish more from fewer, although the extent to which this distinction might be assumed to depend on counting seems likely to vary with the specifics of the sets with respect to which the distinction is to be made—to determine which of two groups of stones contains more stones requires a much less sophisticated grasp of the concept of number if one group contains 1,000 stones and the other 10 than if one group contains 1,000 and the other 999.

A related point may be made with respect to the ability to determine whether two small sets have the same number of items. Piaget (1941/1952) discovered that when very young children are asked to say whether two small sets are equal in number (and they in fact are), they are likely to give the correct answer if the items are spatially aligned so the one-to-one correspondence is salient, but to give the wrong answer if one of the sets is spread out relative to the other so the one-to-one correspondence is perceptually less apparent. Other investigators have provided evidence that children often make judgments of relative number on the basis of overall perceptual features (length, area, density) of a set of items (Brainerd, 1979). The attribution of such findings to numerical incompetence is called into question, however, by the finding that, when young children (two to four years old) are given the choice of either of two sets of M&Ms, both relatively small in number, a majority consistently pick the more numerous set, regardless of their spatial arrangements (Mehler & Bever, 1967).

There is good experimental evidence that people may perceive the number of objects in a collection directly if the number is sufficiently small; some set the limit at three or four, while others say five or six (Chi & Klahr, 1975; Gallistel, 1988; Jensen, Reese, & Reese, 1950; Kaufman, Lord, Reese, & Volkmann, 1949; Klahr & Wallace, 1973, 1976; Mandler & Shebo, 1982; Miller, 1993; Taves, 1941). Such direct perception of numerosity has been distinguished from counting and called *subitizing* (Kaufman et al., 1949). McCulloch (1961/1965) refers to the numbers 1 through 6 as *perceptibles* and to all others as *countables*. The ability to perceive perceptibles, he suggests, is one that nonhuman species probably also possess. According to this view, the ability to distinguish among collections of up to four objects, if not to five or six, is not evidence of the ability to count, and it seems reasonable to suspect that the discrimination by animals of different small collections may be based on an ability of this sort (Davis & Pérusse, 1988a).

Whether subitizing is really fast counting whereby the items or memory representations of them are serially noted (Folk, Egeth, & Kwak, 1988; Gallistel & Gelman, 1991; Gelman & Gallistel, 1978) or a case of all-at-once preattentive processing (Dehaene, 1997; Dehaene & Cohen, 1995) is a matter of debate. An understanding of subitizing is complicated by small numbers of items sometimes being distinguishable on the basis of patterns they form (two forming a line, three a triangle, four a rectangle) and larger numbers of items being distinguished more readily if arranged regularly (in aligned rows, say) than if distributed randomly (Maertens, Jones, & Waite, 1977; Mandler & Shebo, 1982). Also subsets of items can be identified on the basis of specific characteristics, as illustrated in Figure 2.1, and their numbers combined in determining the number of items in the total set.

The idea that the curve relating the time required to make numerosity judgments has an elbow at the point that divides sets within the subitizing range from larger sets—which has been seen as evidence of the reality of subitizing—has been challenged (Balakrishnan & Ashby, 1992).

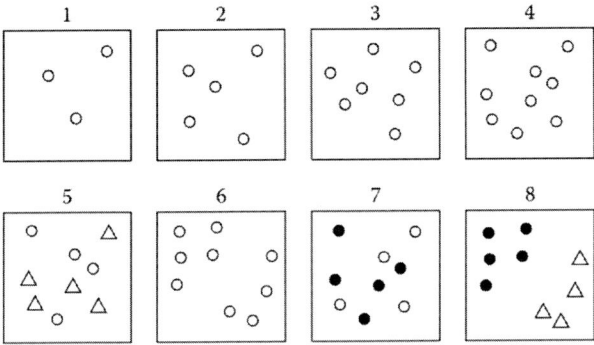

FIGURE 2.1 Readers may find it easy to determine the numbers of items in Box 1 and in Box 2 at a glance, without explicit counting. The numbers in Box 3 and Box 4 are likely to be less easy to determine this way, although explicit counting of every item in each set may be avoided by mentally dividing the items into subsets of three, four, or five items and summing. Boxes 5 through 8 illustrate that clustering may be facilitated by a variety of features, such as shape, spatial arrangement, color, and combinations of them.

Researchers disagree on whether the ability to subitize—assuming there is such a distinctive process—develops before the ability to count (von Glasersfeld, 1982; Klahr & Wallace, 1976; Klein & Starkey, 1988) or after it (Beckwith & Restle, 1966; Mandler & Shebo, 1982; Silverman & Rose, 1980).

   Uncertainties of these sorts have led to questioning the continuing usefulness of the concept of subitizing (Terrell & Thomas, 1990). Some argue that *subitizing* has been used only as a descriptive term and that the concept lacks explanatory value—that simply labeling the process of rapidly discriminating small numbers of visually presented stimuli as subitizing does not shed light on the underlying mechanism (Miller, 1993; Thomas & Lorden, 1993). On the other hand, there is the view that much of what has been interpreted as counting by very young children could be the result of subitizing (Sophian, 1998). Clements (1999a) argues that children should be taught to subitize as a means of facilitating the development of other numerical ideas such as those of addition and subtraction. Whatever the status of subitizing, that the process of determining numerosity is sensitive to the number of items in a display is seen in the time required to name the number, and the frequency of errors, both increasing with the number of items, especially for collections larger than three or four (Bourbon, 1908; Jensen et al., 1950; Logan & Zbrodoff, 2003; Mandler & Shebo, 1982).

   Counting and subitizing are both to be distinguished from estimation, which is also a type of numerosity assessment that has been much studied (Siegler & Booth, 2005). Generally, estimation is required when the number of items in a set is too large to be subitized and the time for inspection is too short to permit counting. One common finding is that adults generally underestimate the number of items in a brief visual display that contains more than a few items, the magnitude of the underestimation increasing with the number of items in the display (Krueger, 1982). Both children and adults differ considerably in their ability to estimate numerical quantities, such as the results of a computation, like 63×112, and that ability tends to be positively correlated with indicants of general cognitive ability, like IQ (Reys, Rybolt, Bestgen, & Wyatt, 1982).

Children learn to say number words—one, two three, ...—in the correct sequence at a fairly early age, but as already noted, this can be done independently of any concept of quantity. In English, *counting* is used to refer to verbalization of this sort as well as to the act of enumeration, which makes for confusion. Fuson and Hall (1983) recommend use of the terms *sequence words* and *counting words* to differentiate the one case from the other.

For present purposes, I want to argue that the main evidence of the ability to count, at least in a mathematically relevant sense, is the ability to put integers into one-to-one correspondence with a set of objects (or with events in time)—to assign to each object (or event) in the set a unique integer name in the conventional order of the integers (1, 2, 3, ...) and to equate the number of items in the set with the integer name assigned to the last object (or event) in the set. This is close to Gelman and Gallistel's (1978) description of counting and essentially what Briars and Siegler (1984) refer to as the word–object correspondence rule, though not perhaps precisely as they would express it. The eminent American mathematician George Dantzig (1930/2005) argues that the principle of one-to-one correspondence is one of two that permeates all of mathematics (the other being the idea of ordered succession). The main point I want to make with this brief consideration of the question of what it means to count, however, is that the answer is not as obvious and simple as one might assume at first thought. We will return to the topic of children learning to count in Chapter 14.

☐ Can Animals Count?

The literature relating directly to the counting, or counting-like, abilities of animals is very large. Here we can only scratch the surface. For the reader who would like more information, there are numerous readily available reviews of work on the topic, among them, those of Honigmann

(1942), Salman (1943), Wesley (1961), Davis and Memmott (1982), Davis and Pérusse (1988a, 1988b), and Rilling (1993).

There have been many claims that specific animals have had the ability—either naturally or as a consequence of training—not only to count but also to solve mathematical problems that are beyond most humans. The claimed abilities of these animals (usually, but not always, either horses or dogs) have been exhibited before large audiences, and in a few cases the animals' abilities have been the focus of scientific study. Perhaps the most famous of the calculating animals was Clever Hans, an Arab stallion whose mathematical prowess—evidenced by his tapping out the answers to mathematical questions with his hoof—was exhibited around Germany by his owner, Wilhelm von Osten, a Russian high school teacher, beginning late in the 19th century. Hans's performance was good enough to convince the 13 members of a commission (including two zoologists and a psychologist) established by Germany's Board of Education that his apparent mathematical knowledge was genuine. Subsequent investigations by a psychologist, Oskar Pfungst (1911/1965), revealed that Hans was reacting to subtle involuntary cues produced by his questioner (usually, but not always, his owner). Hans did poorly on questions to which the questioner did not know the answer. If the accounts of the nature of the cues to which the horse was responding are accurate, they were subtle indeed—too subtle to be detected by numerous human observers, or to be deliberately suppressed by Pfungst when he served as the questioner—and Clever Hans was unquestionably appropriately named, his inability to count or do mathematics notwithstanding.

Clever Hans is well known among psychologists because of the work of Pfungst. There were many other examples at about the same time as Clever Hans of animals that were reputed to be able to count and calculate (Rosenthal, 1911/1965). Among the more spectacular instances were several other horses, owned by Karl Krall, a wealthy merchant of Elberfeld, Germany. (Krall purchased Clever Hans shortly before von Osten's death.) Claims of what some of these "Elberfeld horses" could do (e.g., immediately produce cubic or fourth roots of seven-digit numbers) were nothing short of amazing. The Belgian poet-essayist Maurice Maeterlinck (1862–1949) was sufficiently impressed with their abilities and sufficiently convinced of their authenticity to feature them, and to attempt to explain them in paranormal terms, in his *The Unknown Guest* (Maeterlinck, 1914/1975). Brief accounts of many of the calculating animals are provided by Tocquet (1961), who attributes their performance to the ability to respond to cues unconsciously provided by questioners or observers, which is not to deny the possibility of deliberate deception and fraud in some instances.

In sum, reports, of which there are many, of abilities of animals to do higher forms of mathematics appear to have been quite thoroughly debunked. Clever Hans (or Clever Hans and his owner) was indeed clever enough to convince many observers of his mathematical prowess, but the cleverness rested on abilities other than mathematics. The same conclusion appears to be warranted for all the other instances of computing animals that have been carefully investigated. Few, if any, contemporary researchers believe that nonhuman species can do complicated mathematics.

On the question of whether nonhuman species can count, or perhaps do some rudimentary arithmetic, there is far less agreement. The answer to the question of whether they can count seems likely to depend on how one defines counting. One might object that my definition in the preceding section is too narrow and that with only a slightly less restrictive definition, we would have to conclude that they can. They are able, for example, to discriminate between patterns of dots on the basis of the number of dots the patterns contain, at least for patterns containing relatively small numbers of dots. And some organisms naturally engage in repetitive behavior, repeating some act the same number or times, or approximately so, on different occasions. Davis and Memmot (1982, p. 549) claim, for example, that when a cow chews a cud, it moves its jaw almost precisely 50 times between each swallow. I suspect that few of us would take that as evidence that cows can count to 50, but the observation points up the importance of being clear about what we will take to be evidence of counting.

In the context of an informal discussion of animal intelligence, Sir John Lubbock (1885) recounts an observational report of a wild crow that gave evidence of being able to "count" to four or five. The story is that a man wished to shoot the crow, and to do so he planned to deceive it by having two people enter a watch house and only one leave. The implication is that the crow would make itself scarce if it believed a person was in the vicinity. The man discovered that the crow was not fooled if two men entered and only one left, or if three men entered and two left; only if five or six men entered and four or five left was the crow's ability to keep track exceeded. The accuracy of the account is unknown. The results of recent experiments with ravens, close cousins to crows, suggest that these birds have considerable ability to solve problems that appear to require some logical reasoning (Heinrich & Bugnyar, 2007).

There is at least suggestive evidence that honeybees use landmarks to estimate the distance to a goal. Chittka and Geiger (1995) trained bees to fly from a hive to a food source on a course that passed a given number of landmarks; they found that by changing the number of landmarks after training, they disrupted the bees' ability to find the food source. These investigators interpreted their findings as evidence that the bees were capable of "proto-counting"—something less than *bona fide* counting, but close to it.

Although the terms *numerosity* and *numerousness* are used more or less interchangeably in the literature, Stevens (1951) makes a distinction between them, using *numerosity* to connote the property of a collection that one determines by counting, and *numerousness* to indicate a property of a collection that is perceived without actual enumeration. Dantzig (1930/2005) makes a somewhat similar distinction, in this case between having a number sense and being able to count; nonhuman species have the former, he argues, but not the latter, which ability he sees as exclusively human. It is possible, Dantzig contends, "to arrive at a clear-cut number concept without bringing in the artifices of counting" (p. 6). He illustrates what he means by a "number sense" with behavior exemplified when a mother wasp of a particular species—genus *Eumenus*—which lays eggs in individual cells and provides each cell with several live caterpillars to serve as food for hatchlings, puts 5 caterpillars in cells with eggs destined to become male grubs and 10 in those with eggs destined to become females (which grow to be larger than males).

Some might take this behavior, and other similar number-based distinctions that nonhuman species can make, as evidence of the ability to count, but Dantzig reserves the concept of counting for a more demanding type of process, which rests not only on the ability to perceive differences in quantity but also on a grasp of the concept of ordered succession:

> To create a counting process it is not enough to have a motley array of models, comprehensive though this latter may be. We must devise a number *system*: our set of models must be arranged in an ordered sequence, a sequence which progresses in the sense of growing magnitude, the *natural sequence*: one, two, three, ... Once this system is created, *counting a collection* means assigning to every member a term in the natural sequence in *ordered succession* until the collection is exhausted. (p. 8)

In effect, Dantzig requires that behavior that is to count as counting give evidence of appreciation of the principles of both cardinality and ordinality: "matching by itself is incapable of creating an art of reckoning. Without our ability to arrange things in ordered succession little progress could have been made. Correspondence and succession, the two principles which permeate all mathematics—nay, all realms of exact thought—are woven into the very fabric of our number system" (p. 9).

The question of whether animals count in their natural habitat in a sense that meets Dantzig's criteria seems doubtful. More generally, I believe that the majority of students of animal behavior would not contend that animals naturally count, except in a relatively rudimentary sense of the word. Noting the existence of anecdotal reports of animal behavior in the wild that, if taken at face value, seem to suggest the ability to count, Davis and Memmott (1982) contend that there is no solid evidence that animals count in their natural state. That animals use number (numerosity or numerousness) discrimination in their natural habitat seems pretty well established, but exactly how to relate that to counting appears to be a topic of continuing debate.

But do animals have the ability to *learn* to count, if trained to do so? This question has motivated a considerable amount of research (Davis & Pérusse, 1988a, 1988b; Gallistel, 1990; Rilling, 1993). German zoologist–animal behaviorist Otto Koehler (1937, 1943, 1950) made some of the earliest attempts to train animals, more particularly birds, to make discriminations on the basis of number. An example of the kind of task given to his birds was to select from five boxes the one whose lid contained a pattern matching another pattern lying in front of the row of boxes. The patterns differed with respect to the number of spots they contained, and this varied from two to six.

Koehler taught his birds to respond to numerosity (or numerousness, if one prefers) in other ways as well. For example, he was able to teach them to eat only a fixed number of (up to four or five) items (e.g., peas) from a larger number that was available. Unfortunately, details regarding Koehler's experiments are sparse, and investigators of counting behavior by animals have been cautious about interpreting his results as firm evidence of counting ability. Koehler himself was unwilling to attribute to his subjects the ability to count. They could not count, he argued, because they lacked words. What they could do, he suggested, was "think unnamed numbers." The possibility that Koehler's birds could have been making their discriminations on the basis of cues confounded with numerosity has not gone unnoted (Thomas & Lorden, 1993; Wesley, 1961).

Other investigators have demonstrated that pigeons can be taught to discriminate between sets containing different numbers of objects, within limits. Rilling and McDiarmid (1965) trained pigeons through operant conditioning to discriminate between 50 pecks and 35 pecks, but as noted above, discriminating more from fewer could conceivably be made on the basis of gross perceptual features and not require counting. Being able to discriminate, say, 15 from 14 would be much more compelling evidence of the ability to count.

Watanabe (1998) trained pigeons to respond to a set of four objects while refraining to respond to a set of two objects, and to do this when the objects comprising the sets of four and two were varied in size and shape. He interpreted his results as evidence that the birds were able to abstract "twoness" and "fourness," but not as evidence of their ability to count sequentially from one to four. Other experiments purporting to demonstrate the ability of pigeons to make distinctions on the basis of numerosity include those of Rilling (1967), Honig (1993), Xia, Siemann, and Delius (2000), and Xia, Emmerton, Siemann, and Delius (2001).

Pepperberg trained an African grey parrot, Alex, to produce different vocalizations to collections of from two to six objects, the objects in the collections and their arrangements being varied to invalidate cues that might be used other than number. The same investigator obtained evidence that the bird had acquired a concept of absence or zero, in the sense of being able to vocalize "none" appropriately when questioned about properties or objects that were missing from a display. Pepperberg's 30-year odyssey with this remarkable bird, which included the acquisition of many cognitive abilities other than those involving numbers, is documented in several technical publications (Pepperberg, 1987, 1988, 1994, 1999) as well as in a popular book-length account (Pepperberg, 2008).

Researchers have trained rats by operant conditioning to press a bar a specified small number of times (e.g., eight) in order to obtain food (which is withheld if the number of bar presses is not correct, or nearly so). Mechner (1958; Mechner & Guevrekian, 1962) trained them to press one lever, A, the desired number of times (up to 16) in order to get food by pressing a second lever, B. The number of presses of lever A was generally approximate to the target number, the spread around that number increasing with the number's size. Davis and Bradford (1986) showed that rats can quickly learn to select which of six tunnels to enter in order to obtain food, when cues other than ordinal position of the tunnel are controlled.

On the basis of numerous operant conditioning experiments with rats over three decades, Capaldi (1964, 1966; Capaldi & Miller, 1988a, 1988b) concluded that the animals count reinforcing events. On the question of whether what animals can do really amounts to counting, Capaldi's (1993) answer is unequivocal: "Animals, at least animals as highly developed as the rat, count routinely, I suggest" (p. 193). *Routinely* is worth emphasis here; Capaldi explicitly dismisses the contention that if animals can count, they can do so only under highly contrived circumstances designed to maximize the opportunity for them to learn: "Counting is not some esoteric activity engaged in by rats when no other means of solution is open to them, as suggested by Davis and Memmott (1982) and Davis and Pérusse (1988). Rather, rats count routinely. By this I mean that it is reasonable to assume that rats count in a wide variety of conventional learning situations" (p. 206).

Fernandes and Church (1982) trained rats to press one lever in response to a sequence of two sounds of fixed duration and to press a different lever in response to a sequence of four sounds of the same fixed duration. Because the total sound time was redundant with number of sounds in this task, Fernandes and Church tested the rats on a transfer task in which the durations of the individual sounds was modified so the total sound time was the same for both the two-sound and the four-sound stimuli, and the rats made the discrimination on the basis of the different numbers of sounds.

In a subsequent experiment, Meck and Church (1983) trained rats to respond differentially to stimuli for which duration and number were confounded (e.g., one stimulus was two events in two seconds and another was four events in four seconds). After training, the rats responded correctly either to number (number of events varied with duration held constant) or to duration (number of events held constant with duration varied), showing that they had encoded both number and duration during training. Similar results were obtained by Roberts and Mitchell (1994) with pigeons.

Church and Meck (1984) also trained rats to press either a left or a right lever in response to two tones or two lights (left) or to four tones or four lights (right). When tested, the rats pressed the left lever in response to one tone and one light in combination (two events) and the right lever in response to a combination of two tones and two lights (four events), indicating that they had learned to respond on the basis of number, independently of the signals' sensory mode. These and other experiments (Church & Gibbon, 1982; Gibbon & Church, 1990; Meck, Church, & Gibbon, 1985) provide strong evidence that rats can learn to respond either to duration or to the number of sequential events when possibly confounding temporal variables are adequately controlled. Following a review of experiments of the sort just described, Broadbent, Church, Meck, and Rakitin (1993) conclude that "a substantial body of evidence indicates that timing and counting have a shared mechanism" (p. 185). What that mechanism is remains to be determined, but "it is clear," they contend, "that any model that provides an explanation of counting should also explain the data that link counting and timing" (p. 185).

The ability to learn to make numerousness or numerosity discriminations has also been demonstrated with various species of monkeys (Brannon, 2005; Brannon & Terrace, 2000, 2002; Hicks, 1956; Rumbaugh & Washburn, 1993; Thomas & Chase, 1980; Thomas, Fowlkes, & Vickery, 1980; Washburn & Rumbaugh, 1991), chimpanzees (Boysen, 1992, 1993; Boysen & Berntson, 1989, 1996; Ferster, 1964; Matsuzawa, 1985; Matsuzawa, Asano, Kubota, & Murofushi, 1986; Tomonaga & Matsuzawa, 2002), raccoons (Davis, 1984), and dolphins (Kilian, Yaman, von Fersen, & Güntürkün, 2003).

Using sets controlled for shape, size, and surface area, Brannon and Terrace (1998, 2000) obtained evidence that monkeys represent numerosities one through nine at least ordinally. The animals learned to order sets of one through four in ascending or descending order, and what they learned transferred to sets of five through nine in ascending order, but not in descending order. Similar transfer has been obtained with a squirrel monkey and a baboon (Smith, Piel, & Candland, 2003). More recently Brannon, Cantlon, and Terrace (2006) were able to get some transfer to testing on 3→2→1 when the animals were trained first on 6→5→4.

Some of the results obtained with monkeys suggest that the animals represent numerical values as analog magnitudes. Cantlon and Brannon (2005), for example, trained monkeys to choose the pattern with the larger number of items when the patterns were superimposed on a blue background, and to choose the pattern with the smaller number when the background was red. They found that choice time varied inversely with magnitude of the difference between the numbers of items on the two displays being compared. Similarly, Beran (2007) found that when monkeys had to select the more numerous of two sets varying in size from 1 to 10 items, percent correct varied directly with the magnitude of the difference between the set sizes.

Cantlon and Brannon (2006) trained monkeys to respond to pairs of patterns on the basis of number of items contained in them—to press the pattern with the smaller number of items first and then the one with the larger number. After training with patterns containing from 1 to 9 items, patterns with 10, 15, 20, and 30 items were added to the mix. What the animals learned from training with the less numerous sets transferred spontaneously to the more numerous sets. Response time varied inversely and percent correct directly with the ratio of the numbers of items in the larger and smaller of sets being compared. These results are reminiscent of the finding by Moyer and Landauer (1967, 1973) of evidence of analog representation of numbers by humans. Cantlon and Brannon also suggest that most, if not all, of the quantities monkeys are able to discriminate are probably represented only approximately, although Thomas and Chase (1980) were able to train one monkey to tell the difference between collections with eight elements and those with nine, which seems to require a more exact representation.

A few efforts have been made to give animals a symbolic representation of number. Washburn and Rumbaugh (1991) demonstrated that monkeys are capable of learning to select, after much training, the (numerically) larger of two Arabic numerals, even when the two numerals had not been paired previously. Other successful efforts to train animals to associate symbols with numerosities have been made by Olthof, Iden, and Roberts (1997) with monkeys, by Ferster (1964) and Boysen and Berntson (1989; Boysen, 1993) with chimpanzees, and by Xia et al. (2000, 2001) with pigeons.

Matsuzawa and colleagues trained a chimpanzee to select an Arabic numeral reflecting the number of objects—up to nine—in a display. This is perhaps the most sophisticated behavior relating to counting that has been demonstrated with animals; it clearly reveals the ability to make discriminations on the basis of number (assuming all cues that could be correlated with number are ruled out by counterbalancing), but it does not prove that animals can count in the fullest sense. One could learn to make these discriminations and to associate the different patterns with a set of symbols (numerals or letters of the alphabet) without having an appreciation of the idea that the quantities involved constitute a progression. Tomonaga and Matsuzawa (2002) agree with Murofushi (1997) that the chimpanzee's performance also does not provide compelling evidence of the ability to count in the sense of using a one-to-one mapping process as distinct from estimating numerosity.

Woodruff and Premack (1981) trained a chimpanzee to match objects on the basis of fractional parts, for example, to select 1/2 an apple (rather than 3/4 of an apple) to correspond to 1/2 a glass of water, and to select (more often than not) a combination of 1/2 of one thing and 1/4 of another to correspond to 3/4 of something else, suggesting a rudimentary form of fraction addition.

That animals can make discriminations based on numerosity seems now beyond doubt. But can they *count*? Obtaining an answer to this question is hindered by the fact that what constitutes counting has not yet been established to everyone's satisfaction, though a variety of distinctions relating to the question have been made. Honig (1993) distinguishes among numerosity discrimination, number discrimination, and counting: "A *numerosity discrimination* involves a discrimination between nonadjacent numbers, or between different ranges of numbers of elements. A *number discrimination* involves differential responding to adjacent numbers of items, such as 3 and 4, 7 and 8, and so forth. *Counting* is a number discrimination in which different responses are made to each of a series of adjacent numbers of items" (p. 62). On the basis of his own work with pigeons, Honig concludes that the birds are capable of numerosity discrimination "but number discriminations are more difficult" (p. 62); he did not use a paradigm to test for counting.

Davis and Memmott (1982) distinguish a continuum of number-related abilities ranging from simple number discrimination through counting, to a concept of number, and the ability to perform operations on numbers. With respect to what constitutes counting, they give two requirements: "(a) the availability of some form of cardinal chain and (b) the application of that chain in one-to-one correspondence to the external world" (p. 565). They consider number discrimination to be within the normal ability of many animals. The concept of number, on the other hand, they consider to be probably beyond the ability of most infrahuman species. On the basis of a review of several efforts to teach animals to count involving both operant and Pavlovian conditioning paradigms, they conclude that although for many studies in which counting has been claimed an alternative explanation is possible, nevertheless the evidence is compelling that animals can learn to count, by their definition, at least up to three. Davis and Pérusse (1988a) distinguish between counting and counting-like behavior that falls short of true counting. The latter, which they call proto-counting, is what they believe animals are capable of doing.

Gallistel (1993) contrasts numbers as categories and numbers as concepts:

> A mental category, in the usage I propose, is a mental/neural state or variable that stands for things or sets of things that may be discriminated on one or more dimensions but are treated as equivalent for some purpose. The mental category corresponding to 3 is a mental state or variable that can be activated or called up by any set of numerosity 3 and is activated or called up for purposes in which the behaviorally relevant property of a set is its numerosity. On the other hand, a concept, in the usage I propose, is a mental/neural state or variable that plays a unique role in an interrelated set of mental/neural operations, a role not played by any other symbol. The numerical concept 3 is defined by the role it plays in the mental operations isomorphic to the operations of arithmetic, not by what it refers to or what activates it. (p. 212)

Do animals have numerical categories? Do they have numerical concepts? Gallistel's answer is that they have the former, and perhaps the latter as well. Their possession of numerical categories is seen in the fact that they respond to sets on the basis of their numerosity, even though the process by which they categorize sets on this basis is a noisy one. The noisiness is seen, for example, in that when rats have to press a lever n times to get some reinforcement, they typically learn to press it approximately the right number of times, and the accuracy of the approximation varies inversely with n. Gallistel's conclusion that animals probably also have numerical concepts rests on the results of experiments suggesting that "common laboratory animals order, add, subtract, multiply, and divide representatives of numerosity" (p. 222).

Rumbaugh and Washburn (1993) argue that "to conclude that an animal counts, one must, among other things, demonstrate (a) *how* it enumerates items of sets, things, or events; (b) that it partitions the counted from the uncounted; (c) that it stops counting appropriately; and (d) that it can count different quantities and kinds of things" (p. 95). Again, "to count, an organism must know each number's ordinal rank and that each number's cardinal value serves to declare the total quantity counted at each step in the counting process (e.g., the item assigned 'three' is the third one and, also, that three items have been enumerated)" (p. 96). I think it safe to say that very few studies of animals' dealings with numerosity have clearly demonstrated all of these criteria. So by this definition, while animals unquestionably can make many impressive discriminations based on numerical properties, a large majority of those discriminations fall short of being conclusive demonstrations of counting. Rumbaugh and Washburn (1993) point, however, to some of their own work (Rumbaugh, 1990; Rumbaugh, Hopkins, Washburn, & Savage-Rumbaugh, 1989; Rumbaugh, Savage-Rumbaugh, Hopkins, Washburn, & Runfeldt, 1989) as evidence that a chimpanzee (an extensively language-trained chimpanzee) can be taught to count, which is to say, "to respond to each of three Arabic numbers in a differential and relatively accurate manner—in a manner that we term *entry-level counting*" (p. 101).

Davis (1993) makes a distinction between *relative* and *absolute* numerosity abilities and argues that, while animals have the former, the latter are unique to humans. Animals are able, he contends, to demonstrate many forms of numerical competence under supportive conditions, such as those typically provided in research laboratories, but the supportive conditions are essential. "In short, I do not believe that demonstrations of numerical competence come easily. They are no mean feat, and for each of the successes you have read about, there are untold failures. Although the nondissemination of negative evidence is the way science normally progresses, it is particularly unfortunate in the case of numerical competence in animals because it clouds the question of how general or easily established this ability is" (p. 110). Davis argues that much of what has been interpreted as counting behavior by animals can be explained in other ways, and that the search for rudiments of human abilities among other species may have had the cost of reducing the probability of discovering abilities that other species have that humans do not. Von Glasersfeld (1993) makes the same point and characterizes the "often unconscious supposition that the way we tend to solve problems of a certain kind is the only way of solving them" as "a widespread manifestation of *anthropocentrism*" and one that "seems almost unavoidable in cases where we have not thought of an alternative solution" (p. 249).

I have already noted that Terrell and Thomas (1990) suggest that the concept of subitizing may have outlived its usefulness. Miller (1993) dismisses as untenable the belief that subitizing can account for much of the empirical data on animal perception of numerosity. Thomas and Lorden (1993) also argue that the idea of proto-counting, as proposed by Davis and Pérusse (1988a), is unjustified. They hold that some discriminations that might appear to result from counting can be made on the basis of pattern detection. "Prototype matching is a well-established process to explain the acquisition and use of class concepts in general, and we suggest that numerousness concepts are not an exception" (p. 141). Similar positions have been expressed by Mandler and Shebo (1982) and by von Glasersfeld (1982, 1993). Von Glasersfeld (1993) argues that "it is one thing to recognize a spatial or temporal pattern as a pattern one knows and has associated with a certain name, and quite another to interpret the pattern as a collection of unitary items that constitute a certain numerosity" (p. 233).

Many studies have addressed the question of whether animals have the ability to do, or to be taught to do, simple arithmetic. The literature on the subject is large and I will not attempt to review it here. For present purposes it suffices to note that the types of arithmetic capabilities that have been observed in animal studies are roughly comparable to those that have been observed in studies of human infants and prelingual children (Boysen & Berntson, 1989; Brannon, Wusthoff, Gallistel, & Gibbon, 2001). A study by Rumbaugh, Savage-Rumbaugh, and Hegel (1987) is illustrative of what has been done. These investigators gave chimpanzees a choice between two trays, each of which had a pair of food wells containing a few (in combination not more than eight) chocolate chips. Over time, the chimps learned to choose the tray whose food wells, in combination, held the larger number of chips, suggesting that they were adding the contents of the two food wells, or something equivalent to that. The finding was replicated and extended slightly by Pérusse and Rumbaugh (1990).

In sum, studies have shown with reasonable certitude that animals, including not only primates but rats and birds, can be taught to make limited distinctions based on number. Precisely what animals are doing when they are making such distinctions is still a matter of debate, as are the questions of how numerical quantities are represented by animals (or infants) in the absence of a verbal code and whether the distinctions that are made require a specifically numeric sensitivity or can be done with more generic capabilities (Dehaene & Changeux, 1993; Gallistel & Gelman, 1992, 2000; Gelman & Gallistel, 1978; Meck & Church, 1983; Mix, Huttenlocher, & Levine, 2002a,b; Simon, 1997, 1999). In the aggregate, the evidence suggests that, whatever the representation, it produces a relationship between quantities that is described, to a first approximation and at least for quantities greater than 3, by Weber's law, according to which the just discriminable difference between two quantities is a constant proportion of their size; so it is easier, for example, to discriminate between 4 and 5 than between 8 and 9, and easier to discriminate between 8 and 9 than between 16 and 17. Some theorists hold that small quantities (one to three) are represented discretely, whereas larger quantities are represented in a more continuous form (Cordes & Gelman, 2005).

Few if any experiments have yielded incontestable evidence that animals can learn to count in the sense of doing something equivalent to enumerating the items in a set and consistently telling the difference between sets of modest size that differ only by a single item. A common criticism of the methods used to study counting, or counting-like, behavior by animals is that number often is confounded with other variables (density, area covered by ensembles, brightness, patterns, interelement distances, durations or timing of events, rhythm, and as in the case of Clever Hans, cues too subtle to be detected by most human observers). Moreover, in controlling for one confounding variable, one can easily increase the salience of other confounds, or introduce new ones. Concern about the possibility of responses to correlates of numerosity being misinterpreted as responses to numerosity, per se, is unquestionably well founded. On the other hand, as Miller (1993) points out, absent empirical data, one should not assume that nonnumerical cues will always overshadow numerical cues. Two things are clear. First, the question of whether animals can count, either naturally or as a consequence of careful training, remains a matter of definition and of lively debate. Second, the most impressive examples of the numerical abilities of animals provide only the vaguest hint of the type of numerical competence that human beings somehow developed.

☐ *Homo sapiens* Learns to Count

Mathematics is an astoundingly broad subject. The concepts involved range from those that are sufficiently intuitively simple that they can be comprehended by very young children to those that are sufficiently complex that only a handful of people in the world are likely to be completely in command of them. As is true of other conceptually rich subjects, many of the concepts are related in a hierarchical fashion, in the sense that understanding those at a given level of the hierarchy is unlikely, if not impossible, unless one understands those that are closely related at lower levels in the hierarchy. Understanding the operation of raising to a power, for example, requires an understanding of multiplication. Performance of simple arithmetic operations presumably depends on the ability to count. The ability to count presupposes the ability to distinguish discrete objects.

This view of mathematics as hierarchically structured motivates efforts to determine the *level* at which children of a specific age or mental development are capable of functioning. It also is conducive to the idea that the acquisition of mathematical competence by humankind as a species necessarily progressed up a similar, if not the same, ladder—counting before calculating, concrete concepts before abstract ones, arithmetic before algebra, and so on. From this perspective it is not surprising to find an interest in the development of mathematical competence in children being combined with an interest in the history of mathematics, as is evident, for example, in the work of Piaget (1928, 1941/1952; Piaget, Inhelder, & Szeminska, 1948/1960), who, as Resnick and Ford (1981) put it, "thought it possible ... to understand the development of the species' intellectual capacities by studying the intellectual development of individuals as they grew into adults" (p. 156).

Sadly, the early history of the development of mathematics is not known in any detail. All we have is guesswork aided by a few clues to the progression of the ability to count, represent quantities, and calculate in prehistoric times. One thing is clear, however; the ability to count, whenever and however it was obtained, not only constituted a major step in the development of mathematics but was an extraordinarily useful ability in its own right. Imagine a shepherd who could not count trying to keep track of his sheep. If his flock were sufficiently small, he might satisfy himself that they were all accounted for by checking them one-by-one against a mental list (assuming he could recognize the individual sheep), but if the flock were large, this would be an impossible task. Knowing the *number* of sheep in his flock and being able to count at least up to that number greatly simplifies his task.

There is considerable dispute among mathematicians as to whether ordinality or cardinality is the more fundamental concept, and whether either is more fundamental than the natural numbers. There is some evidence that the individual child develops a concept of, or is taught, ordinality before cardinality, and that the former but not the latter precedes and is fundamental to the development of number competence. Dantzig (1930/2005) notes that inasmuch as cardinality is based on matching only, while ordinality requires both matching and ordering, there is a temptation to assume that cardinality preceded ordinality in the history of the development of the number concept, but he argues that investigations of primitive cultures have not revealed such precedence: "Wherever any number technique exists at all, both aspects of number are found" (p. 9).

Psychologists have been on both sides of the debate regarding the priority of ordinality or cardinality. Some hold that the priority of cardinality is seen in the ability of children to make discriminations on the basis of number (manyness) among collections of objects that are few in number before they can count in any meaningful sense (Nelson & Bartley, 1961). Whether what is being discriminated in these cases is really number, as distinct from some property—such as density, the spatial closeness of the objects—has not always been clear. The priority of ordinality has been proposed by Brainerd (1973a, 1973b, 1979), who contends that most children have a good grasp of ordinality, but not of cardinality, before they begin school. Cardination, he argues, is generally not a relatively stable concept until children are roughly 9 or 10 years old, which is considerably after the time they are expected to begin learning arithmetic.

Brainerd (1979) describes his ordinal theory of the acquisition of number concepts as postulating a process with three overlapping phases: "What the theory actually says is that most children will have made considerable progress with the notion of ordination before they make much progress with arithmetic, and most children will have made considerable progress with arithmetic before they make much progress with cardination" (p. 168). Following a brief review of some work of Piaget (1952), Dodwell (1960, 1961, 1962), Hood (1962), Beard (1963), Siegel (1971a, 1971b, 1974), Wang, Resnick, and Boozer (1971), and Beilin and Gillman (1967), Brainerd concludes that "the available evidence on ordination, cardination, and arithmetic is sketchy and not very conclusive. Insofar as the developmental relationship between ordination and cardination is concerned, there is very little solid evidence" (p. 126). There is much more to be said regarding the question of when and how children acquire an appreciation of the ordinal and cardinal properties of number, but the acquisition of these properties is complicated in that many of the numbers to which children are exposed in their early years—TV channel numbers, numbers on athletes' jerseys, numbers on buses or trains, license plate numbers, street address numbers, telephone numbers—generally are neither ordinals nor cardinals but serve only a nominal function.

A distinction that is important to a full understanding of numbers, in addition to that between ordinality and cardinality, is the distinction between *count* numbers and *measure* numbers (Munn, 1998). Count numbers apply to the numeration of discrete entities; measure numbers are used to represent continuous variables, such as length, weight, and temperature. For present purposes, let us note that the question, still not completely settled, has obvious implications for the teaching of elementary mathematics, a topic to which we will return in Chapter 15.

Are there people (cultures) who do not count, or is counting an activity that is common to all cultures? Apparently there are cultures—probably very few—in which the ability to count has not been developed much beyond distinguishing among one, two, and many. Flegg (1983) cites the example of the Damara of Namibia, who would "exchange more than once a sheep for two rolls of tobacco but would not simultaneously exchange two sheep for four rolls" (p. 19). Another instance of a group that lacks words for specific numbers beyond 2 is a small hunter–gatherer tribe, the Pirahã, that lives in relative isolation along the Macai River in the Amazon jungle (Gordon, 2004).

It appears that children in all cultures that count use their fingers to facilitate the process (Butterworth, 1999), but beyond this, various methods for counting and for representing the results thereof have been developed and used in different cultures over the ages. Conant (c. 1906/1956) speculates that every nation or tribe has developed some method of numeration before having words for numbers. Detailed accounts of several systems, used for both counting and calculating, based on finger positions and other body parts, are readily available (Dantzig, 1930/2005; Ellis, 1978; Flegg, 1983; Ifrah, 2000; Menninger, 1992; Wassmann & Dasen, 1994). A system described by the Venerable Bede, who lived in the eighth century, could represent quantities as great as 1 million (Figure 2.2). Apparently people—merchants, traders, and ordinary folk—in most parts of the world have used finger counting at one time or another, and there are places where its use is still common (Flegg, 1983).

Saxe (1981, 1982, 1985) describes a system of tallying and measuring used by the Oksapmin of Papua New Guinea that associates numbers with 27 different locations on the hands, arms, shoulders, and head (see also Lancy, 1978; Saxe, 1991; Saxe & Posner, 1983). Saxe, Dawson, Fall, and Howard (1996) note that, unlike the notational system (Hindu-Arabic) used today throughout Western society, which is specialized for computation, the Oksapmin system is specialized for counting and certain forms of measurement and is not particularly conducive to computation.

Flegg (1983) notes that an abstract understanding of numbers is not needed in order to be able to count. He contends that the activity of counting, including the development of finger systems to facilitate the process, predated the concept of numbers in the abstract and that the intellectual step from the former to the latter is a large one that came relatively late in human history. Exactly when this step was taken is not known; Flegg speculates that an awareness of numbers in the abstract developed sometime around 3,000 BC or a bit later, but that it did not become an influential part of mathematical thinking until about the time of Pythagoras. But the species did learn to count—one way or another.

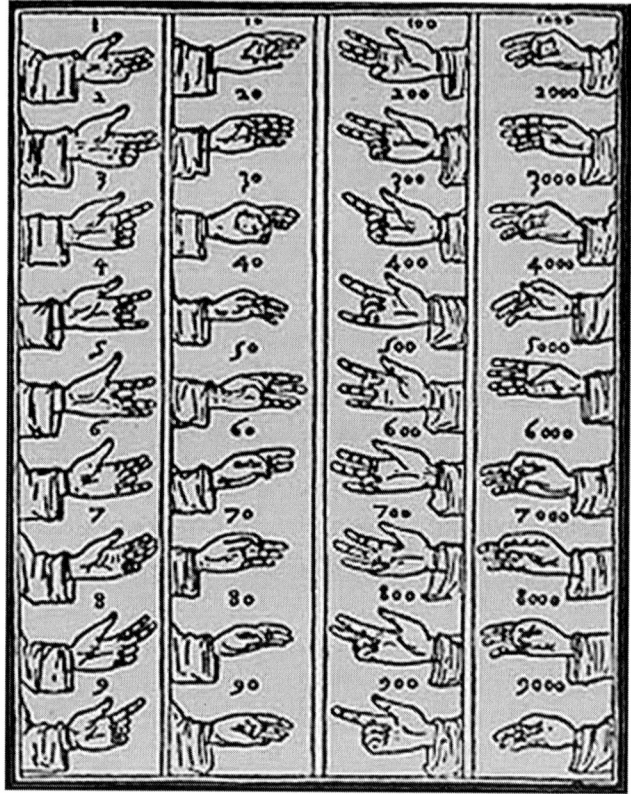

FIGURE 2.2 The Venerable Bede's system for the manual representation of numbers as rendered on a woodcut originally published in *Summa de Arithmetica* by Italian mathematician Luca Pacioli near the end of the 15th century.

☐ And to Calculate

Calculating—performing operations on numbers such as adding, subtracting, multiplying, dividing, and so on—is a considerably more abstract process than counting. For some of the operations that can be performed on numbers there are fairly obvious analogs that can be performed on concrete objects. What one can do with numbers, however, is much less constrained than what one can do with concrete objects. One can add 6 apples and 6 apples and get 12 apples, just as one can perform the abstract operation $6 + 6 = 12$. But one does not multiply 6 apples by 6 apples to get 36 apples, and while 6^2 apples makes sense, (6 apples)^2 does not. The physical analog of the multiplication operation for the natural numbers is successive addition. To stick with the apples example, one gets 36 apples by adding 6 sets of apples, each of which contains 6 apples, and one gets 6^2 apples by computing $6^2 = 36$ and counting out that number of apples.

The origin of calculating, like that of counting, is hidden in the mists of prehistory. No one knows who first noticed that when one item is combined with one item the result is invariably two items. Or that when two items and three items are put together, their combination always is five items. Or how long it took for the concept of an addition operation to emerge.

Much of the history of the development of counting and calculating competence is recorded in the techniques and artifacts that were invented to aid either process or both. Ancient physical aids to counting and calculating include counting boards, abaci, and related devices. Traders, merchants, and tax collectors used such devices for counting and computing before they made use of, or perhaps even had, written systems of numerals.

A system of "reckoning on the lines" was used widely, especially for commerce, from the 12th century in Europe and was sufficiently popular to motivate strong opposition to the adoption of the Hindu-Arabic numerals and their use in calculation (see Flegg, 1983, p. 157 for an example of this representation). A number was represented by the placement of dots with respect to a column of horizontal lines, each line representing a different power of the base. Each dot on a line represented one instance of the associated power of the base; a dot between lines was equivalent in value to five dots on the line immediately below it. Arithmetic calculations with this representation were relatively straightforward.

The Incas had developed a system of knotted cords (quipus or khipus) that was in use for purposes of accounting and numerical record keeping at the time of the Spanish conquest of Peru (Asher & Asher, 1981; Salmon, 2004). Similar systems were used to represent quantities elsewhere, including among North American natives and in parts of Africa, as well.

A system for doing "finger multiplication," described by Robert Recorde in *Grounde of Artes* in 1542, was effective for multiplication and, by using 10s-complement arithmetic, required one to memorize a multiplication table only up to 5×5 (which involves only 15 products, compared with 78 in a table up to 12×12).

The history of the development of procedures of computation and of the invention of devices that compute or facilitate computing is a fascinating one, but one that I will not attempt to relate here. I have touched on the subject briefly elsewhere (Nickerson, 1997, 2005), and there are many sources of extensive information on the topic. It suffices for present purposes to note that *Homo sapiens* not only learned to count, but to compute, and that, arguably, this ability has been second in importance only to that of language in the development of civilization as we know it.

Numbers

It is widely held that no other idea in the history of human thought even approaches number's combination of intellectual impact and practical ramifications. (Brainerd, 1979, p. 1)

The real numbers are taken as "real." Yet no one has ever seen a real number. (Lakoff & Núñez, 2000, p. 181)

"What," Warren McCulloch (1961/1965) asks, "is a number, that a man may know it, and a man, that he may know a number?" Wynn (1998) puts essentially the same question somewhat more prosaically, but no less emphatically: "The abstract body of mathematical knowledge that has been developed over the last several thousand years is one of the most impressive of human achievements. What makes the human mind capable of grasping number?" (p. 3). The concept of a number line extending from 0 to infinity in both directions is among the more powerful representations in mathematics, or any other domain of thought. But what exactly is a number? The question is easy to ask, but the answer is neither obvious nor simple:

The concepts of counting and number have taxed the powers of the subtlest thinkers and problems of number theory which can be stated so that a child can grasp their meaning have for centuries withstood attempts at solution. It is strange that we know so little about the properties of numbers. They are our handiwork, yet they baffle us; we can fathom only a few of their intricacies. Having defined their attributes and prescribed their behavior, we are hard pressed to perceive the implications of our formulas. (Newman, 1956c, p. 497)

The reader who is inclined to doubt such an assertion is referred to Russell's (1903, 1919) Herculean effort to answer the question and to the many objections it evoked (Cassirer, 1910/1923). The puzzle is that we all understand instinctively what a number is—until we try to define it, as mathematicians have been trying to do, with dubious success, for a very long time. Perhaps, as some have claimed, number is a primitive concept, and not definable.

In any case, numbers have a history, and the final chapter has yet to be written. We do not know who first used tokens or symbols to represent quantities. Nor do we know whether this use of tokens or symbols was invented once or many times. In either case, it was a momentous invention. Whether or not counting is unique to humankind, the use of symbols to represent numerical concepts certainly is.

☐ Beginnings

The origins of systems for representing numbers are obscure. Theories rest on scattered archeological evidence from a few sites and a considerable amount of speculation. Some of this evidence suggests that the earliest systems predated the development of written language and were very concrete. Notched bones may have been used to record phases of the moon, the number of animals killed by hunters, and other matters of interest possibly as many as 20,000 to 30,000 years ago. Token systems, of small objects, possibly coded by shape, size, and marking, appear to have been used in Western Asia and the Middle East to record and communicate the number of entities in a collection—number of animals in a herd—at least as many as 11 millennia ago, before a system of written numerals began to emerge.

According to one theory, these token systems were immediate precursors to the development of writing, which evolved from them in an interesting way. Merchants adopted the practice of enclosing in sealed clay containers tokens representing merchandise that was being transported from one place to another and using them as bills of lading. By breaking open a container, a recipient of a shipment of goods could determine from the number and types of tokens inside what the shipment was supposed to contain. At some point, users of this system began making inscriptions on the outside of containers to represent the tokens they contained. In time, it is hypothesized, the redundancy of this system was noticed, the tokens were done away with, and users began to rely solely on the inscriptions (Schmandt-Besserat, 1978, 1984).

Archeological records show that by the latter part of the fourth millennium BC the use of inscriptions on clay tablets was common throughout Sumerian and Elamite sites. The inscriptions, which are believed to be primarily accounts and receipts, contain many number representations (Friberg, 1984).

☐ Number Abstraction

Numerosity initially was probably thought of as a property of objects, so the symbol for a given quantity would differ depending on what was being counted. Thus, the concepts "three sheep" and "three fish" were distinct, represented in different ways, and not thought of as sharing the abstract property of threeness. Perhaps we see some remnants of the one-time concreteness of numbers in the special names that we have for representing the same quantity in different contexts: couple, duo, duet, dual, pair, brace, twice, twin, twain, and so forth. Rucker (1982) points out that Renaissance mathematicians hesitated to add x^2 and x^3 because the first represented an area and the second a volume.

We have no idea how or when abstraction of the concept of number from the thing counted occurred, how or when the realization dawned that three stones and three fingers have something very interesting—their numerosity—in common. The recognition of number as something distinguishable from objects counted and independently manipulatable—that *three* could be a noun as well as an adjective—was a major breakthrough in the intellectual odyssey of humankind. Today we use numbers as both adjectives and nouns, and failure to distinguish between the two uses can make for confusion in beginning arithmetic. Two cats plus three cats makes five cats, but we do not say that two cats times three cats makes six cats; we would say that two cats times three makes six cats. And, of course, without reference to what is being counted, we say without fear of contradiction that $2 + 3 = 5$ and $2 \times 3 = 6$.

Recognition that number is a property of a set and not of the items that comprise the set—that three fingers, three apples, and three ducks have in common their threeness—requires an abstraction. Realization that two fingers, three apples, and four ducks have in common their numerability—that each is a set that can be counted and characterized by a number, though not the same one—represents a higher-level abstraction.

The history of the development of the number concept is a study of a progression of abstractions. The number universe, populated initially with the positive integers ("natural" numbers), effective for counting discrete entities but severely limited as a basis of reckoning and partitioning wholes into parts, expanded over time to include fractions (rationals), negative numbers, irrationals (radicals), imaginaries, and more. The rationals and irrationals in combination comprise the reals. Rational numbers can be expressed as fractions in which both numerator and denominator are whole numbers; irrational numbers cannot be represented this way. When expressed in decimal form, rational and irrational numbers differ in that the former either get quickly to a string of zeros or to a sequence of digits that repeats endlessly, whereas the latter do neither but continue indefinitely with no repeating pattern. The irrationals include both algebraic numbers and transcendentals. Algebraic numbers are solutions of algebraic equations with integer coefficients; the square root of 2, for example, though irrational is algebraic, because it is the solution of $x^2 - 2 = 0$.

Transcendental numbers are a subset of irrationals; they are numbers that are not solutions—roots—of polynomial equations with integer coefficients. In addition to $\sqrt{2}$, they include π, e, ϕ, the golden ratio (about which more in Chapter 11), and many trigonometric and hyperbolic functions. (For one writer's list of the 15 most famous transcendental numbers, see Pickover, 2000.) Although Euler believed that transcendentals exist, their existence was first proved in the 19th century by French mathematician Joseph Liouville. Johann Heinrich Lambert, a German mathematician and physicist, surmised in 1761 that both e and π are transcendental. French mathematician Charles Hermite proved e to be transcendental in 1873, and German mathematician Carl Ferdinand von Lindemann proved π to be so in 1882. German mathematician Georg Ferdinand Ludwig Philipp Cantor proved the transcendentals to be abundant, despite the difficulty of identifying individual cases, and in doing so, evoked much unwelcome and unsettling criticism from other mathematicians.

The early (pre-Pythagorean) Greeks recognized only positive whole numbers as numbers; 2 and 3 were numbers in their view, but not –2 or –3, or 2/3. The Pythagoreans recognized rationals—quantities that could be expressed as the ratio of two whole numbers—as real, but were scandalized by the discovery that the length of the diagonal of a unit square cannot be expressed as such a number. This is an early example—there are many more in the history of mathematics—of mathematical thinking hitting a barrier, the crossing of which required the invention or discovery of new concepts.

The admission to the family of numbers of quantities that cannot be represented as the ratio of whole numbers, such as $\sqrt{2}$ and π, was a slow and painful process; acceptance of $\sqrt{-1}$, or i, and complex numbers was even more so. Today all these types of representations are familiar members of the number lexicon, but there are more recent additions—transfinites (Cantor), hyperreals, hyperradicals and ultraradicals (Kasner & Newman, 1940; Robinson, 1969), surreals (Conway, 1976; Tøndering, 2005), inaccessible, hyperinaccessible, indescribable, and ineffable cardinals (among other esoteric kinds) (Rucker, 1982)—that are considerably less so. A compelling argument can be made that to the question "What really is a number?" there is no universally agreed-upon answer. If Lakoff and Núñez (2000) are right, "not only has our idea of number changed over time, but we now have equally mathematically valid but mutually inconsistent versions of what numbers are" (p. 359).

From the vantage point of the 21st century, it is easy to fail to appreciate the sometimes slow and difficult path to acceptance of many of the abstractions that we take for granted, but two facts pertain to nearly every extension of the number system from natural numbers to rationals, to real numbers, to negative numbers, to complex numbers, and beyond: Each has been motivated to meet a need (to make some mathematical operation possible that was not possible before), and each has occurred against considerable opposition. Negative numbers, for example, were found to be essential to the performance of even the most fundamental of arithmetic operations, such as the subtraction of a larger from a smaller number. One might expect too that they would naturally arise in the treatment of debt in accounting. Nevertheless, most European mathematicians of the 16th and 17th centuries did not accept negative numbers, which they learned of from the Hindus, as *bona fide* numbers—as opposed to "fictitious" ones. McLeish (1994) credits Indian mathematician Brahmagupta, who lived during the seventh century, with the first systematic treatment of negative numbers, including rules for multiplying them, and of their legitimacy as roots of quadratic equations.

Sixteenth- and 17th-century European mathematicians, French polymath René Descartes among them, were unhappy with the idea that negative numbers could be roots of equations. It was not until late in the 17th century that English mathematician John Wallis represented them as quantities to the left of zero on the number line, negative *n* being the same number of units to the left as positive *n* is to the right, a representation that is commonly used today. The use of negative numbers was eventually accepted for practical reasons before a logical foundation had been provided for them, although some textbook authors rejected the possibility of multiplying two negative numbers even as late as the 18th century (Boyer & Merzbach, 1991).

Nahin (1998) recounts the following argument, made by Wallis in 1665:

> Since $a/0$, with $a > 0$, is positive infinity, and since a/b with $b < 0$, is a negative number, then this negative number must be *greater* than positive infinity because the denominator in the second case is less than the denominator in the first case (i.e., $b < 0$). This left Wallis with the astounding conclusion that a negative number is simultaneously both less than zero and greater than positive infinity, and so who can blame him for being wary of negative numbers? (p. 14)

Among other impediments to the acceptance of negative numbers was that in his introduction of analytic geometry, Descartes originally considered equations only in the first quadrant of the x,y plane; as the development of analytic geometry progressed, however, use of the other quadrants, each of which involved negative numbers, was such an obvious extension that this may have facilitated acceptance over time.

☐ Tallies to Ciphers

Realization that symbols like |, ||, and ||| could be used to represent one, two, or three of anything—sheep, fish, children—was an enormously important insight. It was essential to the development of mathematics as we know it. It would be wonderful to know the story of how this happened, but we are unlikely ever to do so. These symbols are used here only for the sake of illustration, but one might defend the notion that the stroke or line (tally mark) is a powerful and profound symbol. It is trivially easy to make with almost any medium (bone, stone, clay, papyrus). The one-to-one correspondence between a sequence of strokes and the set of objects whose numerosity is being represented is direct and salient. Evidence of the notching of bone, presumably to represent tallies, goes back perhaps 30,000 years (Flegg, 1983).

The single vertical stroke was used to represent 1 by several ancient number systems, and some of them use two or three strokes to represent 2 and 3 as well. Apparently around 2000 BC the residents of Harappa, in the Indus Valley (parts of which are in modern India, Pakistan, and Bangladesh), used vertical strokes to represent the numbers from 1 to 7 and abandoned the tally representation only beginning with the number 8, their symbol for which bears some resemblance to our own (Fairservis, 1983).

How long it took from the first uses of tallies in which only a single symbol was used repeatedly to represent a quantity and the beginnings of the use of different symbols to represent different quantities, "cipherization," is not known. Among the oldest known cipherized systems are those of the ancient Sumerians, Babylonians, Egyptians, Greeks, Romans, Aztecs, and Mayans.

☐ The Hindu-Arabic System

Today the Hindu-Arabic system (alternatively referred to as the Indo-Arabic, or simply Arabic, system) for representing numerical concepts is used throughout the civilized world. It is so familiar to us that we may have difficulty in perceiving it as the convention that it is. We tend to think of a string of Hindu-Arabic symbols, such as 267, not as a representation of a number—a numeral—but as the number itself. In fact, this system is a relatively recent chapter in the history of the development of number systems. The Hindu-Arabic system is generally referred to as such because it was widely believed to have been developed by the Hindus in India, and its familiarity to the West owes much to the work of Arab scholars. It appears that this system—in principle but not in detail the one used in India by the seventh century—was adopted by the Arabs probably during the eighth century and was known here and there in Europe for several hundred years before it was appropriated there for general use. Decimal systems with different symbols were in use in other parts of the world, including China, centuries earlier (Ifrah, 2000). (For a brief account of the origin of the characters that constitute our decimal integers, see Friend, 1954, or Pappas, 1989.)

Precisely how and by whom the Hindu-Arabic symbols were introduced to medieval Europe is uncertain. According to Ifrah (1987), the earliest known European manuscript that contains the first nine Hindu-Arabic numerals is the *Codex Vigilanus*, which dates from 976. Ifrah credits the French monk Gerbert of Aurillac, who became Pope Sylvester II, with being the first great scholar to spread the use of these numerals in Europe, although the system was not widely used in Europe until several centuries later. Schimmel (1993) identifies a Latin translation by Robert of Chester in about 1143 of a book, *Concerning the Hindu Art of Reckoning*, written in the ninth century by Mohammed ibn-Musa al-Khwarizmi (whose last name is also the origin of our *algorithm*) as the vehicle that introduced this number system, as well as the concept of algebra, to the West. It was perhaps because of an Arab's (al-Khwarizmi's) role in publicizing the Hindu numerals in Europe that the system became known as Hindu-Arabic. A Syrian reference to the Hindu numerals is known from 662, and an Indian plate survives from 595 in which the date 346 appears in decimal place notation form (Boyer & Merzbach, 1991).

Leonardo of Pisa, a scholarly and widely traveled merchant also known to posterity as Fibonacci (a condensation of Filius Bonaccio, or son of Bonaccio), promoted the use of the Hindu-Arabic system in Europe in his book *Liber abaci*, which was circulated in manuscript form early in the 13th century, and was made more widely available when it was published in Latin some six centuries later. At the beginning of the book, Fibonacci identifies the nine figures of the Indians as 1 2 3 4 5 6 7 8 9 and notes that with these nine figures, and the sign 0, any number can be written.

Europe was slow to adopt the Hindu-Arabic system. Some have speculated that this was due in part to the widespread use of the abacus for computation at the time and the advantages of the Hindu-Arabic notation over Roman numerals being less apparent with this method of computation than when calculations are done with pen and paper. Merchants in Florence were forbidden to use the Hindu-Arabic system for fear that their customers could easily be deceived by it (Ellis, 1978). A struggle between *Abacists*, those committed to the use of the abacus and old traditions, and *Algorists*, who advocated reform, went on for three or four centuries. In retrospect, scholars have viewed the eventual adoption of the Hindu-Arabic system of number representation as one of the defining events in the history of Europe and, by extension, of the world. We are so familiar with the system we use today to represent numbers that it is difficult for us to see it as the thing of beauty and power that it is.

The history of the evolution of systems for representing numbers has been told, at least in part, by numerous scholars, among them Smith and Ginsburg (1937), Menninger (1958/1992), Flegg (1983), Ifrah (1987), Barrow (1992), Kaplan (1999), and Seife (2000). It is instructive to compare the Hindu-Arabic system with its predecessors. When one does so, one sees a variety of similarities and differences that have implications for its use. It is more abstract than several of its predecessors, and therefore probably more difficult to learn, but also more compact, more readily extendable, and more conducive to computational manipulation.

The Hindu-Arabic system was not the first that was able to represent any number with a fixed small set of symbols—the Babylonian system, which predated the Hindu-Arabic system by many centuries, could do so with only two symbols, approximated by I and <, as could the Mayan system, which used only a dot (.) and a dash (—). There were other systems that used more than 2 but less than 10 symbols. But the Hindu-Arabic system had an economy of expression and certain other properties, not shared by most other systems, and especially the Roman system in use throughout Europe before its introduction, that made it especially useful for purposes of computation and that ensured its eventual widespread adoption.

Several of the principles on which the Hindu-Arabic number system is based appear to have been discovered independently by different cultures, including the Sumerian or Babylonian, the Egyptian, the Chinese, and the Mayan. The enormously important trick of using a symbol's position relative to other symbols comprising a number to convey information, for example, was used by the Babylonians, the Chinese, and the Mayans, as well as by the Indians and Arabs. Exactly how this trick was discovered and why it was discovered when it was—possibly independently by each of these different cultures—we do not know.

We can only wonder what the current status of mathematics and science would be if it had never been discovered.

The adoption of place notation is generally recognized as of monumental importance to the subsequent development of mathematics. Dantzig (1930/2005) speaks of the event in superlative terms:

> The greatly increased facility with which the average man today manipulates number has been often taken as proof of the growth of the human intellect. The truth of the matter is that the difficulties then experienced were inherent in the numeration in use, a numeration not susceptible to simple, clear-cut rules. The discovery of the modern positional numeration did away with these obstacles and made arithmetic accessible even to the dullest mind.... One who reflects on the history of reckoning up to the invention of the principle of position is struck by the paucity of achievement. This long period of nearly five thousand years saw the fall and rise of many a civilization, each leaving behind it a heritage of literature, art, philosophy and religion. But what was the net achievement in the field of reckoning, the earliest art practiced by man? An inflexible numeration so crude as to make progress well-nigh impossible, and a calculating device so limited in scope that even elementary calculations called for the services of an expert. And what is more, man used these devices for thousands of years without making a single worth-while improvement in the instrument, without contributing a single important idea to the system! (pp. 27, 29)

Gauss once expressed surprise and disappointment that Archimedes had failed to invent a place notation scheme for number representation. He admired Archimedes as one of the greatest mathematical minds of all time and believed that he could have discovered the place notation principle. Gauss realized the importance to his own mathematical accomplishments of a notational system that was conducive to computation, and he believed that had Archimedes made this discovery, many of the subsequent accomplishments in mathematics and science might have occurred centuries before they did. Bell (1945/1992) saw the failure of the Greeks to appreciate the advantages of the arithmetic and algebra of the Babylonians, which were largely a matter of representation, as the cause of them missing an opportunity to anticipate future developments by 2,000 years. "The algebraic-analytic method in mensuration and geometry was well within the capacity of the Greek mathematicians, and they could have developed it with any degree of logical rigor they desired. Had they done so, Apollonious would have been Descartes, and Archimedes Newton" (p. 83). Dantzig (1930/2005) argues that the use by the Greeks of every letter of their alphabet to represent numerals was a serious impediment to their discovery of positional notation. Bernstein (1996) speculates that the reason the classical Greeks failed to discover the laws of probability, calculus, or even algebra might have been the "clumsy" nature of the number system on which they had to depend.

That several of the same principles can be seen in independently developed systems suggests that number representation did not evolve in an arbitrary way, but that its development was constrained by specific requirements or desiderata. It is as though there are certain principles that must characterize any representational system that is to be an effective vehicle for counting, computing, and thinking quantitatively. Why these principles came to be encoded precisely as they are in the Hindu-Arabic system is an interesting question that deserves more attention from researchers than it has received. I do not mean to suggest that the Hindu-Arabic system has no arbitrary features. The shapes of the individual numerals are obviously arbitrary, as is the reading of them from left to right instead of from right to left. What appear not to be arbitrary are the use of such principles as one-for-one mapping, adoption of a standard quantity to serve as a base or radix (10 in the Hindu-Arabic system, but not in all its predecessors), one-for-many substitution, use of a single symbol to represent a multiple quantity (e.g., 3 instead of, say, 111), the idea of a symbol for representing an empty set, and the use of position to carry information.

Perhaps the least significant aspect of the Hindu-Arabic system is that its base is 10. This is a completely arbitrary feature, very possibly a consequence of our happening to have 10 fingers—10 digits, from the Latin *digitus*, meaning finger. Although many, if not most, counting systems used by various cultures are base 10 systems, other bases would serve the purpose very well. A base 2 system is used almost universally in computer technology. There are a few examples of number systems currently in use in the world that have some base other than 10; the Huku of Uganda and the Huli of New Guinea use a base 4 and a base 15 system, respectively (Cheetam, 1978; Zaslavsky, 1973). Among the many predecessors to the Hindu-Arabic system, about which more presently, there are several with bases other than 10. Arguments have been made that bases other than 10 would be advantageous in specific ways. Some of the advantages of 12 and 16 are tacitly acknowledged by the custom of dividing the foot into 12 inches and the inch into fractional powers of 2—halves, quarters, eighths, sixteenths, and so forth (Nickerson, 1999). Several people, including Georges Buffon, Herbert Spenser, and H. G. Wells, have actively promoted the adoption of a base 12 system—in part because 12 is divisible by 2, 3, 4, and 6, while 10 is divisible only by 2 and 5—but without much success.

In any case, it is easy to see the advantage that a system with a base of a modest number has over systems that mark the extremes of the trade-off between the number of different symbols in the system and the average number needed to represent a given quantity. A system with only a single symbol—say the symbol 1—could represent any number as a tally, but for the representation of all numbers but very small ones would be excessively long. At the other extreme, one can imagine a system in which every number has a unique symbol, in which case the representation of any number would require only a single symbol, but there would have to be as many different symbols as there are numbers to be represented. Clearly both of these extremes would place severe restrictions on people's ability to count and calculate.

The Hindu-Arabic system for representing numbers has been widely recognized as an event of enormous practical importance. "The Indian place-value numeration with a zero sign ranks among humanity's fundamental discoveries. Through the centuries it has been propagated even more widely than the alphabet of Phonecian origin, and it has now become the only real universal language" (Ifrah, 1987, p. 459). British logician Philip Jourdain (1913/1956) points to the Hindu-Arabic system as an instance of "that great spirit of economy which spares waste of labor on what is already systematized, so that all our strength can be concentrated either upon what is known but unsystematized, or upon what is unknown" (p. 13). This system is in large part responsible, he suggests, for the fact that many arithmetical problems that were formidable challenges to the ancients seem easy to us.

☐ Predecessor Systems

Two significant properties of the Hindu-Arabic system—extensibility and economy of expression—can be illustrated by contrasting this system with the hieroglyphic notation used by the Egyptians from at least 3000 BC. Like the Hindu-Arabic system, it was a base 10 system. Unlike the Hindu-Arabic system, however, the Egyptian system used a different symbol to represent each power of 10—thus 1s were represented by │ s, 10s by ∧ s, 100s by ⌒ s—and the number of times the quantity represented by a given symbol was to be counted in determining the value of the number was indicated by the number of times the symbol representing that quantity appeared in the representation of the number. Thus, in Egyptian the equivalent of 342 was represented as

$$⌒⌒ ⌒∧ ∧ ∧ ∧ │ │$$

This principle, sometimes referred to as the additive principle, was used by many systems other than that of the Egyptians, including those of the Cretans (Linear B), Hittites, Greeks, Romans, and Aztecs (Ifrah,

1987). A notable property of a number system built on the additive principle was its lack of a need for a symbol to perform the function that zero performs in the Hindu-Arabic system. Such a system also had the advantage that the value of a number was determined strictly by the symbols that comprised it; the order in which the symbols occurred was irrelevant. A disadvantage of the system was the need to have a unique symbol for every power of its base. For example, to add 1 to

$$ꝯꝯꝯꝯꝯ\ ꝯꝯꝯꝯ∩∩∩∩∩∩∩∩∩|||||||||$$

in the Egyptian system necessitates the introduction of a symbol not used in the representation of any smaller number, and the same need arises each time one wishes to extend the system to another power of the radix, which is 10 in this case. In contrast, the Hindu-Arabic system is capable of representing any number, no matter how large, with its basic set of 10 symbols.

The lack of compactness of the Egyptian system is seen in its taking 36 symbols to represent the Egyptian equivalent to the Hindu-Arabic 9,999, 45 to represent 99,999, and so on. To be sure, it takes only one Egyptian symbol to represent the equivalent of the Hindu-Arabic 10,000 or any integral multiple of 10; however, Egyptian representations are considerably longer than their Hindu-Arabic counterparts on average.

The Roman system for representing numbers made use of the additive principle, but shortened the number of symbols required to represent some quantities both by introducing symbols—V, L, D—to represent quantities halfway between successive powers of 10—5, 50, 500—and by introducing a subtractive principle according to which the values of a symbol preceding one representing the next higher power of 10 was to be subtracted from, rather than added to, the latter. Use of the halfway symbols made it possible to write 5 as V rather than IIIII, 50 as L rather than as XXXXX, 500 as D instead of CCCCC. Use of the subtractive principle enabled such shortening as VIIII to IX and LXXXX to XC. This feature is gained at the cost of the loss of the independence of the value of a number from the order of the symbols comprising it: XC and CX represent different quantities.

The Hindu-Arabic system gains both compactness and extensibility at the cost of being more abstract and less redundantly encoded than the Egyptian system and others of its predecessors. The principle of one-for-many substitution has a straightforward application in arithmetic operations in systems that used the additive principle. For example, because in the Egyptian system one symbol representing a given power of 10 is equivalent in value to 10 symbols representing the immediately lower power of 10, adding two numbers amounts simply to merging the two representations and then, for any symbol that occurs 10 or more times, replacing 10 of them with one symbol representing the next higher power of 10. Thus, to add

$$\text{ꝯꝯ} \wedge\wedge\ \wedge\ \wedge\ \wedge\ |||$$

and

$$\text{ꝯ}\wedge\wedge\ \wedge\ \wedge\ \wedge\ \wedge||$$

one would merge the two representations, getting

$$\text{ꝯꝯ ꝯ}\wedge\wedge\ \wedge\ \wedge\ \wedge\ \wedge\wedge\wedge\wedge\wedge\wedge\ |\,|\,|\,|\,|$$

And inasmuch as there are 11 $\bigwedge$ s in the sum, one would replace 10 of them with one ꝯ and get

$$\text{ꝯꝯꝯ ꝯ}\wedge|\,|\,|\,|\,|$$

The same observation could be made with respect to any system that is based on the additive principle. In the Hindu-Arabic system the analogous operation is usually referred to as carrying, but the substitutionary principle that underlies it is not so obvious.

The operation that we refer to as borrowing, which is used in subtraction when a digit in the subtrahend is larger than the corresponding digit in the minuend, is the inverse of carrying and is based on a many-for-one substitution principle that is also obvious with the Egyptian representation. If, for example, one wished to subtract

$$\text{ꝯ}\wedge\wedge\wedge\wedge||$$

from

$$\text{ꝯꝯꝯ}\wedge\wedge\wedge|\,|\,|\,|$$

one would first make a many-for-one substitution in the minuend, replacing one ꝯ with 10 $\bigwedge$ s. With that done, the problem becomes

$$\text{ꝯꝯ}\wedge\wedge\ \wedge\ \wedge\ \wedge\ \wedge\wedge\wedge\wedge\wedge\wedge\ |\,|\,|\,|$$
$$-\qquad\qquad\qquad\text{ꝯ}\wedge\ \wedge\ \wedge\wedge||$$
$$\overline{\qquad\qquad\qquad\text{ꝯ}\ \wedge\ \wedge\ \wedge\ \wedge\ \wedge\ \wedge\wedge||}$$

and the answer is apparent. In effect, all one needs to do is to cross out a symbol in the minuend for every corresponding symbol in the subtrahend and the remainder is, literally, what remains.

At some point, perhaps around 2000 BC, the Egyptians invented a second system of notation that, like the hieroglyphic system described above, was a base 10 system, but that did not use the principle of symbol repetition. This system, known as the hieratic or sacred system, used a different, but single, symbol to represent each of the quantities between 1 and 9 or any power of 10 thereof. The power of 10 that was intended was indicated by the position of the symbol, so this was a true place notation system and a great advance in economy of expression over the hieroglyphic system, but an advance gained at the cost of an increase in abstractness of notation.

☐ Zero

Unlike a system based on an additive principle, a place notation system requires a symbol to represent zero. Initial recognition of the need for a symbol to represent "nothing" is generally credited to the Babylonians (Kaplan, 1999; Seife, 2000), although, according to Kaplan, the zero marker was used by the Babylonians only in the middle of a number, never at its end. The Greeks learned of the Babylonian's use of this symbol and their astronomers began, sometime after their invasion of Babylonia during the fourth century BC, to use a symbol similar to our 0 or the lowercase omicron of the Greek alphabet, to represent zero in their astronomical tables (the Greeks, like the Hebrews, used letters of their alphabet to represent numbers), but according to Seife (2000), they were not comfortable with a symbol for representing nothing and used it as little as possible. "Zero," Seife contends, "conflicted with the fundamental philosophical beliefs of the West, for contained within zero are two ideas that were poisonous to Western doctrine. Indeed, these concepts would eventually destroy Aristotelian philosophy after its long reign. These dangerous ideas are the void and the infinite" (p. 39). It was the Greeks' philosophical problems with zero and the infinite, Seife argues, that kept them from discovering the calculus. This view is consistent with the idea that the Greeks were more interested in the philosophy of mathematics than in the doing of mathematics (Brainerd, 1979).

Indian mathematicians also learned of zero from the Babylonians, either directly or by way of the Greeks, and they changed its role from that of a place holder to that of a number, although Dantzig (1930/2005) contends that the term the Hindus initially used for zero—*sunya*— meant *empty* or *blank*, and the symbol was used only to indicate a column that contained no marks when recording the results of counting board calculations. The distinction between a symbol to represent the absence of a number and a symbol that is considered a *bona fide* number whose value is null is an important one. As Kaplan (1999) puts it, "The first keeps it estranged from numbers, merely part of the landscape through which they move; the second puts it on a par with them" (p. 46). In his captivating discussion of how the transformation from place holder to number might have been effected over time in India and later in Europe, Kaplan gives 876 AD (as does Friend, 1954) as the date of the earliest known inscription of 0 on an Indian artifact.

Although it is less easily learned by children than are other digits (Wellman & Miller, 1986) and is read more slowly by adults (Brysbaert, 1995), today zero is universally recognized as an indispensable item in the mathematician's tool box. Dantzig (1930/2005) refers to its discovery as "one of the great single achievements of the human race" (p. 35). But there is still some uncertainty about how best to think of it. Is it a symbol to represent "nothing," the empty set? Clearly it is that, and it serves this critical function in the Hindu-Arabic representation of numbers. But it is more than that, serving as it does as a *bona fide* number in all sorts of contexts. But as a number, it is peculiar in certain ways. Regarding every other natural number, it is easy to say whether it is odd or even, positive or negative, but for 0 it is not so easy. Although no less a figure than Euler held that anything divided by 0 equals infinity, today mathematicians disallow the use of 0 as a divisor, because dividing by 0 leads to paradoxical or nonsensical results.

Let $a = b = 1$. Since a and b are equal, the following equality also holds:

$$b^2 = ab$$

Subtracting both terms from a^2, we have

$$a^2 - b^2 = a^2 - ab$$

which can be rewritten as

$$(a + b)(a - b) = a(a - b)$$

Dividing both sides by $(a - b)$ gives

$$a + b = a$$

Subtracting a from both sides gives

$$b = 0$$

and since b was set equal to 1 at the outset, we have

$$1 = 0$$

from which by multiplying, and adding to, both sides of the equation, we can make anything equal anything. For example, by first multiplying by 3 and then adding 4, we would have

$$7 = 4$$

There is nothing wrong with the math here, except that, inasmuch as a and b are equal, dividing by $(a - b)$ is dividing by 0.

More briefly, and perhaps more transparently, we can see that allowing division by 0 would make it possible to prove anything by simply noting that since

$$7 \times 0 = 4 \times 0$$

we could multiply both sides of this equation by $\dfrac{1}{0}$ and obtain

$$\frac{7 \times 0}{0} = \frac{4 \times 0}{0}$$

from which it follows from canceling the zeros that $7 = 4$.

Zero also causes problems that require special treatment when it appears in such contexts as 0! and X^0. Both of these expressions are defined to equal 1. This may seem a bit arbitrary. But there is a rationale for each. Consider 0! first. Recall that $n!$ is the product of all the numbers 1 through n, that is, $n! = n(n - 1)(n - 2) \ldots 3 \bullet 2 \bullet 1$. It follows from this definition that $n! = n(n - 1)!$. Thus, $4! = 4 \bullet 3!$, $3! = 3 \bullet 2!$, and so on. If this relationship is to include 1!, then $1! = 1 \bullet 0!$, but $1! = 1$, and if this is to be consistent with $1! = 1 \bullet 0!$, then it must be that $0! = 1$ also. One might ask, then, does it not also follow that $0! = 0 \bullet (-1!)$? No, because factorials are defined only for positive numbers, of which 0, for some purposes, is one.

Now what about X^0? Conventionally the exponent in the expression X^n indicates the number of times X is to be used as a factor in a succession of multiplications of X by itself; X^3, for example, is a shorthand way of representing $X \bullet X \bullet X$. But how does one apply this interpretation to the expression X^0? How can one use X as a factor 0 times in a succession of multiplications? The interpretation seems nonsensical in this case.

But why should X^0 be defined to equal 1? The answer is that this defini-tion follows from the rules of doing simple arithmetic with exponentials. According to these rules, $X^n/X^m = X^{n-m}$, so, for example, $7^5/7^3 = 7^2$. If this is to be true in general, it must hold for the case in which $n = m$, which is to say when the quotient is X^0. The quotient X^0 is obtained from divisions of the form X^n/X^n (e.g., $7^5/7^5$), which obviously must equal 1.

There is another way to look at the situation that provides a rationale for giving X^0 the value 1. If X is a positive integer, for $n > 0$, $X^n > 1$, and for $n < 0$, $X^n < 1$. As n approaches 0, from either the positive or negative side, X^n approaches 1 (see Figure 3.1), so it seems right that at $n = 0$, X^n should equal 1.

We see then compelling reasons for deciding that $X^0 = 1$. Obviously, 0 multiplied by itself as many times as one wishes equals 0, so $0^X = 0$. (Raising 0 to a negative power is not allowed by virtue of the prohibition of division by 0.) But now, what about the oddest expression of all, 0^0? If $X^0 = 1$ is to hold generally, then 0^0 must equal 1, but if $0^X = 0$ is to hold generally, then 0^0 must equal 0. We seem to be at an impasse; whether we make the value of 0^0 be 1 or 0, we must give up the generality of one or the other of the relationships considered; they cannot both be general.

Because of this dilemma, the value of 0^0 is sometimes said to be indeterminate or undefined, like 0/0. Several rationales have been given,

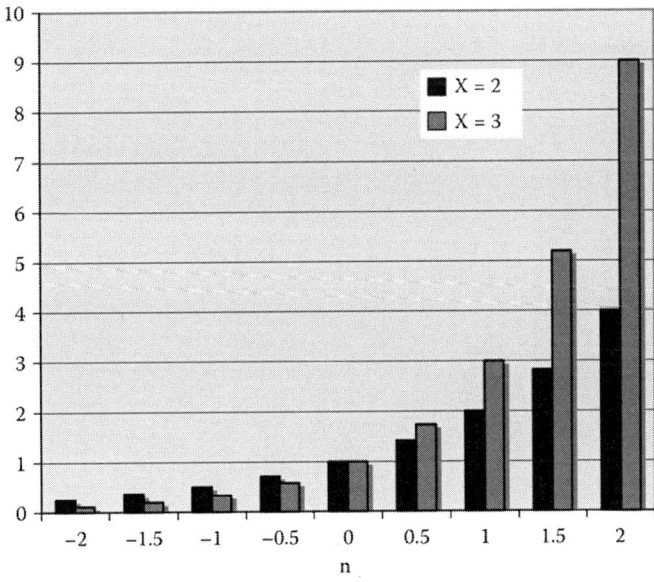

FIGURE 3.1 Illustrating the approach of X^n to 1 as n approaches 0 from either below or above. In this illustration, $X = 2$ (black) and 3 (dark gray).

however, for considering the value to be 1. One such is that defining it as 1 permits the binomial theorem to be general—to not have to treat $X = 0$ as a special case. To my knowledge, there is no widely accepted answer to the question of what the "real" value of 0^0 is, if that is a meaningful question; how it should be treated appears to be a matter of what works in the context in which it is encountered, and perhaps on the predilections of the user.

Defined as the empty set, zero has been used as the basis for set-theoretic definitions of all the natural numbers. And this is considered preferable to using 1 as the basis for such definitions, as was done by Frege and Peano, because 1 is a member of the set that is being defined, which makes the definitions circular. If the definitions based on 0 are not to be seen as circular, 0, in this context, cannot be considered a natural number. But there it is on the number line, and what would the number line be without it? What is zero, really? Lakoff and Núñez (2000) argue that there is no one answer to this question. Each of the candidate answers—empty set, number, point on the number line—"constitutes a choice of metaphor, and each choice of metaphor provides different inferences and determines a different subject matter" (p. 7). Whatever zero is, its importance to mathematical reasoning is beyond dispute; without it, much of mathematics as we know it today could not have been developed.

☐ Fractions

The ancient Egyptians expressed all fractional quantities as sums of different fractions, each of which had a numerator of 1. The fraction 2/3 would be written as 1/2 + 1/6; 3/7 would be 1/3 + 1/11 + 1/231, in Egyptian notation of course. In general, ancient number systems, perhaps with the exception of the Babylonian system, were not ideally suited to representing fractions. One of the strategies used for limiting the need for fractions was that of defining many subdivisions of units of length, weight, and other measures so that calculations could be done in terms of integral multiples of the subdivisions.

The practice of representing fractions by placing the numerator over the denominator and separating them with a horizontal bar was used in Arabia and by Fibonacci in Europe in the 13th century, but it did not come into general use in Europe until the 16th century. The slanted line was suggested by British mathematician–logician Augustus De Morgan in 1845. Hindu mathematicians represented fractions by placing the numerator over the denominator, but with no bar between, as early as the seventh century.

Boyer and Merzbach (1991) note that "it is one of the ironies of history that the chief advantage of positional notation—its applicability to fractions—almost entirely escaped the users of the Hindu-Arabic numerals for the first thousand years of their existence" (p. 255). Although decimal fractions are known to have been used in more than one pre-Renaissance culture on occasion—including China as early as the third century AD—the decimal point was first used to separate the integral from the fractional part of a number sometime close to the end of the 16th century. John Napier, Scottish mathematician-physicist and inventor (among others) of logarithms, was an early advocate of this scheme. The French mathematicians Francois Viète (sometimes Vieta, an amateur to whom Kasner and Newman [1940] refer as "the most eminent mathematician of the 16th century") and Simon Stevin are also credited with playing significant roles in ensuring wide acceptance of decimal fraction notation, which did not occur until about 200 years after Napier's advocacy (Boyer & Merzbach, 1991; Ellis, 1978).

Contemporary conventions for representing numbers differ in some respects in different parts of the world. In America, a period is used to set off the fractional part of a number, while a comma serves the same purpose in Europe. Thus, what Americans would write as 231.05, Europeans would write as 231,05. Americans separate multiples of 1,000 by commas (1,000,000), while Europeans do so with spaces (1 000 000). Especially confusing is that English-speaking Americans and Europeans use the same words to denote different quantities; thus to Americans, one billion is 10^9, whereas to the British, one billion is 10^{12}, that is, 1 million squared.

☐ Mental Number Representation

The many ways in which numbers have been represented in various places and at sundry times is an interesting story, and it raises many questions regarding the relative advantages and disadvantages of specific representational systems from a psychological point of view. Other interesting psychological questions have to do with the way numbers are represented in people's minds. One might assume that numbers are represented as numerals—that 3, for example, is represented in one's mind as the symbol 3. But there is reason to believe that it is not quite that simple.

Moyer and Landauer (1967, 1973) measured the time it takes for people to decide which of two numbers is the smaller and found that it varies inversely with the distance (the difference in value) between the numbers on the number line—the greater the distance (the larger the difference), the faster the decision time. A comparison between 2 and 9, for example, takes considerably less time than one between 9 and 7. This classic result, which was surprising when it was first obtained, has been interpreted as what would be expected if the mental representation of numbers were analog in character, something like an actual number line (Dehaene, Bossini, & Giraux, 1993; Fias, Lammertyn, Reynvoet, Dupont, & Orban, 2003). The "symbolic distance effect," or simply the "distance effect," as it is generally called, has been replicated many times in various forms, including with languages other than English (Banks, Fujii, & Kayra-Stuart, 1976; Dehaene, 1996; Parkman, 1971; Tzeng & Wang, 1983), and with children as well as with adults (Donlan, 1998; Sekuler & Mierkiewicz, 1977; Temple & Posner, 1998), although the effect appears to decrease in magnitude with increasing age (Duncan & McFarland, 1980; Sekuler & Mierkiewicz, 1977).

Dehaene, Dupoux, and Mehler (1990) reported a distance effect with two-digit numbers. Dehaene (1997) gave university students the task of pressing a left- or right-hand key to indicate whether a digit was smaller or larger than 5, and found that the distance effect persisted even after 1,600 trials over several days. He takes the distance effect as evidence of the inadequacy of the popular metaphor of the brain as a digital computer, arguing that the way we compare numbers is more suggestive of an analog machine than of a digital one. But comparing numbers is a relatively restricted type of cognitive activity, and whether the many other types of cognitive activity involved in the doing of complex mathematics are as readily attributed to analog processes remains to be seen.

There are several findings involving the perception of, or decisions about, numbers that are closely related to the distance effect. A case in point is known variously as the "magnitude effect," "size effect," or "problem-difficulty" effect, and it manifests itself in several ways. The time required to read numbers increases with their magnitude, especially for numbers in the range of 1 to about 50 (Brysbaert, 1995; Dehaene, 1992). Given two numbers that differ by a fixed amount, say, 2 and 4 or 8 and 10, the time required to decide which of two is the smaller (or the larger) increases with their size (Antell & Keating, 1983; Strauss & Curtis, 1981). The time required to report the sum, difference, product, or quotient of two numbers, or to judge whether a proposed answer to a computation is correct, increases with the sizes of the numbers involved (Campbell, 1987; Groen & Parkman, 1972; LeFevre et al., 1996; Miller, Perlmutter, & Keating, 1984; Rickard & Bourne, 1996; Stazyk, Ashcraft, & Hamann, 1982). Interestingly, this effect is greater when the numbers involved are represented in printed verbal form (one, two, ...) than as numerals (1, 2, ...) (Campbell, 1999; Campbell & Clark, 1992).

An analog of the finding that the time required to decide that two numbers are different varies inversely with the numerical difference between the numbers has been obtained with nonnumeric stimuli. When people have been asked to indicate as quickly as possible whether two stimuli are the same or different and those that differ can differ to varying degrees, the typical finding has been that same stimuli are identified as "same" faster than different stimuli are identified as "different," on average, but that the time required to identify different stimuli as "different" decreases as the magnitude of the difference between them increases (Bamber, 1969; Egeth, 1966; Hawkins, 1969; Nickerson, 1967, 1972).

A numerical distance effect has also been observed when the task is to report whether a proposed answer to a simple mathematical calculation (e.g., addition of two single-digit numbers) is correct or incorrect and the "distance" involved is the magnitude of the difference between the correct answer and an incorrect proposal: The greater the difference between the correct and the incorrect answer, the faster and more accurately people report that the incorrect answer is incorrect (Ashcraft & Battaglia, 1978; Ashcraft & Stazyk, 1981).

Still another related finding suggests a representation that has numbers proceeding from small to large from left to right. Using left and right keys to respond "smaller" and "larger," respectively, when comparing a visually presented number to a reference number produces faster responses than does assignment of "smaller" and "larger" to the right and left keys, respectively (Dehaene et al., 1990; Dehaene et al., 1993). The advantage of the smaller-left, larger-right mapping has been found both with Hinduu-Arabic numerals and with numbers represented as words (Fias, 2001; Nuerk, Iversen, & Willmes, 2004). The effect has been obtained whether the left and right keys are pressed with the left and right hands (most experiments), with two fingers of the same hand (Kim & Zaidel, 2003), or by crossing the arms and pressing the left and right keys with the right and left hands (Dehaene et al., 1993). Fias and Fischer (2005), who review these and related findings in detail, point out that, in the aggregate, they can be seen as illustrative of a more general class of spatial compatibility effects of the sort noted half a century ago by Fitts and Seeger (1953) and reviewed by Kornblum, Hasbroucq, and Osman (1990). The extent to which this small-to-large left-to-right mapping is peculiar to cultures that read from left to right is, so far as I know, an open question.

In sum, the findings from number comparison studies have been widely interpreted as generally supportive of the idea that numerical quantities are represented, not necessarily exclusively, as locations on an analog number line. Regarding how numbers are spaced on this line—in particular whether the spacing is linear (Gallistel & Gelman, 1992, 2000) or something closer to logarithmic (Dehaene, 1992; Dehaene et al., 1990), so the differences between successive numbers decrease with number size—it appears that the jury is still out. And alternatives to the number-line representation have been proposed that are able to account for many of the findings as well (Zorzi & Butterworth, 1999; Zorzi, Stoianov, & Umilta, 2005).

Several investigators of number processing have used Stroop-like tasks, which require the suppression of some feature of a stimulus in order to respond accurately to some other feature, as for example when one is asked to respond to the color of a word, say *red*, which can be printed in red or blue. A version of the task used in number processing might require a person to say which of two numerals, 2 or 8, was in the larger print. As might be expected from the results of findings with Stroop tests with nonnumeric stimuli, people respond faster and more accurately in determining that 5 is the (physically larger) of 5 and 3 than in determining that 3 is the larger of 5 and 3 (Henik & Tzelgov, 1982; Pansky & Algom, 1999) (Figure 3.2, left), or in reporting that a display contains four digits if the display is of four 4s than if it is of four 3s (Pavese & Umilta, 1998) (Figure 3.2, center). The finding of no Stroop interference when the task is to specify which of two verbal numerals (three or five) is written in the larger font (Figure 3.2, right) has been taken as evidence of separate types of processing of verbal and Arabic numeral representations of numbers (Ito & Hatta, 2003). There is some

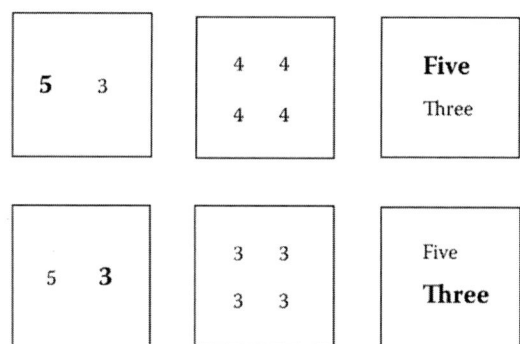

FIGURE 3.2 Illustrating Stroop-like effects with numbers. Left: People are faster to report which numeral is physically larger when that numeral is also numerically larger. Center: People are faster to report the number of numerals when that number corresponds to the numerals shown. Right: The time required to say which number name is written in the larger font appears not to be affected by the numbers named.

evidence that high math anxiety increases the interference one experiences from irrelevant stimulus features on Stroop-like tasks (Hopko, McNeil, Gleason, & Rabalais, 2002).

The finding of a variety of priming effects—in which presentation of a number has an effect on the response to a subsequent occurrence of the same or a different number in the context of some numerical task—has been taken as evidence of automatic activation of number codes and to have implications for models of the nature of those codes. Studies have found a distance-prompting effect, according to which the size of the effect of a prompt* is inversely proportional to the quantitative difference between the prompt and target (the number on which the prompt has its effect). Thus, the effect on one's response to 5 in a numerical task would be greater if prompted with 4 than if prompted with 2 (den Heyer & Briand, 1986). The finding that the magnitude of the effect is strictly a function of the difference between prompt and target and relatively independent of whether the prompt is smaller or larger than the target has been taken as evidence against the idea that numbers are represented mentally on a logarithmically spaced number line (Reynvoet, Brysbaert, & Fias, 2002; Zorzi et al., 2005).

The question of how numbers are represented mentally continues to be a focus of research. One issue is the extent to which the representation of numerical concepts is integrated with, or independent of, the representation of natural language. According to models proposed by McCloskey (1992; McCloskey & Macaruso, 1995) and by Gallistel and Gelman (1992), numerical concepts are represented independently of natural language. A model proposed by Dehaene (1992) and elaborated by Dehaene and Cohen (1995) hypothesizes three separate codes: one that represents magnitudes analogically, one that represents numbers visually as Hindu-Arabic numerals, and one that represents number facts verbally. Details can be found in the cited references and a comparison of the models in Campbell and Epp (2005). In a study of acquisition of the concept of infinity, Falk (in press) obtained evidence that recognition that numbers are endless is facilitated by the ability to separate numbers from their names—children who could not make this distinction had difficulty in seeing that numbers could be increased beyond what could be named.

One aspect of the question of how numeric concepts are represented in the brain that is currently motivating much research is that of identifying specific regions of the cortex that are actively involved in the

---

* What I here refer to as a prompt usually is referred to in the literature as a prime. I have opted for prompt so as to avoid confusion with prime in mathematics to refer to numbers that have no divisors other than 1 and themselves.

processing of such concepts. The application of a variety of neuroimaging techniques has led to a focus on three areas, all in the parietal lobes, that appear to be heavily involved: a horizontal segment of the intraparietal sulcus (HIPS), the angular gyrus, and the posterior superior parietal lobule (PSPL) (Dehaene, Piazza, Pinel, & Cohen, 2005). Evidence that numeric and nonnumeric concepts are represented, at least partially, in different cortical areas comes from studies showing that lesions can impair the processing of numeric concepts while leaving the processing of nonnumeric concepts intact (Anderson, Damasio, & Damasio, 1990; Butterworth, Cappelletti, & Kopelman, 2001; Cappelletti, Butterworth, & Kopelman, 2001), and can have the opposite effect as well (Cohen & Dehaene, 1995; Dehaene & Cohen, 1997).

Studies have also shown dissociation between knowledge of numerical facts and the ability to perform mathematical procedures (Temple, 1991; Temple & Sherwood, 2002), as well as differential impairment of the abilities to do subtraction and multiplication (Dagenbach & McCloskey, 1992; Delazer & Benke, 1997; Lampl, Eshel, Gilad, & Sarova-Pinhas, 1994; Lee, 2000). One interpretation of the latter finding is that multiplication requires access to information stored in verbal form (the typically overlearned multiplication table), whereas subtraction does not (Dehaene & Cohen, 1997; Dehaene et al., 2005). Dehaene et al. (2005) contend that such dissociations suggest the existence of "a quantity circuit (supporting subtraction and other quantity-manipulation operations) and a verbal circuit (supporting multiplication and other rote memory-based operations)" (p. 445). The neuroanatomy of number competence is a lively area of research at the present and promises to continue to be so for the foreseeable future.

☐ Irrational Numbers

"To the Pythagorean mind, ratios controlled the universe" (Seife, 2000, p. 34). More specifically, the controlling ratios were assumed to be ratios of integers. The discovery that there are quantities, such as the square root of 2, that cannot be expressed as the ratio of two integers was a devastating shock to people with this view; the existence of such *irrational* numbers was seen to be a serious threat to the rationality or orderliness of the world. (One wonders whether Pythagoras was aware that $(\sqrt{2}^{\sqrt{2}})^{\sqrt{2}} = 2$.)

One of the implications of the existence of irrational numbers is the failure of the rationals to "fill up" the number line. Given that there are infinitely many rational numbers and that between any two rationals, no matter how close, there are infinitely many other rationals, one might easily assume that the rationals fill up the line. But the existence of irrationals means that there are gaps in the line that are not filled by rationals. Maor (1987) calls the discovery that the rationals leave "holes" in the number line—points that are not rational numbers—"one of the most momentous events in the history of mathematics" (p. 43). This is an interesting claim in view of the distinction between rational and irrational numbers having little, if any, practical importance inasmuch as both can be expressed to whatever level of accuracy (number of significant digits) that is required for computational purposes.

Besides generating reflection on the question of what mystical significance the existence of irrationals might have, their discovery threatened the integrity of geometry by invalidating many of the early geometrical proofs that had been considered beyond question. The restoration of confidence in geometrical methods became a major objective of much mathematical thinking, and according to British mathematician and historian of mathematics Herbert Turnbull (1929), this was eventually accomplished by Eudoxus in the fourth century BC by his establishment of a sound basis for the doctrine of irrationals. Even so, irrationals were not fully accepted as *bona fide* numbers; a distinction was made between numbers and magnitudes—with irrationals considered to be in the latter category—and prevailed for about a millennium (Flegg, 1983).

Notable among irrational numbers are two of the most important constants of mathematics. I refer, of course, to π, the ratio of a circle's circumference to its diameter, and e, the base of natural logarithms. Both are indispensable, appearing in the most mundane as well as the most arcane of computations, and each is a colorful character with an intriguing history. Mathematicians have had techniques for approximating the value of π at least since the time of Archimedes. Non-Western cultures also long ago had methods for approximating this all-important number with a ratio of two integers; Dunham (1991) gives as examples of values produced: $355/113 = 3.14159292$ by Chinese mathematician-astronomer Tsu Ch'ung-chih (sometimes Zu Chongzhi) in the fifth century, and $3,927/1,250 = 3.1416$ by Hindu mathematician-astronomer Bhāskara around 1150. That π is irrational was demonstrated by Swiss mathematician Johann Lambert in the 1760s. A little over a century later, German mathematician Carl Louis Ferdinand von Lindemann showed it to be transcendental.

Why does π appear in so many formulas that have no obvious connection with circles? Why, for example, does it occur in the formula for the probability that a needle tossed on a plane of equally spaced parallel lines will fall across one of the lines: $p = 2n/\pi d$, where n is the length of the needle and d is the distance between the lines? Why does it appear in the formula for the Gaussian, or "normal," probability density, which describes the distribution of so many natural variables? What is it doing in Maxwell's equations describing electromagnetic effects? Or in Einstein's general relativistic equation? Why does it appear in Stirling's formula for bracketing $n!$ for large n? And what business does it have, to ask a qualitatively different question, showing up in such elegant attire as

$$\frac{2}{\pi} = \frac{\sqrt{2}}{2} \bullet \frac{\sqrt{2+\sqrt{2}}}{2} \bullet \frac{\sqrt{2+\sqrt{2+\sqrt{2}}}}{2} \dots ?$$

This equation, discovered by Viète in 1593, is credited by Maor (1987) with being the first explicit expression of an infinite process in a mathematical formula. It is, Maor notes, still admired for its beauty.

Viète's formula was subsequently joined by many others that express π as a function of an infinite arithmetic process and that produce approximations to π that increase in accuracy regularly with the number of terms in the equation. One such is

$$\frac{\pi}{2} = \frac{2}{1} \bullet \frac{2}{3} \bullet \frac{4}{3} \bullet \frac{4}{5} \bullet \frac{6}{5} \bullet \frac{6}{7} \bullet \frac{8}{7} \bullet \dots$$

discovered by 17th-century English mathematician John Wallis. Another is

$$\pi = \cfrac{4}{1 + \cfrac{1^2}{2 + \cfrac{3^2}{2 + \cfrac{5^2}{2 + 7^2 \dots}}}}$$

from 17th-century Irish mathematician Lord William Brouncker, founder and first president of the Royal Society of London. Leibniz and 17th-century Scottish mathematician James Gregory independently discovered

$$\frac{\pi}{4} = \left(\frac{1}{1} - \frac{1}{3} + \frac{1}{5} - \frac{1}{7} + \frac{1}{9} - \frac{1}{11} + \dots \right)$$

In Chapter 1, I mentioned the zeta function of Riemann's famous conjecture,

$$\zeta(x) = \sum_{n=1}^{\infty} \frac{1}{n^x}$$

Euler discovered that, for $x = 2$, this function approximates $\dfrac{\pi^2}{6}$, that is,

$$\frac{\pi^2}{6} = \sum_{n=1}^{\infty} \frac{1}{n^2}$$

Maor (1987) refers to this series as one of the most beautiful results in mathematical analysis, and du Sautoy (2004) calls Euler's discovery "one of the most intriguing calculations in all of mathematics" (p. 80). Intriguing because surprising: The finding "took the scientific community of Euler's time by storm. No one had predicted a link between the innocent sum $1 + \frac{1}{4} + \frac{1}{9} + \frac{1}{16} + \cdots$ and the chaotic number π" (p. 80).

Although all the series mentioned, as well as others, converge so as to yield good approximations to π, they do not all converge at the same rate. The series that was discovered independently by Leibniz and Gregory and that converges to $\pi/4$ requires 628 terms just to produce a value for π that is accurate to two decimal places. In contrast,

$$\frac{\pi^4}{90} = \sum_{n=1}^{\infty} \frac{1}{n^4}$$

also from Euler, converges rapidly.

Other people, including some for whom mathematics was only an avocation, developed other infinite series approximations to π, some of which are highly efficient in that relatively few terms yield approximations that are accurate to dozens of decimal places. Dunham (1991) mentions Abraham Sharp, a British schoolmaster, who found π accurate to 71 places in 1699, and John Machin, a British astronomer, who found it accurate to 100 places in 1706. William Shanks, an amateur mathematician, also British, used Machin's series to approximate π to 707 places in 1873, but more than seven decades later in 1944, another British mathematician, D. F. Ferguson, discovered, with the help of a desktop mechanical calculator, that Shanks's approximation had an error at the 528th place, making the rest of the approximation wrong. In 1947, American mathematician John Wrench Jr. produced an approximation of 808 digits, and again Ferguson found an error, this time in the 723rd place. He and Wrench jointly published a correction to 898 places in 1948. The story of this progression of longer and longer approximations to π, predating the arrival on the scene of the digital computer, is told in detail by Beckman (1971) and Dunham (1991).

The fascination that mathematicians—professional and amateur—have shown for π over the centuries and their willingness to spend countless hours attempting to approximate it to accuracies far beyond any practical usefulness are interesting psychological phenomena. It is one among numerous examples of mathematical entities that have captured the attention of people with a mathematical bent and held it seemingly permanently.

With the arrival of the electronic computer on the scene, the rate at which ever longer approximations to π were found increased exponentially. Among the first to use a computer for this purpose were Daniel Shanks, an American physicist and mathematician, and Wrench, who collaborated on programming an IBM 7090 computer to compute π to a little over 100,000 places in approximately nine hours in 1961. In 2005, Yasumasa Kanada of the Information Technology Center, Computer Centre Division, University of Tokyo, announced having approximated π to over 1 trillion (10^{12}) decimal places; I have no idea how the accuracy of such a computation is verified.

The remarkable constant e, the base of natural logarithms, is a particularly interesting case of the infinitely large and the infinitely small combining to confound our intuitions and to yield an immensely useful construct in the process. It is usually defined as the limit of the expression $(1 + 1/n)^n$, which is sometimes written as

$$\left(1\frac{1}{1}\right)^1, \left(1\frac{1}{2}\right)^2, \left(1\frac{1}{3}\right)^3, \left(1\frac{1}{4}\right)^4, \left(1\frac{1}{5}\right)^5, \ldots$$

or as

$$\left(\frac{2}{1}\right)^1, \left(\frac{3}{2}\right)^2, \left(\frac{4}{3}\right)^3, \left(\frac{5}{4}\right)^4, \left(\frac{6}{5}\right)^5, \ldots$$

As n increases indefinitely, $1/n$ goes to 0, $1 + 1/n$ goes to 1, so, inasmuch as 1^n is 1, we might expect $(1 + 1/n)^n$ to go to 1. On the other hand, inasmuch as x^n increases without limit provided that x is greater than 1, and $1 + 1/n$ is always greater than 1, even if only infinitesimally, we might expect $(1 + 1/n)^n$ to increase without limit. In fact, as n increases indefinitely, $(1 + 1/n)^n$ converges to 2.7182818284 ..., which we have come to know and love as e. As seen in Table 3.1, the sequence has to be extended for about 1,000,000 terms for the converging approximation to be accurate to six decimal places.

TABLE 3.1. Values of $\left(1\frac{1}{n}\right)^n$ for n up to 1,000,000

n	$\left(1\frac{1}{n}\right)^n$
1	2.00000000
2	2.25000000
3	2.37037037
4	2.44140625
5	2.48832000
10	2.59374246
100	2.70481383
1,000	2.71692393
10,000	2,71814594
100,000.	2.71826830
1,000,000	2.71828138

Another sequence that converges to e, and considerably more rapidly, is

$$e = \sum_{k=0}^{n} \frac{1}{k!}$$

This approximation is accurate to seven decimal places when extended to only 10 terms, as shown in Table 3.2. The following series approximation of e is attributed to Euler:

$$e = 2 + \cfrac{1}{1 + \cfrac{1}{2 + \cfrac{2}{3 + \cfrac{3}{4 + \cfrac{4}{5 + \ldots}}}}}$$

TABLE 3.2. The Values of $\sum_{k=0}^{n} \frac{1}{k!}$ **for n From 1 to 10**

n	$\sum_{k=0}^{n} \frac{1}{k!}$
1	2.00000000
2	2.50000000
3	2.66666667
4	2.70833333
5	2.71666667
6	2.71805556
7	2.71825397
8	2.71827877
9	2.71828153
10	2.71828180

The story of how e came to be selected in the 17th century by Napier as the base for natural logarithms has been wonderfully told by Maor (1994) and can be found in part in Coolidge (1950), Péter (1961/1976), and in any general reference on logarithms, infinite series, or number theory. As every student who has finished first-year calculus knows, e^x is unique among mathematical functions in that it is equal to its own derivative, which is to say $\frac{de^x}{dx} = e^x$. This is easily seen when e^x is expressed in power-series form as

$$e^x = \sum_{n=0}^{\infty} \frac{x^n}{n!} = 1 + x + \frac{x^2}{2!} + \frac{x^3}{3!} + \ \cdots$$

which, when differentiated term by term, yields the same expression.

As we have noted, the values of π and e, as well as those of all other irrational constants, cannot be expressed exactly numerically; they can only be approximated. No matter how many digits one uses to approximate them, the representation will fall short of being exact. The concept of an approximation seems to carry the notion that there is some value to be approximated. But what is it in the case of π, say, that is being approximated? What is the *exact* value of this constant, which we can only approximate? The answer appears to be that it has no exact value. One might say that the exact value of π is the ratio of a circle's circumference to its diameter, but one is still left with a value that cannot be expressed exactly. This is a curious situation from a psychological point of view, but one that we seem to accept with no difficulty.

☐ Imaginary Numbers

Nahin (1998) and Mazur (2003) tell the fascinating story of the emergence and slow acceptance of the idea that negative numbers could have square roots and the eventual adoption of i (first used by Euler) as the symbol to represent $\sqrt{-1}$, the square root of minus one. They note that the construct $\sqrt{-1}$ (and multiples of it, such as $\sqrt{-15}$) was used for centuries, because it was found to be useful, before a satisfactory "imagining" of what it might mean was attained. As recently as the 16th and 17th centuries, negative roots of equations were referred to by mathematicians as false, fictitious, or absurd. Writing late in the 18th century, Euler called square roots of negative numbers impossible on the grounds that "of such numbers we may truly assert that they are neither nothing, nor greater than nothing, nor less than nothing" (quoted in Nahin, 1998, p. 31).

Often the use of square roots of negative numbers was covert, which was possible because mathematicians have long been reluctant to expose the sometimes tortuous thought processes that have gone into the solving of problems, preferring to present only polished justifications of results. Dantzig (1930/2005) credits Italian physician–mathematician Girolamo Cardano as the first to use the $\sqrt{-1}$ representation, which he did in 1545. Its appearance in equations was often accompanied by an apology or rationalization for the use of such strange—"meaningless, sophisticated, impossible, fictitious, mystic, imaginary" (Dantzig, 1930/2005, p. 190)— quantities. Cardano himself believed that negative numbers could not have square roots, but considered use of the form in a calculation to be justified if the final answer was a real number. Italian mathematician Raphael Bombelli, a contemporary of Cardano, also used imaginary numbers and defended the practice as essential to the solving of algebraic equations of the form $x^2 + 1 = 0$. Descartes is generally credited with being the first to refer to square roots of negative numbers as "imaginary," presumably to connote something the existence of which was doubted or denied.

Acceptance of imaginary numbers as *bona fide* numbers was slow, but in time it came. Arguably a major factor in their acceptance was the invention in the latter part of the 18th century of complex numbers—numbers of the form $a + ib$, in which a is considered the real part and b the imaginary part—and the convention of representing such numbers geometrically. In this representation, a number is a point in a "complex plane," its real and imaginary parts being indicated by its x and y coordinates, respectively. Accounts of geometrical representations of complex numbers were given by Caspar Wessel in 1799, Jean-Robert Argand in 1806, and William Rowan Hamilton in 1837. This representation gave numbers the capability to represent—now as vectors—both magnitude and direction, and opened up many unanticipated applications in science and engineering. Nevertheless, according to Kline (1980), well into the 19th century "Cambridge University professors preserved an invincible repulsion to the objectionable $\sqrt{-1}$, and cumbrous devices were adopted to avoid its occurrence or use, wherever possible" (p. 158).

Stewart (1990) refers to the agreement to allow −1 to have a square root as "an act of pure mathematical imagination," and to the resulting imaginary numbers as "among the most important and beautiful ideas in the whole of mathematics" (p. 234). Its importance is seen in the impressive range of its uses. As Péter (1961/1976) puts it, "There is no branch of mathematics which does not turn to this i for help, especially when something of a deep significance needs to be expressed" (p. 163).

Today the rules for performing mathematical operations with complex numbers are well developed, and the use of such numbers is ubiquitous in many areas, especially in science and engineering. High school students use such quantities with little realization of the consternation they once caused, although some may have difficulty in accepting them as *bona fide* numbers (Tirosh & Almog, 1989). The modern view recognizes a hierarchy of number concepts in which natural (counting) integers are a subset of the rationals, which are a subset of the reals, which are a subset of the complex numbers (a real number is a complex number with imaginary component $0i$), and so on.

How are we to explain the contrast between the matter-of-fact way in which $\sqrt{-1}$ and other imaginary numbers are accepted today and the great difficulty they posed for learned mathematicians when they first appeared on the scene? One possibility is that mathematical intuitions have evolved over the centuries and people are generally more willing to see mathematics as a matter of manipulating symbols according to rules and are less insistent on interpreting all symbols as representative of one or another aspect of physical reality. Another, less self-congratulatory possibility is that most of us are content to follow the computational rules we are taught and do not give a lot of thought to rationales.

The latter possibility gains some credence from the suspicion that many people who use square roots of negative numbers in computations with apparent ease might be hard-pressed to say what is wrong with the following argument. Clearly $\sqrt{\frac{1}{-1}} = \sqrt{\frac{-1}{1}}$, from which it follows that $\frac{\sqrt{1}}{\sqrt{-1}} = \frac{\sqrt{-1}}{\sqrt{1}}$, which can be written $\sqrt{1}\sqrt{1} = \sqrt{-1}\sqrt{-1}$, and inasmuch as $\sqrt{-1}\sqrt{-1} = i^2 = -1$, we have $1 = -1$. Bunch (1982), from whom I took this example, notes that in order to avoid such unpleasantries, mathematicians prohibit the use of $\sqrt{-1}$ in equations and insist on using i instead. Note this disallowance must extend not only to the stand-alone use of $\sqrt{-1}$, but also to its implicit use in expressions like $\sqrt{\frac{-1}{1}}$ and $\sqrt{\frac{-1}{1}}$. If these forms are permitted and we allow that $\sqrt{\frac{1}{-1}} = \sqrt{\frac{-1}{1}}$ we would still have the problem inasmuch as from $\sqrt{\frac{1}{-1}} = \frac{\sqrt{1}}{i}$ and $\sqrt{\frac{-1}{1}} = \frac{i}{\sqrt{1}}$, we get $\frac{\sqrt{1}}{i} = \frac{i}{\sqrt{1}}$, and $1 = i^2 = -1$.

Mazur (2003) notes that the consternation caused by the early use of square roots of negative numbers did not differ greatly, in kind, from that felt by the use of the rule that the multiplication of two negative numbers should produce a positive number. I venture the conjecture that most of us who learned this rule in elementary school have long since become comfortable with it—which is not to suggest that we necessarily ever lost sleep over it—but many of us would perhaps find it difficult to explain to the satisfaction of an inquisitive child the rationale for the rule.

The reader who has no trouble with this assignment may get more of a challenge from explaining what $\sqrt{-1}^{\sqrt{-1}}$, or i^i, means. Mazur claims that "this concoction can be given a natural enough interpretation, which has a real-number value, as already seen by Euler" (p. 209). I am not the one to explain what it is.

So there are many kinds of numbers, and they all serve us well. But still the question of what exactly a number is—or what numbers are—lingers. Do they exist independently of mankind's knowledge and use of them, or are they products of human inventiveness brought into existence to serve useful purposes? One can take one's pick, and whichever answer appeals, one can find authoritative spokespersons for that view.

☐ A Number Paradox

There are many paradoxes involving numbers. I will mention one because of its relevance to other paradoxes that will be discussed in subsequent chapters. The Berry paradox was made famous by British philosopher-mathematician-logician Bertrand Russell, who attributed the initial version of it to G. G. Berry, an Oxford University librarian. It is a self-referential paradox, similar in important respects to such paradoxes as "This sentence is false" (true or false?) and "All knaves are liars," spoken by a knave.

Presumably there exist, among the infinity of integers, many that are not definable in less than 11 words. (You may prefer to substitute for *definable* some other word, such as *describable, identifiable,* or *referable.*) Consider "the smallest integer not definable in less than 11 words." Inasmuch as "the smallest integer not definable in less than 11 words" is a 10-word definition of the integer of interest, there can be no integer not definable in less than 11 words.

The Berry paradox is illustrative of paradoxes that involve talking about things that cannot be talked about, describing things that cannot be described. There are perhaps numbers that are so large that we cannot name them, and if there are, there must be a smallest one, but, in referring to such a number as the smallest number that we cannot name, we have just named it, have we not? Rucker (1982), who discusses the Berry paradox in various guises at length, notes with some bemusement: "It is curious how interesting it can be to talk about things that we supposedly can't talk about" (p. 107). We shall encounter closely related ideas in the chapter on proofs and, in particular, with reference to proofs of nonprovability, the most famous of which we owe to Kurt Gödel.

☐ Prime Numbers

Prime numbers were mentioned briefly in Chapter 1, but the concept deserves more emphasis and a chapter on numbers seems the appropriate place to provide it. A prime number is a number that cannot be divided evenly (without a remainder) by any numbers other than 1 and itself. All numbers that are not prime numbers are composite numbers, which means that they are integral products of other integers. This distinction is a very old and important one in mathematics, and more will be said about prime numbers in subsequent chapters of this book.

The assertion that every natural number is either prime or the product of a unique set of primes is referred to sometimes as the *fundamental theorem of arithmetic* and sometimes as the *unique factorization theorem.* Seventeen is a prime number, being divisible only by 1 and 17; 30, in contrast, is a composite number, the unique set of primes of which it is composed being 2, 3, and 5. We shall see in subsequent chapters that determining whether a particular (large) number is prime or composite is very difficult, and that this has acquired great practical significance today in its use in making communications systems secure. Many modern encryption systems are based on it being a trivially simple matter to create an extremely large number (say more than 100 digits long) that is the product of two large primes, but exceedingly difficult to determine the factors of such a large composite number, or whether in fact the number is composite. This permits one to encode messages using the composite number, which can be generally known, as the encryption key and be sure that only those who know the prime factors of that number will be able to decode it.

Prime numbers have been of considerable interest to mathematicians over the centuries independently of any practical applications that could be made of the knowledge of them, which were not great until recently. Many conjectures have been advanced about them, some of which have been proved as theorems and some of which still have the status of unproved conjectures. Conjectures have often motivated extraordinary effort from mathematicians to produce the needed proofs (either to show the conjectures to be correct or to show them to be false). One example involving primes is a conjecture by French lawyer and renowned amateur mathematician Pierre de Fermat. It has been known for a long time that every prime number, x, greater than 2 can be represented either by $4n + 1$ or $4n - 1$. (This is not to say that for any n, $4n + 1$ or $4n - 1$ is a prime; for $n = 16$, $4n + 1 = 65$ and $4n - 1 = 63$, both composite.) Fermat claimed that all primes, and only those primes, that can be represented by $4n + 1$ are the sum of two squares. Table 3.3 shows the first few primes that can be represented by $4n + 1$ and the squares that sum to them.

More than a century after Fermat's claim, Euler proved it, but it took him seven years to do so (Singh, 1997). Why it is that some of the primes that can be expressed as $4n + 1$ are equal to the sum of two squares while none of those that can be expressed as $4n - 1$ are remains unexplained, so far as I have been able to determine. The history of work on prime numbers is another illustration of how a particular aspect of mathematics can motivate years of effort by extraordinarily bright people who, in many cases, have little to gain but the satisfaction of solving a problem or at least hopefully getting a bit further on it than others have yet done.

Prime numbers continue to provide endless fascination for mathematicians; the search for patterns among them is the persisting challenge. How is it, one wants to know, that the pattern of primes can be so irregular in one sense (seen in the absence of any structure that would allow the prediction of which numbers in a sequence will be prime) and so remarkably regular in another sense (the increasingly close correspondence between the density of primes and $1/\log_e n$ with increasing n)? The mix of unpredictability on one level with predictability at another is analogous to that seen with many random processes (coin tossing, die rolling, radioactive decay) in which the behavior is random

**TABLE 3.3. The First Few Values of *n* for Which
4*n* + 1 Is a Prime and Also the Sum of Two Squares**

Prime 4*n* + 1	*n*	Squares
5	1	$2^2 + 1^2$
13	3	$3^2 + 2^2$
17	4	$4^2 + 1^2$
29	7	$5^2 + 2^2$
37	9	$6^2 + 1^2$
41	10	$5^2 + 4^2$
53	13	$7^2 + 2^2$
61	15	$6^2 + 5^2$

(unpredictable) at the level of individual events but highly predictable in the aggregate. We will return briefly to the topic of prime numbers and the continuing search for ever larger ones in Chapter 11.

Numbers are fascinating—at least to some people—and for many reasons. Moreover, all numbers are fascinating—there are none that are not. If there are any numbers that are dull—not fascinating—there must be a smallest dull number. But there can be no smallest dull number, because a smallest dull number would be fascinating by virtue of being unique in this respect. Therefore, there can be no dull numbers. (This is not original with me, but I cannot recall to whom I am indebted for it.)

****

This chapter has focused on numbers, and many different kinds of numbers have been noted. I thought it would be good, before leaving the topic, to provide a diagram that shows how the different kinds that have been mentioned relate to each other. Things went well at first, but when my attempt got to about the point shown in Figure 3.3, I began to find it difficult to proceed. The problem is that the set–subset relationships represented in the figure tell only part of the story. To be sure, there are many such relationships: The primes are a subset of the natural (counting) numbers, which are a subset of the integers, which are a subset of

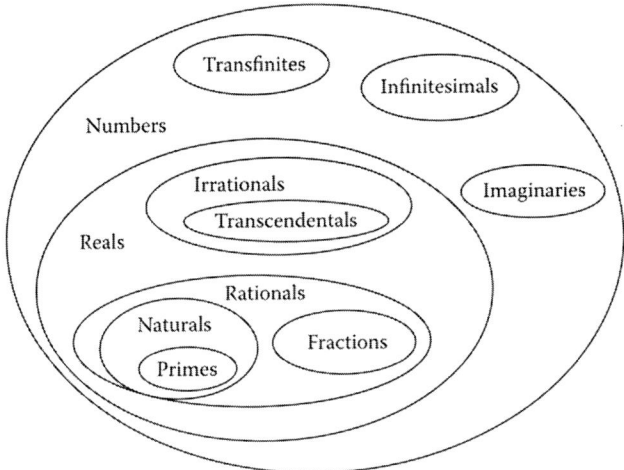

FIGURE 3.3 The universe of numbers; an incomplete representation.

the reals, which are a subset of the complex numbers, and so on. But I was not sure how best to represent the distinction between positive and negative numbers in this scheme. Representation of the distinction between algebraic numbers—numbers that are roots of polynomial equations with rational coefficients—and nonalgebraic numbers was complicated by the fact that some irrationals ($\sqrt{2}$, for example) are algebraic, while others (π, for example) are not. Where should surreals, superreals, and hyperreals appear in such a diagram? Or quaternions, vectors, and matrices? And where does humble, indispensable 0 belong? You, dear reader, may wish to try to construct a representation of the number universe that is more inclusive, and more enlightening, than Figure 3.3. It is one way to get a feel for how richly that universe is populated.

The attentive reader will note that nowhere in this chapter is there a definition of *number*. That is because I do not know how to define it. My sense is that definitions one finds in dictionaries—including even mathematical dictionaries—in effect take an understanding of the concept of *natural number* as given—essentially indefinable—and use it as a basis for defining *types* of numbers that are derivative from it. Whatever a number is, there are few concepts that have been more important to the history of humankind, or more instrumental to the intellectual development of the species.

Deduction and Abstraction

> Mathematics as a science commenced when first someone, probably a Greek, proved propositions about *any* things or about *some* things, without specification of definite particular things. (Whitehead, 1911/1963, p. 54)

A distinguishing task of pure mathematics is to make explicit, by deducing theorems, the relationships that are implicitly contained in the axioms of a given mathematical system. A distinction between mathematics and empirical science can be made on the basis of the different ways in which they ensure the objective validity of thought. "Science disciplines thought through contact with reality, mathematics by adhering to a set of formal rules which, though fixed, allow an endless variety of intellectual structures to be elaborated" (Schwartz, 1978, p. 270). Regarding the difference between mathematical and nonmathematical (commonsense) arguments, Schwartz contends that because of their precise, formal character, the former are sound even if they are long and complex, whereas the latter, even when moderately long, easily become far-fetched and dubious. Davis and Hersh (1981) refer to Euclidean geometry as the first example of a formalized deductive system and the model for all subsequently developed systems.

☐ Postulational Reasoning

> Primarily, mathematics is a method of inquiry known as postulational thinking. The method consists in carefully formulating definitions of the concepts to be discussed and in explicitly stating the assumptions that shall be the bases for reasoning. From these definitions and assumptions conclusions are deduced by the application of the most rigorous logic man is capable of using. (Kline, 1953a, p. 4)

A major difference between the mathematics of Egypt and Babylonia and that of Greece was that the conclusions in the former were established empirically, whereas those in the latter were established by deductive reasoning. Kline (1953a) describes the insistence by the Greeks on deductive reasoning as the sole method of proof in mathematics as a contribution of the first magnitude. "It removed mathematics from the carpenter's tool box, the farmer's shed, and the surveyor's kit, and installed it as a system of thought in man's mind. Man's reason, not his senses, was to decide thenceforth what was correct. By this very decision reason effected an entrance into Western civilization, and thus the Greeks revealed more clearly than in any other manner the supreme importance they attached to the rational powers of man" (p. 30).

As already noted, geometry was once an empirical discipline. Its development was motivated, as its name suggests, at least in part by an interest in measuring land. The idea of building geometry as an axiomatic system took shape gradually over several hundred years and found its first extensive expression in the work of Euclid of Alexandria, who lived in the third century BC. Euclid constructed his geometry from a set of 23 definitions, 5 postulates, and 5 common notions. He considered the postulates and common notions (sometimes referred to collectively as axioms) to be sufficiently obvious that everyone would agree to their truth. The goal was to keep the number of postulates to a minimum, to assume nothing that could be derived. The high place that deductive reasoning has held in Western culture is probably a consequence, to a significant degree, of the success that Euclid and his followers had in deriving deductive proofs in geometry. (Some mathematicians make a distinction between postulates and axioms, some do not. In what follows, the terms are used more or less interchangeably, except when in quotes of others' writings, where, of course, the terms are those used by the quoted authors.)

Euclid's *Elements* is undoubtedly the most durable text on mathematics ever written. The estimated number of editions that have been published since the appearance of the first printed version in Venice in 1482 exceeds 1,000. Boyer and Merzbach (1991), to whom I am indebted for this fact, refer to Euclid's *Elements* as the most influential textbook of all times, an assessment that agrees with Wilder's (1952/1956) surmise that there may be no other document that has had a greater influence on scientific thought. The entire text was composed of 13 books that deal with plane and solid geometry, the theory of numbers, and incommensurables.

Euclid did not claim originality for most of what he wrote; his intention was to produce a text that covered the elements of mathematics as they were understood in his day, and he drew much from the work of others in the pursuit of this goal. The deference shown to his product over two millennia attests to the success of his efforts. What gives *Elements* its lasting importance is not so much the theorems proved in it, but the method Euclid explicated, which involved starting with a few "self-evident" truths and, using deduction as the only tool, making explicit what they collectively imply. Beckman (1971) refers to *Elements* as the first grandiose building of mathematical architecture and describes Euclid's achievement in metaphorical terms as the construction of an edifice, the foundation stones of which are the postulates. "Onto these foundation stones Euclid lays brick after brick with iron logic, making sure that each new brick rests firmly supported by one previously laid, with not the tiniest gap a microbe can walk through, until the whole cathedral stands as firmly anchored as its foundations" (p. 48). Euclid, Beckman contends, is not just the father of geometry but the father of mathematical rigor.

As an aside, we may see in the U.S. Declaration of Independence a reflection of Euclid's approach of constructing a body of geometric truths composed of deductions from self-evident axioms. The writers of this document identified four truths that they considered to be self-evident—that all men are created equal, that they are endowed by their Creator with certain inalienable rights, and so on—and then deduced what they saw to be the implications of these truths, which they concluded justified their declaration of the right of the colonies to be free and independent states. Meyerson (2002), who makes this comparison, contends that the U.S. Constitution also may be seen as basically a set of axioms, and that it is the business of the Supreme Court to pass judgment on what is implicit in those axioms—which explains why the Constitution can be as short as it is and yet have such far-ranging influence.

The impressiveness of Euclid's monumental accomplishment notwithstanding, his logic has not escaped criticism. Even he was not immune to the common problem of making tacit assumptions without recognizing them as such. Critical views as to the tightness of Euclid's logic have been expressed by Russell (1901/1956a, pp. 1588–1589) and Bell (1946/1991, p. 332), among numerous others. Bell (1945/1992) agrees that Euclid's contribution to mathematics was monumental, not because of the postulates he proved, but because of the "epoch-making methodology" of his work:

> For the first time in history masses of isolated discoveries were united and correlated by a single guiding principle, that of rigorous deduction from explicitly stated assumptions. Some of the Pythargoreans and Eudoxus before Euclid had executed important details of the grand design, but it remained for Euclid to see it all and see it whole. He is therefore the great perfector, if not the sole creator, of what is today called the postulational method, the central nervous system of living mathematics. (p. 71)

Dunham (1991) contends that Euclid's sins were sins of omission and that his works are free of sins of commission, which is to say that while some of his proofs are incomplete—by today's standards—none of his 465 theorems has been shown to be false.

But despite the emphasis on deduction, the Greeks saw their mathematics as dealing with truth, though not necessarily truth that depended on empirical observation. Given the Euclidean view of axioms as obviously true statements about reality as the point of departure, and deduction as the method of drawing conclusions from those axioms, the conclusions drawn—the theorems proved—could also be considered true assertions.

This view of the matter persisted until relatively recent times. To German physician-physicist Hermann von Helmholtz, for example, the axioms that comprise the foundation of geometry were unprovable principles that would be admitted at once to be correct by anyone who understood them. Von Helmholtz (1870/1956) wondered, however, about why there should be such self-evident truths and about the basis of our confidence in their correctness. "What is the origin of such propositions unquestionably true yet incapable of proof in a science where everything else is reasoned conclusion? Are they inherited from the divine source of our reason as the idealistic philosophers think, or is it only that the ingenuity of mathematicians has hitherto not been penetrating enough to find the proof?" (p. 649).

Von Helmholtz rejected the view held by German philosopher Immanuel Kant that the axioms of geometry are necessary consequences of an *a priori* transcendental form of intuition, because it is possible to imagine spaces in which axioms different from those of plane geometry would be intuitively obvious to the inhabitants, and even we, who live presumably in a Euclidean space, can imagine what it would be like to live in a non-Euclidean one. The axioms of Euclidean geometry are what they are, according to this view, because of the properties of the space in which we live, but we can imagine them otherwise because we can imagine other types of spaces.

This perspective represents a small departure from that of the Greeks because, while it sees the axioms of geometry as reflections of the way things are, it recognizes the possibility that they could be otherwise. However, it does not abandon the basic assumption of a direct connection between the axioms and the properties of the physical world; in fact, it reinforces that assumption by allowing as to how a world with different characteristics would give rise to a geometry with different axioms.

Modern mathematicians have a very different attitude toward axiomatic systems. The empirical truth of a system's axioms, or postulates, as they are often called, is irrelevant to the mathematical integrity of the system. The axioms are viewed as conventions that mathematicians agree to take as a system's foundation, and the only requirements are that they be consistent with each other and that the logic by which theorems are deduced be valid. It is consistency *within* an axiomatic system that is required; different systems need not be consistent with each other, nor need they be descriptive of the physical world. What constitutes deductively valid logic is taken to be a matter of agreed-upon convention. Given these conditions, one can say that *if* a system's axioms are empirically true, theorems that are validly deduced from them are true also. It is not a requirement, however, that either the axioms or the theorems be empirically true.

Russell (1901/1956a) expresses this attitude with respect to Euclid's system in particular. "Whether Euclid's axioms are true, is a question to which the pure mathematician is indifferent" (p. 1587). Russell further points out that the question of their truth "is theoretically impossible to answer in the affirmative. It might possibly be shown, by very careful measurements, that Euclid's axioms are false; but no measurements could ever assure us (owing to the errors of observation) that they are exactly true" (p. 1587). This is not to deny that they have proved to be sufficiently accurate to be immensely useful for practical purposes in the physical world.

Philosophers have sometimes made a distinction between analytic and synthetic truths. Analytic truths are not verified by observation; true analytic statements are tautologies and are true by virtue of the definitions of their terms and their logical structure. Synthetic truths relate to the material world; the truth of synthetic statements depends on their correspondence to how physical reality works. Mathematics, according to this distinction, deals exclusively with analytic truths. Its statements are all tautologies and are (analytically) true by virtue of their adherence to formal rules of construction.

☐ Mathematical Induction

> Mathematical induction is a method that can be used to prove a countable infinity of statements true in a finite number of steps. (Bunch, 1982, p. 43)

In books on logic and reasoning a distinction is often made between deduction and induction, deduction involving reasoning from more general assertions to more particular ones and induction involving going from the more particular to the more general. In most everyday situations, and probably in most scientific contexts as well, if one entertained the hypothesis "If A then B" and upon observing a large number of instances of A found B to be present in every case, one would undoubtedly consider the hypothesis to be supported strongly and would expect to see B in all subsequent observations of A. This is a form of induction and it works very well most of the time. Mathematicians use induction, sufficiently broadly defined to include guessing, hunch following, intuiting, and trial-and-error experimenting in their efforts to construct, validate, and refine deductive systems, but it is not what is meant by *mathematical induction*.

Kasner and Newman (1940) describe mathematical induction as "an inherent, intuitive, almost instinctive property of the mind. 'What we have once done we can do again'" (p. 35). The form of reasoning involved may be represented in the following way:

If R is true for n, it is true for $n + 1$.
R is true for n.
Therefore, R is true for $n + 1$.

This representation looks very much like a deductive argument, in particular the *modus ponens* form of the conditional syllogism, and so it is. What motivates referring to it as mathematical *induction* is that it must be applied iteratively. To use this form of argument, one must first show that the major premise holds—that R is *necessarily* true for $n + 1$ if it is true for n. With that settled, if one can then demonstrate that R is true for some specific integer value of n, it must be true for all greater integers. One shows this by iterating on the syllogism with specific integers: If I have shown that R is true for 1, for example, I can conclude that it is true for 2; if it is true for 2, it must be true for 3; if true for 3, ... This is a very power-ful form of argument. As Péter (1961/1976) points out, what mathematical induction allows us to do—demonstrate that something holds for *all* natural numbers—would otherwise be impossible for finite brains.

This form of argument is sufficiently different from that in which one simply generalizes from a sample to a population, on the strength of the assumption that inasmuch as the sample reveals no exceptions to a rule, the population is unlikely to have any either, to warrant a different name—Dantzig (1930/2005) has proposed *reasoning by recurrence*—but *induction* continues to be used, although generally with the modifier *mathematical* to accentuate the distinction. That a conjecture has been shown to be correct with respect to every specific instance that has been checked is never taken by a mathematician as proof of its truth. Many formulas have been proposed for generating prime numbers. Some of these have worked impressively well, generating nothing but primes for a long time. One might say they have passed any reasonable test that could be applied to justify the leap from many consistent observations to the statement of a general law. But mathematicians are rightfully wary of such leaps, because some of the prime number generators that have appeared to be so promising have eventually proved to be not quite perfect. The formula $n^2 - 79n + 1{,}601$, for example, produces nothing but primes for all values of n up to 79, but for $n = 80$, the formula gives 1,681, which is 41^2. The numbers 31; 331; 3,331; 33,331; 333,331; 3,333,331; and 33,333,331 are all primes. One might guess on the basis of such regularity that any sequence of 3s followed by a 1 would yield a prime number, but in fact 333,333,331 is not prime; it is the product of 17 and 19,607,843. Until 1536 people believed that any number of the form $2^p - 1$ was prime if p was prime, but in that year Hudalricus Regius discovered that $2^{11} - 1 = 2{,}047$, the product of 23 and 89. (As far as I can tell, this is the only thing for which Hudalricus Regius is remembered. A search of indexes of mathematicians failed to find him listed, and all the numerous references to him I was able to find referred to him simply as "a mathematician" and mentioned him only in reference to this one finding.) Before Fermat's last theorem ($x^n + y^n \neq z^n$ for $n > 2$) was proved to be true in the 1990s, it had been shown to be true for all n up to 4 million, but this did not suffice to guarantee it to be true for all n. An obscure conjecture in number theory known as Merten's theory was verified to be true for the first 7.8 billion natural numbers before it was shown to be false in 1983 (Devlin, 2000a, p. 74).

☐ The Trend Toward Increasing Abstraction

> The longer mathematics lives the more abstract—and therefore, possibly, also the more practical—it becomes. (Bell, 1937, p. 525)

The history of mathematics reveals a progression from the more concrete to the more abstract, beginning with the earliest knowledge we have of the origins of counting and number systems and continuing to the present day. This progression is seen in numerous ways—in the emergence of the distinction between numbers as properties of things counted and numbers as interesting entities in their own right (between *three* as an adjective and *three* as a noun), in the evolution of systems for representing numbers, in the early attention to mathematics for the sake of practical applications (trade, surveying, construction) spurring exploration of increasingly esoteric avenues of mathematical thought for its own sake, in the invention or discovery of increasingly abstract and counterintuitive ideas (negative numbers, irrational numbers, imaginary numbers, infinities of different order, proofs of nonprovability). It is of at least passing interest that students of cognitive development in children report a similar progression from more concrete to more abstract thinking in the normal course of cognitive maturation—"away from the material and toward the formal, or away from ideas rooted in the here and now and toward ideas addressing events that are distant in time and space" (White & Siegel, 1999, p. 241).

The progression from the more concrete to the more abstract in the history of mathematics is seen clearly in the early development of methods of counting and representing quantities. Schmandt-Besserat (1978, 1981, 1982, 1984) notes markers of this trend:

- Animal bones and antlers bearing series of notches and dating from prehistoric times, considered to be reckoning or tally devices and illustrative of the use of the principle of one-to-one correspondence.

- Three-dimensional tokens of various shapes used for counting, also on a one-to-one basis, dating from about 8000 BC—the shape of the token indicating what was represented and the number of tokens indicating the number of units.

- Ideograms impressed on clay tablets. Such ideograms were less concrete than the three-dimensional tokens, but were still relatively concrete inasmuch as they were indicative of the things they represented and not just of the numbers of such things.

- Symbols, appearing around 3100 BC, that represented numbers, or quantities, that were independent of the things whose quantities were being represented.

As noted in Chapter 3, a trend toward increasing abstractness is seen in the evolution of systems for representing numbers over the centuries. The Hindu-Arabic system that is used almost universally today encodes the basic principles on which the representational scheme is based in a more abstract way than did many of its predecessors. The greater abstractness has provided greater computational convenience, perhaps at the cost of making the system's rationale somewhat more obscure.

Abstraction in mathematical operations is seen already even in elementary arithmetic. Alfred North Whitehead (1911/1963a) puts it this way:

> Now, the first noticeable fact about arithmetic is that it applies to everything, to tastes and to sounds, to apples and to angels, to the ideas of the mind and to the bones of the body. The nature of the things is perfectly indifferent, of all things it is true that two and two make four. Thus we write down as the leading characteristic of mathematics that it deals with properties and ideas which are applicable to things just because they are things, and apart from any particular feelings, or emotions, or sensations, in any way connected with them. This is what is meant by calling mathematics an abstract science. (p. 52)

Algebra represents a higher level of abstraction than do number and arithmetic. Number and arithmetic are abstractions from the entities numbered, or added and subtracted and otherwise manipulated; algebra, in modern terms, is an abstraction not only from particular numbers but from the concept of number itself. As Bell (1937) puts it,

> Once and for all Peacock [in his *Treatise on Algebra*, published in 1830] broke away from the superstition that the $x, y, z, \ldots$ in such relations as $x + y = y + x$, $xy = yx$, $x(y + z) = xy + xz$, and so on, as we find them in elementary algebra, necessarily 'represent numbers'; they do not, and that is one of the most important things about algebra and the source of its power in applications. The $x, y, z, \ldots$ are merely arbitrary marks, combined according to certain operations, one of which is symbolized as $+$, and other by $\times$ (or simply as xy instead of $x \times y$), in accordance with postulates laid down at the beginning. (p. 438)

The abstract nature of mathematics is seen also in the debated question of what it means for some of its simplest constructs to exist. Geometry, for example, deals with shapes, only approximations to which are found in the physical world; the entities that populate its axioms and theorems are idealizations, figments of the mathematician's imagination.

> For even though mathematics teaches us that there are cubes and icosahedrons, yet in the sense that there are mountains over 25,000 feet high, that is, in the sense of physical existence, there are no cubes and no icosahedrons. The most beautiful rock-salt crystal is not an exact mathematical cube, and a model of an icosahedron, however well constructed, is not an icosahedron in the mathematical sense. While it is fairly clear what is meant by the expressions 'there is' or 'there are' as used in the sciences

dealing with the physical world, it is not at all clear what mathematics means by such existence statements. On this point indeed there is no agreement whatever among scholars, whether they be mathematicians or philosophers." (Hahn, 1956, p. 1600)

The question of the existence, or nonexistence, of mathematical entities has motivated much discussion and debate, and proposed answers distinguish several schools of thought, about which more is in Chapter 13. For the present it suffices to recognize that such debates have seldom, if ever, gotten in the way of the doing of mathematics, and to note that whatever the sense in which mathematical entities may be said to exist, it differs from the sense in which physical objects may be said to exist. As Kasner and Newman (1940) put it, "A billiard ball may have as one of its properties, in addition to whiteness, roundness, hardness, etc., a relation of circumference to diameter involving the number π. We may agree that the billiard ball and π both exist; we must also agree that the billiard ball and π lead different kinds of lives" (p. 61).

Geometry illustrates the increasingly abstract nature of mathematics in another way as well. Presumably geometry initially grew out of practical concerns about measuring physical areas and making calculations that could be useful for purposes of building physical structures. Once geometry was framed by Euclid as deductions from a set of axioms, it became possible to explore the consequences of changing the axioms, although more than 2,000 years were to pass before such exploration occurred. Why it took this long for the idea to surface is an interesting question. Apparently, the assumption that the theorems of geometry were descriptive of the physical world was sufficiently strong to preclude the consideration of other perspectives. The development of non-Euclidian geometries in the 19th century, perhaps more than any earlier event, challenged the prevailing idea that the axioms of mathematics are examples of truths that would be recognized universally as such by all rational persons. It demonstrated that geometry (one should say any particular geometry) could be treated as an abstract deductive system in which one states a set of axioms and investigates what logically follows from them—any correspondence to the physical world, or lack thereof, being irrelevant to the enterprise. Instead of being assertions of obvious truths about the physical world, the axioms of geometry now were better viewed as "conventions," as Poincaré (1913) put it, and thus at once abstract and arbitrary. Non-Euclidean geometries have been around long enough now that we easily accept them, but their introduction was greeted with great skepticism and angst.

That mathematics generally can be viewed strictly as symbol manipulation, without any reference to what, if anything, the symbols represent is a relatively recent idea. Among the more explicit statements of this view is one by De Morgan, who proclaimed that (with the exception of =) the symbols he used in mathematical expressions had no meaning whatsoever. Algebra to him was nothing more or less than the business of manipulating symbols according to specified rules. French polymath Louis Couturat (1896/1975) describes what the mathematician does as the laying down of symbols and the prescribing of rules for combining them. He treats the conventions by which mathematical entities are created as arbitrary and equates the process with that by which chessmen and the rules that govern their moves are defined.

Lakoff and Núñez (2000) contend that even such fundamental concepts as point, line, and space have been transformed in what they refer to as the "discretization" of mathematics. Space, once thought of as continuous as attested by our ability to move smoothly within it, became conceptualized as a set of points. The latter, less intuitively natural, conceptualization is a reconceptualization of the former, they suggest, constructed to suit certain purposes. According to the earlier conceptualization, lines and planes exist independently of points; according to the more recent one, lines and planes are composed of points, and a point, defined as a line of zero length, is about as abstract a concept as one can imagine.

Lakoff and Núñez illustrate the abstractness of the concept of a point with the question of whether points on a line touch. One answer that might be expected is that of course they touch, else the line would not be continuous. Another is that of course they do not touch, because, if they did, there would be no distance between them and they would therefore be the same point. The latter seems to follow from the definition of a point as a line of zero length, but that does not make the definition invalid. Lakoff and Núñez contend that other counterintuitive ideas emerge from the discretization of mathematics, and that one just has to get used to it. "In thinking about contemporary discretized mathematics, be aware that your ordinary concepts will surface regularly and that they contradict those of discretized mathematics in important ways" (p. 278). (We will return to this topic in Chapter 9.)

The progression from the more concrete to the more abstract can be seen over the entire recorded history of mathematics, but the rate of change appears to have increased substantially in relatively recent times. Most of math was empirically based up until the time of Galileo, with math concepts straightforward abstractions from real-world experience. But things had begun to change rapidly by the 17th century. Wallace (2003) puts it this way:

> By 1600, entities like zero, negative integers, and irrationals are used routinely. Now start adding in the subsequent decades' introductions of

complex numbers, Naperian logarithms, higher-degree polynomials and literal coefficients in algebra—plus of course eventually the 1st and 2nd derivative and the integral—and it's clear that as of some pre-Enlightment date math has gotten so remote from any sort of real-world observation that we and [Ferdinand de] Saussure can say verily it is now, as a system of symbols, 'independent of the objects designated,' i.e. that math is now concerned much more with the logical relations between abstract concepts than with any particular correspondence between those concepts and physical reality. The point: It's in the seventeenth century that math becomes primarily a system of abstractions from other abstractions instead of from the world. (p. 106)

Arguably it is the increasing tendency of mathematical concepts to be abstractions from other mathematical concepts, themselves abstractions, that makes much of higher mathematics opaque to nonmathematicians. Devlin (2002) maintains that "the only route to getting even a superficial understanding of those concepts is to follow the entire chain of abstractions that leads to them" (p. 14). Follow the chain, that is, if you are able. But in many cases of contemporary higher math, one is likely to find that to be a tall order. Devlin (2000a) also contends that an inability to deal effectively with abstraction is the single major obstacle to doing well at mathematics.

Bell (1945/1992) describes the magnitude of the change toward abstractness and generality that occurred during the first half of the 19th century this way: "By the middle of the nineteenth century, the spirit of mathematics had changed so profoundly that even the leading mathematicians of the eighteenth century, could they have witnessed the outcome of half a century's progress, would scarcely have recognized it as mathematics" (p. 169). In 1900, German mathematician David Hilbert proposed a program, the aim of which was to axiomatize all of mathematics. The trend continues and new areas of mathematics tend to be more abstract than those from which they emerged. Innovative work in mathematics today is often so abstract, as well as dependent on considerable specialized background knowledge, that few but specialists in the areas of development can follow it. By 1925 Whitehead (1925/1956) could, without much fear of contradiction, describe mathematics as the science of the most complete abstractions to which the human mind can attain. The history of mathematics can be viewed as a continuing attempt to extend the limits of what is attainable in this regard.

In focusing on the tendency of mathematics to become increasingly abstract over the course of its existence and realizing that the criteria that a modern mathematical system must satisfy do not include making true statements about the physical world as we understand it, one is led to wonder whether mathematics would have been of interest—whether it would ever have been developed—if it were not so obviously applicable to the world of the senses. We should note too that essentially all of the early abstractions were abstractions from the perceived physical world.

Surprisingly, perhaps, that mathematics has become increasingly abstract does not mean that it has become increasingly useless. To the contrary, it can be argued that its utility, as well as its beauty, has only been enhanced by its tendency to eschew the concrete in its preference for the domain of pure thought. Bell (1945/1992) describes the abstractness of mathematics as "its chief glory and its surest title to practical usefulness" (p. 9). Kline (1980) has a similar assessment. "Though it [mathematics] is a purely human creation, the access it has given us to some domains of nature enables us to progress far beyond all expectations. Indeed it is paradoxical that abstractions so remote from reality should achieve so much. Artificial the mathematical account may be; a fairy tale perhaps, but one with a moral. Human reason has a power even if it is not readily explained" (p. 350). We will return to the topic of the usefulness, some would say the surprising usefulness, of mathematics in Chapter 12.

Implicit in much of the foregoing account of the increasingly abstract nature of mathematics is the equating of increasing abstraction with progress. Few would question the assertion that over the centuries mathematics has become more and more abstract and mathematics has made progress. It seems only natural to yoke these two observations in a causal way: Mathematics has made progress because it has become increasingly abstract, or it has become increasingly abstract because it has made progress—or becoming increasingly abstract and making progress are the same thing.

The equation can be challenged. Smith and Confrey (1994) argue that making increasing abstraction the universal standard for mathematical progress has two unfortunate tendencies. "First, it tends to treat abstraction as an ahistorical concept, that is, it assumes that we can interpret historical mathematical events in terms of some timeless concept of abstraction. Second, it encourages the creation of an historical record in which only those events that are viewed as part of the story of increasing abstraction are considered important, while events that do not fit into this framework are often considered superfluous or wrong" (p. 177). The second tendency has the effect of defining progress in terms of increasing abstraction, and simply ignoring, or not building on, innovations that do not fit that definition.

Smith and Confrey argue that progress lies, to some extent, in the eye of the beholder. "As we look backwards, it is often easy to see what we now understand but which those before us seemingly did not.... However, what is much more difficult to see is what they did understand that we do not and perhaps cannot, because we cannot enter the historical and social world in which they lived" (p. 178). We may, they argue, see the understanding of forerunners as confused, when, in fact, the confusion lies in our own inability to imagine the world as they saw it.

The development of a new concept or area of mathematics arguably follows a common trajectory, and increasing abstraction appears to be a major aspect of it. Devlin (2000) describes the trajectory as one that begins with the identification and isolation of new key concepts and that later is followed by analysis and attempts at axiomatization, which generally means increased abstraction, which in turn leads to generalizations, new discoveries, and greater connections to other areas of mathematics. It must be pointed out, however, that increasing abstraction does not mean to all mathematicians an abandoning of the concrete. Some argue that no matter how abstract some aspects of mathematics may become, there will always be a place for the concrete. Kac (1985) forcefully defends this position:

> By its nature and by its historical heritage, mathematics lives in the interplay of ideas. The progress of mathematics and its vigor depend on the abstract helping the concrete and on the concrete feeding the abstract. To isolate mathematics and to divide it means in the long run to starve it and perhaps even destroy it.... The two great streams of mathematical creativity [the concrete and the abstract] are a tribute to the universality of human genius. Each carries its own dreams and its own passions. Together they generate new dreams and new passions. Apart, both may die—one in a kind of unembodied sterility of medieval scholasticism and the other as a part of military art. (p. 153)

The existence of abstract mathematics is something of an enigma. What is there in evolutionary history that can explain not only the intense desire to acquire knowledge, even that which has no obvious practical utility, but the apparent ability of people to do so? From where, in particular, comes the fascination with and propensity for abstract mathematics? As Davies (1992) puts it, "It is certainly a surprise, and a deep mystery, that the human brain has evolved its extraordinary mathematical ability. It is very hard to see how abstract mathematics has any survival value" (p. 152). The question of the basis for mathematical ability touches on that of what it means to be human.

☐ Freedom From Empirical Constraints

> The mathematician is entirely free, within the limits of his imagination, to construct what worlds he pleases. (Sullivan, 1925/1956, p. 2020)

That the geometry that Euclid systematized stood alone for 2,000 years is testimony both to the influence of his work and to the strength of the connection in people's thinking between geometry and the perceived properties of the physical world. The ability to think of geometry as an abstract system, rather than as a description of physical reality, was essential to the development of geometries other than that of Euclid, and it was a long time in coming. It emerged from efforts of mathematicians to deal with the "parallel postulate" of Euclid's geometry, which had been a challenge and frustration to them for centuries. I have put "parallel postulate" in quotes because, as expressed by Euclid, the postulate did not mention explicitly parallel lines, but rather referred to the angles made by a line falling across two straight lines; what is commonly cited today as Euclid's parallel postulate is a rephrasing of what Euclid said by Scottish mathematician John Playfair:

> Given a line *l* and a point *P* not on *l*, there exists one and only one line *m*, in the plane of *P* and *l*, which is parallel to *l*.

Many believed this postulate to be derivable from the others, and countless hours were spent on efforts to prove it to be so. Proofs were published from time to time, but invariably they were shown, sooner or later, to be invalid.

The idea that a geometry might be developed that did not contain the equivalent of Euclid's parallel postulate, either as a postulate or as a theorem, was not seriously entertained for a long time, because of the prevailing conception of geometry as descriptive of real-world relationships, and it was obvious to anyone who thought about it that parallel lines could never meet. The strength of this conviction is illustrated by the work of Italian logician-theologian-mathematician Giovanni Girolamo Saccheri. Saccheri demonstrated that Euclid's geometry was not the only one possible, but refused to accept his own findings. Bell (1946/1991) refers to Saccheri's success in convincing himself of the absolutism of Euclid's geometry as "one of the most curious psychological paradoxes in the history of reason. Determined to believe in Euclid's system as the absolute truth, he constructed two other geometries, each self-consistent and as adequate for daily life as Euclid's. Then, by a double miracle of devotion, he disbelieved both" (p. 344). Kac (1985) speculates that the reason that Sacherri failed to accept his findings as a basis for the development of a geometry in which the parallel postulate did not hold was that, being convinced that the postulate was correct, he was searching for a contradiction; he never found one, but he apparently never became convinced that there was not one to be found. One of the new geometries that Saccheri comtemplated and dismissed was the one that Russian mathematician Nikolai Lobachevsky developed 97 years later. Saccheri's work did not come to the attention of other mathematicians until more than 150 years following his death.

When the possibility of non-Euclidean geometries began to be considered seriously during the 19th century, several of them were introduced within a relatively short period of time by such eminent mathematicians as Hungarian János Bolyai, German Carl Friedrich Gauss, Nikolai Lobachevsky, and German Georg Friedrich Bernhard Riemann, although some of these individuals had difficulty accepting this development at first. Bolyai, Gauss, and Lobachevsky explored the implications of a geometry in which the parallel postulate was replaced by one that held it to be possible to draw more than one parallel to a straight line; Riemann considered the implications of an axiom that permitted no parallels to be drawn to a given line through a point not on the line. In the geometry resulting from the first approach—hyperbolic geometry—the sum of the angles of a triangle is always less than 180 degrees; in Riemann's geometry—elliptical geometry—the sum of the angles of a triangle is always greater than 180 degrees.

These strange new ideas opened the door to a new era of innovation in mathematics. The non-Euclidean—hyperbolic and elliptical— geometries that emerged at the beginning of this era appeared at first to be products of the rarified kind of thinking in which mathematicians indulge for their own intellectual amusement and to have no connection with the physical world. As has been true time and time again, however, these products eventually proved to be powerful new tools for furthering our understanding of the universe; in this case their usefulness was first recognized in the area of theoretical physics and notably in Albert Einstein's work on relativity, the geometry of which is that of Riemann.

The creation of non-Euclidean geometries forced a major rethinking of the nature of mathematics. Kasner and Newman (1940) refer to the development of these geometries as a "sweeping movement" that has never been surpassed in the history of science, and contend that it "shook to the foundations the age-old belief that Euclid had dispensed eternal truths" (p. 134). Kline (1953) also describes its significance in similarly superlative terms: "It is fair to say that no more cataclysmic event has ever taken place in the history of all thought" (p. 428). What it did was compel mathematicians, scientists, and others "to appreciate the fact that systems of thought based on statements about physical space are different from that physical space" (p. 428). It demonstrated, as no prior development had done, the independence of mathematics from the material world.

The change in perspective forced by this demonstration was to many not only profound but profoundly unsettling. "Prior to the coming of non-Euclidian geometry, there was a unity, a confidence, and a certainty to our knowledge of the world. Afterwards, it was not enough to know that God is a geometer. The one unassailable truth about the nature of the physical world had been eroded and so, along with it, had centuries of confidence in the existence and knowability of unassailable truths about the Universe" (Barrow, 1992, p. 14). British mathematician-philosopher William Kingdom Clifford (1873/1956) compared the revolution represented by the invention of non-Euclidean geometries to that which Copernicus brought about on Ptolemaic astronomy; in his view, the consequence in both cases was a change in our conception of the cosmos. That is not to say that the change was immediate. Lobachevsky's work attracted little attention until about 30 years after it was published.

The freeing of geometry from considerations of real-world constraints was advanced considerably by the work of Hilbert, who developed a system of geometry based on a small set of undefined terms and relations and 21 axioms, which he referred to as assumptions. Far more important than Hilbert's particular system—which became known as *formalism*—was his insistence that it is not necessary that the constructs of such a system represent anything at all in the real world, and that all that is necessary is that the system be internally consistent, which is not to suggest that internal consistency among a set of more than a very few axioms is readily established.

The cost of this newfound freedom was great, and not everyone was willing to pay it. Gauss delayed publication of his own work on non-Euclidean geometry in the interest of avoiding controversy. Kline (1953) describes the effect of the development of non-Euclidean geometry as that of not only depriving mathematics of its status as a collection of truths, but perhaps of robbing man of the hope of ever attaining certainty about anything. On the other hand, it also gave mathematicians *carte blanche* to wander wherever they wished, which is precisely what they proceeded to do. Kline's assessment of this result is not positive.

The development of non-Euclidean geometries demonstrated also the inappropriateness of characterizing mathematics as an axiomatic system. Mathematics is not *an* axiomatic system; it encompasses *many* such systems. While it is required of any axiomatic system that it be self-consistent, mathematics, as a whole, need not and does not meet that requirement; it contains many axiomatic systems, each of which is intended to be self-consistent, but it is not essential that the theorems derived from one set of axioms be consistent with those derived from another, or even that the axioms themselves be consistent across systems. The arbitrariness of the axioms of any mathematical system is seen perhaps most starkly in the interchangeability of the notions of *point* and *line* in projective geometry. Given the axioms of this discipline, these two constructs are entirely interchangeable in the sense that either can play the role of the fundamental element, and if every mention of *point* were replaced with *line* and conversely, the system would remain intact.

Polish-American mathematician Nathan Court (1935/1961) captures the modern mathematician's indifference to the correspondence between mathematical statements and physical reality in somewhat whimsical terms. "If a mathematician takes a notion to create a mathematical science, all he has to do is to set up a group of postulates to suit his own taste, postulates which he by his own fiat decrees to be true, and involving things nobody, including the mathematician himself, knows about, and he is ready to apply formal logic and spin his tale as far and as fast as he will" (p. 24). And, more soberly, "The postulates of a mathematical science may be laid down arbitrarily. The rest of the doctrine is developed by pure logic and the test of its validity is that it must be free from contradictions" (p. 26).

The essence of an axiomatic system is that all that can be said about the system—the total collection of assertions that can be made—is implicit in the axioms. The challenge to the mathematician is to make what is implicit explicit by applying agreed-upon rules of inference. That is it in a nutshell. As Whitehead (1898/1963b) puts it, "When once the rules for the manipulation of the signs of a calculus are known, the art of their practical manipulation can be studied apart from any attention to the meaning to be assigned to the signs" (p. 69). There are many axiomatic systems in mathematics. Euclidean geometry is one such. Among many others are hyperbolic geometry, elliptical geometry, probability theory as axiomatized by Russian mathematician Andrey Nikolaevich Kolmogorov, and the set theory of German mathematician Ernst Zermelo and Israeli mathematician Abraham Fraenkel. The push within mathematics to axiomatize has been very strong. The vision of many mathematicians over the ages has been the development of a single axiomatic system that would provide a foundation for all of mathematics. This dream appears to have been shown by Kurt Gödel to be unattainable (more on that subject in subsequent chapters). Suffice it to note here that, while Gödel's work, which showed it to be impossible to have an axiomatic system that was complete even for arithmetic, was seen by some to be devastating to the mathematical enterprise, it appears not to have slowed mathematical activity appreciably, if at all. If total axiomatization is not possible, there appear to be plenty of challenges that do not require it. As Moore (2001) puts it: "Whatever the appeal of axiomatic bases, we must not regard them as sacrosanct. After all, people were engaged in arithmetic for thousands of years before any attempt was made to provide it with one" (p. 182).

Proofs

> In practice, proofs are simply whatever it takes to convince colleagues that a mathematical idea is true. (Henrion, 1997, p. 242)

> There is an infinite regress in proofs; therefore proofs do not prove. You should realize that proving is a game, to be played while you enjoy it and stopped when you get tired of it. (Lakatos, 1976, p. 40)

If we are to understand mathematical reasoning at all, we must understand, at least from the perspective of our culture and time, something about the nature of mathematical proof and the processes involved in proof construction. What constitutes a proof? What do mathematicians mean when they use the term? Where and when did the idea of a proof originate? How do proofs get built? How can one be sure that a proposed proof is valid? Who is qualified to judge the validity of a proof?

A proof in mathematics is a compelling argument that a proposition holds without exception; a disproof requires only the demonstration of an exception. A mathematical proof does not, in general, establish the empirical truth of whatever is proved. What it establishes is that whatever is proved—usually a theorem—follows logically from the givens, or axioms. The empirical truth of a theorem can be considered to be established only to the extent that the axioms from which it is derived can be considered to be empirically true—to be accurately descriptive of the real world.

☐ Origin and Evolution of the Idea of Proof

> The concept of proof perhaps marks the true beginning of mathematics as the art of deduction rather than just numerological observation, the point at which mathematical alchemy gave way to mathematical chemistry. (Du Sautoy, 2004, p. 29)

The origin of the notion of proof is obscure. Apparently the ancient Egyptians lacked it. They also did not make a sharp distinction between exact relationships and approximations (Boyer & Merzbach, 1991). They did use demonstrations of plausibility such as noting, in the context of claiming that the area of an isosceles triangle is half its base times its height, the possibility of seeing an isosceles triangle as two right triangles that can be rearranged to form a rectangle with the same height as the triangle and a width of half its base, but they did not prove theorems in a formal way.

Like the Egyptians, the Babylonians appear to have dealt primarily with specific cases and not to have attempted to produce general formulations of unifying mathematical principles. They too failed to distinguish sharply between exact and approximate results. As Boyer and Merzbach point out, however, that statements of general rules have not been found on surviving cuneiform tablets is not compelling evidence that no such rules were recognized; the many problems of similar types that are found on Babylonian tablets could be exercises that students were expected to work out using recognized rules and procedures. It cannot be said with certainty that pre-Hellenic peoples had no concept of proof. The mathematics of many ancient cultures—Egyptian, Babylonian, Chinese, and Indian—display a mixture of accurate and inaccurate results, of primitive and sophisticated methods, and of the simple and the complex.

Bell's treatment of the question of whether the Babylonians had the concept of a proof is puzzling. In his *Men of Mathematics*, published in 1937, he refers to the Babylonians as "the first 'moderns' in mathematics" and credits them with "recognition—as shown by their work—of the necessity for *proof* in mathematics" (p. 18). He calls this recognition "one of the most important steps ever taken by human beings" (p. 18) and notes that "until recently," it had been supposed that the Greeks were the first to have it. However, in his *The Magic of Numbers*, which was first published in 1946, Bell, after noting that some historians of mathematics considered the Babylonian algebra of 2000–1200 BC to be superior to any other produced before the 16th century and their work in geometry and mensuration to be almost as good, says that the work has "no vestige of proof" (p. 27).

Whatever the status of the concept of proof, or of precursors to this concept, in the pre-Hellenic world, there is little doubt that the concept was explicitly articulated by the classical Greeks. Bell (1937) credits Pythagoras with importing proof into mathematics and calls it his greatest achievement, but probably the name that is most closely associated with the concept of proof is that of Euclid. To be sure, the idea of what constitutes an acceptable proof has changed considerably since Euclid's time. Many of the proofs in his monumental *Elements* that went unchallenged for centuries do not meet current standards. Some theorems use undefined terms that have not been identified as such; some depend on unstated assumptions or postulates; definitions that are given are sometimes exceedingly vague; and so on. But there is no denying the enormous influence that Euclid's work had in giving the idea of a deductive proof center stage in mathematics and focusing the efforts of subsequent generations of mathematicians on the activity of proof making. That contemporary mathematicians find it easy to point out the deficiencies in Euclid's proofs may be seen as evidence not so much of weakness in Euclid's thinking as of changes in the standards of precision, rigor, and proof since his day, indeed changes that were largely consequences of the thinking that his own work and that of his contemporaries set in motion. Moreover, whatever the shortcomings of Euclid's proofs, even if many, it took two millennia of mathematizing to improve much upon them.

One of the major changes in perspective that has implications for the nature of proofs was the change from thinking of geometry as self-evidently descriptive of the way the world is to thinking in terms of an axiomatic system. As was noted in the preceding chapter, according to the prevailing modern perspective, whether (any particular) geometry describes the physical world is incidental; what is important from a mathematical point of view are the implications of the axioms that constitute the system. As Devlin (2000a) puts it, "When it comes to establishing the theorems that represent mathematical truths, the axioms are, quite literally, all there is" (p. 163).

But even when it is not required of mathematical theorems that what they assert is descriptive of the physical world, what constitutes a proof may be in some dispute, as may be the validity of specific proofs. Bell (1937) credits Gauss with being the first to see clearly that a proof that can lead to absurdities is no proof at all, and with being responsible for imposing a rigor on mathematics that was not known before his time.

☐ A Proof as the "Final Word"

Proof has a ring of finality to it. To say that something has been proved is to say, or so it would appear, that we can be certain it is true—in the sense of being derivable from the system's axioms. Once an assertion has been proved to be true—once it has attained the status of a theorem—from that point on one can take it as a given and need no longer worry about it. German-American philosopher of science Carl Hempel (1935/1956a) expresses essentially this idea in contrasting the status of the theories of empirical science with the theorems of mathematics in the following way: "All the theories and hypotheses of empirical science share this provisional character of being established and accepted 'until further notice,' whereas a mathematical theorem, once proved, is established once and for all; it holds with that particular certainty which no subsequent empirical discoveries, however unexpected and extraordinary, can ever affect to the slightest extent" (p. 1635).

The centrality of the role of proofs distinguishes, probably more than anything else, mathematics from the empirical sciences. Du Sautoy (2004) argues that a major reason that proofs play a central role in mathematics but not in the empirical sciences is that the subject matter of mathematics is ethereal while that of the empirical sciences is tangible.

> In some respects, the ethereal nature of mathematics as a subject of the mind makes the mathematician more reliant on providing proof to lend some feeling of reality to this world. Chemists can happily investigate the structure of a solid buckminsterfullerene molecule; sequencing the genome presents the geneticists with a concrete challenge; even the physicists can sense the reality of the tiniest subatomic particle or a distant black hole. But the mathematician is faced with trying to understand objects with no obvious physical reality such as shapes in eight dimensions, or prime numbers so large they exceed the number of atoms in the physical universe. Given a palette of such abstract concepts the mind can play strange tricks, and without proof there is a danger of creating a house of cards. In the other scientific disciplines, physical observation and experiment provide some reassurance of the reality of a subject. While other scientists can use their eyes to see this physical reality, mathematicians rely on mathematical proof, like a sixth sense, to negotiate their invisible subject. (p. 31)

Du Sautoy goes on to note too that perhaps the most compelling reason for the emphasis on proofs in mathematics is that proofs are possible in this domain, and that is not true of the empirical sciences. "In how many other disciplines is there anything that parallels the statement that

Gauss's formula for triangular numbers will *never* fail to give the right answer? Mathematics may be an ethereal subject confined to the mind, but its lack of tangible reality is more than compensated for by the certitude that proof provides" (p. 32). Many other definitions or descriptions of *proof* could be quoted that would give it a similar ring of finality.

☐ The Relativity of Proofs

To some, the idea that a proof is the "final word" on a mathematical question is an aspect of mathematical reasoning that sets it apart from, and in a nontrivial sense above, reasoning in other domains. Knorr (1982) refers to mathematicians as intellectual elites among members of both the sciences and the humanities, and surmises that the basis of this status is the incontrovertible nature of properly reasoned mathematical arguments. The degree of consensus that is attainable in mathematics is found nowhere else. At least this is a common claim. But not everyone, not even every mathematician, holds this view.

Apparently Eric Temple Bell saw things differently. Kline (1980) quotes him as follows: "Experience has taught most mathematicians that much that looks solid and satisfactory to one mathematical generation stands a fair chance of dissolving into cobwebs under the steadier scrutiny of the next.... The bald exhibition of the facts should suffice to establish the one point of human significance, namely, that competent experts have disagreed and do now disagree on the simplest aspects of any reasoning which makes the slightest claim, implicit or explicit, to universality, generality, or cogency" (p. 257). Regarding how drastically views as to what constitutes a proof can change from one generation to another, Bell (1946/1991) contends that "a proof that convinces the greatest mathematician of one generation may be glaringly fallacious or incomplete to a schoolboy of a later generation" (p. 66).

Mathematical proofs are relative in several ways. First, any mathematical proof exists within some specific axiomatic system. As already noted, the truth that it establishes is relative to the axioms of that system. That is to say, the proof of a theorem establishes the theorem to be true only to the extent that the axioms of the system are held to be true. To be completely consistent with currently prevailing ideas about mathematics, we probably should not use the concept of truth at all. What a proof purports to show is that a theorem follows from the axioms of the system of which it is a part. Inasmuch as it is required of the axioms of a system only that they be consistent with each other, and not that they be true in the sense of accurately reflecting properties of the physical world, it cannot be required of proved theorems that they be true in this sense either. I shall continue to speak of proved theorems as true as a matter of convenience, as do mathematicians, but it must be borne in mind that truth in this context has the specific meaning of "following from the axioms."

Second, the history of mathematics is replete with proofs that, after standing for a considerable time, have been shown to be inadequate in retrospect. I have already noted that many of the proofs in Euclid's revered *Elements* fail to meet contemporary standards. Here is the assessment by one prominent mathematician from the vantage point of more than 2,000 years of intervening work: "Any impartial critic may convince himself in less than an hour—as many did when European geometers began to recover from their uncritical reverence for the Greek mathematical classics—that several of Euclid's definitions are inadequate; that he frequently relies on tacit assumptions in addition to the postulates to which he imagined he had restricted himself; that some of his propositions, as he states them, are false, and that the supposed proofs of others are nonsense.... If it were worth anyone's trouble, the entire logical structure of the geometrical portions of the *Elements* might be destructively analyzed for inexplicit assumptions and defective proofs" (Bell, 1946/1991, p. 332). Hersh (1997) has similarly harsh words for what he refers to as the myth of Euclid, and for those who perpetuate it: "Today advanced students of geometry know Euclid's proofs are incomplete and his axioms are unintelligible. Nevertheless, in watered-down versions that ignore his impressive solid geometry, Euclid's *Elements* is still upheld as a model of rigorous proof" (p. 37). Even among mathematicians, Hersh claims, the Euclid "myth" was universal until well into the 19th century. An abiding challenge to researchers of human cognition is to figure out how it is that what can appear to be a compelling proof to some mathematicians can be unconvincing—and in some cases even appear to be nonsensical—to others.

It has not been unusual for generations of great mathematicians to overlook specific problems in proofs. As von Mises (1951/1956) puts it, "All followers of the axiomatic method and most mathematicians think that there is some such thing as an *absolute* 'mathematical rigor' which has to be satisfied by any deduction if it is to be valid. The history of mathematics shows that this is not the case, that, on the contrary, every generation is surpassed in rigor again and again by its successors" (p. 1733). The problems that mathematicians find in the arguments of predecessors often have less to do with what was said than with what was not said—less likely to lie with what the authors of the arguments knew they had assumed than with what they unconsciously assumed. "Each generation criticizes the unconscious assumptions made by its parents. It may assent to them, but it brings them out in the open" (Whitehead, 1925/1956, p. 406).

Third, even proofs that are considered sound differ considerably in their ability to convince. There are many theorems that have been proved in a variety of ways. Several hundred different proofs have been offered of the Pythagorean theorem, which relates the length of the hypotenuse of a right triangle to the lengths of its other two sides (Loomis, 1968). It seems a safe bet that the reader who will take the trouble to check out a few of them will find some more compelling, or more readily grasped, than others. The amount of attention this theorem has received, as indicated in the great variety of proofs of it that have been constructed, gives credence to Dunham's (1991) reference to it as "surely one of the most significant results in all of mathematics" (p. 47).

Fourth, proofs, again even when considered sound, vary also in their simplicity (beauty, elegance, attractiveness to mathematicians). Complex, ugly, inelegant proofs stand as challenges to mathematicians to find ways to improve upon them. Joseph Bertrand's conjecture that one can always find a prime between any number, n, and its double, $2n$, was proved by Russian mathematician Pafnuty Chebyshev seven years after it was proposed. Chebyshev's proof, though accepted as valid, was not seen by all number theorists to be as elegant as possible. Indian mathematician Srinivasa Ramanujan found a way to improve upon it, as did Hungarian mathematician Paul Erdös independently some years later (Du Sautoy, 2004). There are countless other examples of "proof perfecting," and many remaining challenges.

Fifth, the concept of proof is itself evolving. Hungarian philosopher-mathematician Imre Lakatos (1976) points out that changes in the idea of what constitutes a rigorous proof have engendered revolutions in mathematics. The Pythagoreans, for example, held that rigorous proofs have to be arithmetical, but upon discovering a rigorous proof that the square root of 2 is irrational, they had to change this criterion for rigor. As a consequence, geometrical intuition took the place once held by arithmetical intuition. Newton, Leibniz, Euler, LaGrange, and Laplace, all of whom were great analysts, had little conception of what is now acceptable as a proof involving infinite processes. And what is now acceptable depends on whom one asks; some mathematicians recognize only constructive proofs, for example, and rule out indirect proofs such as those that employ the *reductio ad absurdum* argument. (A constructive proof is one that is derivable—constructible—from integer arithmetic. To prove a proposition indirectly, one shows that assuming the proposition to be false leads to a contradiction, which, according to Aristotelian logic, means that the proposition must be true.) German mathematician-logician Leopold Kronecker was a staunch advocate of recognizing only constructive proofs and used this restriction to discredit Cantor's work. Mathematicians disagree about the legitimacy of the "law of the excluded middle," according to which a mathematical statement is either true or false; Dutch mathematician Luitzen Brouwer, for example, rejected it, whereas Hilbert considered it an essential mathematical tool and likened the prohibition of its use to prohibiting a boxer from using his fists (Hellman, 2006).

That the concept of proof has changed over time should temper harsh judgments of the inadequacy of proofs that were considered to be sound by earlier generations. It is hardly fair to judge the proofs of Euclid by standards that were developed gradually over hundreds of years after his death. It should also make us less than completely certain of the finality of current ideas on the matter. Kline's (1980) observation that "the proofs of one generation are the fallacies of the next" (p. 318) undoubtedly applies to our own generation not only as one that follows those that have already passed, but also as one that precedes those that are to come. It would be presumptuous to assume that today's ideas about what constitutes adequacy of proof will be regarded much more highly by succeeding generations than those of preceding generations are regarded by our own.

Sixth, even within a specific time frame, *proof* can have different connotations in different contexts. Hersh (1997), for example, distinguishes two meanings of the term as it is used today in mathematics, one that applies in practice and the other in principle. "Meaning number 1, the *practical* meaning, is informal, imprecise. *Practical mathematical proof is what we do to make each other believe our theorems.* It's argument that convinces the qualified, skeptical expert.... What is it *exactly*? No one can say. Meaning number 2, theoretical mathematical proof, is formal.... It's transformation of certain symbol sequences (formal sentences) according to certain rules of logic (*modus ponens*, etc.). A sequence of steps, each a strict logical deduction, or readily expanded to a strict logical deduction" (p. 49).

Finally, even among mathematicians who agree on the rules, disputes regarding the validity of specific proofs continue to arise. Such disputes are resolved—if they are resolved—by consensus. "Real proofs aren't checkable by machine, or by live mathematicians not privy to the mode of thinking of the appropriate field of mathematics. Even qualified readers may differ whether a real proof (one that's actually spoken or written down) is complete and correct. Such doubts are resolved by communication and explanation" (Hersh, 1997, p. 214). Hersh emphasizes the social nature of mathematics and the importance of the influence of the culture in which it is done—and of the tentativeness of the proofs that mathematicians develop. "The mathematics of the research journals is validated by the mathematical community through criticism and refereeing. Because most mathematical papers use reasoning too long to survey at a glance, acceptance is tentative. We reconsider our claim if it's disputed by a competent skeptic" (p. 224). Fortunately, as a general rule, mathematicians agree relatively quickly on the merits of most proposed proofs, but sometimes a consensus as to the adequacy of a complicated proof can be a long time in forming, if it forms at all.

To nonmathematicians, formal proofs can be intimidating and sometimes incomprehensible. Rucker (1982, p. 274) illustrates how cumbersome the process of writing out a formal proof can be with an example of such a proof of $(\forall y)$ $[0 + y = y]$, that is, for all y, $0 + y = y$. The proof takes 17 steps and uses on the order of—I am estimating—400 to 500 symbols. Rucker argues that, despite their "nitpicking, obsessive quality," fully formalized proofs "are satisfyingly solid and self-explanatory. Nothing is left to the imagination, and the validity of a formal proof can be checked simply by looking at the patterns of symbols. Given the basic symbols, the rules of term and formula formation, the axioms and axiom schemas, and the rules of inference, one can check whether or not a sequence of strings of symbols is a proof in a wholly mechanical fashion" (p. 275). (Note the difference between Rucker's claim that a formal proof can be verified mechanically and Hersh's insistence, mentioned above, that real proofs are not checkable this way.)

Casti (2001) argues that proofs differ in quality, and proposes that three grades be recognized:

> The first, or highest quality type of proof, is one that incorporates *why* and *how* the result is true, not simply that it is so. ... Second-grade proofs content themselves with showing *that* their conclusion is true, by relying on the law of the excluded middle. Thus they assume that the conclusion they want to demonstrate is false and then derive a contradiction from this assumption. ... In [third, or lowest-grade proofs] the idea of proof degenerates into mere verification, in which a (usually) large number of cases are considered separately and verified, one by one, very often by a computer. (p. 137)

Casti points to Appel and Haken's (1977b) proof of the four-color theorem as an example of the third type. Obviously, from this perspective, the goal is to produce highest quality proofs; second or third level is to be settled for only when the first one proves to be out of reach.

All this being said, the only absolute conclusion we can draw is that proofs are relative. What a mathematical proof gives us, in Kline's (1980) words, is "relative assurance. We become quite convinced that a theorem is correct if we prove it on the basis of reasonably sound statements about numbers or geometrical figures which are intuitively more acceptable than the one we prove" (p. 318). Intuitively acceptable statements in this context must include statements that are acceptable by virtue of themselves having already been proved. The chain of inferences that gets one back to basic givens, or statements that are intuitively acceptable without proof, is, in some instances, very long. The history of proof making and proof discrediting dictates caution in accepting new proofs as infallible.

☐ A Proof in Process

The nature of proof and the process of proof making, as well as those of proof challenging and proof repairing, have been explored in a delightfully readable work by Lakatos (1976), published two years after his untimely death at age 52. The aim of Lakatos's "case study," as he called it, of the methodology of mathematics was, in his words, "to elaborate the point that informal, quasi empirical mathematics does not grow through a monotonous increase of the number of indubitably established theorems, but through the incessant improvement of guesses by speculation and criticism, by the logic of proofs and refutations" (p. 5). (For a concise description of "the method of proofs and refutations," which he also calls "the method of lemma incorporation," see Lakatos, 1976, p. 50.)

Lakatos's essay recounts a lengthy discussion among a group of students and a teacher in a classroom. The classroom and the participants in the discussion are fictitious, but the discussion tracks the development of mathematical thinking over several centuries as it relates to the problems on which the class focuses. The problem that has the class's attention at the outset is the question of whether there is a relationship between the number of vertices (V), the number of edges (E), and the number of faces (F) of a polyhedron. The students discover by trial and error that for regular polyhedra these variables are related according to the formula $V - E + F = 2$. (Both Euler and Descartes had observed this relationship, Descartes in 1640 and Euler in 1752. Euler expressed the relationship with the formula just mentioned.)

The teacher proposes a proof that this relationship holds for *all* polyhedra. The students challenge the validity of the proof (which actually was believed to be valid by several notable 19th-century mathematicians) by questioning the truth of some of the claims that comprise it. This they do by finding counterexamples to one or more of these claims. (Lakatos makes a distinction between a local counterexample, which refutes a lemma—a proposition subsidiary to a theorem, proved, or assumed to be true, in order to simplify the proof of the theorem—of a proof but not necessarily the main conjecture that one is trying to prove, and a global counterexample, which refutes the main conjecture itself. A global counterexample shows the main conjecture to be false, whereas a local counterexample shows only that some element of the proof is false, but does not rule out the possibility that the conjecture itself is true.) The teacher concedes that the students have indeed shown the proof to be invalid, but rather than discard it, he attempts to improve it so that it will be able to stand up to the criticisms.

As the dialogue proceeds, alternative proofs are offered and challenged with counterexamples of one or another type. Counterexamples are sometimes challenged or made to be irrelevant by the "method of monster-barring," whereby the original conjecture is modified, or the class of interest (in this case polyhedron) is redefined in such a way that the counterexample becomes a "monster" with respect to that class, which is to say no longer a member of it. This leads to a discussion of the importance of definitions, and also to the recognition that definitions can often be a focus of debate and disagreement. Lakatos points out that short theorems in mathematics are sometimes obtained at the expense of long definitions; the definition of *ordinary polyhedron* in the 1962 edition of *Encyclopedia Britannica*, for example, takes up 45 lines.

Sometimes the response to a demonstration that a conjecture is false is a modification of the conjecture. For example, when shown not to be true, the original conjecture "For all polyhedra, $V - E + F = 2$" is replaced with "For all *simple* polyhedra, $V - E + F = 2$," where *simple* is meant to rule out polyhedra of the type represented by a picture frame (a polyhedron with a hole in it). When that conjecture is shown to be false, it, in turn, is replaced by "For a simple polyhedron, *with all its faces simply connected, $V - E + F = 2$.*" And so on.

Definitions are crucial because of the vagueness of language, especially in its everyday use. Lakatos contends that one can always find a sufficiently narrow interpretation of the terms of a proposition to make it be true as well as a sufficiently wide interpretation to make it be false. The rigor of a proof of a theorem may be increased through redefinition or the incorporation of new lemmas at the expense of decreasing the inclusiveness of the theorem's domain.

Lakatos refers to the idea that the path of discovery is a simple progression from facts to conjecture, and from conjecture to proof as "the myth of induction" (p. 73), noting that basic concepts often become modified substantially in the making, criticizing, and revising of proofs: "Naive conjectures and naive concepts are superseded by improved conjectures (theorems) and concepts (proof-generated or theoretical concepts) growing out of the method of proofs and refutations. And as theoretical ideas and concepts supersede naive ideas and concepts, theoretical language supersedes naive language" (p. 91).

Thus as a consequence of the process of attempting to prove a conjecture, the conjecture itself may be modified, as may the concepts that comprise it, so what ends up being proved is something other than the conjecture that motivated the proof-making effort. The conjecture, as modified, is likely to be considerably more precise, and perhaps narrower in scope—as a result of the introduction of precise definitions and the delimitation of conditions under which it is claimed to hold—than as originally conceived. Casti (2001) cites Lakatos's work as a prime illustration of an increasing tendency to acknowledge the empirical component in the practice of mathematics. He notes that Lakatos's view—that "the practice of mathematics constitutes a process of conjecture, refutation, growth, and discovery"—has much in common with Karl Popper's ideas about the nature of the scientific enterprise.

As an example of how a proof can evolve as generations of mathematicians work on it, Barrow (1991) points to the case history of the *prime number theorem*, which derives from a conjecture of Gauss and French mathematician Adrien-Marie Legendre—known naturally as either the prime number conjecture or the Gauss-Legendre conjecture—regarding the proportion of numbers less than any given value that are primes. The conjecture was that the number of primes less than n was approximated by $n/\log n$ ever more closely with increasing size of n. The first proof of the theorem, given by French mathematician Jacques Hadamard (1865–1963) and Belgian mathematician Charles de la Vallée-Poussin in 1896, involved complex analysis and was very difficult. Somewhat simpler proofs were produced later by German mathematician Edmund Landau and American mathematician Norbert Wiener. In 1948, Erdös and Norwegian mathematician Atle Selberg gave what Barrow says could be considered an elementary proof of some 50-plus pages in length, which was refined and made truly elementary by Norman Levinson. Barrow argues that this is characteristic of the evolution of mathematics, and that *bona fide* original ideas of the caliber of Cantor's diagonal argument and Gödel's proof of undecidability (about both of which more later) are very rare.

☐ Refutations

In the course of political, scientific and everyday disputes, in the process of a court investigation and analysis, in attempts to solve various problems, one must learn not only to prove, but also to refute. (Bradis, Minkovskii, & Kharcheva, 1938/1999, p. 2)

The give and take in Lakatos's (1976) account of the role of refutations in proof making and proof improving illustrate that a refutation of a proof is itself a proof of sorts; it is a proof that the proof that is being refuted is not valid—is not a proof after all. As we have noted, the history of mathematics has many examples of "proofs" that have survived for some time only eventually to be considered to be faulty. French mathematician Jean le Rond D'Alembert, Euler, and Lagrange all produced proofs of the fundamental theorem of algebra—according to which every polynomial equation has at least one root—which were later determined to be wrong (Flegg, 1983). (The fundamental theorem is also expressed as follows: Any real polynomial of degree n can be factored into n linear factors.)

The importance of refutations is also seen in the history of attempts to solve some of the famous problems that have tantalized mathematicians, and many would-be mathematicians as well, over the centuries—squaring the circle, trisecting an angle, proving the four-color conjecture or Fermat's last theorem. Proposed solutions or proofs have to be shown to be wrong if they are to be dismissed.

In another delightful book, published originally in 1938, Russian mathematicians V. M. Bradis, V. L. Minkovskii, and A. K. Kharcheva give numerous "proofs" of mathematical absurdities. Examples of what is "proved" include: $45 - 45 = 45$; $2 \times 3 = 4$; every negative number is greater than the positive number having the same absolute value; all triangles are of equal area; the length of a semicircle is equal to its diameter; $\pi/4 = 0$; $1/4 > 1/2$; and every triangle is a right triangle. I will reproduce here two of the faulty proofs from Bradis et al., the first algebraic and the second geometric. (If it is not obvious where these proofs go wrong, the reader is referred for explanations to Bradis et al., pp. 115, 123.)

- Proof of the equality of two arbitrary numbers (p. 80). Take two arbitrary numbers a and $b > a$, and write the identity:

$$a^2 - 2ab + b^2 = b^2 - 2ab + a^2 \tag{5.1}$$

where the algebraic sums in the right- and the left-hand members differ from one another only by the order of the terms.

Equation (5.1) we rewrite in the shorter form, making use of the formula for the square of a difference:

$$(a - b)^2 = (b - a)^2 \tag{5.2}$$

Extracting the square root from both members, we obtain:

$$a - b = b - a \tag{5.3}$$

whence, upon transferring some terms, simplifying and dividing both members by 2, we have:

$$a + a = b + b, \; 2a = 2b, \; a = b$$

- Proof that the segments of parallel straight lines bounded by the sides of an angle are equal. Take an arbitrary angle and intersect its sides by two arbitrary parallel straight lines. Let AB and CD be the segments of the parallels included between the sides of that angle, and E its vertex (Figure 5.1).

  As is well known, parallel straight lines intersect proportional segments of the sides of the angle. Consequently,

$$AE : CE = BE : DE$$

and

$$AE \times DE = BE \times CE \tag{5.4}$$

  Multiplying both members of Equation (5.4) by the difference $AB - CD$, we carry out the following transformations:

$$AE \times DE \times AB - AE \times DE \times CD = BE \times CE \times AB - BE \times CE \times CD$$

$$AE \times DE \times AB - BE \times CE \times AB = AE \times DE \times CD - BE \times CE \times CD$$

$$AB(AE \times DE - BE \times CE) = CD(AE \times DE - BE \times CE) \tag{5.5}$$

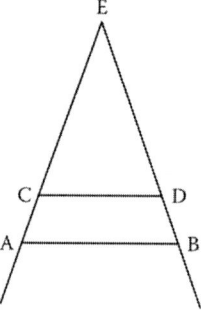

FIGURE 5.1 Supporting the "proof" that the segments of parallel straight lines bounded by the sides of an angle are equal.

Dividing both members of the last equality by the difference $AE \times DE - BE \times CE$, we obtain the equality $AB = CD$. Thus, the segments of parallels confined between the sides of a given angle are always equal (p. 123).

For each of the faulty proofs, Bradis et al. provide an explanation of where it goes wrong. I strongly suspect that most readers, including those with a considerable knowledge of mathematics, will have to work a bit to find the faulty steps on their own in some cases. Bases for faulty proofs that Bradis et al. identify include incorrect usage of words, inaccurate formulations, neglect of the conditions of applicability of theorems, hidden execution of impossible operations, and invalid generalizations—as in passing from a finite set to an infinite one. What these authors demonstrate is the ease with which such errors can be made and go undetected. One cannot read their book and ever again accept even the simplest and most transparent of proofs as "obviously infallible" without some thought of how it might be wrong.

The point of course is that one does well to be wary of accepting proofs too quickly. Kline (1980) makes a stronger statement: "No proof is final. New counterexamples undermine old proofs. The proofs are then revised and mistakenly considered proven for all time. But history tells us that this merely means that the time has not yet come for a critical examination of the proof" (p. 313). Proofs are accepted, Kline argues, by virtue of being endorsed by the leading specialists of the day or because of employing currently fashionable principles. But tentative or no, proofs remain the objective, and the ticket for recognition among mathematicians. As someone has said—I cannot remember where I read it—"Of scientists one asks, what did they discover; of mathematicians, what did they prove."

☐ Unproved Conjectures

One of the lessons that the history of mathematics clearly teaches us is that the search for solutions to unsolved problems, whether solvable or unsolvable, invariably leads to important discoveries along the way. (Boyer & Merzbach, 1991, p. 595)

Throughout the history of mathematics, mathematicians have been tantalized by certain conjectures that are believed to be true, but that have resisted all past attempts to prove them to be so. Some conjectures have been proved true only after existing as conjectures for a very long time. A case in point is the conjecture that every positive whole number is the sum of no more than four squares. This was known, as a conjecture, to the classical Greeks, and existed as a conjecture until Lagrange proved it in 1770.

Some conjectures have been known and have remained unproved for a long time despite countless hours of intense work by many mathematicians seeking the elusive proofs. Among the better known examples, which were mentioned in Chapter 1, are Goldbach's conjecture that every even number is the sum of two primes, Gauss's conjecture that the number of primes between 1 and n is approximated by $n/\log_e n$, the four-color conjecture according to which four colors suffice to color any map without using the same color for any two bordering areas, and Fermat's "last theorem," according to which integers cannot be found to solve the equation $x^n + y^n = z^n$ for $n > 2$ and xyz not equal to 0. Many proofs of all of these conjectures have been proposed over the years, only eventually to have been shown to be invalid.

As of 2000, Goldbach's conjecture, which was put forward by Prussian mathematician Christian Goldbach in 1742, had been shown to be valid for all numbers up to 400,000,000,000,000, but it has yet to be proved either true for all numbers or not true for all numbers. Gauss's conjecture was proved (by French mathematicians Jacques Hadamard and Charles Jean de la Vallée-Poussin) approximately 100 years after it was stated by Gauss.

The four-color conjecture has it that four colors are sufficient to color any conceivable map without making any bordering countries have the same color. (A shared point, like that shared by Arizona, Utah, Colorado, and New Mexico, does not count as a common border.) According to Rouse Ball (1892), the problem was mentioned by German mathematician-astronomer August Ferdinand Möbius in lectures in 1840, but received little attention until it was communicated to De Morgan around 1850. De Morgan learned of the problem from a student, Frederick Guthrie, whose brother Francis, a South African botanist and mathematics professor, conceived the question as a consequence of actually coloring a map. May (1965) claims that the evidence about the origin of the problem is tenuous.

In any case, the problem became widely known, and despite many efforts to prove that four colors are enough, which most mathematicians appear to have believed, the conjecture remained unproved for 150 years, although several people believed, at least for a short time, that they had succeeded. One proof, published by British mathematician Arthur Kempe in 1879, stood for 11 years until another British mathematician, Percy Heawood, who himself worked on the problem over 60 years, showed Kempe's proof to be flawed. A proof of the conjecture that in time became widely accepted as such was produced by American mathematicians Kenneth Appel and Wolfgang Haken with assistance from John Koch and much use of a computer. It was described for mathematicians in the *Illinois Journal of Mathematics* in two parts (Appel & Haken, 1977a; Appel, Haken, & Koch, 1977) and for a more general audience in *Scientific*

American (Appel & Haken, 1977b). An engaging historical account of the development of the proof is given in Appel and Haken (1978), in which the authors credit Kempe with producing an extremely clever argument containing most of the basic ideas that eventually led to their proof. Appel and Haken point out that much of what became known as graph theory, which now has numerous practical applications, grew out of the work done in the countless efforts to prove the four-color conjecture.

Mathematicians labored in vain for over three centuries to generate a proof of Fermat's last theorem. According to Aczel (1996), the year following the announcement in 1908 of the Wolfskehl Prize of 100,000 German marks for a proof of it, 621 proposed solutions were submitted, none of them sound. Some have believed they have succeeded; however, every "proof" that had been advanced prior to the one announced and improved upon by American mathematician Andrew Wiles in the 1990s was subsequently found to be invalid. And some have believed that they were very close to producing a proof, but discovered later that they were unable to complete the feat. In 1847, shortly after the French Academy of Sciences established a gold medal and monetary prize for anyone who first produced a *bona fide* proof, French mathematicians Gabriel Lamé and Augustin Cauchy each announced to the assembled academy members that he was on the verge of doing so, but it would be another century and a half before Wiles actually produced one.

It is interesting that we read much about the energy that was put into finding a proof of the theorem but relatively little about attempts to show the theorem to be false, when all that was necessary for the latter was to find one set of values for which $x^n + y^n = z^n$ is true. And given that there are an infinite set of values for which $x^3 + y^3 + z^3 = w^3$ is true, one might be excused for thinking that perhaps there should be at least one set for which $x^n + y^n = z^n$ is true. Apparently, many mathematicians—perhaps this is testimony to the esteem in which Fermat was held—believed from the beginning that no such set exists. Over time evidence supporting the assumption accumulated in the form of demonstrations that it was true for values of n up to a specified limit, which by 1993 was over 4 million.

☐ Failed Proofs and Mathematical Progress

Attempts to find a general proof of Fermat's last theorem led serendipitously to many other mathematical developments. Indeed, it can be argued plausibly that major unproved conjectures that have motivated intensive search for proofs (or disproofs) over long periods of time have provided a great service to the advancement of mathematics because of the unanticipated discoveries that have been made as a consequence of these efforts. We will consider proofs of the four-color theorem and Fermat's last theorem again below because they provide further evidence of how difficult it can sometimes be to decide when a proof is a proof.

Prominent among failed attempts to prove something are the numerous efforts over several centuries to prove that Euclid's parallel postulate—the fifth of his geometry—is deducible from the other Euclidean postulates. Generations of mathematicians have been uneasy about it, believing it to be unnecessary by virtue of being implicit in the other postulates, but no one was able to prove that to be the case. As we have already noted, after many centuries of failed efforts to demonstrate the parallel postulate to be redundant, several mathematicians began to consider the possibility of geometries in which it did not hold, and such non-Euclidian geometries were formulated in the 19th century.

The classical Greeks described three geometrical construction problems that have challenged professional and amateur mathematicians alike for more than two millennia. The problems may be stated as follows: Using only a straightedge and compass, (1) construct a square whose area is equal to the area of a given circle, (2) construct a cube the volume of which is twice that of a given cube, and (3) divide an angle into three equal angles. Countless hours were devoted to efforts to solve these problems and many "solutions" were proposed, all of which were shown eventually to be invalid. Numerous constructions have been developed over the centuries, some by the early Greeks themselves (see Beckman, 1971), but none that requires only straightedge and compass. (Incidentally, not until 1672 did someone—Danish mathematician Georg Mohr—point out that, given that a line is determined when its two endpoints are specified, any plane construction that can be effected by straightedge and compass can be effected by compass alone [Boyer & Merzbach, 1991].)

Eventually, during the 19th century, all three of these problems were shown to be insoluble (Jones, Morris, & Pearson, 1991). Does this mean that all of the time spent working on them should be considered wasted? In fact, some of this work produced very useful results. The conic sections as well as numerous other mathematical phenomena, including many that have proved to have great practical utility, were discovered as a consequence of attempts to solve these problems. Work on analytic curves, cubic and quartic equations, Galois theory, and transcendental numbers has been attributed, at least in part, to these attempts (Paulos, 1992). More generally, efforts to solve problems that eventually were shown to be insoluble have often led to unanticipated advances in mathematics. Stewart (1987) refers to Fermat's last theorem as an example of a problem that is so good that even the failures to solve it have enriched mathematics greatly. A similar observation might be made regarding other conjectures that have remained unproved for a long time. Failed attempts to prove Riemann's hypothesis have frequently led to mathematical discoveries. (Du Sautoy [2004] notes that mathematicians are sufficiently confident that Riemann's hypothesis is true, despite the nonexistence of a proof, that some have used it in the production of proofs of other theorems; this is a risky strategy, however, because if the hypothesis is eventually proved to be false, theorems that have been based on it will fall.)

This is not to suggest that all such work is productive in any meaningful sense. Determining a cost-benefit ratio for work on insoluble mathematical problems is undoubtedly an insoluble mathematical problem, for practical if not for theoretical reasons. It is interesting to note that in 1755 the French Academy resolved to examine no more manuscripts purporting to "square the circle," even though at the time of the academy's decision, the impossibility of solving this problem had not yet been proved. Apparently the academy had decided that what was to be gained by continuing to receive such submissions was not worth the effort of reviewing them. Before Wiles presented his proof of Fermat's last theorem in 1993, the Göttingen Royal Society of Science had received over 5,000 proposed proofs, all of which had to be evaluated by a qualified mathematician (Casti, 2001).

In 1900 David Hilbert presented to the International Congress of Mathematics 23 unsolved problems that represented, in his view, the greatest challenges to mathematicians at the dawn of the 20th century. (Two additional problems not on the list of 23 were mentioned in his introductory remarks.) The Riemann hypothesis and Goldbach's conjecture in combination—generally referred to as problems of prime numbers—constituted the eighth item in Hilbert's list. Accounts of the status of Hilbert's problems at the end of the 20th century are given by Gray (2000) and Yandell (2002). Some of the problems turned out to be too vague to admit a precise solution, but of those that were sufficiently precise, all but one—the Riemann hypothesis—had been solved by the end of the century.

But there appears to be no end of challenging problems. At a 1974 symposium at which progress on Hilbert's problems was discussed by experts in the various relevant areas, a new 23-item set of problems was described (Browder, 1976). The mathematical world was ushered into the 21st century with the announcement, in May 2000, of a $1 million prize to anyone who could solve any of seven problems then considered by the offerers of the prize to be among the most difficult mathematical problems still unsolved. Prize money—$7 million—was provided by Landon Clay, founder of the Clay Mathematics Institute. Among the Millennium

Problems, as the set of seven is known, is the Riemann hypothesis, the only carryover from Hilbert's list. A description of all seven problems, written expressly for the interested layperson who is not an expert mathematician, is provided by Devlin (2002); this is not to say that it is an easy read.

☐ Proofs as Convincing Arguments

> In practice, proofs are simply whatever it takes to convince colleagues that a mathematical idea is true. (Henrion, 1997, p. 242)

Devlin (2000a) gives a definition of a proof very similar to Henrion's just quoted, but with the qualification that who needs to be convinced is "any sufficiently educated, intelligent, rational person" (p. 51). This is an important qualification inasmuch as to be "sufficiently educated" to understand some proofs (e.g., the four-color theorem, Fermat's last theorem) means knowing a great deal of rather esoteric mathematics.

For a proof of a theorem to be compelling, every assertion must be either an axiom or a statement that follows logically from the system's axioms either directly or indirectly through other already proved theorems. As Nozick (1981) puts it, "A proof transmits conviction from its premises down to its conclusion, so it must start with premises ... for which there already is conviction; otherwise, there will be nothing to transmit" (p. 14). Accepting the axioms as givens is one necessary condition for accepting a proof as a whole; another is believing the assertions that are derived from them to be valid inferences. But while this combination is essential, it does not suffice to satisfy all inquiring minds.

Polya (1954b) notes the possibility that a mathematician may be convinced that every step in a proof is correct and still be unsatisfied if he does not feel he understands the proof as a whole. "After having struggled through the proof step by step, he takes still more trouble: he reviews, reworks, reformulates, and rearranges the steps till he succeeds in grouping the details into an understandable whole. Only then does he start trusting the proof" (p. 167). Kline (1980) makes essentially the same point in noting that an intuitive grasp of a proof can be more satisfying than logic. "When a mathematician asks himself why some result should hold, the answer he seeks is some intuitive understanding. In fact, a rigorous proof means nothing to him if the result doesn't make sense intuitively" (p. 313).

Penrose (1989) similarly emphasizes the importance of being able to see the truth of a mathematical argument in order to truly be convinced of its validity, and insists that mathematical truth is not ascertained merely by use of an algorithm. Gödel's theorem, more about which will presently be discussed, shows us the necessity for external insights for deciding the validity of algorithms. Indeed, our ability to be persuaded by Gödel's argument is itself evidence of this need. "When we convince ourselves of the validity of Gödel's theorem, we not only 'see' it, but by so doing we reveal the very non-algorithmic nature of the 'seeing' process itself" (p. 418).

In an algorithmic approach to mathematics, one applies useful techniques to solve problems, but does not worry much about how the techniques were derived or why they work. This approach, effective though it may be for practical purposes, is unlikely to satisfy the mathematician who wishes to understand mathematics at a deeper level. Just so, a proof that is accepted as such because no fault can be found with the sequence of steps that comprise it (a strictly algorithmic proof) is not likely to be as satisfying to a mathematician as an insightful proof that provides, or at least facilitates, an understanding of why the proved relationship is what it is.

Casti (2001) holds that a good proof has three characteristics: It is convincing, surveyable, and formalizable. By convincing, he means convincing to mathematicians—if a proof is a good one, most mathematicians will believe it. To be surveyable means to "be able to be understood, studied, communicated, and verified by rational analysis" (p. 70). "Formalizability means we can always find a suitable formal system in which an informal proof can be embedded and fleshed out into a formal proof" (p. 70). These three characteristics, Casti argues, represent, respectively, the anthropology, epistemology, and logic of mathematics. From a psychological point of view, the first of these characteristics is paramount. If a proof is not convincing to people who are sufficiently knowledgeable to follow it, there is not much else to be said for it.

Some proofs are relatively straightforward in the sense that they involve only a few inferences that most people probably would have little trouble following. Consider, for example, Euclid's proof that there is no largest prime number. I have seen two versions, and will give both. Version 1: Suppose there were a largest prime number, p. Suppose further that we pick a number, p^*, that is equal to the product of all the primes, plus 1. Obviously, p^* cannot be evenly divided by any prime—it was constructed so as to ensure that division by any prime would leave a remainder of 1—so p^* itself must be prime. This contradicts the assumption that p is the largest prime. And inasmuch as the same reasoning can be applied to p^* and to any larger prime that is found, it follows that there is no largest prime, or, equivalently, that there are an infinity of primes. Version 2: Let p represent any prime. Construct $p! + 1$. The result clearly is not divisible by p or any number smaller than p. Either it is not divisible or it is divisible by a prime between p and $p! + 1$; either possibility implies the existence of a prime larger than p. Euclid's strategy—proof by contradiction, or *reductio ad absurdum,* in which one shows that the assumption that the assertion to be proved is false leads to a contradiction and that the assertion therefore must be true—has been widely used in mathematics. Euclid's proof that there is no largest prime is often held up as an example of an elegant proof.

Another example is the proof traditionally attributed to the Greeks that the square root of 2 is not a rational number. It starts by assuming that $\sqrt{2}$ *is* rational—that it can be represented as the ratio of two integers. If the two integers have any common factors, those factors can be eliminated by dividing each of the integers by them and expressing the ratio in its lowest terms, say $\frac{p}{q}$. If $\frac{p}{q} = \sqrt{2}$, then $\frac{p^2}{q^2} = 2$ and $p^2 = 2q^2$, from which it follows that p and q must both be even numbers, and therefore have at least the common factor 2, which contradicts the assumption that $\frac{p}{q}$ is the ratio in its lowest terms.

For a charmingly presented collection of richly illustrated and relatively simple proofs, see Polster (2004). Polster begins with the observation that "proofs should be as short, transparent, elegant, and insightful as possible" (p. 2), and then provides numerous proofs that meet these criteria, some, of course, better than others.

☐ Cantor's Proofs

Some especially elegant and ingenious proofs were produced by Cantor in his work on infinity—more accurately, infinities—during the latter part of the 19th century. We will consider first his proof that the rational numbers are countable—can be put in one-to-one correspondence with the natural numbers—and then his proof that the real numbers, which include not only the rationals but also the irrationals (nonrepeating infinite decimals), are not. This distinction is the basis of his concept of different infinities.

The proof that the rationals can be put in one-to-one correspondence with the natural numbers requires the construction of a table of fractions. All the fractions in a given row have the same numerator, and the numerator for each row is increased by 1 in successive rows. All the fractions in a given column have the same denominator, and the denominator for each column is increased by 1 in successive columns. So the upper left portion of this table is as follows:

1/1	1/2	1/3	1/4	1/5	. . .
2/1	2/2	2/3	2/4	2/5	. . .
3/1	3/2	3/3	3/4	3/5	. . .
4/1	4/2	4/3	4/4	4/5	. . .
5/1	5/2	5/3	5/4	5/5	. . .
.	.	.	.	.	
.	.	.	.	.	
.	.	.	.	.	

One has to imagine the table being continued indefinitely both to the left and down, which is to say that it contains an infinity of rows and an infinity of columns. Cantor pointed out that the fractions in this table can be put into one-to-one correspondence with the natural numbers by simply progressing through the table in an orderly fashion—in a diagonal-by-diagonal pattern. By starting with the upper left fraction, 1/1, and proceeding as shown below, one will not miss any numbers. This demonstrates that rational numbers can be put into one-to-one correspondence with the natural numbers.

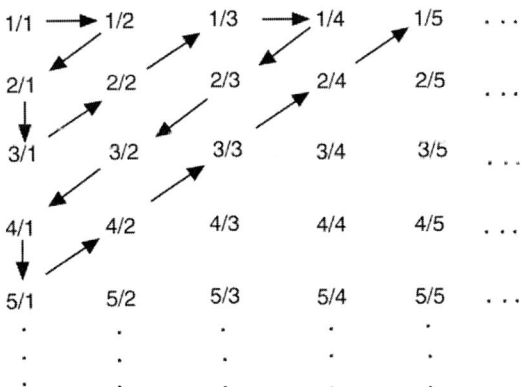

To construct the proof that the set of real numbers is not only infinite but, unlike the set of rational fractions, uncountable, Cantor (1911/1955) used the concept, which he originated, of a *diagonal number*. Suppose, he argued, that there were a finite number of decimals between 0 and 1. Imagine that we listed them all, in no particular order, as follows:

.77358436 …

.84663925 …

.16486902 …

.53932175 …

.35487250 …

.94882604 …

.04327419 …

.36498105 …

A diagonal number may be composed from this set of numbers by making its first digit correspond to the first digit of the first number in the set (7), its second digit to the second digit of the second number (4), and, in general, making the nth digit in the diagonal number correspond to the nth digit of the nth number in the set. Thus, the first eight digits of the diagonal number defined on the above set are

.74437615 …

Now, suppose we construct a new number, say

.85548726 …

that differs from the diagonal number with respect to every digit—we change the first digit from 7 to 8, the second from 4 to 5, and so on. We can be sure that the resulting number differs from *every* number in the original set with respect to at least one digit and so is not a member of that set. Inasmuch as it would always be possible to define such a number, no matter how many numbers there were in the original set, our supposition that there are a finite number of decimals between 0 and 1 must be false. It follows from Cantor's demonstration that the real numbers cannot be put in one-to-one correspondence with the integers. Therefore, Cantor concluded that the "power" of the set of reals is greater than that of the integers, although the sets are infinite in both cases. Cantor's diagonal-number proof of the uncountability of the reals is another example of the *reductio ad absurdum*, in which one shows that the assumption that something is true (in this case that the reals are countable) leads to a contradiction, which permits one to conclude that the assumption is not true after all.

A propos the conception of a proof as a convincing argument, we may note that French mathematician-physicist-philosopher Jules Henri Poincaré did not find Cantor's proof of the uncountability of the real numbers just described to be convincing. He did not accept the inability to devise a way to match the natural numbers with the reals in a one-to-one fashion as compelling evidence that the latter were more numerous than the former. There are many other examples of mathematicians who have rejected proofs that have been accepted by most of their colleagues. As already noted, there is a sense in which what constitutes mathematical truth is determined by consensus, though most mathematicians do not promote this aspect of the discipline.

Cantor's genius at proof making is seen (among other places) in his proof that all the points of a plane can be mapped onto the points of a line. Inasmuch as every point on a plane is defined by two coordinates, whereas every point on a line has only a single coordinate, the idea that the points of a plane can be mapped—in one-to-one fashion—to the points of a line seems impossible on the face of it. And so it seemed to many mathematicians, including Cantor, for a long time. Cantor's stroke of genius was to see that one could merge the two coordinates of any point on the plane in such a way that the resulting number represented a unique position on the line. Imagine a unit square, with both x and y coordinates going from 0 to 1, and a unit line starting at 0 and terminating at 1. Consider any point on the square, say the point at $x = .379253 \ldots$ and $y = .849016\ldots$. If we merge these coordinates, taking the first number from x, the second from y, the third from x, and so on, we generate the blended number .387499205136 ..., which identifies a unique point on the line. We can do this (in our imagination) for every point on the plane, and every blend will identify a unique position on the line. (And the process works for spaces of more than two dimensions.)

Cantor's proof is reminiscent of a paradox involving infinity described by logician-philosopher Albert Ricmerstop (Albert of Saxony) in his book *Sophismata*, which was published in Paris in 1489, a century after his death. Consider a beam of unit width and height and of infinite length. Imagine cutting the beam into $1 \times 1 \times 1$ blocks. Since there would be an infinite number of these blocks, there would be enough of them to completely fill an infinite three-dimensional space.

☐ Proof of Impossibility and Nonprovability

> Now to establish a formula is one thing, but to establish the *nonexistence* of a formula is a chore of quite a different magnitude. (Mazur, 2003, p. 228)

All that is necessary to show that a conjecture in the form of a universal assertion is false is to find a single case for which it does not hold. There are many famous conjectures in mathematics that have stood for some time only to have been shown to be false when someone eventually found a counterexample. A case in point is Fermat's conjecture that numbers of the form $2^{2^n} + 1$ are always prime. This is interesting because Fermat made it on the basis of knowing of only a very few values of n for which it held. Euler showed it to be false by demonstrating that $2^{2^5} + 1 = 4,294,967,297$ is factorable. That he was able to find the factors of a number this size without the assistance of modern computational devices is impressive. A conjecture that is consistent with some, but not all, relevant cases may remain tenable for a long time if a case that would reveal it to be false is exceptionally difficult to find.

There is a great difference between confessing to being unable to find an answer to a question and demonstrating conclusively that there is no answer to be found. Beginning in the 19th century, mathematicians became interested in the idea that it may be impossible to solve some problems, or to prove some assertions in mathematics even if they are true (and not axiomatic), and a new challenge became that of developing proofs of impossibility or nonprovability. The proof of Fermat's last theorem is an example of a demonstration that no integral solution can be found for an equation of the form $x^n + y^n = z^n$ for $n > 2$. A proof by Norwegian mathematician Niels Abel that a formula for fifth degree equations cannot be found, and one by French mathematician Évariste Galois (1811–1832) that extends the principle to equations of higher degree are other well-known examples of demonstrations of impossibility.

The impossibility of proving Euclid's parallel postulate by deduction from his other postulates was demonstrated chiefly through the work of Gauss, Bolyai, Lobachevsky, and Riemann. In 1882 Lindemann proved π to be a transcendental number, from which it follows that the circle cannot be squared by the methods of Euclidean geometry. Only in the 20th century was it proved that none of the three straightedge and compass construction problems posed by the Greeks are solvable. No one knows the amount of time that was devoted to searches for solutions to these problems—especially the squaring of the circle—by countless mathematicians, professional and amateur, during the 2,000 years that they were not known to be unsolvable, but the consensus seems to be that it was enormous.

Arguably the most significant proof of nonprovability in all of mathematics was the demonstration by Austrian mathematician-philosopher Kurt Gödel in 1931 that mathematics cannot be proved to be internally consistent—that one cannot prove from the axioms of a mathematical system that contradictions will never occur within that system. Other famous proofs of nonprovability that followed hard on Gödel's were the proof by British mathematician Alan Turing of the impossibility of solving the "halting problem"—determining, of any computer program, whether it will produce a result and stop—and the "undecidability" proof by American mathematician Alonzo Church that no general procedure can be specified that will invariably tell whether the arithmetic truth or falsity of any statement can be determined; both were given in 1936.

Gödel (1930, 1931) published two theorems demonstrating certain limitations inherent to any axiomatic approach to the construction of a deductive system and, in particular, showed the impossibility of guaranteeing the logical consistency and completeness of a system as complex as elementary arithmetic. The first of the theorems showed that arithmetic (as axiomatized by Italian mathematician Guiseppe Peano) is incomplete; the second showed that it could not be proved to be consistent. Generally reference is made to Gödel's incompleteness theorem, singular, the two being treated as one inclusive demonstration that *any* formal theory sufficiently complicated to include arithmetic that was consistent would necessarily be incomplete, which is to say that for any such system there are true statements that cannot be derived from the system's axioms. It is important to note that Gödel did not prove arithmetic to be inconsistent; he only showed that it cannot be proved to be consistent, but this was enough to be very unsettling to the mathematical world.

Central to the proof is the demonstration that a system as complex as elementary arithmetic is capable of producing assertions of the sort "This assertion cannot be proved"—shades of Russell's antinomies. To prove the statement to be true would prove it to be incapable of being proved. To prove it to be false would show it to be provable—and therefore, by its own claim, not capable of being proved. As Stewart (1987) puts it, "Gödel showed that there are true statements in arithmetic that can never be proved, and that if anyone finds a proof that arithmetic is consistent, then it isn't!" (p. 214).

Much has been written about the significance of Gödel's theorems. Here are just a few of the claims that have been made:

- "Mathematics was forced to face an ugly fact that all other sciences had come to terms with long before: it is impossible to be absolutely certain that what you are doing is correct" (Stewart, 1987, p. 218).

- "In the axiomatization game, the best you can do is to *assume* the consistency of your axioms and *hope* that they are rich enough to enable you to solve the problems of highest concern to you" (Devlin, 2000a, p. 84).

- [Gödel's theorem is] "the deathblow for the Hilbert program" (Singh, 1997, p. 142). Hilbert had challenged mathematicians to provide mathematics with a foundation that was free of doubt and inconsistency.

- "Gödel's theorem is one of the most profound results in pure mathematics. When it was first published, in 1931, it had a devastating impact" (Moore, 2001, p. 172).

- [Gödel's theorem is] "probably the single most profound conceptual result obtained by mankind in the twentieth century" (Ruelle, 1991, p. 143).

- "What it seems to say is that rational thought can never penetrate to the final, ultimate truth" (Rucker, 1982, p. 165).

- "Gödel showed that provability is a weaker notion than truth, no matter what axiomatic system is involved" (Hofstadter, 1979, p. 19).

- [Gödel's theorem] "says something about the possibilities of a certain kind of knowledge; yet it is expressed within that body of knowledge itself. This is the mystery of Gödel's theorem—that within the context of logical thought one can deduce limitations on that very thought" (Byers, 2007), p. 282).

Barrett (1958) argues that Gödel's theorem had repercussions far beyond mathematics, which, since the time of Pythagoras and Plato, had been considered "the very model of intelligibility" and "the central citadel of rationalism." Although commentaries tend to emphasize its implications for mathematics—and sometimes the limitations of human knowledge more generally—the theorem is also lauded for its esthetic appeal. Moore (2001), for example, refers to its sheer beauty as enough to take one's breath away. As for its implications for mathematics, Rucker (1982) sees the incompleteness of arithmetic as not entirely a bad thing, because, if it were complete, there would be no more need for mathematicians; "we could build a finite machine that would answer every question we could ask about the natural numbers" (p. 277). Moore (2001) makes a similarly positive observation in noting that "the infinite richness of arithmetical truth is beyond the reach of any finite collection of arithmetical axioms" (p. 182).

In demonstrating unmitigated faith in the infallibility of the queen of the sciences to be unjustified, Gödel's theorem provided support for the increasingly experienced existential sense of alienation, estrangement, and irrationality. Readily available simplified—not to say simple—expositions of Gödel's proof include those of Nagel and Newman (1958/2001), Hofstadter (1979), Bunch (1982), Rucker (1982), Moore (2001), and Meyerson (2002).

To my eye, a particularly enlightening—and entertaining—explication of Gödel's proof is the beautifully illustrated book-length treatment by Hofstadter (1979). Hofstadter presents the argument as an example of a "strange loop"—a phenomenon that occurs "whenever, by moving upwards (or downwards) through the levels of some hierarchical system, we unexpectedly find ourselves right back where we started" (p. 10). He draws analogies between the strange loop that constitutes Gödel's argument and other strange loops that he sees in the compositions (especially fugues) of J. S. Bach and the etchings of M. C. Escher.

Hofstadter describes Russell and Whitehead's *Principia Mathematica* as a mammoth attempt to exorcise strange loops from logic, set theory, and number theory. The exorcising strategy was to propose the notion that sets form a hierarchy of *types*, and that any given set can be a member only of a set that is higher than itself in the hierarchy; no set can be a member of a set at its own level, so, in particular, no set can be a member of itself. Although this strategy works in the sense of getting rid of (disallowing) self-referential paradoxes, it does so, Hofstadter argues, "at the cost of introducing an artificial-seeming hierarchy, and of disallowing the formation of certain kinds of sets—such as the set of all run-of-the-mill sets. Intuitively, this is not the way we imagine sets" (p. 21). Russell and Whitehead's theory of types banishes all forms of self-reference. Hofstadter describes this as overkill, inasmuch as the effect is the treatment of many perfectly good constructions as meaningless.

Hofstadter sees Gödel's proof as the translation of the Epimenides paradox, also known as the liar paradox, into mathematical terms. A modern form of the paradox is "This statement is false." "The proof of Gödel's Incompleteness Theorem hinges upon the writing of a self-referential mathematical statement, in the same way as the Epimenides paradox is a self-referential statement of language" (p. 17). Gödel invented a coding system in which every possible symbol and statement (sequence of symbols) in number theory would be represented by a unique number. In this system, numbers could represent not only statements of number theory but statements about statements of number theory.

The relationship of Gödel's work to that of Cantor and that of Turing illustrates the continuity of mathematical innovation involving different mathematicians. In his proof of the incompleteness theorem, Gödel made use of Cantor's diagonal number proof of the uncountability of the real numbers (see Moore, 2001, p. 175). Casti (2001) notes the close correspondence between Gödel's theorem and Turing's halting theorem, contending that the latter is really the former expressed in terms of computers instead of the language of deductive logic.

Although there can be no doubt that Gödel's accomplishment was momentous, destroying forever, as it did, the idea of the perfect certainty of mathematics, Du Sautoy (2004) urges caution not to overemphasize its significance. "This was not the death knell of mathematics. Gödel had not undermined the truth of anything that had been proved. What his theorem showed was that there's more to mathematical reality than the deduction of theorems from axioms" (p. 182). Du Sautoy argues that Gödel's result exposed the dynamic character of mathematics and the importance of the intuitions of mathematicians in its continuing evolution. Byers (2007) makes a similar point in noting that while Gödel's theorem marked an end from one point of view, it could be seen as a beginning from another. "The collapse of the hope for a kind of ultimate formal theory can be seen as a kind of liberation. It is a liberation from a purely formal or algorithmic view of mathematics and opens up the possibility of a view of mathematics that is more open and filled with creative possibilities" (p. 273). In any case, there is little, if any, evidence that awareness of the theorem and its implications has deterred many mathematicians from the doing of mathematics.

I would guess that Gödel's demonstration of the incompleteness of mathematics is considerably better known among the general public, if not among mathematicians, than Turing's demonstration of the insolvability of the halting problem, or what he referred to as the *Entscheidungsproblem* (decision problem). It has been argued, however, that Turing, in fact, did more than Gödel and that his approach was more fundamental:

> Turing not only got as a corollary Gödel's result, he showed that there could be no decision procedure. You see, if you assume that you have a formal axiomatic system for arithmetic and it's consistent, from Gödel you know that it can't be complete, but there still might be a decision procedure. There still might be a mechanical procedure which would enable you to decide if a given assertion is true or not. That was left open by Gödel, but Turing settled it. The fact that there cannot be a decision procedure is more fundamental and you get incompleteness as a corollary. (Chaitin, 1995, p. 30)

Another example of proof of nonprovability is one constructed by Paul Cohen addressed to the first of the 23 problems that David Hilbert identified in 1900. Cantor had proved that the set of real numbers (which includes irrationals) is larger than the set of rationals, even though both sets are infinitely large. Hilbert's question was whether there is an infinite set that is between these two in size. Cohen proved that it is impossible to tell, which is to say that it is impossible to construct either a proof that such a set exists or a proof that such a set does not exist.

That it is possible to show that something is not provable in mathematics is an interesting and important fact. As it happens, it is also possible to show that a theorem is provable without actually proving it. Early in the 20th century, Finnish mathematician Karl Frithiof Sundman proved that the famous and recalcitrant "three-body problem"—determining the behavior of three bodies under mutual gravitational attraction—has a solution, although he did not provide the solution itself (Bell, 1937). More recently, several collaborating computer scientists and mathematicians from Canada, Israel, and the United States have promoted the idea of "zero-knowledge proofs," and shown how it can be demonstrated that a theorem has been proved without providing details of the proof itself (Peterson, 1988, p. 214).

Finally, a word of caution regarding proofs of nonprovability. Austrian philosopher Ludwig Wittgenstein (1972) reminds us that to say that a proposition cannot be proved is not to say that it cannot be derived from other propositions. Any proposition can be derived from others, but it may be that the propositions from which the one to be proved is derived are no more certain than the one that is derived from them, in which case, the derivation would be a "proof" in a peculiar sense and unlikely to be of much interest.

☐ The Quest for Generality, Unity, and Simplicity

The development of mathematics has been characterized by a striving for ever higher and higher degrees of generality. (Boyer & Merzbach, 1991, p. 483)

It is one thing to demonstrate the possibility of constructing with straightedge and compass a regular polygon of 17 sides, as Gauss, at 18, was the first to do; it is quite another to discover a rule that will distinguish all regular polygons that can be constructed from those that cannot be, which Gauss also did, though not at 18. What he discovered was the constructability of all regular polygons whose number of sides is expressible as $2^{2^n}+1$. The demonstration—no mean feat—pertains to a single regular polygon, whereas the rule relates to all—an infinity—of them. Mathematicians strive for generality. They seek theorems that state properties of all the members of classes of interest. It is not enough to know that Goldbach's conjecture—that every even number is the sum of two primes—holds for all even numbers up to some astronomically large number; one wants a proof either that it holds for *all* even numbers or that it does not.

The search for generality is seen also in the preference for *if and only if* relationships over simple *if* relationships. For example, it is interesting and useful to realize that if a number is a prime, other than 2, it can be represented either as $4n + 1$ or as $4n - 1$, but it would be much more interesting and useful to be able to say that if and only if a number is a prime, other than 2, it can be represented either as $4n + 1$ or as $4n - 1$. The latter statement is false, however, inasmuch as it is not the case that all numbers that can be represented either as $4n + 1$ or as $4n - 1$ are primes. Nine, for example, which is $4 \times 2 + 1$, is not prime, nor is 15, which is $4 \times 4 - 1$. Misinterpretation of *if* (conditional) assertions as *if and only if* (biconditional) assertions is a common source of error in mathematical and logical reasoning.

The power of general relationships, which motivates the search for them, is seen in the following comment about the highly abstract concept of a *group*:

> Having proved, *using only the group axioms*, that group inverses are unique, we know that this fact will apply to every single example of a group. No further work is required. If tomorrow you come across a quite new kind of mathematical structure, and you determine that what you have is a group, you will know at once that every element of your group has a single inverse. In fact, you will know that your newly discovered structure possesses *every* property that can be established—in abstract form—on the basis of the group axioms alone. (Devlin, 2000a, p. 193)

The abstractness of the concept of a group is captured in a definition offered by Newman (1956a). "The theory of groups is a branch of mathematics in which one does something to something and then compares the result with the result obtained from doing the same thing to something else, or something else to the same thing" (p. 1534). Browder and Lane (1978) refer to the notion of a group as having become in the 20th century "the fundamental conceptual and formal tool for mathematical descriptions of the physical world" (p. 343).

There are many examples in mathematics of relationships that have been discovered that are tantalizingly close to being general, but still fall a little short. Fermat's "little theorem" is a case in point. One would like a formula that will distinguish all prime numbers from all composites. Fermat claimed that a number, x, is not prime if $2^{n-1} \neq 1 \pmod{n}$. So the inequality identifies (some) nonprimes. However, from $2^{n-1} = 1 \pmod{n}$, one cannot conclude that n is prime, because the equality holds for some composite numbers as well as for all primes. Another statement of the theorem is:

> If n is a whole number and p is a prime number that is not a factor of n, then p is a factor of $n^{p-1} - 1$.

Fermat claimed to have a proof for this theorem, but he did not publish it. Euler constructed a proof of it nearly 100 years later.

Closely related to the search for generality in mathematics is the search for unity. Just as scientists seek to unify the forces of nature by accounting for them within a single cohesive theoretical structure, so mathematicians seek a conceptual framework within which all of mathematics can be viewed. Interest in unifying different mathematical areas has been a continuing one, and the history of mathematics records many efforts to join various combinations of the "branches"—arithmetic, geometry, algebra, analysis, and probability theory, among others—in terms of which the mathematical tree has developed. Among the reasons for the attention that topology has received in the recent past is the prospects that some see for unifying much of mathematics within it.

A major unifying concept in mathematics, introduced at the turn of the 20th century, is that of a *set*. As Rucker (1982) puts it: "Before long, it became evident that all of the objects that mathematicians discuss—functions, graphs, integrals, groups, spaces, relations, sequences—can be represented as sets. One can go so far as to say that mathematics is the study of certain features of the universe of set theory" (p. 41). Whether all mathematicians would agree with this assessment of the unifying nature of set theory, Rucker's statement illustrates the interest that potentially unifying concepts hold.

Many proofs in mathematics are simple and comprehensible; many are not. Undoubtedly some proofs are complex because simpler ones are impossible, but a difficult proof is often taken as a sign that the essential nature of the problem has not been understood and as a challenge to construct a simpler one. "Cleaning up old proofs is an important part of the mathematical enterprise that often yields new insights that can be used to solve new problems and build more beautiful and encompassing theories" (Schecter, 1998, p. 59). Schechter quotes Italian-American mathematician-philosopher Gian-Carlo Rota's observation that "the overwhelming majority of research papers in mathematics is concerned not with proving, but with reproving; not with axiomatizing, but with reaxiomatizing; not with inventing, but with unifying and streamlining" (p. 59), and he notes that Pafnuty Chebyshev's difficult proof of a conjecture by French mathematician Joseph Bertrand that there is always at least one prime number between every number and its double stood for 80 years before being tidied up by the teenage Paul Erdös.

As Casti (2001) points out, there is no reason to expect that every short, simple, true statement should have a short, simple, true proof. On the other hand, mathematicians attempt to produce proofs that are as simple as they can be, and they strive to simplify existing complex proofs whenever they believe it possible to do so. One might wonder why one should believe, in any particular case, that a simpler proof than all those that have already been developed might be found. One suspects that Stewart (1990) expresses a belief that is deeply held by many mathematicians when he claims that "the best mathematics is always simple, if you look at it the right way" (p. 73). Unhappily, for most of us much of mathematics seems exceedingly complex, no matter how we look at it.

Usually anyone with the necessary training will have no difficulty understanding each of the steps in a complicated mathematical proof or chain of deductions; grasping the entire proof or deductive chain in its entirety is another matter. A proof can be so complex that understanding it in its entirety may not be possible. Appel and Haken's proof of the four-color theorem represented the beginning of a new era in proof making. It involved an excessive amount of computation, which was done by a computer, and the result was so complicated as to be incomprehensible. Stewart (1987) notes that a full proof, if written out, would be so enormous that nobody could live long enough to read it, and he asks the obvious question of whether such a monstrosity is really a proof. On that question opinions are divided. To accept it as a proof, which many—but not all—mathematicians do, requires an act of faith, inasmuch as nobody could check it without the help of computers, even in principle.

Another proof the complexity of which challenges comprehension is the more recent one by Andrew Wiles of Fermat's last theorem. If, as has been claimed, not more than 1 mathematician in 1,000 can understand Wiles's argument, perhaps we have to say that the theorem has been proved to a few people—in particular mathematicians conversant with elliptic equations, modular forms, especially the ground-breaking work of Japanese mathematicians Yutaka Taniyama and Goro Shimura, and numerous other mathematical arcana—but that for the rest of us to consider it proved requires an act of faith, namely, faith in the competence of those few mathematicians who claim they understand it well enough to vouch for its adequacy.

If it is true that very few mathematicians understand Wiles's proof in its entirety, it does not follow that only those few are in a position to challenge its authenticity; one need not follow an argument in its entirety to find flaws in parts of it if such flaws exist. Doubts were raised about specific aspects of Wiles's proof soon after it was published (Cipra, 1994a, 1994b), but it appears that the necessary repair work was done in collaboration with Richard Taylor, and as of May 1995, it had been sufficiently accredited among mathematicians who are deemed capable of passing judgment to warrant an entire issue of *Annals of Mathematics* (Taylor & Wiles, 1995; Wiles, 1995). This does not ensure that no questions of the adequacy of this proof will ever be raised in the future. It can take a long time for the mathematical community as a whole to make up its mind as to whether to accept a complicated proof or to keep looking for a fatal flaw. Anyone who is disappointed that Fermat's last theorem has finally been proved, because he or she had hoped to be the first to that goal, may still aspire to a place beside Wiles in the mathematical pantheon by inventing a proof of the theorem that is within the grasp of garden-variety minds—or the one that Fermat might have wished he could have written in the margin of his book.

The complexity of some proofs and the energy that can be put into evaluating them are illustrated also by Russian mathematician Grigori Perelman's recently announced proof of a conjecture by Poincaré involving the topological equivalence of a three-dimensional manifold and a three-dimensional sphere under certain conditions. Poincaré made the conjecture 100 years ago, but neither he nor any of the many mathematicians who tried between then and now were able to prove it until Perelman came along. Perelman was offered, and declined to accept, the prestigious Fields Medal for the accomplishment, and appears to have stopped doing mathematics, at least as a professional, since 2003, being disillusioned by the perceived unseemly behavior—apparently relating at least in part to issues of credit—of some fellow mathematicians. As of 2006, it was not clear whether, if offered it, he would accept the $1 million prize provided by the Clay Institute for the proof of Poincaré's conjecture (Nasar & Gruber, 2006).

Possibly the most monstrous proof ever is one in group theory—the classification theorem for finite simple groups. Authored by hundreds of mathematicians and published in pieces in a variety of journals, it has been estimated to be around 15,000 pages long and, not surprisingly, is known among mathematicians as simply the enormous theorem (Conway, 1980; Davis & Hersh, 1981; Stewart, 1987). Although it is doubtful that anyone understands this proof in its entirety in any very deep sense, many mathematicians rely on it for various purposes. A considerable effort is being devoted to finding a way to shorten the proof to the point where it could be understood by a single individual; although progress to this end has been reported, the task is proving to be formidable (Cipra, 1995).

The idea that credence would be given to a mathematical proof that relatively few mathematicians understand may come as a surprise to nonmathematicians. What may be even more surprising is that such proofs are not rare exceptions. Mathematics is not the homogeneous discipline that it is commonly believed to be; it is a very large territory and individual mathematicians spend their lives exploring small parts of it. Frontier-extending work often is understood by very few people other than those who are doing it. As Davis and Hersh (1981) put it, "The ideal mathematician's work is intelligible only to a small group of specialists, numbering a few dozen or at most a few hundred. This group has existed only for a few decades, and there is every possibility that it may become extinct in another few decades" (p. 34). That what appears clear to one mathematician may prove to be abstruse to another is illustrated by the fact that French mathematician Siméon-Denis Poisson, for whom the famed Poisson distribution of probability theory is named, found the paper in which Galois presented what later became widely recognized as the beautiful Galois theory to be incomprehensible (Aczel, 1996).

Why should anyone accept as a mathematical proof an argument that few, if any, people can understand in its entirety? Because "the strategy makes sense, the details hang together, nobody has found a serious error, and the judgment of the people doing the work is as trustworthy as anybody else's" is Stewart's answer (1987, p. 117). Regarding proofs that depend on many computer computations, Appel and Haken (1978) acknowledge that there is a tendency among mathematicians to prefer proofs that can be checked by hand over those that can only be checked by computer programs, but argue with respect to long computationally complicated proofs that, even when it is feasible to check them by hand, the probability of human error is likely to be higher than the probability of computer error. That most known proofs of the past are reasonably short they attribute to the lack of tools for producing extraordinarily long proofs: "If one only employs tools which will yield short proofs that is all one is likely to get" (p. 178).

On the other hand, Paulos (1992) points out that even if the probability that any specific bit of a proof as complicated as the four-color theorem is in error is miniscule, there are so many opportunities for the small probability event to occur that the most one should conclude is that such a proof is probably true, which is not the same, in his view, as being "conclusively proved." At best, such proofs are less than completely satisfying psychologically, because making sure a conjecture is true is only one reason for wanting proof in mathematics; another is to understand *why* it is true. Of course, what it means to understand a proof (or anything else, for that matter) is a nontrivial question in its own right; suffice it to say here that proofs differ in the degree to which they make us believe we understand, and most of us probably have a preference for those that do a good job of that. Stewart (1987) refers to an *illuminating* proof as the kind of proof that mathematicians really like, but notes that what constitutes such a proof is more a matter of taste than of logic or philosophy.

So although mathematicians greatly prefer simple proofs to complicated ones, they often have to settle for the latter. But the search for simplifications continues. Insofar as possible, one wants proofs that are simple enough to be understood in their entirety and simple enough so that one can have relatively high confidence that they do not conceal undetected faults.

☐ Some Notable Proofs

There are millions of proofs in the mathematical literature—some simple, some complex, some elegant, some ugly. Most of us will never see the vast majority of those proofs, and probably neither will most professional mathematicians. There are a few proofs, however, that receive special attention in books on the history of mathematics. What determines that a proof is considered worthy of such attention is an interesting question. Possible factors include simplicity, elegance, novelty (as when a proof constitutes an entirely new way of looking at a relationship), difficulty, and implications (as when a proof opens up a new area of mathematics).

We have already encountered some proofs that have received a great deal of attention, among them Georg Cantor's (1911/1955) proof that the real numbers are not countable, Kenneth Appel and Wolfgang Haken's (1977a) proof of the four-color theorem, and Andrew Wiles's (1995) proof of Fermat's last theorem. Dunham (1991) describes, and provides the historical context of, several proofs of what he characterizes as the great theorems of mathematics. These include:

- Hippocrates' proof of a procedure for quadrature (squaring the area) of a lune

- Euclid's proof of the Pythagorean theorem

- Euclid's proof that there are an infinity of prime numbers

- Archimedes' proof of a procedure for determining the area of a circle to any desired accuracy

- Heron of Alexandria's proof of a procedure for determining the area of a triangle, given only the lengths of its three sides

- Cardano's proof of a rule for solving a cubic equation

- Newton's proof of a method for approximating π

- The brothers (Johann and Jakob) Bernoulli's proof that the harmonic series, $\sum_{n=1}^{\infty} \frac{1}{n}$, diverges

- Euler's proof that the sum of the reciprocals of the squares, $\sum_{n=1}^{\infty} \frac{1}{n^2}$, converges

- Euler's proof that Fermat's conjecture that all numbers of the form $2^{2^n} + 1$ are primes is false

- Cantor's proof that any set is smaller than its power set (the power set of a set is the set of all its subsets)

☐ The Transitory Nature of Understanding

Understanding a proof, or any other argument, must depend, in part, on having some familiarity with the concepts of which it is composed. But this cannot be the whole story. Relevant background knowledge may be a necessary condition for understanding, but it is not a sufficient one. Moreover, why one knowledgeable individual finds an argument compelling and another does not—whether in mathematics or elsewhere—is not always clear. Perhaps even more perplexing is why the same individual will find the same argument persuasive on one occasion and not on another. Bertrand Russell (1944/1956b) reports being able to remember the precise moment "one day in 1894, as I was walking along Trinity Lane," when he "saw in a flash (or thought I saw) that the ontological argument is valid. I had gone out to buy a can of tobacco; on my way back, I suddenly threw it up in the air and exclaimed as I caught it: 'Great Scot, the ontological argument is sound'" (p. 386).

What happened in Russell's head that convinced him at that particular moment that this argument, with which he was thoroughly familiar and that, up until that time, he had not found compelling, was valid? Had his brain been busy weighing the strengths and weaknesses of the argument and suddenly found the scale tipping in favor of the pros? Russell wrote these words many years after the incident he described and long after he no longer considered the ontological argument to be sound. So we have an example of the interesting case of believing an argument to be unsound, then believing it to be sound, and then again believing it to be unsound.

Even outstanding mathematicians can be unsure of whether an ostensible proof is really a proof. Cantor sought for years for a compelling proof of his "continuum hypothesis," and he many times thought he had found it, only soon to become convinced that he had not, or even to be convinced that he had now found a proof that it was false (Aczel, 2000). Cantor's repeated unsuccessful attempts to prove the continuum hypothesis have been considered by some to be the cause of, or at least a factor contributing to, his bouts of depression, which tended to occur when he was working on the problem. Long after his death, the hypothesis was demonstrated to be not decidable.

Probably most of us have had the experience of being convinced on one occasion that a particular argument is compelling and equally convinced on another that it is not. This should make us cautious about taking dogmatically rigid stands on controversial issues even when we are quite sure (at the moment) that we are right. It also should make us sympathetic to the notion that all proofs are tentative, in a sense. One can never be certain that what one takes to be an indisputable argument at a given time will be perceived by all others as indisputable, or even that it will still be perceived so by oneself at a later time.

This type of uncertainty is one of the reasons why some theorists discount feelings of conviction as at all relevant to the question of the soundness of proofs or logical arguments. Hempel (1945), for example, contends that both the idea that the feeling of plausibility should decide the acceptability of a scientific hypothesis and the opinion that the validity of a mathematical proof should be judged by reference to convincingness have to be rejected on analogous grounds:

> They involve a confusion of logical and psychological considerations. Clearly, the occurrence or non-occurrence of a feeling of conviction upon the presentation of grounds for an assertion is a subjective matter which varies from person to person, and with the same person in the course of time; it is often deceptive, and can certainly serve neither as a necessary nor as a sufficient condition for the soundness of the given assertion. A rational reconstruction of the standards of scientific validation cannot, therefore, involve reference to a sense of evidence; it has to be based on objective criteria. (p. 9)

But how is one to judge the merits of Hempel's own argument regarding the need for objective criteria, except by appealing to one's intuitions—to one's feelings of soundness or convincingness? And, assuming one accepts the argument, how is one to decide what the objective criteria should be? Objectivity is the abiding goal of empiricism, but it is elusive. We want objective criteria, and groups of people may be able to agree as to what those criteria should be in specific instances, but the opinions from which such agreement is derived must be based, in the final analysis, on feelings or convictions of correctness that are subjective through and through.

6

Informal Reasoning in Mathematics

A mathematician's work is mostly a tangle of guesswork, analogy, wishful thinking and frustration, and proof, far from being the core of discovery, is more often than not a way of making sure that our minds are not playing tricks. (Rota, 1981)

The characterization of mathematics as a deductive discipline is accurate but incomplete. It represents the finished and polished consequences of the work of mathematicians, but it does not adequately represent the *doing* of mathematics. It describes theorem proofs but not theorem proving. Moreover, the history of mathematics is not the emotionless chronology of inventions of evermore esoteric formalisms that some people imagine it to be. It has its full share of color, mystery, and intrigue.

That the process of mathematical discovery is not revealed in the finished proofs that mathematicians publish was pointed out by Evariste Galois, the brilliant French mathematician who, after inventing group theory, died in a duel at the age of 21. It has been convincingly documented by Polya (1954a, 1954b) and Lakatos (1976). In addition to deducing the implications of axioms, mathematicians also invent new axiomatic systems, and this cannot be done by deductive reasoning alone. As Polish-American mathematician Stanislav Ulam (1976) puts it, "In mathematics itself, all is not a question of rigor, but rather, at the start, of reasoned intuition and imagination, and, also, repeated guessing" (p. 154). Rucker (1982) makes essentially the same point: "In the initial stages of research, mathematicians do not seem to function like theorem-proving machines. Instead, they use some sort of mathematical intuition to 'see' the universe of mathematics and determine by a sort of empirical process what is true. This alone is not enough, of course. Once one has discovered a mathematical truth, one tries to find a proof for it" (p. 208).

Fundamental to an understanding of mathematical reasoning is the distinction between mathematics as axiomatic systems and the thinking done by mathematicians in the process of defining and refining those systems. The systems themselves are better viewed as the results of mathematical thinking than as illustrations of it. Theorems make explicit what is implicit in a discipline's axioms and its theorems, but they do not reveal much of the nature of the reasoning that resulted in this explication. In somewhat oversimplified terms, the distinction contrasts proofs with proof making. Polya (1945/1957) captures this distinction in noting the difference between mathematics, when presented with rigor as a systematic deductive science, and mathematics in the making, which he describes as an experimental inductive science. Penrose (1989) makes a similar distinction in noting that a rigorous argument typically is the last step in the mathematician's reasoning and that generally many guesses—albeit constrained by known facts and logic—precede it. Casti (2001) contends that mathematics—the doing of mathematics—is an experimental activity, and points to Gauss's notebooks as supportive evidence of the claim.

☐ Knowledge in Mathematics

> A great many problems are easier to solve rigorously if you know in advance what the answer is. (Stewart, 1987, p. 65)

To be a good historian one must know a great many historical facts; to be a competent lawyer one must be familiar with much law and numerous legal cases; physicians are expected to know much of what there is to know about the human physiological system or at least that part of it on which they specialize. I suspect that similar assumptions are not made about mathematicians; what is assumed to be required to do well at mathematics is to be able to reason well, not necessarily to know a lot of facts. Mathematics, being the relatively abstract and content-free type of entity that it is, is not usually thought of as a knowledge-intensive discipline.

The well-known fact that mathematicians tend to make their major discoveries while relatively young, often too young to have amassed a large amount of mathematical knowledge, generally supports this view.

It may be that if the various cognitively demanding professions were ordered in terms of the extent to which success depends on accumulating a large body of factual knowledge, mathematics, especially theoretical mathematics, would be at, or close to, the bottom of the list. There is perhaps no other profession in which one can accomplish so much on the basis of abstract thought alone.

But this is not to say that knowledge is of no use to mathematicians, or that it never plays a role in mathematical discoveries. Extraordinary mathematicians not only are very good thinkers, but often they know a lot—have a large storehouse of facts—about mathematical entities and operations. Ulam (1976) considered a good memory to be a large part of the talent of both mathematicians and physicists. British mathematician John Littlewood once claimed that every positive integer was one of Srinivasa Ramaujan's personal friends. By way of lending credence to this claim, Littlewood's colleague and fellow British mathematician G. H. (Godfrey Harold) Hardy (1940) recounts an occasion on which he informed Ramanujan that the taxicab in which he had just ridden had the "rather dull" number, 1729. Ramanujan informed him that he was wrong about the dullness of the number, because it was the smallest number that could be expressed as the sum of two cubes in two different ways—$12^3 + 1^3$ and $10^3 + 9^3$. On the other hand, because Ramanujan lacked knowledge of what other mathematicians had done, many of his own discoveries were rediscoveries of what others had discovered before him. The indispensable role of knowledge in the development of some cutting-edge proofs is seen clearly in Wiles's proof of Fermat's last theorem, which makes extensive use of the work of numerous preceding mathematicians (see Chapter 5).

Often history credits an individual with a major mathematical innovation, and one can get the impression that the contribution was made single-handedly in a vacuum. When one looks more closely at the circumstances under which the innovation was made, however, one is likely to find that the innovator benefited greatly from the work of forerunners with which he or she was thoroughly familiar. The importance of knowledge to mathematical discovery is also illustrated by the numerous occasions in the history of mathematics on which two or more people have come up with the same innovation independently at the same, or approximately the same, time. (For lists of simultaneous, or near simultaneous, discoveries or inventions in mathematics and science, see Ogburn [1923, pp. 90–102] and Simonton [1994, pp. 115–122].) One example of simultaneous, or near simultaneous, discoveries in mathematics is that of logarithms by Scottish mathematician John Napier, German mathematician Michael Stifel, and Swiss clockmaker–mathematician Joost Bürgi.

Another is that of hyperbolic geometry by Lobachevsky, Gauss, and Bolyai.

Perhaps the best known instance of a major mathematical innovation that has been credited to two people working independently is the infinitesimal calculus by Gottfried Leibniz and Isaac Newton. We will have occasion to revisit this event later. The point I wish to make here is that such occurrences would be remarkable coincidences if it were not the case that mathematical innovations spring from the soil of current mathematical knowledge. Individuals who have added significantly to that knowledge have done so by building on what already exists.

Mathematicians, like lawyers, often find it convenient to refer to similar cases with which they are familiar and for which they know the solution or disposition. A large mathematical knowledge base permits the mathematician to classify problems as to types, to anticipate the form that a solution is likely to take, and to select an appropriate approach. It may permit one also to judge whether a problem is reasonable, whether it is well or poorly formed, and whether it is important or trivial.

Because mathematical knowledge accumulates and mathematicians build on the work of their predecessors, much of the work done by mathematicians in any given age could not have been done at an earlier time. There are some notable exceptions to this rule, but, in the main, it stands. It follows also that the opportunities for mathematical discoveries were never greater than they are now, because mathematicians never had more on which to build. Indeed, we are living in a period of unusual mathematical inventiveness; by some estimates, more mathematics has been created during relatively few decades in the recent past than in all previous time. Fifty years ago Kline (1953a) pointed out that the elementary school graduate of that day knew far more mathematics than a learned mathematician of Medieval Europe. This is a sobering and exciting thought.

☐ Intuition

We have noted that current attitudes regarding the intuitive status of the axioms of a mathematical system, say geometry, are different from the attitudes of the ancients and from those of even philosophers of only a couple of centuries ago. No longer is it held that the empirical truth—correspondence to physical reality—of such axioms must be intuitively obvious.

Does this mean that intuition plays no role in modern mathematics, or in the thinking of modern mathematicians? In fact, intuition is a very important factor in the psychology of mathematics, in the sense that mathematicians spend a great deal of time exploring guesses and checking out hunches in their efforts to discover and prove new theorems. Proofs and proof making are at the core of mathematics, but what tells the mathematician what to try to prove? Hersh (1977) refers to intuition as "an indispensable partner to proof," and argues that one sees it everywhere in mathematical practice.

One evidence of intuition at work that has been noted by many writers is that mathematicians have typically made effective use of concepts and relationships before they have been proved. British mathematician–logician Philip Jourdain (1913/1956) puts it this way: "In mathematics it has, I think, always happened that conceptions have been used long before they were formally introduced, and used long before this use could be logically justified or its nature clearly explained. The history of mathematics is the history of a faith whose justification has been long delayed, and perhaps is not accomplished even now" (p. 35). Dantzig (1930/2005) notes that in the history of mathematics the *how* has always preceded the *why*, which is to say that the invention of problem-solving techniques has preceded the development of theoretical explanations of why they work. He argues that this is particularly true of arithmetic, the technique of counting and the rules of reckoning being established hundreds of years before the emergence of a philosophy of number. Dantzig notes too that the same history reveals that progress in mathematics has been erratic and that intuition has played a predominant role in it.

Kline (1980) contends that the essential idea of new mathematical work is always grasped intuitively before it is explicated by a rational argument, and that mathematical creation is the special province of those "who are distinguished by their power of intuition rather than by their capacity to make rigorous proofs" (p. 314). Nasar (1998) emphasizes the importance of intuition in the work of contemporary American mathematician–economist John Nash, which she likens to that of other great "mathematical intuitionists," Riemann, Poincaré, and Ramanujan: "Nash saw the vision first, constructing the laborious proofs long afterward. But even after he'd try to explain some astonishing result, the actual route he had taken remained a mystery to others who tried to follow his reasoning" (p. 12).

Intuition also figures in the work-a-day world of mathematicians in the monitoring and evaluating of their own work. Hadamard (1945/1954) points out, for example, that although good mathematicians frequently make errors—he claims to have made many more of them than his students—they usually recognize them as errors and correct them so no trace of them remains in the final result. But how do they recognize the errors as such? Hadamard refers to a "scientific sensibility," which he calls insight, but perhaps might as appropriately be called intuition, that warns the mathematician that the calculations do not look as they ought to look. It is not clear how best to account for this in current psychological terms; possibly what clues one that things are not right is a realization of incompatibility between the "answer" in hand, or the process by which it was obtained, and other relevant things one knows.

But intuition also plays a much more fundamental role in mathematics, considered as a deductive system. If, as Whitehead and Russell (1910) argue, mathematics is reducible to logic, our acceptance of the basic rules of the former rests on our acceptance of the basic rules of the latter. And our acceptance of the rules of logic is a matter of intuition because there is no way to justify them without appeal to the very laws one is seeking to justify.

One might protest that intuition is an unreliable basis for anything, because intuitions change over time. Ideas that are intuitively acceptable at one time may be unacceptable at another, and conversely. Intuitions do change as a consequence of learning and developing the ability to see things from new perspectives. But changes of perspective can occur only to the extent that one becomes convinced, intuitively, of their justification. As Kline (1980) puts it, "We are now compelled to accept the fact that there is no such thing as an absolute proof or a universally acceptable proof. We know that, if we question the statements we accept on an intuitive basis, we shall be able to prove them only if we accept others on an intuitive basis" (p. 318).

Undoubtedly, intuition can lead one into paradoxes and other types of mathematical quicksand. On the other hand, what one is willing to accept as a resolution of a paradox, or solutions of other types of problems, with the help of definitions and formalisms must ultimately depend on how intuitively compelling one finds such a resolution or solutions to be, given the definitions and formalisms proposed. One may find oneself in the situation of saying yes, given such and such a definition or formalism, I accept the conclusion that follows, while recognizing that the acceptance is provisional on the givens, and one may harbor some reservations about the givens as representing the way things are as distinct from being the specifications of an arbitrary abstract game. *Defining* an infinite set as any set the elements of a proper subset of which can be put into one-to-one correspondence with the elements of the set, and *defining* equality as one-to-one correspondence answers the question of how it is that the set of all even integers can be said to equal the set of all integers, but only for one who is willing to accept the definitions.

☐ Insight

Numerous mathematicians have reported the experience of suddenly realizing the solution to a problem on which they had labored unsuccessfully for some time, but about which they were not consciously thinking when the solution came to mind. Poincaré (1913) reports realizing, as he was about to enter a bus, the solution of a problem relating to the theory of Fuschsian functions with which he had struggled to no avail for a fortnight. Immediately prior to the insight he had been traveling, and by his own account, the incidents of the travel had made him forget his mathematical work. Poincaré experienced other moments of insight or "sudden illumination" at unexpected times, and the experiences left him convinced of the importance of unconscious work in mathematical invention.

Many other mathematicians have claimed to have experienced sudden insights at unexpected moments. Gauss reported seeing, "like a sudden flash of lightning," the solution to a problem on which he had worked unsuccessfully for years, and confessed to not being able to say "what was the conducting thread which connected what I previously knew with what made my success possible" (quoted in Hadamard, 1945/1954, p. 15). We have already noted that, when only 18, Gauss discovered how to construct a 17-sided regular polygon with straightedge and compass. According to his account of this discovery, the enabling insight occurred to him (after concentrated analysis) before getting up from bed one morning while on vacation. The insight involved the expression

$$-\frac{1}{16}+\frac{\sqrt{17}}{16}+\frac{\sqrt{34-2\sqrt{17}}}{16}+\frac{\sqrt{17+3\sqrt{17}-\sqrt{34-2\sqrt{17}}-2\sqrt{34+2\sqrt{1}}}}{8}$$

(Kaplan & Kaplan, 2003)—not your run-of-the-mill initial wake-up thought.

Irish mathematician-physicist William Hamilton tried unsuccessfully for 10 years to extend work that he had done on what later became known as complex numbers to three dimensions by using ordered number triples instead of couples. One day, during a stroll with his wife, it suddenly occurred to him to try using quadruples instead of triples. By doing so, and dropping the commutative law of multiplication, he created a new type of number—the quaternion. Though still very useful in certain contexts, the quaternion has, for most purposes, been eclipsed by the invention of vectors and matrices, which have many capabilities in common with it.

Spurred by an interest in Gödel's incompleteness theorem, and Hilbert's dream of a machine that could decide, of any statement, whether it could be proved, Alan Turing became intrigued by the question of whether it was possible to prove that such a machine could not be built. His answer rested on an insight he had while jogging in Cambridge. The insight involved seeing a connection between the proof he was seeking and the approach Cantor had taken to prove that irrational numbers outnumber rational numbers. From there he was able to construct the proof he sought.

Hungarian-American mathematician Paul Halmos (1985) gives an engaging account of his struggle with, and eventual solution of, a problem in algebraic logic:

> I lived and breathed algebraic logic during those years, during faculty meetings, during my after-lunch lie-down, during concerts, and, of course, during my working time, sitting at my desk and doodling helplessly on a sheet of yellow paper. ... The theorem that gave me the most trouble was the climax of Algebraic Logic II. ... I remember the evening when I got over the last hurdle. It was 9 o'clock on a nasty, dark, chilly October evening in Chicago; I had been sitting at my desk for two solid hours, concentrating, juggling what seemed like dozens of concepts and techniques, fighting, writing, getting up to walk across the room and then sitting down again, feeling frustrated but unable to stop, feeling an irresistible pressure to go on. Paper and pencil stopped being useful—I needed a change—I needed to do something—I pulled on my trenchcoat, picked up my stick, and mumbling "I'll be back," I went for a walk, out toward the lake on 55th street, back on 56th, and out again on 57th. Then I saw it. It was over. I saw what I had to do, the battle was won, the argument was clear, the theorem was true, and I could prove it. Celebration was called for. ... (p. 211)

Andrew Wiles pays tribute to the role of the subconscious in helping one get by what appears to be an insurmountable impasse in a problem-solving effort, but he stresses also the importance of prolonged concentrated work on the problem first: "When you've reached a real impasse, when there's a real problem that you want to overcome, then the routine kind of mathematical thinking is of no use to you. Leading up to that kind of new idea there has to be a long period of tremendous focus on the problem without any distraction. You have to really think about nothing but that problem—just concentrate on it. Then you stop. Afterwards there seems to be a kind of period of relaxation during which the subconscious appears to take over, and it's during that time that some new insight comes" (quoted in Singh, 1997, p. 208). Regarding his own concentrated effort to prove a conjecture (the Taniyama-Shimura conjecture that all elliptic equations are modular) that was essential to his proof of Fermat's last theorem: "I carried this thought around in my head basically the whole time. I would wake up with it first thing in the morning, I would be thinking about it all day, and I would be thinking about it when I went to sleep. Without distraction I would have the same thing going around and around in my mind" (quoted in Singh, 1997, p. 211). Hadamard (1945/1954) also stresses the role that unconscious thought plays in mathematical invention or discovery: "Strictly speaking, there is hardly any completely logical discovery. Some intervention of intuition issuing from the unconscious is necessary at least to initiate the logical work" (p. 112). Hadamard recognizes four phases of mathematical discovery that British psychologist Graham Wallas (1926/1945) had articulated—preparation, incubation, illumination, and verification—and points out that at least the first three of them had been discussed by others, notably Helmholtz and Poincaré, before Wallas's treatment of them. Whether in mathematics or elsewhere, invention and discovery take place, in Hadamard's view, by combining ideas. But most of the countless combinations that could be made are not useful and are effectively filtered out by a process of which we are not aware, so most of those combinations of which we are conscious are fruitful or at least potentially so. While a strong believer in the effectiveness of unconscious thought processes, Hadamard cautions that the feeling of certitude that often accompanies an unexpected inspiration can be misleading, so it is essential that what appear to be insights be verified in the light of reason.

☐ Origins of Mathematical Ideas

In mathematics, as in science, it is difficult to trace ideas to their origins. We typically associate one or a few specific names with each major development, but a close look generally reveals other lesser-known contributors on whose work those we remember built. What appears often to happen is that an idea emerges, perhaps in more than one place, in a vague and imprecise form and that over a period of time it is explicated and refined sufficiently to become part of the mainstream of mathematical thought. The names that become associated with such ideas in the historical record are as likely to be those of the individuals who helped to sharpen them or to communicate them to the mathematical community as those of the people who were more responsible for their initial emergence.

The relationship between the hypotenuse and the other two sides of a right triangle that is stated in the theorem that bears the name of Pythagoras was widely known among the Babylonians long before the time of Pythagoras, and perhaps also among the Hindus of Iran and India, and in China, as well. According to McLeish (1994), it was known even by the builders of megalithic structures in Britain also long before the time of Pythagoras. It is not clear, however, that the relationship was "known" in the same sense in every instance; knowing that it held in certain cases—as evidenced by the "Pythagorean triples" shown on the Babylonian tablet, Plimpton 322—does not require recognition of the generality of the relationship. And knowing that a rope marked off in, say, 3, 4, and 5 units could be used to construct right angles, as it is believed that the Egyptians knew, does not require explicit recognition of the Pythagorean relationship at all.

"Pascal's triangle" of binomial coefficients was described by Arab and Chinese mathematicians several centuries before it was published by French mathematician-philosopher Blaise Pascal in 1665. The 13th-century Chinese mathematician Yang Hui, for example, knew of it (Dunham, 1991). McLeish (1994) credits its discovery to Halayudha, a 10th-century Jaina mathematician who predated Yang Hui. (There are discrepant accounts of when he lived; but most references I found claimed 10th century AD.) It appeared in several works published in Europe during the 16th century. In short, as a discoverer of Pascal's triangle, Pascal had many predecessors, none of whom, incidentally, acknowledged the others. It could be, of course, that most of them did not know of the others. It is true of many mathematical discoveries, as with this one, that we cannot be sure who first made them—we cannot rule out the possibility that they were made by people whose names will never be known; we can be sure, however, that the people whom history has credited with them were not the first in many cases, because the evidence of predecessors is clear. That Pascal's name has been attached to the triangle is not without some justification. Flegg (1983) describes his 1665 exposition of the triangle's properties as standing out from all the others for its elegant and systematic thoroughness.

Western histories of mathematics are likely to emphasize the contributions of the Greeks, Indians, and Arabs to the early development of mathematics; apparently the Chinese too did much early work. McLeish (1994) sees the insights of the Greek and Arab mathematicians as mainly derivative from the discoveries of the Chinese, Indians, and Babylonians, which is not to deny the genius of specific Greek and Arab individuals. He credits the Chinese with inventing "the decimal system, the place-value concept, the idea of zero and the symbol we use for it, the solution of indeterminate, quadratic and higher-order equations, modular arithmetic and the remainder theorem" (p. 71) centuries, if not millennia, before they were taken up, or independently discovered, by European scholars. According to Flegg (1983), simultaneous linear equations were solved in China from as long ago as there are surviving mathematical records, and methods of solution using bamboo rods are known to have been in existence as early as 1000 BC. McLeish notes also that with respect to arithmetic, the Maya, who flourished in the region of the Yucatan during the European middle ages, were far ahead of Europe in some respects; their use of zero, for example, predated the adoption of that symbol in Europe by a long time. Quite possibly a great deal of mathematics was done in various parts of the world that is not generally described in Western accounts of the history of math.

The extent to which specific ideas passed from one culture to another, as opposed to arising independently in different places at different times, is unclear in many cases. The use of a symbol to serve the purpose that zero now serves is an especially interesting case of an invention that probably occurred independently several times; such a symbol was used by the Babylonians, the Indians, the Chinese, and the Mayans, and conceivably was invented independently in each case. Very readable histories of this enormously important concept and symbol have recently been provided by Kaplan (1999) and Seife (2000).

That Newton and Liebniz quarreled over the question of which of them had precedence in the development of the infinitesimal calculus is well known. Newton's thinking about the calculus presumably benefited from lectures of his teacher, British mathematician Isaac Barrow, who had developed methods for finding areas and tangents to curves. Much less widely known than the work of Newton and Liebniz is that the calculus was invented at about the same time in Japan by Kowa Seki (sometimes Seki Kowa), who did not publicize the invention (Davies, 1992). Moreover, Fermat anticipated them all with a method of differentiation that is essentially the one that is used today, but he did not see his method as a general one that was suitable for a whole class of problems and that deserved further investigation and development (Hadamard, 1945/1954). Most of Fermat's work in this area, as well as what he did pre-Descartes in analytical geometry, was not published until after his death.

The calculus encompasses a number of inventions for which several mathematicians deserve credit. "By even the simplest accounting, royalties would need to be shared by a good dozen mathematicians in England, France, Italy, and Germany who were all busily ramifying Kepler and Galileo's work on functions, infinite series, and the properties of curves" (Wallace, 2003, p. 126). The idea of approximating the properties of curved figures by aggregations of successively smaller rectangles, which is fundamental to the calculus, goes back at least to the "method of exhaustion" used by the Greek mathematicians-philosophers Antiphon, Eudoxus, and Archimedes. The method is illustrated by the approximating of the area of a circle by determining the area of an inscribed many-sided regular polygon; what is exhausted is the difference between the area of the circle and that of the inscribed polygon as the number of sides of the polygon is increased. By starting with hexagons and doubling the number of sides four times to arrive at polygons of 96 sides, Archimedes determined the value of π to be between 3 10/71 and 3 1/7 (Beckman, 1971).

We think primarily of Lobachevski, Bolyai, Riemann, and perhaps Gauss as the originators of non-Euclidean geometry, but Giovanni Saccheri, George Klügel, and Johann Heinrich Lambert all did prior work in this area; even George (Bishop) Berkeley expressed doubts about the absoluteness of the truths of Euclidean geometry. The plane on which complex numbers are represented as points is known today as the Gaussian plane, but although Gauss did promote this form of representation, the method was described in publications by both Norwegian-born surveyor Caspar Wessel and French accountant Jean-Robert Argand several years before Gauss published about it.

Looking back over the history of mathematics, it is hard to understand why some of the ideas that we take for granted today took so long to become established. Some observers have argued that the acceptance of certain ideas or attitudes at critical times precluded or postponed the development or wide acceptance of other ideas, and that the history of both mathematics and science could have been quite different if this had not been the case. Bell (1945/1992) argues that the ancient Greeks, despite their spectacular accomplishments, delayed the development of mathematics and science for centuries by virtue of the Pythagoreans' adoption of number mysticism with its abhorrence of empiricism and by their failure to appropriate the algebra of the Babylonians. "Had the Pythagoreans rejected the number mysticism of the East when they had the opportunity, Plato's notorious number, Aristotle's rare excursions into number magic, the puerilities of medieval and modern numerology, and other equally futile divagations of pseudo mathematics would probably not have survived to this day to plague speculative scientists and bewildered philosophers. ... If on the other hand the early Greeks had accepted and understood Babylonian algebra, the time-scale of mathematical development might well have been compressed by more than a thousand years" (p. 54). In Bell's view the Greeks more than made up for these sins against the future, however, by two monumentally great and lasting contributions to thought: "explicit recognition that proof by deductive reasoning offers a foundation for the structures of number and form" and "the daring conjecture that nature can be understood by human beings through mathematics, and that mathematics is the language most adequate for idealizing the complexity of nature into apprehensible simplicity" (p. 55).

Where do individual mathematicians get their ideas for original work? The same question may be asked, of course, regarding creative endeavor in any field. Where do creative thinkers get their ideas? Halmos (1985), in considering this question, emphasizes the catalytic role of specific concrete problems. He categorically rules out the possibility that good ideas generally come from a desire to generalize, and claims that just the opposite is true: "The source of all great mathematics is the special case, the concrete example. It is frequent in mathematics that every instance of a concept of seemingly great generality is in essence the same as a small and concrete special case. Usually it was the special case that suggested the generalization in the first place" (p. 324).

Kac (1985) distinguishes between two kinds of mathematical creativity. One, which he likens to the conquering of a mountain peak, "consists of solving a problem which has remained unsolved for a long time and has commanded the attention of many mathematicians. The other is exploring new territory" (p. 39).

Byers (2007) sees ambiguity as a major source of mathematical creativity. "Ambiguity is not only present in mathematics, it is essential. Ambiguity, which implies the existence of multiple, conflicting frames of reference, is the environment that gives rise to new mathematical ideas. The creativity of mathematics does not come out of algorithmic thought; algorithms are born out of acts of creativity, and at the heart of a creative insight there is often a conflict—something problematic that doesn't follow from one's previous understanding" (p. 23). Byers makes ambiguity the central theme of his book-length treatment of the question of how mathematicians think. Logic is necessary too, as Byers sees it, but no more so than ambiguity: "Logic moves in one direction, the direction of clarity, coherence, and structure. Ambiguity moves in the other direction, that of fluidity, openness, and release. Mathematics moves back and forth between these two poles. ... It is the interactions between these different aspects that give mathematics its power" (p. 78).

Lakoff and Núñez (2000; Núñez & Lakoff, 1997, 2005) emphasize the role that metaphor plays as a source of mathematical ideas. Indeed, they see metaphor to be central to all thought: "the basic means by which abstract thought is made possible" (p. 39). They argue that conceptual metaphor is not just facilitative of understanding mathematics, but essential to it. Much of the abstraction of higher mathematics, they contend, "is a consequence of the systematic layering of metaphor upon metaphor, often over the course of centuries" (p. 47).

All of mathematics, Lakoff and Núñez (2000) argue, is derivable, psychologically, from four *grounding metaphors* (4Gs)—object collection (forming collections), object construction (combining objects to form new collections), the measuring stick (using devices to make measurements), and motion along a line (moving through space). These grounding metaphors are assumed to give rise to additional metaphors, such as a mental rotation metaphor that extends the number system to include negative numbers. The 4Gs are considered grounding metaphors because they represent direct links between mathematics and sensory-motor experience, and it is in sensory-motor experience, Lakoff and Núñez contend, that all mathematics finds its roots. Metaphors that are derived from the grounding four are more abstract than the grounding four themselves, and the longer the chain of derivation, the more abstract they become, but ultimately they all are traceable, in Lakoff and Núñez's view, to bodily grounding.

In addition to grounding metaphors, Lakoff and Núñez describe *linking methaphors*, the function of which is to support the conceptualization of one branch of mathematics (arithmetic) in terms of another branch (set theory). "Linking metaphors are different from grounding metaphors in that both the source and target domains of the mapping are within mathematics itself" (p. 142). As examples of classical branches of mathematics that owe their existence to linking metaphors, Lakoff and Núñez give analytic geometry, trigonometry, and complex analysis. Insistence that all mathematics is derivative from a few metaphors that are grounded in sensory-motor experience is novel, and whether it will have a lasting effect on the philosophy of mathematics is yet to be determined. We will encounter the idea again in Chapter 8.

☐ Strategies and Heuristics

Both investigators of problem solving by human beings and developers of problem-solving programs for computers stress the importance of strategies and heuristic procedures, especially in the context of problems for which algorithmic solutions are impractical or unknown. Mathematicians use such procedures in their efforts to solve mathematical problems. One finds many rule-of-thumb approaches described in the mathematical literature.

Considering Analogous Problems

Finding a solution to a difficult problem can sometimes be facilitated by considering a problem that is analogous to the one for which a solution is sought but that is more tractable, because either one is familiar with it or it is inherently simpler. "Quite often [mathematicians] do not deliver a frontal attack against a given problem, but rather they shape it, transform it, until it is eventually changed into a problem that they have solved before" (Péter, 1961/1976, p. 73). One is more likely to be able to do this, of course, if one has solved many problems, and many types of problems, than if one has not.

Bell (1945/1992) sees the approach of transforming problems for the purpose of making them more tractable as a defining characteristic of mathematical thinking. The methodology of transforming problems and reducing them to standard forms is seen, he suggests, in all the greater epochs of mathematics. "A relatively difficult problem is reduced by reversible transformations to a more easily approachable one; the solution of the latter then drags along with it the solution of the former and of all problems of which it is the type" (p. 36). Bell credits the ancient Babylonians, who appear to have solved cubic equations by reducing them to a canonical form, with being the first to use this approach.

Specialization and Generalization

One often sees an interplay between specialization and generalization in mathematical thinking. Mason, Burton, and Stacey (1985) emphasize the importance of this interplay and make it central to their approach to the teaching of mathematical problem solving. In specializing, one considers concrete examples of abstract problems. The hope in doing this is that one will find solutions to the concrete problems that will be extendable to the general case of which they are particular instances. If, for example, one is trying to solve a problem having to do with a certain property of parabolas, consideration of several specific parabolas may help one to see what the nature of the general solution must be. Conversely, sometimes turning one's attention from a specific concrete problem to the general problem type of which the specific problem is an example can facilitate progress.

Considering Extreme Cases

This trick, which is a special case of specialization, is often used to advantage in mathematical problem solving. It is nicely illustrated by Polya (1954a), who strongly advocated its use.

> Two men are seated at a table of usual rectangular shape. One places a penny on the table, then the other does the same, and so on, alternately. It is understood that each penny lies flat on the table, and not on any penny previously placed. The player who puts the last coin on the table takes the money. Which player should win, provided that each plays the best possible game? (p. 23)

Polya reports having watched a mathematician to whom this puzzle was posed respond by supposing that the table is so small that it is covered by one penny. Obviously in this case the first player will win (but, of course, only the one penny that he put down). Now imagine the size of the table being gradually increased. If the first player places the first penny precisely in the center, as soon as the table is large enough to hold a penny beside the first penny on any side, it will be large enough to hold another penny on the opposite side as well. By generalizing this argument, we can see that, irrespective of the size of the table, if the first player puts his first penny in the middle and after that always precisely matches what the second player does, but on the opposite side of the table, he will invariably win. It is not necessary to imagine the extreme case to solve this problem, because one could make an argument from symmetry straightaway, but use of the extreme-case heuristic can help one see the argument that can be made.

Visualization

The ability to visualize is considered by some to be of great benefit to both scientific and mathematical thinking (Ulam, 1976; Zimmerman & Cunningham, 1991). There are differences of opinion, however, regarding just what the role of visualization in mathematics is. Kline (1953a) emphasizes the limitations of visualization, arguing that many of the relationships with which mathematicians deal are inherently unvisualizable: "Anyone who insists on visualizing the concepts with which science and mathematics now deal is still in the dark ages of his intellectual development" (p. 447).

On the other hand, even inherently unvisualizable concepts or relationships may be developed with the help of visualization. We know, for example, that to develop his theory of electromagnetism—which was so abstract that it could not be stated in ordinary language—James-Clerk Maxwell used visualizations of rotating vortices interconnected and transmitting their rotation by means of cogs and wheels.

There is also the belief that sometimes mathematics can get in the way of visualization, to the detriment of creative thinking. Freeman Dyson (1979) reports a conversation with Richard Feynman in which the latter argued that Einstein's failure in later life to match his early spectacular successes was because "he stopped thinking in concrete physical images and became a manipulator of equations" (p. 62).

Perhaps the safest conclusion to draw regarding visualization in mathematics is that some mathematical relationships and problems lend themselves to visualization while others do not, and that some, but not all, mathematicians find the ability to visualize to be a great asset in their work. Unfortunately, the role that visualization plays in the development of mathematical proofs is generally obscured in mathematical publications, which typically focus on the proofs themselves and not on the reasoning that went into their development. We will encounter the question of the importance of visualization ability to the learning of mathematics in Chapter 16.

Diversion

Hadamard (1945/1954) reports frequently abandoning a problem for a while with the intention of returning to it again later on and says it is something he always recommends to beginners who consult him. "One rule proves evidently useful: that is, after working on a subject and seeing no further advance seems possible, to drop it and try something else, but to do so provisionally, intending to resume it after an interval of some months. This is useful advice for every student who is beginning research work" (p. 55). Apparently Hadamard believed that an especially good diversion was sleep. "One phenomenon is certain and I can vouch for its absolute certainty: the sudden and immediate appearance of a solution at the very moment of sudden awakening" (p. 8).

Experimentation

As already noted, a major difference between mathematics and science is that science looks to experimentation and empirical observation for validation of its theories, whereas mathematics does not. The relationship represented by the equation $2 + 2 = 4$ is not validated by checking to see if two apples plus two apples equals four apples, two oranges plus two oranges equals four oranges, and so on. The equation is not a theoretical assertion about the physical world that is in principle falsifiable and that launches investigators on a search for a disconfirming case that will bring the theory down.

Experimentation does play a role in mathematical thinking nevertheless, albeit a role that is different from the one it plays in science. Peterson (1988) points out, for example, that experimentation is important in research in number theory because so many of the key ideas in this area are resistant to definitive analysis. "Although their work differs from the experimental research associated with, say, test tubes and noxious chemicals, number theorists often collect piles of data before they can begin to extract the principles that neatly account for their observations" (p. 23).

I have mentioned a few of the strategies that are used in mathematical reasoning. There are many more. Some are used by many, if not most, mathematicians; some are used mainly, or only, by their inventors. For some problems, strategies may be helpful but unnecessary; for others, finding a solution apart from a strategy would be very difficult, if not impossible.

☐ Computing Aids and Devices

Probably for as long as people have counted and computed they have found ways to facilitate doing so. The development of aids to computing strikes me as an especially interesting aspect of the history of cognitive technology (Nickerson, 1997). The abacus of the Middle East, the soroban of Japan, and the knotted cords (quipu) of the Incas are among the better known of the devices that have been used by different cultures to facilitate counting and computing. A device that served Europe and America very well for about three recent centuries was the logarithmic slide rule. Invented, perhaps independently, by British mathematicians Edmund Wingate and William Oughtred around 1630, this device became an essential tool for engineers and others whose work required frequent calculations. Many variations on the basic logarithmic rule were invented for specific purposes; in his *History of the Logarithmic Slide Rule and Allied Instruments*, Cajori (1910/1994) gives a list of 256 different slide rules that were made between 1800 and 1910, the time of his writing. Uses included gauging, ullaging, and the computation of taxes and tariffs. Special rules were designed for many different purposes: "The change from one system of money, weight, or other measure, to another system, or the computation of annuities, the strength of gear, flow of water, various powers and roots. There are stadia rules, shaft, beam, and girder scales, pump scales, photo-exposure scales, etc." (p. 73). Rules other than slide rules—wantage rules, lumber rules, shrink rules, and so on—were designed with special-purpose scales that, in effect, enabled a user to carry out one or another type of computation in the process of making a measurement.

Less well known, but no less useful for specific purposes, are countless devices that have been designed to provide users with answers to computational questions without having to do the actual calculations. Many of these devices carry product advertisements and have been distributed by companies for promotional purposes. I have a small collection of such devices, mostly old, a few of which are described in Nickerson (2005):

- A circular celluloid device distributed by Sunkist Oranges for calculating the costs and selling prices of oranges (lemons on the flip side), per dozen, given the cost per box and the number of oranges (or lemons) per box, and assuming a specified markup.

- A similar device produced by Post Cereals for computing the per-package retail sale price of a product, given the wholesale price of a box of packages, the number of packages per box, and the desired profit percentage.

- A device (copyrighted in 1924) advertising Mead's dextri-maltose, a dietary supplement for babies, that permits one to find the recommended mixture of milk, water, and dextri-maltose for each feeding and number of feedings in 24 hours, given the baby's age and weight.

- A shop-cost calculator produced by General Electric (reflecting costs in 1953) that calculates the labor costs for operating a shop, given an hourly wage, number of operations performed per minute, and an overhead rate.

- A device distributed by the Esso Corporation (predecessor of Exxon) that can be used to calculate distance traveled by an airplane, given speed and time in flight; gallons of fuel consumed, given gallons consumed per hour and time in flight (or fuel consumed per hour, given total consumed and time in flight); and speed, given distance traveled and time in flight. It allows for making a correction in air speed, given the temperature and altitude, and can compute drift angle and ground speed, given course heading, wind velocity and direction, and air speed.

The list of examples could be greatly extended. My collection also includes devices that, in effect, calculate: (1) the appropriate torque setting on an adjustable torque wrench for a wrench extension of a given length, (2) a correction factor for a steam flow system, given a calibrated pressure and an operating pressure, (3) the feed rate for a turning, boring, or milling machine, given certain parameters of the machine, stock, and desired product, (4) relative humidity, given dry-bulb and wet-bulb temperatures, (5) certain motor data, given motor horsepower, and (6) conduit data, given wire size and composition (copper or aluminum)

Perhaps devices of the sort just noted are better described as devices that make computation unnecessary than as aids to computation. In that respect, they are similar to many of the developments in mathematics—especially perhaps in the development of notational systems—that have had the effect of easing one type of cognitive burden on doers of mathematics, freeing cognitive capacity for application to other demands of the problem-solving situation. Unburdening shortcuts have always been welcome to the mathematician, provided that they work.

☐ Computers in Mathematical Thinking

The arrival of high-speed digital computers on the scene has affected the doing of mathematics in several ways. It is somewhat ironic that while computers, including pocket calculators, have arguably made obsolete the learning of certain algorithmic procedures that used to be taught in basic mathematics courses, they have also increased the utility of an algorithmic approach to some types of mathematical problems by providing the computational power that is necessary to carry them out.

Examples of the use of computers in proof making are given in Chapter 5. Some of the problems on which mathematicians work today could not be approached without their involvement. Number theory provides many examples of problems that are beyond the capabilities of manual techniques. The problem of determining whether very large numbers (numbers with thousands or tens of thousands of digits) are prime is one case in point. The problem of factoring large composite numbers—which has practical implications for cryptography and computer system security—is another. Sixteenth-century German mathematician Ludolph van Ceulen spent most of his life determining the value of π to 35 places; as of 2003, thanks to the use of computers in its calculation, π was known to 1.2 trillion decimal places (Gibbs, 2003), although what "known" means in this context is open to some question.

Chaitin (1995) describes the effect of computers on the doing of mathematics this way:

> The computer has enormously and vastly increased mathematical experience. It's so easy to do calculations, to test many cases, to run experiments on the computer. The computer has so vastly increased mathematical experience, that in order to cope, mathematicians are forced to proceed in a more pragmatic fashion, more like experimental scientists. This new tendency is often called "experimental mathematics" ...
>
> It's often the case that when doing experiments on the computer, numerical experiments with equations, you see that something happens, and you conjecture a result. Of course it's nice if you can prove it. Especially if the proof is short. I'm not sure that a thousand-page proof helps too much. But if it's a short proof it's certainly better than not having a proof. And if you have several proofs from different viewpoints, that's very good.
>
> But sometimes you can't find a proof and you can't wait for someone else to find a proof, and you've got to carry on as best you can. So now mathematicians sometimes go ahead with working hypotheses on the basis of the results of computer experiments. Of course, if it's physicists doing these computer experiments, then it's okay; they've always relied heavily on experiments. But now even mathematicians sometimes operate in this manner. (p. 44)

Research on mathematical chaos, the behavior of nonlinear systems, fractal geometry, and cellular automata, among other areas, is very much dependent on the use of computers. Much of this work is greatly facilitated by computer graphics. Thomas Hale's proof of Kepler's conjecture that the face-centered cubic lattice is the densest of all possible three-dimensional sphere packings runs to 250 pages of text in addition to about 3 gigabytes of computer programs and data (Devlin, 2000).

The use of computers in mathematical problem solving has led to the distinction between problems that are likely to be tractable with the help of practically feasible amounts of computing power and those that are not. Most problems grow in complexity with their size, but some grow considerably faster than others. Consider the problem of determining the minimum number of people who must be invited to a party in order to guarantee that either at least m of the guests will all know each other or n will be mutual strangers. For $m = n = 3$, the problem is relatively simply solved: There must be a minimum of six guests in order to guarantee that at least three know each other or that at least three are mutual strangers.

To get an intuitive feel for the problem, try to satisfy the condition (either three people who know each other or three mutual strangers) in a party with five guests. To decide that this is impossible, one must

TABLE 6.1. A Party of Five, Lacking 3 Mutual Friends and 3 Mutual Strangers

	B	C	D	E
A	X	X	O	O
B		O	X	O
C			O	X
D				X

X in a cell indicates that the individuals represented by the associated row and column know each other; O indicates that they are strangers

convince oneself that the negation of the condition is true (that it is possible for there to be both fewer than three who know each other *and* fewer than three mutual strangers). Suppose the guests are *A, B, C, D,* and *E*. It is easy, of course, to have one of these conditions, but is it possible to have both (fewer than three mutual friends and fewer than three mutual strangers) with a group of five? Suppose the friendship relationships are as shown in Table 6.1, where X means *knows* and O means *does not know,* and the relationships are reciprocal: If *A* knows *B, B* knows *A*.

According to the table, *A* knows *B* and *C*, but not *D* and *E*; *B* knows *A* and *D*, but not *C* and *E*; and so on. With only five guests, it is easy enough to consider exhaustively all possible combinations of three to see if there are either three mutual friends or three mutual strangers. This is done in Table 6.2, where it is seen that, with this particular set of relationships, there are no combinations either of three mutual friends or of three mutual strangers.

As *m* and *n* are increased only a little, however, the problem becomes very difficult. The problem has yet to be solved for $m = n = 5$. The answer has been known for over 20 years to be between 43 and 49, but as of 1998, the exact number remained to be determined, and this despite that computers have been used with abandon in efforts to solve the problem. According to Hoffman (1998), "The most complex party problem that has been solved with the aid of computers, 110 of them running in sync, is the case of the minimum guest list needed to guarantee a foursome of friends or a fivesome of strangers. In 1993, the answer was found to be 25" (p. 54).

The computational complexity of some problems increases as an exponential function of their size, whereas that of others increases as a polynomial function of their size. Letting *N* represent the size of a

TABLE 6.2. Given the Relationships Represented in Table 6.1, There Are No Instances Either of 3 Mutual Friends or 3 Mutual Strangers

Combination	3 Mutual Friends?	3 Mutual Strangers?
ABC	No	No
ABD	No	No
ABE	No	No
ACD	No	No
ACE	No	No
ADE	No	No
BCD	No	No
BCE	No	No
BDE	No	No
CDE	No	No

problem, say, the number of nodes in a network that is to be analyzed, the complexity of a problem that grows exponentially with size would be given by C^N, whereas that of one that grows as a polynomial function would be given by N^C, where C is constant and relatively small in both cases. For a given small C, it is easy to see that C^N increases with N much faster than does N^C. Given $C = 2$, for example, C^N increases from 2 to more than 1,000,000 as N goes from 1 to 20, whereas N^C increases from 1 to 400 as N increases over the same range. This type of distinction led to the idea of *NP-complete* problems, introduced by Stephen Cook (1971), an NP-complete problem being one for which a computer algorithm can be constructed but cannot be run to solution in realizable time. When NP-complete problems are encountered in practical contexts such as scheduling, bin packing, or route planning, the goal of an optimal solution must yield to one of a good-enough solution or one that can be considered an approximation to optimal.

Although the use of computers in proof construction is controversial (Kleiner & Movshovitz-Hadar, 1997), their use in proof (or conjecture) refutation is not. In mathematics, it takes only one counterexample of a general assertion to show the assertion to be false. Using a computer to find a counterexample to an assertion that has been widely believed to be true—because no one has yet found a counterexample despite trying hard to do so—is not controversial, and it has been done on several occasions, especially in the area of number theory.

CHAPTER

Representation in Mathematics

> The history of mathematics shows that the introduction of better and better symbolism and operations has made a commonplace of processes that would have been impossible with the unimproved techniques. (Kline, 1953a, p. 240)

Thinking is greatly aided by representational systems used for purposes of communication. Nowhere is this more apparent than in the area of mathematics. Although physicists, chemists, musicians, and architects, among others, all have highly specialized representational systems that facilitate their work, no field has a longer or more impressive history than does mathematics with respect to the development and use of symbol systems.

It is very easy for us, who have been introduced to current mathematical symbols and notational conventions in a matter-of-fact way, to assume unthinkingly that things are as they are because that is the way they should be and always have been. In fact, current symbology and notation are the results of a long history of developments. The basic arithmetic operations—addition, subtraction, multiplication, and division— that we are likely to consider to be simple and straightforward, requiring only the memorization of a few elementary rules, were not always so simple and straightforward. Both the representational schemes and operational algorithms that we were taught as children are products of many centuries of inventions and evolutionary change. An engaging account of much of this history has been provided by Flegg (1983).

Throughout the history of mathematics, the emergence and refinement of new ideas have been accompanied by the invention of new ways of representing those ideas. The introduction of new notational conventions has provided significant economies of expression and greatly facilitated the performance of mathematical operations. And the notational systems invented to represent new mathematical ideas have stimulated and made possible further advances in mathematical thinking. So central are representations to mathematics that, according to one view, "mathematics can be said to be *about* levels of representation, which build on one another as the mathematical ideas become more abstract" (Kilpatrick, Swafford, & Findell, 2001, p. 19).

Mathematicians who have developed new areas of mathematics have often found it essential to invent new notational schemes in order to make progress. Diophantus, Descartes, Euler, and Leibniz are all remembered for their original contributions to mathematics; each of them also introduced new notational conventions and did so because the existing ones were not adequate to represent the thinking they wished to do. Jourdain (1913/1956) claims that Leibniz, who is remembered for numerous contributions to philosophy, science, and mathematics, attributed all his mathematical discoveries to his improvements in notation.

As discussed in Chapter 4, mathematical ideas have progressed from the more concrete to the more abstract. The emergence of new notational conventions often has been forced by the need to represent a new level of abstraction. This progression is illustrated by the symbols 3, x, and $f(x)$, which represent the increasingly abstract ideas of number, variable, and function. The concept *three*, as distinct from three stones or three sheep, is an abstraction; threeness is the property that three sheep and three stones have in common. The concept *number* is a further abstraction; numberness is what 3, 17, and 64 have in common. Algebra is more abstract than arithmetic because, whereas arithmetic deals with entities (numbers) whose values are constant, algebra deals with entities (variables and functions) whose values can vary.

☐ Beginnings of Algebraic Notation

The limitations of the mathematical accomplishments of the early Greeks have been attributed, in part, to the limitations of their notational system. The Greeks had a good notation for geometry, but less effective ones for arithmetic and algebra. Their notation was relatively effective for representing relationships among various parts of a figure, which are static in nature, but not for representing relationships among variable quantities, which are dynamic. Maor (1994) argues that the lack of an adequate representational system—in particular, the language of algebra—helps explain why, despite that Archimedes managed to apply Eudoxus's "method of exhaustion," which came close to the modern integral calculus, to the finding of the area of the parabola, the Greeks failed to discover the calculus.

A classification of representational conventions in algebra, dating from the middle of the 19th century and attributed to G. H. F. Nesselmann (1842), distinguishes three phases of development: rhetorical algebra, syncopated algebra, and symbolic algebra. Rhetorical algebra means algebra expressed in ordinary language; syncopated algebra made use of abbreviations, and symbolic algebra is essentially the system we currently use. Progressing from rhetorical algebra through syncopated algebra to symbolic algebra had the effect of reducing very considerably the amount of cognitive effort that one has to put into the process of solving any given mathematical problem. On the other hand, it also weakened the connection between the variables in equations and the real-world entities they may be used to represent; as we have noted, modern algebra really is symbol manipulation, and whether the symbols represent real-world entities is irrelevant.

Diophantus of Alexandria made a start during the third century toward the development of a notational system for algebra. He represented variables with symbols other than words; a limitation of his notation was its use of the same symbol to represent different variables. In the sixth century, Hindu mathematician Aryabhatta suggested the use of letters to represent unknowns. Like Diophantus, the Hindus and Moslems also used some form of what has been called *additive juxtaposition* to represent successive powers of the unknown. A specific letter or abbreviated word was used to represent the square, perhaps a different one to represent a cube, and concatenations of these to represent higher powers. Although both Hindu and Moslem mathematicians made rudimentary advances toward an operational symbolism, the Moslems eventually veered from this path and chose to write out everything, including the names of numbers.

During the 13 centuries between the time of Diophantus and that of French mathematician and counselor to the king Francois Viète, notational innovations were made, but none with revolutionary effects. Medieval scholar Jordanus Nemorarius, about whom little is known with certainty, used letters to represent numbers in his book *Arithmetica* (published in the 13th century), but he sometimes represented a given number by two letters (suggesting the endpoints of a line segment), and sometimes he used only one (suggesting one endpoint of a line segment, the other of which was understood).

Before the convention was adopted of using a letter to represent the unknown in algebraic expressions of one unknown, it was represented by a word, often the equivalent in the language of use of *thing* in English. Italian friar-mathematician Fra Luca Pacioli, for example, used *co*, short for *cosa* (*thing*), to represent variables of unknown quantity in equations. An economy of expression was realized by 15th-century French mathematician Nicolas Chuquet, who adopted the convention of omitting an explicit representation of the unknown altogether, using only coefficients and exponents of the various terms. Thus, what we would now represent as $6x$ and $4x^3$, he would have represented as $.6.^1$ and $.4.^3$.

A form of notation used to represent algebraic equations by some 15th-century mathematicians in Europe, Pacioli among them, is illustrated by the following expression (Scott, 1958; reproduced in David, 1962, p. 48):

4.p.R6 4.m.R6 Productum 16.m.6 10

which in modern notation would be written as

$$(4 + \sqrt{6})(4 - \sqrt{6}) = 16 - 6 = 10$$

A major advance was made by Viète, who introduced a notational convention for distinguishing unknowns from constants: He used letters to represent coefficients in polynomial equations, and introduced the convention of using a vowel to represent the unknown and a consonant to represent any quantity that was assumed to be known, thereby making explicit the distinction between variables and parameters. Although they credit Viète with making a very significant advance in algebraic notation, Boyer and Merzbach (1991) classify his system as "fundamentally syncopated rather than symbolic" because, along with the letters for variables and parameters and German symbols for the operations of addition and subtraction, he used words and abbreviations for other constructs (e.g., *A quadratus* and *A cubus* for A^2 and A^3, and *aequalis* for =). Viète is also credited with expressing proofs in strictly algebraic terms, which represented a departure from the then prevailing use of geometric proofs.

Although a form of exponential notation was used over 2,000 years ago by the Greek geometer Apollonius of Perga, more cumbersome schemes were used to represent powers more recently. Among 16th- and 17th-century ways of representing A raised to the seventh power were $AAAAAAA$ and $Aqqc$ (for A squared squared cubed). Sometimes a Roman numeral or an encircled number was placed directly above the coefficient of a term of a polynomial to indicate the power to which the variable in that term was to be raised.

The notation of symbolic algebra did not come into general use until the middle of the 17th century, and then through the influence of John Napier, René Descartes, and John Wallis. Galileo made no use of this or any other special notation. In *La géométrie*, in which Descartes (1637) presented his invention of analytic geometry, he introduced the convention of denoting variables by letters toward the end of the alphabet (x, y, z) and constants by letters toward the beginning of it (a, b, c). Descartes also promoted the notation for powers that we now use, x^3, x^4 x^5, ..., except he inexplicably chose to represent what we now write as x^2 by xx. Viète had used superscripts to represent powers, but Descartes denied that he had read Viète (Watson, 2002). Livio (2002) credits French mathematician Albert Girard with the introduction in 1634 of the use of subscripts to represent numerical position in a sequence. x_3, for example, would represent the third term, x_n the nth, and x_{n+k} the kth term following the nth.

Despite its advantages, which are obvious from our perspective, Descartes's notation met considerable resistance from mathematicians of the day and did not become the norm throughout Europe for several decades. Writing 150 years after Descartes, Laplace notes in his *Théorie analytique des Probabilités* the importance of notation in mathematics and points especially to Descartes's way of representing powers. Flegg (1983) explains the reluctance of mathematicians to adopt quickly the new notational scheme this way:

> The answer would seem to lie in the habit of expecting mathematical abbreviations to retain some obvious link with what they were signifying and especially with the spoken word. The older Greek works, which inspired so much of the mathematics of the Renaissance period and after, were purely rhetorical. Mathematical ideas were explained in words; mathematical arguments were written in words. To adopt abbreviation of words is therefore a natural step; the change to abstract symbolism demands an intellectual leap of extraordinary magnitude. It is precisely this requirement to write and hence to think in terms of symbols which makes mathematics a difficult subject in the classroom today unless attention is paid to this particular intellectual demand. (p. 224)

The invention of symbolic algebra represented a very significant step forward, not only for mathematics, but for the history of thought. An algebraic equation provides a means of packing a large amount of information into a few symbols. Jourdain (1913/1956) puts it this way: "By means of algebraic formulae, rules for the reconstruction of great numbers—sometimes an infinity—of facts of nature may be expressed very concisely or even embodied in a *single* expression. The essence of the formula is that it is an expression of a *constant* rule among *variable* quantities" (p. 38).

The wide adoption of a standard notational scheme greatly facilitated communication among mathematicians and the building of any given mathematician on the work of others. A standard symbology was also a practical necessity for the printing of mathematical works, and the emergence of print technology was an impetus to the development of one.

Algebraic notation not only makes possible great economies of expression, but it also facilitates computation. Indeed, one may see the history of improvements in algebraic symbolism as a shifting of an ever-greater portion of the burden of computation and inference from the person to the symbolism. The symbolism encoded much of what its developers had learned about mathematical inference and preserved that knowledge so that it would not have to be rediscovered anew each time it was needed. Inferences that would be very difficult to make without the use of this, or some comparable, notation may become, with its use, matters of straightforward mechanical symbol manipulation.

In many cases the need to make inferences was replaced with the ability to apply an algorithmic procedure. Dantzig (1930/2005) argues that the symbol is not a mere formality but rather the essence of algebra: "Replaced by a symbol the object becomes a complete abstraction, a mere *operand* subject to certain indicated operations" (p. 83). Descartes considered algebra to be a means of mechanizing mathematical operations, thus relieving the mathematician of much mental effort. Kaplan (1956), who says that in algebra the notation is everything, also argues that the power of algebra consists in its allowing the symbolism to think for us. Mathematical notation is but one of many examples of the development of tools designed to aid human cognition, making certain tasks less cognitively demanding, thus increasing the capacity to deal with other tasks (Nickerson, 2005; Salomon, 1993).

These observations may help account for the much discussed and disheartening finding that students who are able to solve textbook algebra problems that are already formulated as equations often cannot set up appropriate equations when given verbal descriptions of the problems to be solved. The setting up of an equation requires some thinking and understanding of the problem, whereas the solving of a preformulated equation may require only the application of memorized rules.

Barrett (1958) makes the point that, as a general matter, people today live on a level of abstraction way beyond that of their forebears. "When the contemporary man in the street with only an ordinary education quickly solves an elementary problem in arithmetic, he is doing something which for a medieval mathematician—an expert—would have required hours" (p. 26). He cautions, however, that this does not necessarily mean a higher level of understanding. "No doubt, the medieval man would have produced along with his calculation a rigorous proof of the whole process; it does not matter that the modern man does not *know* what he is doing, so long as he can manipulate abstractions easily and efficiently" (p. 27). Barrett may well be overestimating the medieval man's penchant for developing rigorous proofs of the validity of his reckoning, but there can be little doubt that many people today are capable of putting mathematics to practical use without a deep understanding of the rationales of the processes they are using.

☐ Mathematical Constants and Variables

Among the more important distinctions in mathematics is that between constants and variables. The words are their own definitions: A constant is an entity whose value does not change; a variable is an entity whose value may differ from occasion to occasion.

There are certain constants that play sufficiently important roles in mathematics that each has come to be represented by a single symbol, usually a Greek or English letter. Perhaps the two most common ones, already discussed in Chapter 3, are π (the ratio of a circle's circumference to its diameter) and e (the base of Napierian, or natural, logarithms). Napier, among others, invented logarithms in the early part of the 17th century, but Euler was the first to use e to represent their base over a century later. The delay in giving this constant a special symbol was short, however, compared to the case of π; although there is evidence that the significance of the ratio of a circle's circumference to its diameter had begun to be appreciated as early as 2000 BC, the symbol π was not used to represent it until the 18th century AD. The symbols π, e, i, Σ, and $f(x)$ all came from Euler, to whom Boyer and Merzbach (1991) refer as the most successful notation builder of all time.

We are so used to the idea of a variable that we are likely to be oblivious to the enabling power it represents. Tarski (1941/1956), noting the necessity of variables for the writing of equations, characterizes their significance this way:

> It is to the introduction of variables that we are indebted for the development of so fertile a method for the solution of mathematical problems as the method of equations. Without exaggeration it can be said that the invention of variables constitutes a turning point in the history of mathematics; with these symbols man acquired a tool that prepared the way

for the tremendous development of the mathematical sciences and for the solidification of its logical foundations. (p. 1909)

Paulos (1992) likens variables to pronouns—variables are to mathematics what pronouns are to natural language. Just as the same pronoun—*she*—can represent different individuals on different occasions, the same variable—*x*—can represent different values in different contexts. Extending the metaphor, one may think of constants as nouns; 5 is 5 and 6 is 6, no matter what the context in which one encounters them.

Clegg (2003) credits 17th-century English mathematician John Wallis as the first to use ∞ to represent infinity. It appeared in his *De sectionibus conicis* [*On conic sections*], written in 1655, and again in his *Arithmetica infinitorum*, written in 1656.

☐ Operators

In addition to symbols to represent variables, functions, and special constants, there is a need for symbols or notational conventions to represent mathematical operations: addition, subtraction, multiplication, division, exponentiation, and so on. The origins of all the many such symbols that are used today are obscure; the main design requirements for them, however, are fairly obvious. They should be convenient to write and sufficiently distinct not to be easily confused with each other. In some cases, a given operation can be represented in more than one way; the division of a by b, for example, can be represented as a/b, $a \div b$, or ab^{-1}, the choice being strictly a matter of convenience. Having more than one way to represent the same operation is useful because it sometimes happens that one representation is convenient for some contexts and not for others, but it also increases what the user of the symbology must learn in order to keep things straight.

Not only can some operations be represented in more than one way, it is also the case that some symbols have more than one meaning. The + sign, for example, is used to represent the operation of addition and to identify positive numbers; similarly, – is used to represent subtraction and to identify negative numbers. By convention, unsigned numbers are assumed to be positive. The dual usage of these signs can be algorithmically convenient; one learns by rote, for example, that two juxtaposed – signs are equivalent to one +, so subtracting a negative number is equivalent to adding a positive one. It can make for conceptual difficulties, however; $a - b$ does not distinguish between whether one is subtracting a positive number, $a - (+b)$, or adding a negative one, $a + (-b)$.

One suspects that what it means to subtract a negative number, $a - (-b)$, is less than crystal clear to many people who are able to apply the algorithm, $a + b$, correctly.

Early uses of the plus (+) and minus (−) signs in Europe occurred during the 15th or 16th century. Boyer and Merzbach (1991) give *Rechnun uff allen Kauffmanshafften*, published in 1489 by Johann Widman, a German mathematician and lecturer at Leipzig, as the oldest book in which + and − appear in print, but they note that these symbols were used first to indicate excess and deficiency in warehouse measurements and only later became associated with arithmetic operations. The + and − signs are also seen in German mathematician Michael Stifel's *Arithmetica integra*, which was published in Nuremberg in 1544, and in German mathematician Christoff Rudolff's *Die Coss* of 1525 (David, 1962). Bell (1945/1992) takes Widman's use of + and − as the event that marks the beginning of algebra becoming more operationally symbolic than it had been for Diophantus and the Hindus. He notes, too, that there is some evidence that sometime between 700 and 1100, the Hindus may have indicated subtraction with a sign similar to our plus sign written after the subtrahend. The × was first used to represent multiplication by William Oughtred, who introduced it in his *Clavis mathematicae* in 1631. Oughtred also invented the symbol :: to represent proportion (Maor, 1994). Today we use ! to represent the factorial function: n! = 1•2•3• … •n. In the 19th century, what we now write as n! was sometimes written as ⌐n (a vertical line followed by n).

Today we represent roots either by fractional exponents ($x^{1/2}$, $5^{1/3}$) or by use of the sign $\sqrt{}$, either by itself to represent square root (e.g., $\sqrt{x}$) or with a preceding superscript to represent a root other than square (e.g., $\sqrt[5]{x}$). The use of fractional exponents as a way of representing roots began to be adopted early in the 17th century. Before the current symbology evolved, words were sometimes used to represent the intended operations. In 16th-century Italy, for example, square and cube roots were identified, respectively, by the terms *lato* and *lato cubico*. *Lato* is Italian for side, so the idea seems to have been that of equating the square root of a number with the length of a side of a square, the area of which would represent the number the square root of which is being taken. Similarly, *lato cubico* would represent the length of a side of a cube, the volume of which would represent the number the cube root of which was desired (Mazur, 2003).

Another early form of representation of roots, also noted by Mazur (2003), used a symbol something like R (or R with a small slash on the right leg, like the sign that is used for medical prescriptions). R.q. (short for *radice quadrata*) and R.c. (short for *radice cubica*) would represent square root and cube root, respectively. Mazur speculates that the modern $\sqrt{}$ might have evolved from the letter r written cursively.

Leibniz, after some experimentation with other possibilities, fixed on dx and dy to represent what we now refer to as differentials, and on ∫, a large stylized s for sum, to indicate the operation of integration. The use of the notation f'(x), f''(x), …, fn(x), … to represent first-, second-, and nth-order derivatives comes from French mathematician Joseph Lagrange. Boyer and Merzbach (1991) call Leibniz "one of the greatest of all notation builders, being second only to Euler in this respect," and point out that in addition to giving us the notation of the calculus, he was the first prominent mathematician to use systematically the dot for multiplication and to write proportions in the form $a:b = c:d$. He also developed a way of representing sets of simultaneous equations that anticipated the invention of determinants. Leibniz used something similar to ∩ and ∪ to represent multiplication and division, respectively; whether or not this representational scheme has anything else to recommend it, it does convey the idea that multiplication and division are inverses of each other. We owe to Leibniz also the use of ~ for "is similar to" and ≅ for "is congruent to."

Leibniz published, when only 20 years old, a treatise entitled *Dissertatio de Arte Combinatoria*. Much of it deals with philosophical matters, but it contains some discussion of the problem of finding the number of combinations of n things taken m at a time. To represent a set of things the members of which are to be taken two, three, or four at a time, he uses, respectively, the designations com2natio (combinatio), com3natio (conternatio), com4natio, and so on (Todhunter, 1865/2001, p. 32). French mathematician Pierre Rémond de Montmort represented the combination of n things taken m at a time by a small rectangle with n above it and m below it.

Euler, possibly the most prolific mathematician who ever lived, recognized the importance of convenient notation and contributed substantively to its development. He was the first to use Σ to indicate summation, f(x) to stand for function of x, and $\left[\frac{n}{m}\right]$—which became $\binom{n}{m}$—to represent $\frac{n!}{m!(n-m)!}$, the number of combinations of n things taken m at a time.

A symbol that we take for granted today that did not come into general use until about the 16th century is the equals sign (=). It was introduced by Welshman Robert Recorde, physician to Edward VI and Mary Tudor and amateur mathematician and astronomer, in *The Whetstone of Witte* (1557). A sign composed of a pair of parallels of one length was chosen by Recorde to represent equality because, as he put it, "no two things could be more equal" (Struik, 1969, p. 4). The signs > and < to representing, respectively, "greater than" and "less than" were introduced by British mathematician Thomas Harriot in a book, *Praxis*, published posthumously in 1631.

The equals sign has come to have different meanings in different contexts, and failure to recognize the differences can make for confusion. Consider the following expressions:

$$x^2 + 3x - 10 = (x + 5)(x - 2)$$

$$x^2 + 3x - 10 = y$$

$$x^2 + 3x - 10 = 0$$

$$\sum_{n=1}^{\infty} \frac{1}{2^n} = 1$$

The first expression is a tautology; the right side is simply a restatement of the left. The second one expresses a functional relationship, showing the dependence of the value of one variable on that of another. The third expression is a constraint equation and can be solved for the values of x. The right side of the expression equals the left in all three instances, but not in the same sense. In the fourth equation, the sum does not actually equal 1; 1 is the limit the sum approximates ever more closely as the value of n increases.

We can add to the confusion by noting that many computer programming languages permit expressions of the sort

$$x = x + 1$$

which is nonsensical mathematically, but in the context of programming can have the perfectly reasonable interpretation "give the variable x a new value, namely one more than its current value."

The multiple uses of the equals sign is an unfortunate example of lack of precision in mathematical notation. Presumably, this ambiguity causes mathematicians no difficulties, because they know immediately from the context what is meant, but it is quite possible that failure to make a clear distinction among these uses may cause problems for students just learning algebra. Kieran (1989) contends that many students learn to interpret = as "and the answer is."

Although the operation of multiplication is called for several times in the first three equations above, an explicit symbol representing multiplication is not used. This is consistent with the familiar convention of representing multiplication by juxtaposition. A recent finding of difficulties that some people have in interpreting very simple equations involving multiplication raises a question as to the advisability of this convention.

It is apparent that there is some degree of arbitrariness about many of the symbols that have been chosen to represent mathematical operations. Surely, it would have made very little difference to the development of mathematics if − had been used to represent addition, and + subtraction, or if multiplication had been denoted by ÷ and division by ×. How these particular assignments of symbols to operations came about is not fully known. One conjecture about ÷ is that the dots above and below the line signified the placement of divisor and dividend in the $\frac{a}{b}$ representation; $a ÷ b$ has an obvious advantage over $\frac{a}{b}$ for purposes of typesetting.

That the selection of a representational convention is not completely arbitrary, however, is illustrated by the fates of the notational systems proposed by Newton and Leibniz to represent the differential calculus. Newton's notation was inferior to that of Leibniz in at least two respects: It did not explicitly identify the independent variable involved in the functional relationship, and it was not suitable for representing derivatives of higher degree; for these reasons, it was less conducive to the solving of differential equations. It has been claimed that the progress of mathematics in England was delayed relative to advances made elsewhere in Europe by more than a century as a consequence of the failure of the British to see or acknowledge the superiority of Leibniz's notation (Gleick, 2004; Jourdain, 1913/1956; Turnbull, 1929).

☐ Functions and Function Graphs

The concept of a mathematical function emerged relatively recently in the history of mathematics, but having emerged, its importance is widely acknowledged. Dubinsky (1994a) refers to it as perhaps "one of the most important ideas that students must learn in all of their studies of mathematics" (p. 235). Kasner and Newman (1940) make the even more extravagant claim that the word *function* "probably expresses the most important idea in the whole history of mathematics" (p. 5).

The functional notation, $y = f(x)$, indicates that the value of the variable y depends on—is a function of—the value of the variable x; similarly $y = f(x,z)$ indicates that the value of the variable y depends jointly on the values of the variables x and z. Typically in algebra functional relationships are represented without the use of the $f()$ notation; one simply writes $y = 3x^2$ to represent y as a particular function of x. An instance in which the $f()$ notation is used to advantage is that of representing probabilities. The convention of using p to represent probabilities in

19th-century textbooks led to difficulties that are avoided by the current convention of representing probabilities as functions of one or two arguments: $p(A)$ or $p(A|B)$, the latter meaning the probability of A given B.

The concept of a function is closely associated with analytic geometry invented by Descartes (and Fermat, whose work preceded Descartes's but was published later). The application of algebraic symbols to geometry permits one to think of figures in a "cartesian space" in terms of equations that define the value of one coordinate (y) as a function of the value of the other coordinate (x). The graphical representation of functional relationships is used so extensively in mathematics today that we are likely to overlook how powerful it is and how long it took for the idea to emerge and to be adopted widely. By blending geometry and algebra, analytic geometry makes it possible to solve algebraic problems geometrically and geometric problems algebraically. Stewart (1990) calls the cartesian coordinates of analytic geometry "a trick to convert geometry into algebra" (p. 41).

Some of the groundwork for the development of analytic geometry was done in the 14th century by French polymath and cleric Nicole (sometimes Nicholae) Oresme. His work spurred considerable interest in the graphical representation of functions, which he referred to as the "latitude of forms" (Boyer & Merzbach, 1991; see also Boyer, 1959; Kaput, 1994). But it is Descartes who is generally credited with clearly articulating the correspondence between plane curves and equations in two variables, and thereby coupling geometry with algebra in a way that had not been done before.

A function graph is a plot of the values of an independent variable for all values of a dependent variable over some range of the latter. It provides explicitly and graphically information contained implicitly in a functional equation. Most commonly, function graphs are drawn on a cartesian plane with linear dimensions, which is to say the dimensions are at right angles to each other and are divided into units of equal length. For some purposes, it is more convenient to use dimensions that are not at right angles or that are divided unequally (e.g., in logarithmic or other nonlinear units).

The use of graphs did not become widespread quickly after the possibility was initially noted; in fact, it took a rather long time for their use to become common. According to Wainer (1992), the only European journal that contained any graphs during the entire 18th century was the *Mémoires de l'Académie Royal des Sciences et Belle-Lettres,* and it contained very few. British theologian and philosopher Joseph Priestly, the discoverer of oxygen, "found it necessary [in 1765] to fill several pages with explanation in order to justify, as a natural and reasonable procedure, representing time by a line in his charts" (p. 12).

In addition to the connotation on which the foregoing has focused, the term *graph* has a quite different connotation as well. It sometimes refers to a set of nodes connected by lines. The nodes represent the elements of some set, and the lines show how the elements are connected. Graphs of this type differ from function graphs in that they are nonmetric and represent connectedness in a topological sense. Figure 7.1 illustrates the distinction between function graphs and nonmetric graphs, showing the graph of the function $y = \frac{x}{2} + 1$ on the left and a graph like one that might be used to represent the semifinals and finals of an elimination tournament on the right.

☐ Equations

The power of equations can seem magical. Like the brooms created by the Sorcerer's Apprentice, they can take on a power and life of their own, giving birth to consequences that their creator did not expect, cannot control, and may even find repugnant. (Wilczek, 2003, p. 132)

An equation is a mathematical expression composed of two parts separated by an equals (=) sign; it expresses what would appear to be the most straightforward of all relationships—equality. This equals that; the part to the left of the sign is said to equal, or be equivalent to, the part to the

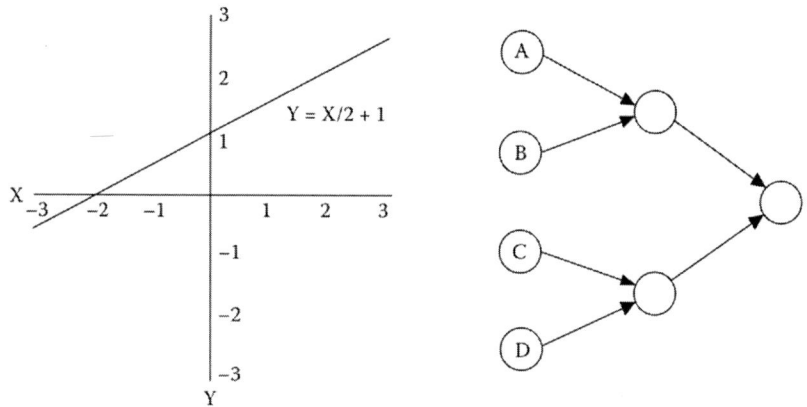

FIGURE 7.1 A graph of the function $y = \frac{x}{2} + 1$ (left) and a nonmetric graph like one that might be used to represent an elimination tournament.

right of it. What could be simpler and more mundane? And yet the concept is an enormously powerful one. If one has a valid equation—valid in the sense that the two parts are indeed equal or equivalent—then the parts will remain equal or equivalent no matter how many legitimate mathematical operations one performs on them, so long as one always performs the same operation on both parts.

Examples of simplifications in representations in the interest of economy of expression are easy to find in any field with a history. In mathematics the invention of symbolic algebra represented an enormous economy of expression over the syncopated algebra that preceded it and the rhetorical algebra that preceded that. Consider how the Greeks expressed the idea of a constant ratio in the time of Plato:

> Whenever among three numbers, whether solids or any other dimension, there is a mean, so that the mean is to the last term as the first term is to the mean, and when (therefore) the mean is to the first term as the last term is to the mean, then, the mean becoming both first and last, and the first and last both becoming means, all things will of necessity come to be the same, and being the same, all will be one. (Bell, 1946/1991, p. 170)

Today, we would express the relationship between the three terms, letting x, y, and z stand for first, second, and last-mentioned numbers, respectively, as

$$x/y = y/z$$

Tarski (1941/1956) illustrates that it is possible to represent in a relatively terse mathematical equation relationships that would require many words to describe by contrasting the way an elementary theorem of arithmetic is represented using the conventional notation of algebra:

For any numbers x and y, $x^3 - y^3 = (x - y)(x^2 + xy + y^2)$

with the way the same theorem might be represented without the use of this notation:

> The difference of the third powers of any two numbers is equal to the product of the difference of these numbers and a sum of three terms, the first of which is the square of the first number, the second the product of the two numbers, and the third the square of the second number. (p. 1908)

The importance of the emergence of an effective algebraic symbolism for the further development of mathematics has been stressed by Bell (1945/1992):

Unless elementary algebra had become "a purely symbolical science" by the end of the sixteenth century, it seems unlikely that analytic geometry, the differential and integral calculus, the theory of probability, the theory of numbers, and dynamics could have taken root and flourished as they did in the seventeenth century. As modern mathematics stems from these creations of Descartes, Newton and Liebniz, Pascal, Fermat, and Galileo, it may not be too much to claim that the perfection of algebraic symbolism was a major contributor to the unprecedented speed with which mathematics developed after publication of Descartes' geometry in 1637. (p. 123)

Even to the casual observer with little knowledge of mathematics, it is obvious that equations can differ greatly in complexity. What is perhaps less obvious is that even simple equations can differ qualitatively with respect to the type of equality they connote. We have already noted the distinction among tautologies, functional relationships, and constraint equations. More subtle distinctions can also be made. Both $2 + 3 = 5$ and $(x + y)^2 = x^2 + 2xy + y^2$ are tautologies, but of rather different kinds. The equation $2 + 3 = 5$ states a relationship among specific quantities; it asserts that adding the quantity 3 to the quantity 2 yields the quantity 5, always, everywhere, without exception (assuming, of course, use of the decimal number system, or at least one with a radix greater than 5). In contrast, $(x + y)^2 = x^2 + 2xy + y^2$ says no matter what quantities x and y represent, one will get the same result if one takes the sum of the square of x, the square of y, and twice their product as one will if one adds x and y and takes the square of their sum.

Now consider the familiar $C/D = 3.14159 \ldots$, where C and D represent the circumference and diameter of the same circle (measured in the same units). This says that the ratio of a circle's circumference to its diameter is the same—a constant—for all circles. As already noted, this particular constant is so ubiquitous and useful in mathematics that it is represented by its own universally recognized symbol, π, mention of which prompts notice of another type of equation, namely, a definitional one, illustrated by $\pi = 3.14159\ldots$ and $e = 2.71828. \ldots$ A definitional equation does no more than give a name (π and e in the examples) to a value. The equation $h^2 = a^2 + b^2$, where h represents the hypotenuse of a right triangle and a and b represent the other two sides—the celebrated Pythagorean theorem—expresses another universal geometric truth; unlike $C/D = \pi$, however, there are no constants involved. What is asserted is that the length of the hypotenuse of a right triangle is a function of the lengths of the other sides forming the triangle, which is to say that if one knows the lengths of the sides forming the right angle, one can infer the length of the remaining side, and this is true for all right triangles. This type of functional relationship is represented by many equations familiar from high school geometry and algebra. Examples:

$A = LW$, where A, L, and W represent, respectively, area, length, and width of a rectangle

$D = ST$, where D, S, and T represent, respectively, distance, speed, and time

$F = MA$, where F, M, and A represent, respectively, force, mass, and acceleration

Although the types of equations mentioned so far find many applications to the world of real processes and physical entities, none of them depends on real-world observations or measurements for its authenticity. The concepts and relationships involved are matters of definition and can be treated in the abstract without reference to the world of tangible objects and events. $C/D = \pi$ whether or not the world contains any truly circular objects, and $D = ST$ even in a world where no one goes anywhere.

On the other hand, equations that have been found to be descriptive of physical relationships have enormous practical value for many purposes. The use of equations is so common today that it is easy to forget that they were invented only a few hundred years ago. The importance of this development to the advance of mathematics is difficult to overstate.

☐ Representational Systems as Aids to Reasoning

> Most thinking in which human beings engage, even in highly mathematical fields like physics or economics, is not rigorous in the sense in which logicians and pure mathematicians use that term. Words, equations and diagrams are not just a machinery to guarantee that our conclusions follow from their premises. In their everyday use, their real importance lies in the aid they give us in reaching the conclusions in the first place. (Simon, 1995, p. xi)

One of the major benefits that the use of mathematical symbols provides is the economizing of thought that it makes possible. Jourdain (1913/1956) makes this point and notes that mathematical symbology permits us to represent many observations "in a convenient form and in a little space," to remember or carry about "two or three little formulae instead of fat books full of details" (p. 6). Again, "it is important to realize that the long and strenuous work of the most gifted minds was necessary to provide us with simple and expressive notation which, in nearly all parts of mathematics, enables even the less gifted of us to reproduce theorems which needed the greatest genius to discover. Each improvement in notation seems, to the uninitiated, but a small thing: and yet, in a calculation, the pen sometimes seems to be more intelligent than the user" (p. 13). Jourdain stresses especially the importance of the representational systems of analytical geometry and the infinitesimal calculus, and argues that their ability to make thinking more efficient is responsible, to no small degree, for their usefulness as instruments for solving geometrical and physical problems. The secret of these systems, Jourdain argues, is that they make it possible to solve difficult problems almost mechanically.

Diagrams are often used to elucidate mathematical proofs. Understanding Cantor's proof that the set of real numbers is uncountable, for example, is facilitated by the visual representation of a list of numbers, as shown in Chapter 5. Cantor's argument can be made without the use of the representation, because the logic does not depend on it and can be expressed without it. In general, it is probably not correct to say that such elucidating representations *constitute* proofs. Moreover, diagrams can be misleading. Ogilvy (1984) cautions against their use in proof making: "If we set out to prove something in mathematics, we must prove it. We are not allowed to say, 'Well, it's so because I can see it in the diagram'" (p. 90). That diagrams can facilitate comprehension of an argument, however, is beyond doubt.

The power of representational systems as vehicles of thought is seen not only in mathematics. Natural language, in both its spoken and written forms, is of course the most obvious example of a representational system without which thinking would be very different indeed. But there are numerous examples of systems that have evolved to meet special needs—music notation, logic diagrams, chemical transformation equations, blueprints, circuit diagrams, geopolitical maps, and so forth. Such systems are used to great advantage in countless contexts.

The details of any representational system must be constrained by what the system is intended to represent, by the representational medium, and by the capabilities and limitations of human beings as information processors. Symbols must correspond, in some fashion, to what they symbolize; they must be producible with the media at one's disposal (stone and chisel, clay and stylus, papyrus or paper and pen), and they must be discriminable and—at least potentially—interpretable by human observers. But within these broad constraints, there is much latitude for arbitrariness. We may ask with respect to any representational system why it is what it is and not something else. Why, for example, do we represent numerical concepts the way we do? How did the system most commonly used for representing Western music come to be what it is? Where did the conventions that rule the production of geopolitical maps come from, and why are they what they are?

The representation of numerical and mathematical concepts is of special interest in part because of the enormous range of applicability of these concepts and in part because of the obvious importance of symbols and notational conventions for the doing of mathematics. A better understanding of the role of representation in mathematics is one avenue to a better understanding of mathematical thinking and, quite possibly, to important insights into the nature of thinking more generally as well.

☐ Representations as Aids to "Seeing" Relationships

According to a well-known theorem in number theory,

$$\sum_{k=1}^{n} (2k - 1) = n^2$$

which is to say that the sum of the first n odd positive integers is equal to n^2. Thus,

$$1 = 1 = 1^2$$

$$1 + 3 = 4 = 2^2$$

$$1 + 3 + 5 = 9 = 3^2$$

$$1 + 3 + 5 + 7 = 16 = 4^2$$

and so on.

When one first learns of this theorem, it may seem to involve a curious relationship. Why should there be a connection between odd numbers and squares? How can one be sure the relationship holds in general?

A simple pictorial representation dispels the mystery and makes the reason for the relationship perfectly clear. If we represent n^2 with an $n \times n$ square arrangement of dots, we see immediately that in order to go from this arrangement to an $(n + 1) \times (n + 1)$ arrangement, representing $(n + 1)^2$, we need to add an odd number of dots, in particular $2n + 1$ dots, to the picture. Starting at the beginning, we represent 1^2 with a single dot:

•

In order to go from this arrangement to an arrangement representing 2^2, we have to add three dots:

To go from this to an arrangement representing 3^2, we have to add five dots:

and so on:

In general, in order to expand an $n \times n$ arrangement of n^2 dots into an $(n + 1) \times (n + 1)$ arrangement of $(n + 1)^2$ dots, one must add a row and a column with $n + 1$ dots in each, but inasmuch as one dot is common to the row and column, the total number of added dots is $2(n + 1) - 1$ or $2n + 1$.

The relationship can be demonstrated, of course, without the use of a diagram:

$$\text{If} \quad \sum_{k=1}^{n} (2k - 1) = n^2,$$

$$\text{then} \quad \sum_{k=1}^{n+1} (2k - 1) = (n + 1)^2,$$

because $\displaystyle\sum_{k=1}^{n+1}(2k-1) = \sum_{k=1}^{n}(2k-1) + (2k+1) = n^2 + (2n+1) = (n+1)^2.$

Given this relationship, if $\displaystyle\sum_{k=1}^{n}(2k-1) = n^2$ holds for any value of n, it holds for $n+1$ and, by induction, for all subsequent integer values of n. The equation holds for $n = 1$; therefore, it holds for all positive integer values of n.

For most of us, I suspect, the diagram provides an intuitively more compelling demonstration of why the sum of successive odd positive integers beginning with 1 is a square than does the series of equations.

Consider the following elegant equality:

$$\sum_{k=1}^{n} k^3 = \left(\sum_{k=1}^{n} k\right)^2 \qquad (7.1)$$

which says that the sum of the cubes of the first n integers equals the square of the sum of those integers:

$$1^3 + 2^3 + 3^3 + \cdots + n^3 = (1 + 2 + 3 + \cdots + n)^2$$

Again, it is not immediately obvious from the equation why this relationship should hold. We see from Table 7.1 that it does indeed hold, at least for the values considered.

This is, of course, no proof that the relationship holds in general, but simply a demonstration that it holds for a few values of n, which is enough to make us wonder whether it might hold indefinitely. But consider again the sequence of odd numbers

1, 3, 5, 7, 9, 11, 13, 15, 17, 19, 21, 23, 25, 27, 29

We have seen already that the sum of the first n odd integers is equal to n^2. Greek mathematician Nicomachus of Gerasa, who lived around the end of the first century AD, noticed that when the odd integers are grouped so that the jth group contains j integers—the first one, the second two, the third three, and so on—the sum of the integers in each group equals the cube of the number of integers in that group. Thus, the integers in the first five groups,

(1) (3, 5) (7, 9, 11) (13, 15, 17, 19) (21, 23, 25, 27, 29)

sum to 1, 8, 27, 64, and 125, or 1^3, 2^3, 3^3, 4^3, and 5^3, respectively.

TABLE 7.1. Showing the Sum of the Cubes of the First *n* Integers Equals the Square of the Sum of Those Integers

n	1	2	3	4	5	...
$\displaystyle\sum_{k=1}^{n} k$	1	3	6	10	15	...
n^3	1	8	27	64	125	...
$\displaystyle\left(\sum_{k=1}^{n} k\right)^2$	1	9	36	100	225	...
$\displaystyle\sum_{k=1}^{n} k^3$	1	9	36	100	225	...

Is there a way to represent the sequence of odd numbers that will make it apparent why the rule noted by Nicomachus holds? Suppose we were to replace each of the numbers in each group with the mean of the numbers in that group. By definition, the sum of *n* numbers equals *n* times the mean of those numbers, so we know that the sums of the numbers in the groups will not be affected by this substitution.

(1) (3, 5) (7, 9, 11) (13, 15, 17, 19) (21, 23, 25, 27, 29) ...

(1) (4, 4) (9, 9, 9) (16, 16, 16, 16) (25, 25, 25, 25, 25) ...

What one notices immediately is that each of the numbers in this new arrangement is a square, and in particular each of the numbers in the *j*th group is j^2. So the arrangement can be represented as

(1^2) $(2^2, 2^2)$ $(3^2, 3^2, 3^2)$ $(4^2, 4^2, 4^2, 4^2)$ $(5^2, 5^2, 5^2, 5^2, 5^2)$ ...

And, inasmuch as there are *j* squares in the *j*th group, we can represent the sums as

(1×1^2) (2×2^2) (3×3^2) (4×4^2) (5×5^2) ...

or equivalently,

$$(1^3)\ (2^3)\ (3^3)\ (4^3)\ (5^3)\ \ldots$$

Again, not a proof that the relationship holds in general, but the representation helps one see why it might. There are other intriguing relationships in the sequence of odd numbers (see Backman, 2007), but to explore them would take us too far afield from the main point of this discussion, which is to illustrate that the "seeing" of a mathematical relationship may be facilitated by the way in which the relationship is represented.

This focus on the equivalence between the sum of the first n cubes and the square of the sum of the first n whole numbers prompts mention of a report by Tocquet (1961, p. 16, footnote 1) of the empirical discovery by 81-year-old "lightning calculator" Jacques Inaudi (about whom more is in Chapter 10) that the sum of the first n cubes could be found with the formula

$$S = \left(\frac{n(n+1)}{2}\right)^2$$

Inasmuch as $\frac{n(n+1)}{2}$ is the sum of the first n integers, this formula is the equivalent to the right-hand term in Equation (7.1).

The ancient Greeks, even when solving problems that would be classified as algebraic today, tended to think in geometric terms. In his discussion of algebraic problems in the *Elements*, for example, Euclid represents numbers as line segments. Although this may have been constraining in some respects, it also may have helped them see the reason for certain relationships more clearly than if they had thought strictly in algebraic terms. For example, comprehension of the distributive law of multiplication, according to which $a(b + c + d) = ab + ac + ad$, would have been easy for the Greek scholar who would have interpreted it to mean that the area of the rectangle on a and the sum of line segments b, c, and d is equal to the sum of the areas of the rectangles formed by a and each of the line segments individually, as shown in Figure 7.2.

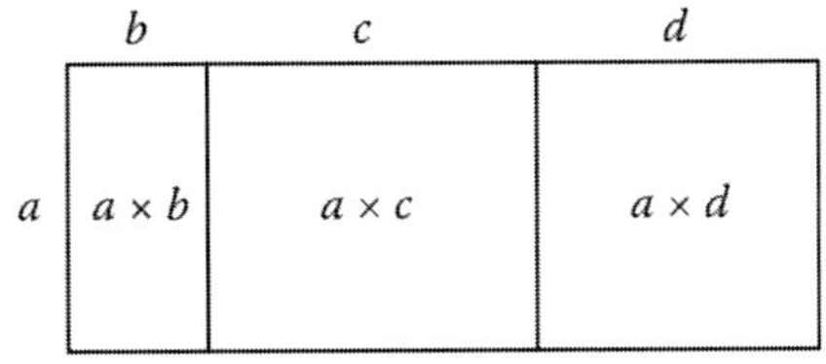

FIGURE 7.2 Illustrating geometrically the distributive law of multiplication, that is, that $a(b + c + d) = ab + ac + ad$.

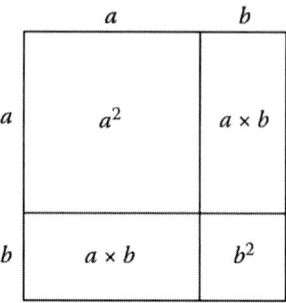

FIGURE 7.3 Illustrating geometrically that $(a + b)^2 = a^2 + 2ab + b^2$.

Similarly, the relationship $(a + b)^2 = a^2 + 2ab + b^2$ becomes obvious when one thinks in terms of a geometrical representation of it, as shown in Figure 7.3. Other algebraic relationships could be represented geometrically in a similar way.

The following problem also illustrates how representations can be useful, not only in aiding problem solving, but in helping to explain why a solution is a solution and in clarifying relationships that otherwise might be difficult to see.

> There are two containers, A and P. Container A has in it 10 ounces of water from the Atlantic Ocean; P contains 10 ounces of water from the Pacific. Suppose that 2 ounces of Atlantic water is removed from container A and added to the contents of container P, and then, after the water in container P is thoroughly mixed, 2 ounces of the mixture is removed and added to the contents of container A. Which container now has the greater amount of foreign water, the Atlantic water being foreign to P and the Pacific to A?

The answer is that both have the same amount. One way to demonstrate that this is true is to track the interchange of liquids step by step. After 2 ounces of the Atlantic water from A is added to the 10 ounces of the Pacific in P, P contains 12 ounces of water in the proportion 10 parts Pacific to 2 parts Atlantic. When 2 ounces of this thoroughly mixed mixture is transferred to A, P is left with $2/12 \times 10$, or 1.67, ounces of foreign (Atlantic) water. Container A, on the other hand, receives 2 ounces mixed in the proportion 10 parts Pacific to 2 parts Atlantic, so the amount of foreign (Pacific) water that goes into A will be $10/12 \times 2$, or 1.67, ounces. Thus, each container ends up with the same amount of foreign water.

Another way of viewing the problem makes the answer obvious. Suppose we represent the situation, after the exchanges have been made, with a 2 × 2 table in which the rows represent the two containers and the columns the two liquids, as shown in the left-most table in Figure 7.4.

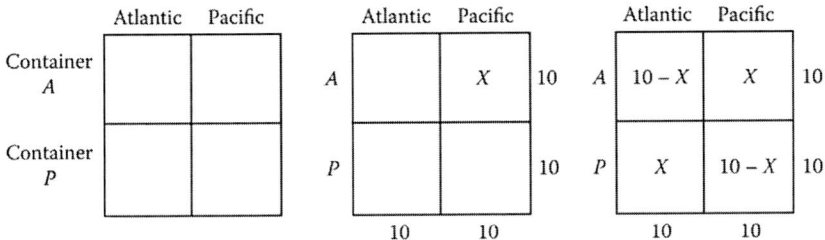

FIGURE 7.4 A way of representing the two-containers problem.

What we want to show in the cells of this table is how much of each liquid is in each container. We can fill in the row and column totals right away (as shown in the middle table of Figure 7.4), because we know that, assuming no liquid was lost in the exchanges, we end up with the same amount of liquid of both types (Atlantic and Pacific) as we began with (10 ounces), and because we took 2 ounces out of container A and then put 2 back, we also end up with the same amount of liquid (10 ounces) in each container. Now suppose we had not done the calculation to discover how much Pacific water ended up in A. We know that it is some amount, but not how much, so we represent the amount in our table by X, as also shown in middle table of Figure 7.4.

It should be clear at this point that all the other cells of the table are determined: If there are X ounces of Pacific water in A, A must contain $10 - X$ ounces of the Atlantic; if X ounces of Pacific water is in A, that means the remaining $10 - X$ ounces of the Pacific must be in P; and if P has $10 - X$ ounces of Pacific water, it must have X ounces of the Atlantic. In other words, given that the total amounts of Atlantic water and Pacific water do not change, and that both containers end up with the same amount of liquid that they had initially, it follows that whatever amount of the Atlantic is missing from A (and therefore is in P) must have been replaced by an equal amount of the Pacific (which is missing from P), as shown in the rightmost table of Figure 7.4.

This example illustrates that problems often can be approached in radically different ways. Both of the solutions given above are correct, but they differ in some important respects. Most of the people I have watched try to solve this problem have taken the first approach of tracking the results of the individual transactions. This approach, although tedious, produces a solution that suffices to answer the specific question that was asked, but it does not generalize readily to related cases. Moreover, the answer that is obtained seems to lack intuitive force; one's belief in its accuracy rests on one's confidence that the sequence of calculations was performed without error.

The second approach produces a solution that has considerable generality. Viewing the problem in this way makes it clear that the "equal amounts of foreign water" answer holds independently of how many transfers are made, what amounts are involved in each transfer, or how thoroughly the mixing is done, provided each container holds the same amount of liquid in the end as in the beginning. This representation of the problem solution also is intuitively compelling; one sees the relationships involved and why the answer has to be what it is. The drawback to this approach is that people typically do not think initially to look at the problem this way; something of an insight seems to be required to put one on this track.

Here is a third example of how a simple representation can help to make clear a relationship that some people may have difficulty seeing without it. The problem is as follows:

> One morning, exactly at sunrise, a monk began to climb a mountain. A narrow path, a foot or two wide, spiraled around the mountain to a temple at the summit. The monk ascended at varying speeds, stopping many times along the way to rest. He reached the temple shortly before sunset. After several days at the temple, he began his journey back along the same path, starting at sunrise and again walking at variable speeds with many pauses along the way. His average speed descending was, of course, greater than his average climbing speed. Does there exist a particular spot along the path that the monk will occupy on both trips at precisely the same time of day? (Adapted from Adams, 1974)

Some people have trouble with this problem, imagining first the upward journey and where the monk would be at different times of the day and then the downward trek and how that might progress. The difficulty seems to be in somehow getting the two journeys into the same frame of reference.

A natural way to represent the situation diagrammatically is with a graph showing position (say, distance from the bottom of the mountain) as a function of time of day. Thus, the upward and downward journeys might be represented as shown in Figure 7.5.

This representation makes it clear that there indeed will be some spot that the monk will occupy at the same time of day. The spot could be at any of many places, depending on the relative speeds of the ascending and descending journeys, but there obviously must be at least one such spot. (There could be more than one, if he did any backtracking on either trip.) There is no way to draw one line from the bottom to the top of the graph and another from the top to the bottom without having the lines cross.

This story illustrates the power of a graphical representation to facilitate the understanding of a relationship that might be difficult to

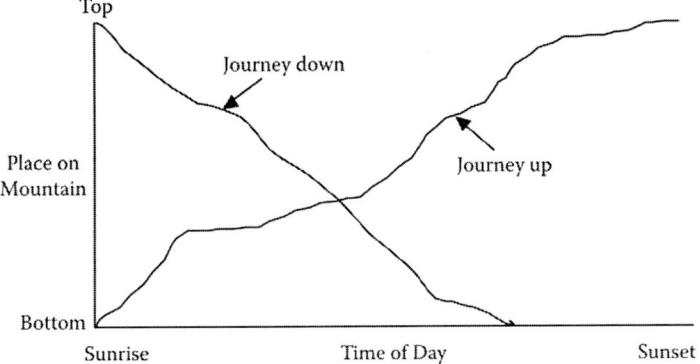

FIGURE 7.5 Illustrating that there must be at least one spot along the path that one takes going up and down a mountain on different days that one will occupy on both trips at precisely the same time of day, assuming one starts out at the same time both days.

see otherwise. The problem also presents another opportunity to note the effectiveness of the heuristic of finding an analogous problem that is easier to solve, in the hope that solving the easier problem will provide some useful hints regarding the solution to the more difficult one. A situation that is analogous to the one just considered is that of two monks, one starting at the bottom and climbing to the top and the other starting at the top and descending to the bottom, both beginning at the same time and completing their journeys on the same day. In this case, it is intuitively obvious that the two monks' paths must cross, so they will be at the same place at some point in the day. To accept this answer as appropriate also for the original problem, one must, of course, be convinced that the two situations are indeed analogous. Most readers, I suspect, will see the situations as sufficiently similar in the right ways to support the conclusion that what holds in the one case holds also in the other. What is likely to be the more serious limitation of the heuristic of finding analogous but easier problems is the difficulty most of us may have in coming up with them when we need them.

☐ Representations in Problem Solving

Whatever the details, most would agree that *some* idea of representation seems to be at the heart of understanding problem-solving processes. (Kaput, 1985, p. 381)

Numerous studies have shown the important role that representations play in problem solving, in mathematics and other domains (Hayes & Simon, 1974; Koedinger & Anderson, 1995; Larkin, 1980, 1983; Larkin & Simon, 1987; Mayer, 1983; Paige & Simon, 1966). Books giving "how to" advice on problem solving almost invariably stress the finding of a useful representation of the problem as an indispensable early step in the process—and the finding of a different representation if the one in hand does not appear to be leading to a solution.

Studies of the differences in the performance of expert and novice problem solvers highlight the greater use by experts than by novices of qualitative representations (e.g., diagrammatic sketches) to ensure they understand a problem and to help plan an approach to it before rushing ahead to attempt to compute or deduce a solution. Heller and Hungate (1985) characterize the difference between expert and novice problem solvers in this regard this way: "Understanding is viewed as a process of creating a representation of the problem. This representation mediates between the problem text and its solution, guiding expert human and computer systems in the selection of methods for solving problems. Novices tend to be quite deficient with respect to understanding or perceiving problems in terms of fundamental principles or concepts. They cannot or do not construct problem representations that are helpful in achieving solutions" (p. 89).

In a meta-analysis of a large number of studies of mathematical problem solving by children in grades K through 4, Hembree (1991; also summarized in Hembree & Marsh, 1993) found that, of the various techniques for problem solving on which instruction was focused, the most pronounced effect on performance was obtained from the development of skill with diagrams. Hembree and Marsh note, however, that explicit training appeared to be essential inasmuch as performance was not improved as a result of practice without direct instruction.

Sometimes the right representation can greatly reduce the amount of cognitive effort that solving a problem requires. A representation can, for example, transform what is otherwise a difficult cognitive problem into a problem, the solution of which can be obtained on a perceptual basis. A compelling illustration of this fact has been provided by Perkins (2000). Consider a two-person game in which the players alternately select a number between 1 and 9 with the single constraint that one cannot select a number that has already been selected by either player. The objective of each player is to be the first to select three numbers that sum to 15 (not necessarily in three consecutive plays). A little experimentation will convince one that this is not a trivially easy game to play. One must keep in mind not only the digits one has already selected and their sum, but also the digits one's opponent has picked and their running sum. Suppose, for example, that one has already selected 7 and 2 and it is one's turn to play. One would like to pick 6, to bring the sum to 15, but one's opponent has already selected 6 along with 4. So, inasmuch as one cannot win on this play, the best one can do is to select 5, thereby blocking one's opponent from winning on the next play. In short, to play this game, one must carry quite a bit of information along in one's head as the game proceeds.

One could reduce the memory load of this game, of course, by writing down the digits 1 to 9 and crossing them off one by one as they are selected. And one could also note on paper the current sum of one's own already selected digits and that of those selected by one's opponent. Better yet, as Perkins points out, the game can be represented by a "magic square"—a 3 × 3 matrix in which the numbers in each row, each column, and both diagonals add to 15. With this representation, the numbers game is transformed into tic-tac-toe. The player need only select numbers that will complete a row, column, or diagonal, while blocking one's opponent from doing so. There is no need now to remember selected digits (one simply crosses them out on the matrix as they are selected) and no need to keep track of running sums.

Students often have difficulty with mathematical word problems even when they are able to do the computations that the solutions of the problems require, if given the appropriate computational formulas (Hegarty, Mayer, & Green, 1992; Hegarty, Mayer, & Monk, 1995; Lewis & Mayer, 1987). Unquestionably, the ability to solve such problems can be facilitated by representing the relationships between variables in diagrammatic form. The educational challenge is to find ways to teach students how to construct effective diagrams.

Infinity

What is man in nature? A Nothing in comparison with the Infinite, an
All in comparison with the Nothing, a mean between nothing and everything. Since he is infinitely removed from comprehending the extremes,
the end of things and their beginning are hopelessly hidden from him in
an impenetrable secret; he is equally incapable of seeing the Nothing from
which he was made, and the Infinite in which he is swallowed up. (Pascal,
1670/1947, p. 200)

The infinite more than anything else is what characterizes mathematics
and defines its essence.... To grapple with infinity is one of the bravest
and extraordinary endeavors that human beings have ever undertaken.
(Byers, 2007, p. 187)

Some concepts in mathematics constitute more of a challenge to intuition than do others. Some have been accepted only slowly over many
years, proving their practical worth before being widely acknowledged
as fully legitimate. Some continue to baffle, especially when one tries to
understand at a deep level what they "really mean." Among those in this
category are infinity and infinitesimals.

These concepts have given mathematicians—among others, including many of the world's most notable philosophers—great pleasure and
much trouble for a very long time. They are at once fascinating and
extraordinarily perplexing. They draw any thinking person to questions
that engage the mind in exciting excursions that transcend the constraints of the physical world as we know it, and that, in many cases, do
not admit to uncontroversial answers. As Burger and Starbird (2005) put

it, "Long before we reach infinity, we must face ideas beyond our grasp" (p. 233).

Byers (2007) contends that the use of infinity brought mathematics from the domain of the empirical to the domain of the theoretical. "The use of infinity in any specific manner requires considerable mental flexibility. It requires new ways of using the intellect, a certain subtlety of thought, an ease with complex contradictory notions. In the use of the concept of infinity there is always the danger that things will get out of control and slip into the realm of the purely subjective. That is, there is the danger that we will not be doing mathematics anymore" (p. 121).

Mathematicians and philosophers who have struggled with ideas involving the infinitely large and the infinitesimally small include Archimedes, Aristotle, Zeno, Descartes, Pascal, Kant, Leibniz, Gauss, Cantor, as well as countless other lesser lights. Few concepts have captured more attention from inquiring minds over the centuries. It is with some trepidation that I turn to these ideas in this chapter and the next, but their prominence in the history of mathematics makes it imperative that they be considered in any book that purports to be about mathematical reasoning.

A caveat is in order before proceeding. In what follows reference is often made to what can and cannot be done in mathematics. Generally, unless otherwise stipulated, when I claim that something can be done, I mean that it can be done conceptually, but not necessarily actually—that one can conceive of it being done, even if one cannot do it. I borrow an illustration from Moore (2001). Suppose there were no practical constraints on how fast one could work. It would be possible then to write an infinity of natural numbers in a minute. One would take half a minute to write 0, one-quarter of a minute to write 1, one-eighth of a minute to write 2, and so on, halving the time to write each successive integer. One would have written an infinity of them in a minute—obviously impossible, but conceivable. Moore suggests calling stories in which infinitely many things can be done in finite time "super-task stories." One encounters such stories often in discussions of infinity.

☐ Origin of the Idea of Infinity

> This is a most important lesson, namely that the infinite in mathematics is conceivable by means of finite tools. (Péter, 1961/1976, p. 51)

Where did the idea of infinity originate? What prompted its emergence? One possibility is that it originated in the compelling belief that every number has a successor—that there is no largest number. Where does this idea come from? Not from logical necessity and certainly not from the immediate experience of our senses. But logical necessity and experimental evidence are not all there is to the objective world we call reality. Perhaps there is, as Dantzig (1930/2005) argues, "a mathematical necessity which guides observation and experiment, and of which logic is only one phase. The other phase is that intangible, vague thing which escapes all definition, and is called intuition…. The concept of infinity is not an experiential nor a logical necessity; it is a *mathematical necessity*" (p. 256).

The ancient Greeks were suspicious of the concept of infinity, in part because of the unsettling effect of the famous paradoxes of Zeno of Elea that yielded, by presumably self-evidently true assumptions and impeccable logic, startling revelations. One example: If swift Achilles gives a lumbering tortoise a head start in a race, the former will never be able to catch the latter. Another: An arrow can never leave the archer's bow. And, more generally: Movement (or change of any sort) is impossible. The concept is difficult to avoid completely, however, especially by minds as fertile of those of the classical Greeks.

The claimed dislike of infinity by the ancient Greeks is sometimes described in strong terms. Dantzig (1930/2005), for example, refers to the horror this concept held for them. Whether the Greeks actually had such a horror is a contested point; Knorr (1982) calls the idea "a preposterous myth whose demise can only be welcome" (p. 143). Moore (2001) takes the more moderate position that while the Greeks did not make infinity an important object of mathematical study, they embraced the concept in an indirect way, for example, in taking a line to be indefinitely extendable and infinitely divisible.

Whether or not the Greeks were horrified by infinity, there can be no doubt that the concept was problematic for them and continued to be so for others for many centuries. Hopper (1938/2000) uses the phrase *repugnance of the idea of infinity* to describe the attitude during medieval times. Wallace (2003) contends that "nothing has caused math more problems—historically, methodologically, metaphysically—than infinite quantities" (p. 32). Some have argued that serious grappling with the concept has caused, or at least hastened, the descent of more than one brilliant mathematician into madness. Descartes (1644), who declined to become involved in "tiresome arguments" about the infinite, famously took the position that inasmuch as the human mind is finite, we have no business thinking about such matters. Galileo (1638/1914) also considered both infinities and indivisibles to be incomprehensible to finite understanding (the one because of its largeness and the other because of its smallness), but accepted that "human reason does not want to abstain from giddying itself about them."

☐ Paradoxes of Infinity

To the reader whose interest in mathematics is focused exclusively or primarily on its practical applications, a discussion of paradoxes of infinity may seem an unnecessary digression. However, development of the calculus—the mathematics that is basic to the study of real-world phenomena of motion in space and change in time and without which modern technology could not exist—involved confrontation of paradoxes of this sort. Much of the philosophical difficulty people, including mathematicians, had with the calculus when it was in its initial stages of development had to do with what appeared to be absurdities involving infinity that cropped up in efforts to understand time and motion. It would seem, for example, that if time is infinitely divisible, when an object passes from a state of rest to one of movement at some speed, it must pass through an infinity of speeds in a finite time. But if it spends any time at all, no matter how tiny an instant, at each speed, how can an infinity of those instants fit within a span of finite duration? The inability to answer this question seems to force one to the conclusion that the body does not spend any time at all at any of the speeds between rest and the final speed obtained, but this seems equally as absurd as the belief that it spends a very small amount of time at each of the infinity of intermediate speeds. Questions of this sort challenged many of the better minds of the 17th century, including those of Galileo, Pascal, Leibniz, and Newton.

Zeno's paradoxes (about which there is more in the next chapter) have provided entertainment and frustration to generations of mathematicians from his time to ours. They have been resolved to the satisfaction of some, and defined out of existence by some, but they continue to bedevil others. Bertrand Russell (1926) credits them with affording, in one way or another, grounds "for almost all the theories of space and time and infinity which have been constructed from his day to our own" (p. 183). They continue to inspire serious philosophical treatises. Adolf Grünbaum (1967), who considers Zeno's arguments to be fallacious, describes them as "inordinately subtle, highly instructive, and perennially provocative" (p. 40). His own book-length treatment of them cites treatments also by Henri Bergson, William James, Alfred North Whitehead, Bertrand Russell, Gerald J. Withrow, Hilary Putnam, and Max Black, among others. In another place (Grünbaum 1955/2001a), he mentions also Emmanuel Kant, Paul du-Bois Reymond, and Percy W. Bridgman as among the notables who have wrestled with one or another of Zeno's ideas. An anthology edited by Wesley Salmon (2001) contains writings on the topic by several of those just mentioned, including Grünbaum, plus J. O. Wisdom, Paul Benecerraf, James Thomson, and G. E. L. Owen.

Many paradoxes involving infinity have been invented since the time of Zeno. Generally credited to Hilbert, the story of an imaginary hotel with an infinity of rooms has been told numerous times in a variety of versions. The essentials are that Hotel Infinity has an infinity of rooms, and even with an infinity of guests, can always find room for more. If a new guest arrives after infinitely many have already been booked in, the clerk simply moves the guest that is in Room 1 into Room 2, the one that was in Room 2 into Room 3, and so on, thus making a vacant room for the new guest. If an infinity of new guests wish to sign in, the clerk moves each of the already-booked guests into a room the number of which is twice the number of his or her existing room—from Room 1 to Room 2, from 2 to 4, from 3 to 6, and so on, which leaves an infinity of odd-numbered rooms for the infinity of new guests. And so the story goes.

The following paradox is described by Ross (1976). Imagine an infinitely large urn and an infinite collection of balls numbered 1, 2, 3, ... At one minute before 12, balls 1 through 10 are placed in the urn and number 10 is removed. At ½ minute before 12, balls 11 through 20 are tossed in and number 20 is removed. At ¼ minute before 12, balls 21 through 30 go in and number 30 comes out. And so on indefinitely. How many balls will be in the urn at 12 o'clock? The answer is an infinite number, inasmuch as any ball whose number is not some multiple of 10 will still be in the urn. But suppose the in-and-out rule is changed slightly so that when balls 1 through 10 go in, number 1 comes out; when 11 through 20 go in, number 2 comes out; when 21 through 30 go in, number 3 comes out; and so on. Now it appears that the urn is empty at 12 o'clock, because, for any ball, one can say at precisely what time it was removed from the urn. This paradox is discussed also by Paik (1983).

Falk (1994) presents a different version of the problem, which shows that there is still a paradox even if one considers only the case in which balls are tossed out in numerical sequence.

An infinite line of tennis balls, numbered 1, 2, 3, ..., is arranged in front of an empty room. Half a minute before 12 o'clock, balls 1 and 2 are tossed into the room and 1 is thrown out. A quarter of a minute before 12:00, balls 3 and 4 are tossed in and 2 is tossed out. In the next 1/8 minute balls 5 and 6 are tossed in and 3 is tossed out, and so on. The question is how many balls will the room contain at 12:00? Cogent arguments can be made for two diametrically different answers (p. 44):

First answer. There will be infinitely many balls in the room at 12:00. Argument 1: The number of balls in the room increases by one at each tossing event. Hence, for any N you suggest, I can compute an exact time (before 12:00) when the number of balls in the room exceeded N.

Second answer. There will be no ball in the room at 12:00. Argument 2: If you claim that there is any ball there, when you name it, I can tell you the exact time (before 12:00) when it was tossed out.

To resolve the paradox, Falk uses the concept of a limit of an infinite sequence and distinguishes between two possible limits—the limit of the *number* of balls in the room and the limit of the sequence of *sets* of balls in the room. Regarding the definition of the limit of an infinite sequence of sets, Falk cites Ross (1976, pp. 38–39) and Shmukler (1980, pp. 13–14). She argues that both answers are legitimate, that one can interpret the original question either way, and that the answer one gets from a formal analysis depends on which interpretation is used (the number interpretation yields infinity, whereas the set interpretation yields 0).

The basic problem, from Falk's point of view, is psychological:

> In the absence of an authoritative criterion to tell us which is the correct choice, we have to decide which of the two formal interpretations of the problem we endorse. That decision is hard because both interpretations make sense and we are disturbed by the disparity between their conclusions. We face a psychological difficulty inasmuch as we intuitively expect the limit of the cardinal numbers of the sets to equal the cardinal number of the limiting set. The point of difficulty is that cardinality, as a function of sets, is "discontinuous at infinity in the sense that the values of the function are ever increasing but its value at the limit point is zero" (Paik, 1983, p. 222). (Falk, 1994, p. 49)

Falk notes that one can react to this analysis in either of two ways: "to question the acceptability of the definition suggested by Ross (1976) and Shmukler (1980) for the limit of an infinite sequence of sets" (p. 49) or "to admit the existence of a contradiction between the answers according to two interpretations, and to understand that the source of the 'trouble' is that the same question, as phrased in natural language, can be translated into two different (mathematical) questions in formal language" (p. 49).

There are many paradoxes involving infinity in geometry. One such—the Koch curve—involves a shape that has finite area but an infinite perimeter. The curve was described by Swedish mathematician Niels Helge von Koch in 1906. Sometimes referred to as a snowflake curve, the Koch curve is constructed as follows. Start with an equilateral triangle. Replace the middle third of each side with two sides of an equilateral triangle, the base of which is the third of the side that is being replaced. On each successive step repeat this process on each of the straight-line edges of the figure. Figure 8.1 shows the original triangle and three successive applications of the transformation rule.

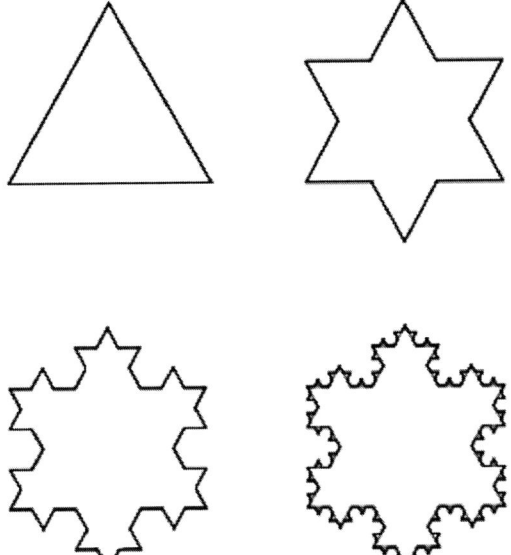

FIGURE 8.1 The first four steps in producing a Koch "snowflake" object, which is a finite area with an infinite perimeter.

It is easy to see that each application of the rule increases the length of the perimeter by one-third. Kasner and Newman (1940, p. 346) give a proof that in the limit the perimeter is infinite. Not only does the perimeter of the Koch curve become infinite in the limit, so does the distance between any two points on the curve; this follows from the length of the entire perimeter, and consequently the distance between any two points on it, increasing by one-third at each iteration.

Perhaps even more surprising than the existence (conceptually) of finite areas with infinite perimeters is the existence (again conceptually) of three-dimensional objects with finite volume but infinite surface area. Rotation of the function $y = 1/x$, $1 \leq x$, about the x axis produces a trumpet-shaped figure—a hyperboloid; it is known both as Gabriel's horn and as Torricelli's trumpet, the latter after its discoverer, 17th-century Italian physicist-mathematician Evangelista Torrecelli. The volume of the figure is finite—in fact it is πx—but its surface area is infinite, which means that a horn of this shape could not hold enough paint to cover its surface.

Draw a triangle on a piece of paper (Figure 8.2, left). (The point I wish to make is usually illustrated with an equilateral triangle, but any triangle will do as well.) Construct an enclosed triangle by joining the

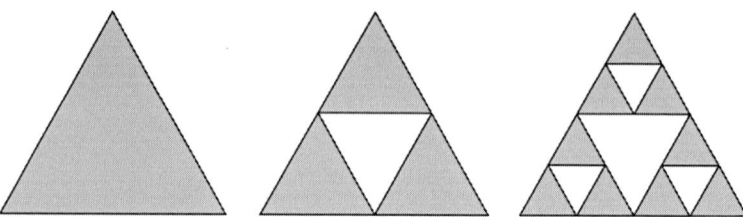

FIGURE 8.2 Illustrating the construction of a triangle with central 1/4 removed, and the process repeating by removing the central 1/4 of the remaining nine triangles.

midpoints of the original triangle's sides as shown in Figure 8.2, center. By this construction you have divided the original triangle into four congruent triangles, each of which represents 1/4 of the original triangle's area. Imagine that you started with the original triangle colored gray, and that now you make the interior triangle white, signifying its removal. You are left with three gray triangles, each of which is a miniature replica of the original one; and, in the aggregate, the three have 3/4 of the original area. Now treat each of these triangles as you did the original one: Construct an enclosed triangle by connecting the midpoints of its sides and whiten the enclosed triangle to signify its removal, as shown in Figure 8.2, right. You now have left nine smaller triangles, which, in the aggregate have 3/4 the area of the three with which you began the second cycle and $(3/4)^2$ or 9/16 the area of the triangle with which you started.

Obviously this process can be continued indefinitely, conceptually. At every step you remove 1/4 of what remains of the original triangle at that point, so after n steps what remains of the original area is $(3/4)^n$. As n increases, this number shrinks quite rapidly and, in the limit, approaches zero, but the perimeter of the holes is infinite. Figure 8.3 shows the result

FIGURE 8.3 The result of application of the area-extraction algorithm six times.

of applying the area-extraction algorithm six times. The remaining area is $(3/4)^6$, or about .18 of the original triangle.

The shape produced by this process is known as Sierpinski's gasket or Sierpinski's triangle, after Polish mathematician Waclaw Sierpinski, who first described it in 1915. It is representative of many fractal shapes (Mandelbrot, 1977) that are constructed by iteratively performing a deletion operation on segments of a figure produced by a preceding application of the same operation. Another such shape described by Sierpinski, and known as Sierpinski's carpet, is constructed by dividing a square into nine equal squares, removing the central one, and then repeating the division and removal operation on each of the remaining squares at each step in the process (see Figure 8.4). The area that remains after n steps is $(8/9)^n$ of the original area. Austrian mathematician Karl Menger described a three-dimensional generalization of Sierpinski's carpet (Menger's sponge) in which successive removal of a fixed proportion of the remaining volume of a cube at each step yields, in the limit, a structure with no volume but infinite area surrounding its holes.

All of these, and other similar fractals, may be seen as generalizations of the simplest illustration of the process, which is its application to a line segment. In this case the center 1/3 of the line is removed, and in the next step each of the end segments is treated as the original line and the center 1/3 of it is removed, and so on. This fractal is known as a Cantor set.

I have argued that intuition plays a critical role in the development and the understanding of mathematics. How does one reconcile these and similar paradoxes with that claim? Is there anything more counter-intuitive than the claim of the existence of a finite area with an infinite perimeter, or a three-dimensional shape with a finite volume and an infinite surface area? One accepts these things, if one does, because one has the choice of either accepting them or rejecting the mathematics that yields them. A person who has no great confidence in the mathematics may well reject them—perhaps should reject them. But one who has

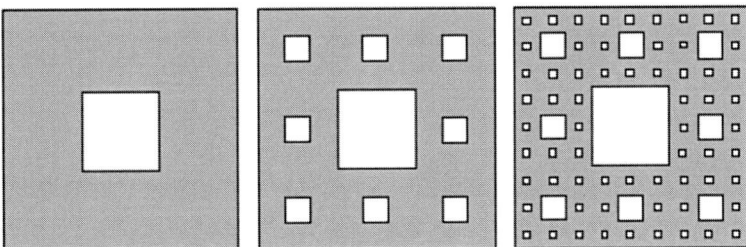

FIGURE 8.4 Sierpinski's carpet, three steps.

confidence in the mathematics may find it more acceptable to educate one's intuition, so the paradoxical result no longer is seen as paradoxical, than to toss out the mathematics.

There is an elegant paradox in the belief that any knowledge about infinity is beyond the capability of human reason: "If we cannot come to know anything about the infinite, then, in particular, we cannot come to know that we cannot come to know anything about the infinite; if we cannot coherently say anything about the infinite, then, in particular, we cannot coherently say that we cannot coherently say anything about the infinite. So if the line of thought above is correct, then it seems that we cannot follow it through and assimilate its conclusion. Yet that is what we appear to have done" (Moore, 2001, p. 12).

☐ Types of Infinity

We have noted that the ancient Greeks had some difficulties with the concept of infinity. Aristotle came to terms with it by making a distinction between the *potentially* infinite and the *actually* infinite and resolving to recognize the reality of only the former. The distinction between potential and actual infinity motivated a great deal of discussion and theorizing for many centuries following Aristotle and indeed to the present day.

In one elaboration of the distinction Moore (2001) partitions concepts relating to infinity into two clusters. "Within the first cluster we find: boundlessness; endlessness; unlimitedness; immeasurability; eternity; that which is such that, given any determinate part of it, there is always more to come; that which is greater than any assignable quantity. Within the second cluster we find: completeness; wholeness; unity; universality; absoluteness; perfection; self-sufficiency; autonomy" (p. 1). Moore notes that the concepts that comprise the first cluster, which he calls *mathematical infinity*, are more negative and convey a sense of potentiality, whereas those in the second cluster, which he associates with *metaphysical infinity*, are more positive and convey a sense of actuality. He contends that the first cluster is likely to inform more mathematical or logical discussions of infinity, while the second is likely to inform more metaphysical or theological discussions of the topic.

Regarding the claim that Aristotle abhorred the idea of the infinite, Moore (2001) argues that "what he abhorred was the metaphysically infinite, and (relatedly) the actual infinite—a kind of incoherent compromise between the metaphysical and the mathematical, whereby endlessness was supposed to be wholly and completely present all at once. It was the mathematically infinite that he was urging us to take seriously. Properly understood, the mathematically infinite and the potentially infinite were, for Aristotle, one and the same. Far from abhorring the mathematical infinite, he was the first philosopher who seriously championed it" (p. 44).

It is not the case, however, that mathematical infinity and potential infinity are the same for everyone. Indeed, the question has been raised as to whether mathematical infinity should itself be considered actual or only potential. There seems to be fairly general agreement that there is no such thing as infinity in the physical world; as Bernstein (1993) puts it, "True infinities never, as far as we know, occur in nature; and if a theory predicts them, it can be taken as an indication that the theory is 'sick' or, at the very least, is being applied in a regime where it is not applicable" (p. 86). But the debate about the actuality of mathematical infinity continues. "In mathematics no other subject has led to more polemics than the issue of the existence or nonexistence of mathematical infinities" (Rucker, 1982, p. 43).

Gauss was unwilling to consider mathematical infinity to be an actuality. "I protest against the use of an infinite quantity as an actual entity; this is never allowed in mathematics. The infinite is only a manner of speaking in which one properly speaks of limits to which certain ratios can come as near as desired, while others are permitted to increase without bound" (quoted in Clegg, 2003, p. 78).

Falk and colleagues (Falk & Ben-Lavy, 1989; Falk, Gassner, Ben-Zoor, & Ben-Simon, 1986) recognize the distinction between potential and actual infinity. Potential infinity is represented by an unending process, seen in the realization that one can increase numbers indefinitely by always adding 1 to any number, no matter how large it is. The set of numbers, in their view, constitutes an actual infinity. Falk and colleagues argue that comprehension of the infinitude of numbers requires three insights: everlasting process (that the process of increasing numbers is interminable), boundless amount (that the set of numbers is actually infinite), and immeasurable gap (that the gap between an infinite set and any finite set is itself infinite, no matter how numerous the finite set is). Falk (in press) presents evidence that by the age of 8 or 9, children generally have acquired the first two insights—they understand that numbers continue indefinitely and that they comprise an infinite set—but that appreciation of the immeasurable gulf between infinity and any finite set does not come until later, if at all.

Lakoff and Núñez (2000) also recognize the distinction between potential and actual infinity, and describe the latter as a metaphorical concept. Potential infinity they see as illustrated by imagined unending processes such as building a series of regular polygons with more and more sides, or writing down more and more decimals of an irrational number like $\sqrt{2}$. Actual infinity, in contrast, is a metaphorical concept, and as such, it allows us to treat a process that has no end—no final result—as though it did have an end and a final result. "We hypothesize that all cases of actual infinity—infinite sets, points at infinity, limits of infinite series, infinite intersections, least upper bounds—are special cases of a single general conceptual metaphor in which processes that go on indefinitely are conceptualized as having an end and an ultimate result" (p. 158). This metaphor, which Lakoff and Núñez refer to as the basic metaphor of infinity (BMI), plays a very prominent role in their treatment of several key mathematical concepts and developments.

An alternative view of the current status of the distinction between potential and actual infinity is given by Barrow (2005). As we have already noted, before the discovery of geometries other than that of Euclid, which was generally considered to represent physical reality, "existence" was more or less equated with physical existence. But subsequent to the discoveries—or inventions—of non-Euclidean geometries, mathematical existence gradually came to be taken to mean no more (and no less) than logical self-consistency. So in this sense, infinity—infinities, thanks to Cantor—could be seen as actually existing, mathematically if not physically. Cantor himself distinguished three types of infinity: *physical* infinity, existent in the physical universe; *mathematical* infinity, existent in the mind of man; and *absolute* infinity, the totality of everything, existent only in the mind of God.

Czech theologian-philosopher-mathematician Bernhard Bolzano (1851/1921) argued for the acceptance of the idea of actual infinity and introduced the scandalous notion (Kasner and Newman [1940, p. 44] call it the "fundamental paradox of all infinite classes") that a part of an infinite collection can be as numerous as the whole of it. German mathematician Julius Richard Dedekind and Cantor built on this foundation.

☐ Infinity and Numbers

Most of us, I suspect, tend to think of infinity as a very large number. And for many applications of the concept, such as its use in limit theorems, it does not inhibit our understanding of a relationship if we think of what happens when some quantity is allowed to become arbitrarily large, as opposed to thinking of it being infinite. But in fact infinity is not a very large number; it is not a number at all, and such phrases as "approaching infinity," "an almost infinite number," and "nearly infinite in extent" are contradictions in terms. Think of the largest number you can imagine. How close is this to infinity? Not close. And it does not matter how large this number is. A googol, so named in 1938, it is claimed, by a nine-year-old nephew of American mathematician Edward Kasner, is 10^{100}. A googolplex is 10 raised to the googolth power: $10^{10^{100}}$. This is a very large number indeed, but larger ones have been expressed. The number $e^{e^{e^{79}}}$, which is approximately equal to and generally represented by $10^{10^{10^{34}}}$, was used by South African mathematician Stanley Skewes in 1933 in a proof regarding the distribution of prime numbers and has been known since as Skewes's (often Skewes or Skewes') number. Graham's number, larger still—large enough to require special notation to be expressed—was once held by Martin Gardner (1977) and the editors of the *Guinness Book of World Records* (1980) to be the largest number ever used in a serious mathematical proof. In the interim even larger numbers have been used. The important point for present purposes is that none of these unimaginably large numbers is close to infinity. No matter what one does to increase the size of the largest number that one can conceptualize or represent, and no matter how large that number becomes, it gets no closer to infinity than the humble 1; between it—our largest number—and infinity there will remain a gulf of infinite extent, and there is nothing one can do to decrease it.

The same point may be made with the observation that *every* number is closer to 0 than to ∞. Consider any number, X. This number is X units from 0. Given X, one can specify another number, Y, that is more than twice as large as X, say $3X$ or $1,000X$. For all such cases, $X < Y - X$, which is to say that X is closer to 0 than to Y and therefore is closer to 0 than to ∞. By similar reasoning one could argue that no number is closer to infinity than any other, or that every number is infinitely far from infinity. In fact, however, the very term *closer to infinity* is contradictory; it makes no sense to describe a point as being close to something that has no location.

Or think of it this way. Presumably there is some number, call it X (of course no one knows what it is), that is the largest number that has ever been, or that ever will be, expressed by a human being. If it were possible to select a number at random from all possible numbers (it is not), the probability of selecting a number smaller than X is essentially 0. In other words, the probability that a number selected at random from all possible numbers would be within the range of all numbers expressed by human beings is 0, so miniscule is that range relative to infinity.

Slote (1986/1990) poses a question that relates to these ruminations in a discussion of rational dilemmas. Imagine a wine connoisseur who has been condemned to an infinite life with only finitely much of his favorite wine. For how many bottles should he ask? The point of the story is to illustrate the possibility of a rational dilemma, because no matter what number is given, one may wonder why it was not bigger. But one might also argue that it really does not matter what the number is; for any finite number it will be the case that an infinite time will be spent without wine.

Imagine participating in a contest in which a very desirable prize is to be given to the contestant who writes the largest integral number, and assume that the number can be written in any interpretable fashion—as a name, a string of digits, in exponential form. As a contestant, what number should one write? No matter what number one writes, one knows that there are infinitely many that are larger than it. The concept of "the largest number that one can think of" is strangely frustrating. "In trying to think of bigger and bigger ordinals, one sinks into a kind of endless morass. Any procedure you come up with for naming larger ordinals eventually peters out, and the ordinals keep coming" (Rucker, 1982, p. 69).

Sometimes one sees references to numbers "selected at random." Such statements require qualification, inasmuch as it is not possible to select a number at random from the infinite set of all possible numbers. Any random selection must be from a finite set, so to select a number at random must mean to select at random a number between X and Y, the values of X and Y being either explicit or assumed. (Suppose one were to claim to have selected a number—a positive integer, to be specific—from the infinite set of positive integers. No matter what integer one selected, there would be an infinite number of larger positive integers but only a finite number of smaller ones, and this is inconsistent with the idea of random sampling from a set.) Ignoring this proviso and proceeding as though random selection from an infinite set were possible leads to paradoxes such as Lewis Carroll's obtuse angle problem (Falk & Samuel-Cahn, 2001).

When we use the expressions "as n approaches infinity," "letting k go to infinity," and the like, we perhaps should qualify them with "so to speak" to remind ourselves that what we really mean is that we are imagining n and k becoming indefinitely large, but that no matter how large they become, they will be no closer to infinity than when they are ever so small. This being said, I need to recognize an important observation made by Falk (1994) in a very insightful discussion of infinity as a cognitive challenge. Noting that "almost infinite" is a self-contradictory expression, she points out that there is a practical sense in which it can be meaningful. She cites an explanation by Asimov (1989) that Newton's theories of motion and gravitation would have been absolutely right only if the speed of light were infinite, but they were very nearly right in the sense that the error in the time required for light to travel a given distance was very small. "Thus if light traveled at infinite speed, it would take light 0 seconds to travel a meter. At the speed at which light actually travels, however, it takes it 0.0000000033 seconds. That is why Einstein's relativistic equations, taking the finite speed of light into consideration, offered only a slight correction of Newton's computations" (p. 56, footnote 1).

Many assertions in mathematics apply to all numbers. We may say, for example, that every integer is either even or odd, or that every integer is either prime or nonprime. But what should such statements be taken to mean, given that it is not possible to produce all integers or even all integers of a given type (odd integers) of which there are infinitely many? In what sense does a very large number that has never been written or thought exist? Is not the idea of an infinite set itself a contradiction in terms? There is no way to identify all the members of such a set. The notion of "all the members" seems not to apply. We can list as many of the members of such a set as we wish, but no matter how many we list, the number of unlisted members will still be infinite. How can anything infinite exist in a finite universe? The reader will think of many other questions of this sort that could be—and probably have been—raised.

☐ Modern Conceptions of Infinity

> The mathematics of the infinite is a sheer affirmation of the inherent power of reasoning by recurrence. (Kasner & Newman, 1940)

What are we to make of the concept of infinity? It seems to be totally beyond our comprehension. The reward for a few minutes' pondering it can be a feeling of abject intellectual inadequacy. One is hardly surprised to learn that the ancients disliked the concept, and that it has continued to frustrate thinkers throughout the ages and to the present time.

Is the characterization of something as infinite—in extent, in number, in duration—simply an admission, as Thomas Hobbes believed, of our inability to conceive of it? Is the claim that the infinite exists a misuse of language, as Wittgenstein insisted? Is infinity a concept that can be grasped by the intellect but not by the imagination, as Leibniz contended? What does it mean to say that it can be grasped by the intellect? Is it the case, as Moore (2001) contends, that "anything whose existence we can acknowledge we are bound to recognize as finite, on pain of contradiction and incoherence"? (p. 217).

Is infinity one concept or several? To make sense of it must we distinguish among physical infinity, metaphysical infinity, mathematical infinity, and perhaps other types? Can there possibly be anything that is infinite in nature? Is the universe itself infinite in space and time? Is there any way of knowing? Is mathematical infinity anything more than a concept invented by mathematicians to enable certain types of computation, a fiction sustained by its mathematical usefulness? In what sense, if any, might it be held to be real? Is an understanding of the concept of infinity simply beyond finite minds, so that attempts to deal with it are bound to end in a quagmire of incoherence?

Austrian philosopher-logician-linguist Ludwig Wittgenstein argued compellingly that the meanings of words are determined by their uses; this is of questionable help in the case of *infinity* because it was, and is, used by different users in different ways. Wittgenstein held that some of the difficulties with the concept stemmed from careless use of language, as, for example, when one uses language to refer to an infinite collection as though it existed in nature rather than only conceptually.

Despite the attention and ink that have been spent on such questions, answers remain elusive. As Moore (2001) puts it, "The same old puzzles and preoccupations are as relevant as they ever were to discussion of the infinite. A survey of the current literature reveals a continuing concern with all the perennials: the distinction between the actual infinite and the potential infinite; the relationship between the infinite and time; Zeno's paradoxes; the paradoxes of thought about the infinite; and so forth" (p. 142). Moore speaks dismissively of the view that the last word regarding what infinitude really is has been spoken in the definition of infinity "as a property enjoyed by any set whose members can be paired off with the members of one of its proper subsets" (p. 198).

Nevertheless, we seem to be able to accept infinity more easily today than did people in previous centuries. That may be in part because the idea has become somewhat more clearly understood as a consequence of the work of Bolzano, Dedekind, and Cantor, among others. It also may be that we simply *accept* the idea without thinking very deeply about it.

Whatever the reason, and despite the difficulties the concept of infinity has caused thinkers for millennia, we use it quite matter of factly and effectively in even relatively simple mathematics. When we use it, we typically consider a trend that can be seen when we let the values of an index variable range over a few numbers, 1, 2, 3, 4, …, and then make the, one might say unconscionable, leap to a conclusion regarding what will happen as this number is allowed to increase indefinitely. And even though we know that nothing "approaches infinity," we act, when computing, as though values do, and we seem to be able to solve problems just the same.

☐ Sets, Subsets, and One-to-One Correspondence

Scholars sometimes saw in the peculiarities of infinity the basis for conclusions about the nature of the physical world. For example, an argument was made during the 13th century, perhaps first by the Franciscan cleric St. Bonaventure, that the world cannot have existed from eternity past because if it had, the number of the moon's revolutions around the Earth and the Earth's revolutions around the sun would both be infinite, but the moon would have revolved 12 times as frequently as the Earth (Murdoch, 1982). This argument was anticipated by seven centuries by the Alexandrian scholar John Philoponus, who observed that, however old the world is, the number of days it has existed is 30 times the number of months, which means that if the world had always existed, one infinity would be greater than another, which he considered to be absurd.

The idea of infinities that, from one point of view, appear to differ in size, but whose items can be put into one-to-one correspondence, has perplexed thinkers for a long time. In the sixth century, Philoponus also pointed out that if there are infinitely many even numbers, there must be as many even numbers as odd and even numbers combined, and this he saw as justification for rejecting the idea of countable infinities. Thomas Bradwardine, English scholar and cleric (once Archbishop of Canterbury), made a similar observation in the 14th century.

In the 16th century, Galileo observed that inasmuch as *every* integer can be squared, there must be as many squares as there are integers. This observation had the surprising implication that an inexhaustive subset of the set of all integers (the subset of integers that are perfect squares) is as numerous as the entire set. Not only are the squares an inexhaustive subset of the integers, they are an infinitesimally small subset because, as Galileo also noted, the proportion of the first n integers that are squares gets increasingly close to 0 as n increases. Galileo held that the infinite is beyond the finite understanding of humans and that the difficulties we have with it come from applying to it concepts that are appropriately applied only to finite entities.

Similar observations were made with respect to other sets and subsets thereof. Consider the simple equation $y = 2x$, where x and y need not be integers. This says that for every x there is a y that has twice the value of that x, or conversely that for every y there is an x that has half the value of that y. Imagine the set of all ys that have values between, say, 0 and 2. According to the equation, there must be, for each of these ys, an x that has a value between 0 and 1. It follows that there must be as many values between 0 and 1 as there are between 0 and 2, even though the second interval is twice as large as the first.

Many illustrations of the possibility of putting the items of an infinite set into one-to-one correspondence with a proper subset of itself were noted over the long period of time that the concept of infinity was taking shape, and that illustrate the truth of a quip by Hoffman (1998) that "in the realm of the infinite, things are often not what they seem" (p. 236), and Seife's (2000) definition of the infinite as "something that can stay the same size even when you subtract from it" (p. 149).

In 1872 Dedekind defined infinity as follows: "A system S is said to be *infinite* when it is similar to a proper part of itself; in the contrary case S is said to be a finite system." Boyer and Merzbach (1991) note that this definition may be rephrased in somewhat more modern terminology as "a set S of elements is said to be infinite if the elements of a proper subset S' can be put into one-to-one correspondence with the elements of S" (p. 566). This is the definition that Cantor also used in his classic work on the concept.

Dedekind's *defining* infinity in terms of the property of being able to put a set into one-to-one correspondence with another set of which it is a subset was a stroke of genius. It resolved the problems by defining them out of existence. Given this definition, it is possible to make the subset as small a fraction of the original set as one likes and still have the subset be infinite. Consider, for example, the function $y = 1,000,000x$, where x and y are integers. For every value of x there is a value of y, despite that x can have any integer value on the number line, whereas y can have only one value in every million. If one wants a sparser subset, one may make the ratio of y to x as large as one pleases. As long as the two sets can be put into one-to-one correspondence, as they can when $y = ax$, no matter what the value of a is, then if x is infinite, y is also. The same is true of any single-valued function of x, such as x^a, a^x, or x^x, despite that with such functions the number of integers between successive values of the function increases without limit.

These ideas about infinity lead to such conclusions as that any two line segments, regardless of length, have the same number of points, and, in particular, that even the smallest line segment imaginable has as many points as a line of infinite length. A well-known demonstration that two lines of different lengths have the same number of points is shown in Figure 8.5. The circumference of the outer circle is larger than that of the inner circle, but any radius drawn on the former will intersect the latter at a unique point, so there must be as many points on the smaller circle as on the larger. A diagram of this sort was used by 13th-century Scottish theologian and philosopher Duns Scotus to support the contention that lines are not composed of an infinity of infinitesimal points, because if they were, the inner and outer circles—having the

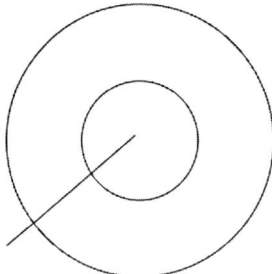

FIGURE 8.5 Every line from the origin through the larger circle also passes through the smaller one, demonstrating that for every point on one circle there is a corresponding one on the other.

same number of such points—should be equal in circumference, which they manifestly are not. Galileo resolved this conundrum to his own satisfaction by assuming that the larger circle had gaps between its points that the smaller one did not.

Today it is generally held not only that all line segments have an equal number of points, but that every line segment has the same number of points as does a two-dimensional plane, or a three-dimensional volume. Even though there are an infinite number of points representing rational numbers on a line segment of any length, the proportion of the total number of points on the segment that represent rational numbers is infinitesimally small. Between any two rationals, no matter how close they are, there are infinitely many irrationals, and between any two irrationals there are infinitely many rationals. And so on. Thinking about it makes one's head spin. It seems the more precisely infinity is defined, the more strange the concept becomes. What could illustrate more clearly the insightfulness of a quip by Wallace (2003): "It is in areas like math and metaphysics that we encounter one of the average human mind's weirdest attributes. This is the ability to conceive of things that we cannot, strictly speaking, conceive of" (p. 22).

☐ Enter Georg Cantor

This is not the end of the story with respect to infinity, of course. One might assume that, given Dedekind's definition, or the modern restatement of it, it should be possible to put any infinite set in one-to-one correspondence with any other. But the definition had barely been written down when German mathematician Georg Cantor showed that this is not the case. The real numbers, for example, cannot be put in one-to-one correspondence with the integers (see Chapter 5). In this respect, the set of reals, which includes irrationals such as e, π, and $\sqrt{2}$, differs from the set of rationals.

The difference illustrated by the possibility of putting the rational numbers into one-to-one correspondence with the integers and the impossibility of doing so with the reals led Cantor to note that some infinite sets are *countable* or *denumerably infinite*—can be put into one-to-one correspondence with the integers—and some are not, and to distinguish infinite sets of different *powers* or *cardinality*. (For some years Cantor had believed, and tried in vain to prove, that all infinite sets are countable.) He argued that the set of rationals or the set of integers should be considered the same power, but that the set of reals should be considered a higher one, although the set was infinite in each case. He went on to establish that there are infinitely many powers of infinite sets, which is to say there is no largest infinite set—given any infinite set, it is possible to describe a larger one.

Suppose there were a largest infinite set, perhaps the set of all sets. Cantor showed that the set of all subsets of any given set A (which would be called the power set of A, or $P[A]$) contains more members than does the set A. From this it follows that there is no largest set, because, given any set, no matter how large it is, there is a larger one, namely, the set of all its subsets. Suppose there is a largest set—call it S_L; its power set, $P[S_L]$, is larger than S_L. Moreover, $P[S_L]$ is itself a set and its power set, $P[P[S_L]]$, is larger than it. And so on, *ad infinitum*. So the supposition that there is a largest set is apparently wrong.

Although the conclusion that there is no largest set seems harmless enough when one first encounters it, it led to a conceptual difficulty involving the set of all sets—the universal set. Moore (1995) describes the problem and Cantor's treatment of it:

> Given Cantor's theorem, this collection [the set of all sets] must be smaller than the set of sets of sets. But wait! Sets of sets are themselves sets, so it follows that the set of sets must be smaller than one of its own proper subsets. That, however, is impossible. The whole can be the same size as the part, but it cannot be smaller. How did Cantor escape this trap? With wonderful pertinacity, he denied that there is any such thing as the set of sets. His reason lay in the following picture of what sets are like. There are things that are not sets, then there are sets of all these things, then there are sets of all those things, and so on, without end. Each set belongs to some further set, but there never comes a set to which every set belongs. (p. 116)

Moore (1991) describes the desire both to and not to admit the existence of a set of all sets as a paradox and contends that the best way to deal with it is to refrain from talking about it. The final resolution of the problem was to axiomatize set theory in such a way as to define the universal set out of existence.

Aczel (2000) notes the correspondence between Cantor's reasoning about the impossibility of the existence of a set containing everything and Gödel's incompleteness theorem. "The impossibility of a set containing everything brought Cantor to the conclusion that there was an *Absolute*—something that could not be comprehended or analyzed within mathematics. Cantor identified the absolute with God ... the impossibility of a universal set, and the unattainable Absolute perhaps lend credence to Gödel's incompleteness principles: there is always something outside, something larger than any given system" (p. 196).

Cantor's work on infinities, which included the discovery of infinities of different sizes (orders of infinities, indeed an infinite order of infinities) and the arithmetic of infinities (transfinite arithmetic), caused more than a little scratching of many heads at the time, including his own. Barrow (2005) contends that Cantor's discovery that there is an infinity of infinities of different sizes and that they can be distinguished unambiguously was one of the greatest discoveries of mathematics, and completely counter to the opinion prevailing when it was made.

Cantor went on to develop rules for doing arithmetic with infinite entities. In transfinite arithmetic the rules for basic operations (addition, subtraction, multiplication, division, exponentiation) differ from the rules that apply to finite arithmetic, and those that apply to transfinite cardinal numbers differ from those that apply to transfinite ordinals. Cantor's impact on the world of mathematics was to many perplexing if not deeply troubling. "It was Cantor's greatest merit to have discovered in spite of himself and against his own wishes in the matter that the 'body mathematic' is profoundly diseased and that the sickness with which Zeno infected it has not yet been alleviated" (Bell, 1937, p. 558).

Cantor speculated, but was unable to prove, that there does not exist an infinite set larger than the counting numbers but smaller than the reals. His inability to prove, or disprove, the conjecture—it is claimed that at various times he thought he had done the one or the other—was a major disappointment to him. The conjecture became known as the *continuum hypothesis* and appeared as the first of the 23 problems that Hilbert identified as the major unsolved problems in mathematics as of 1900. The continuum hypothesis continues to be a significant mathematical challenge. Cohen (1963, 1964) showed that whether it is true or false depends on the assumptions with which one begins, which means that it is independent of the axioms of set theory and thus can be treated as an additional axiom that one is free to accept or reject.

Although Cantor is venerated today as an exceptionally original thinker, his work was severely criticized by some of the mathematical luminaries of the day, notably German mathematician–logician Leopold Kronecker—who believed that reality was represented by the integers and detested the very idea of infinity—and he died in a mental hospital after spending his last years clinically depressed. The cause of Cantor's illness, which came and went, has been the topic of much speculation, and it remains unknown. What is known is that his bouts of depression tended to coincide with periods when he was thinking about his continuum hypothesis. While it would be wrong to infer cause and effect from this fact, one cannot help but wonder. Recent accounts of Cantor's work on infinity that are accessible to the general reader include those of Love (1989), Aczel (2000), and Clegg (2003). For his own groundbreaking articles published in 1895–1997, see Cantor (1911/1955).

☐ One-to-One Correspondence Versus Same Size

What does it mean for two infinite sets to be the same size, or for one to be larger than another? Rucker (1982) notes that the fact that an infinite set can have the same cardinality as a proper subset of itself was so puzzling to pre-Cantorian thinkers that "they generally believed it was hopeless to attain a theory of infinite cardinalities much more sophisticated than: 'All infinities are equal'" (p. 230).

Cantor argued that two infinite sets should be considered the same size if the elements of one can be put into one-to-one correspondence with those of the other. Sometimes referred to as the "correlation criterion" for size comparisons, this is the fundamental assumption, or definition, that underlies the transfinite mathematics that he developed. Not everyone accepted, or accepts, the one-to-one matching procedure as *the* way to compare the sizes of infinite sets. An alternative approach is to consider whether one of the sets is a proper subset of the other and, if it is, to conclude that the set is larger than the subset on the grounds that when the subset is removed from the set, some elements still remain in the latter. Leibniz acknowledged that each even number could be paired with each natural number, but he did not see this as a basis for concluding that there are as many even numbers as natural numbers.

Lakoff and Núñez (2000) argue that according to our usual concept of "more than," there are more natural numbers than even numbers; they resolve the problem by making a distinction between the concepts of *pairabilty* and *same number as*. They object to the characterization of what Cantor did as proving that there are just as many even integers as natural numbers. "Given our ordinary concept of 'As Many As,' Cantor proved no such thing. He proved only that the sets were pairable. In our ordinary conceptual system, there are *more* natural numbers than there are positive even integers.

It is only by use of Cantor's metaphor that it is correct to say that he proved that there are, metaphorically, 'just as many' even numbers as natural numbers" (p. 144). In effect, Cantor *defined* same size in terms of one-to-one correspondence. Lakoff and Núñez contend that Cantor "intended pairability to be a *literal* extension of our ordinary notion of Same Number As from finite to infinite sets" (p. 144). In this, they argue, Cantor was mistaken; it is a metaphorical extension only. "The failure to teach the difference between Cantor's technical metaphorical concept and our everyday concept confuses generation after generation of introductory students" (p. 144).

Byers (2007) makes a similar argument in contending that the notion of equality is ambiguous. According to the notion that two sets are equal in number if the items of the two sets can be put into one-to-one correspondence, the set of squares is equal to the set of counting numbers. But if two sets are considered to be equal if and only if they have identical elements, then the set of squares is not equal to the set of counting numbers, because the latter contains elements that the former does not.

There is much more to the story of the concept of infinity as it has evolved in mathematics than this brief discussion conveys. What is important to note for present purposes is that the concept has indeed evolved, and is still evolving. It has perplexed more than one capable thinker in the past and is likely to remain an intellectual challenge for a long time to come. Despite this, the concept has proved to be an immensely useful one in mathematics, both pure and applied. Indeed, *useful* is not sufficiently strong to describe its importance to much of modern mathematics; *essential* is a more appropriate word: "Without a consistent theory of the mathematical infinite there is no theory of irrationals; without a theory of irrationals there is no mathematical analysis in any form even remotely resembling what we now have; and finally, without analysis the major part of mathematics—including geometry and most of applied mathematics—as it now exists would cease to exist" (Bell, 1937, p. 522).

Just as different systems of geometry have been developed from different starting definitions and rules, so different concepts of infinity are possible, depending on how one chooses to define one's primitives, and the rules of operating on them. Cantor, working with Dedekind's definition of infinity in terms of the one-to-one correspondence between elements of a set and any of its proper subsets, was able to develop a system of operations that led to the identification of different levels of infinity, but the definition was crucial as the point of departure. Given a different definition, the destination would not be the same.

But one may ask: What is the *real* situation. After all is said and done, are even numbers *really* as numerous as whole numbers? Has Cantor proved this to be the case? He has shown that every even number can be paired with a natural number, and conversely, but should we take this as proof positive that the two sets really are equal in size? This is perhaps a good place to remind ourselves that mathematics, according to the modern view, has no obligation to be descriptive of the real world; it is obliged only to be consistent with the definitions and axioms with which one starts. From this perspective, the question of what the *real* situation is does not arise. The conclusion that the even numbers are as numerous as the whole numbers follows from the criterion for the same number being defined as the ability to be put in one-to-one correspondence. If one uses a different definition, one gets a different result.

☐ Infinite Time

So far, I have mentioned only in passing the concept of infinite time—eternity. It too has its share of paradoxes and mind-numbing puzzles. I have always found it easier to imagine a future that has no end than to imagine a past that has no beginning. Looking forward, I see no great conceptual problem in the idea that the universe, in some form, could go on indefinitely. But looking back, the idea that the universe—or time, for that matter—always existed seems incomprehensible. If it always existed—if it stretches back forever—then it has already existed for an eternity. And if it has existed for an eternity, how did it get to be what it is now? Presumably things are changing and are likely to be different in the future than they are now. But if things have been going on forever, why have they not long since attained whatever state an eternity of evolving would have produced?

After writing the preceding paragraph, I learned that puzzlement of this sort has a history. Moore (2001) refers to a "curious asymmetry" according to which "an infinitely old world strikes us as more problematical than a world with an infinite future, though it is very hard to say why" (p. 91). This curious asymmetry was the basis of the "kalam cosmological argument" that the universe must have had a beginning. If its existence extended to the infinite past, the argument goes, getting to the present would have required crossing an infinite gap, which is impossible. But the assumption that the universe had a beginning does not really solve the puzzle, in my view, because if the universe began at a certain point in time, one still has the question of how *that* point in time came to be, if time itself extended to an infinite past; getting from infinity past to the point at which the universe came into existence would also have required the crossing of an infinite gap. Perhaps the answer is that time itself had a beginning. This of course is one interpretation of the big bang theory of the origin of the universe. According to it, asking what was going on before the big bang makes no sense because there was no before "before" the big bang; neither time nor space existed. This may solve the puzzle for some; it does not quite do it for me.

☐ Acquiring the Concept of Infinity

Falk (1994) reviews evidence that individuals acquire the concept of infinity gradually over several years. The idea that there is no largest number— that no matter what number one gives, someone can always give one that is larger—appears to be graspable, if not spontaneously expressible, by a majority of children by the time they are about five to seven years old (Evans, 1983, cited in Falk, 1994). Falk et al. (1986) found that most eight-year-olds recognized the advantage of being the second person to name a number in a game of two players in which the player who names the larger number wins, and were able to verbalize the idea that no matter how large the number named by the first player, the second player can always name a larger one. Eight-year-old children were likely to consider the natural numbers to be more numerous than the number of grains of sand on earth; younger children were likely to consider the grains of sand to be the more numerous. But many of the children, even as old as 12 or 13, who considered the numbers to be more numerous than the grains of sand, thought the latter to be *almost* as numerous as the former. "Roughly speaking, children of ages 8–9 and on seem to understand that numbers do not end, but it takes quite a few more years to fully conceive, not only the infinity of numbers, but also the infinite difference between the set of numbers and any finite set" (p. 40).

Results obtained in two of the studies reviewed by Falk (1994)— Fishbein, Tirosh, and Hess (1979) and Moreno and Waldegg (1991)—suggest that a large majority of primary and junior high school students are likely to believe that the set of natural numbers (1, 2, 3, …) contains more elements than the set of all even numbers (2, 4, 6, …). Falk found in a study of her own that about 55% of approximately 100 college students who had not taken college courses in higher mathematics also considered the set of natural numbers to be the more numerous of the two.

Anticipating the topic of the next chapter, we may note too that high school students are likely to consider numbers to be infinitely divisible (Tirosh & Stavy, 1996). Smith, Solomon, and Carey (2005) found that many elementary school students (third through sixth grades) also are likely to consider numbers to be infinitely divisible, and to consider physical quantities to be infinitely divisible as well. Children who considered infinite divisibility to be the case in either the domain of numbers or that of matter tended to consider it to be possible in both.

****

This discussion of infinity has focused on the concept primarily as it relates to mathematics. Before leaving the topic, we should note that the concept is also highly relevant to science, especially physics, astronomy, and cosmology. Whether the universe is finite or infinite in either space or time is a question that continues to be debated. The widely accepted view that it had a beginning with the big bang and that it is limited in extent is not universally espoused, and even the big bang theory leaves open the possibility of an infinite sequence of oscillations between expansion and contraction and of the existence of an infinity of universes (small u) comprising the Universe—all there is (multiverse?)—about which we know only what little we can learn from the one in which we live.

One may take the position that theorizing about such matters is not science but metaphysics. And well it may be. But it is no less fascinating for that. And not without practical consequences. The argument has been made often that, in an infinite universe, anything that has nonzero probability of happening—you, for example—will happen infinite times. It is something to reflect upon when you are having trouble getting to sleep. Barrow (2005) provides a thoughtful, and thought-provoking, discussion of this topic, including its relevance to ethics. "Unusual consequences seem to follow," he notes, "if we take seriously the idea that there exists an infinite number of possible worlds which fill out all possibilities" (p. 208). The consequences follow, in large part, because beliefs influence behavior, and what we as individuals believe about the universe in which we live conditions how we treat it, including, importantly, people in it.

CHAPTER

Infinitesimals

> The infinitesimal has a fascinating history. At least as far back as Archimedes, it's been used by mathematicians who were perfectly aware that it didn't make sense. (Hersh, 1997, p. 289)

Equally as problematic as the concept of infinity, or the infinitely large, is the idea of infinitesimals or the infinitely small. Among the ancient Greeks, both Eudoxus of Cnidus and Archimedes of Syracuse used the idea of quantities as small as one wished to find areas and volumes. As noted in Chapter 6, Antiphon, Eudoxus, and Archimedes used the method of exhaustion, which involved determining properties of curved figures by approximating them, ever more closely, with increasingly many-sided polygons. Dantzig (1930/2005) goes so far as to identify Archimedes as the founder of the infinitesimal calculus, and to suggest that the failure of other Greeks to extend his work in this direction was due in part to lack of a proper symbolism and in part to their horror of the infinite. (I have already noted that whether the Greeks really had a horror of the infinite is a debated question.)

Bell (1946/1991) credits Zeno's invention of his paradoxes with being "partly responsible for the failure of the Greek mathematicians to proceed boldly to an arithmetic of infinite numbers, an arithmetical theory of the continuum of real numbers, an analysis of motion, and a usable theory of continuous change generally" (p. 224). The "partly responsible" in this comment reflects that Bell, like Dantzig, considered the lack of an efficient symbolism for representing numbers to be another serious limitation of the time.

The extent to which Greek mathematicians were influenced by the philosophers' arguments about infinity and indivisibles is unclear. One gets the impression from some accounts that many of the most challenging problems with which the mathematicians struggled were first articulated by philosophers. However, there is also the view that the mathematicians were not much influenced by the thinking of the philosophers, but that the problems on which they focused reflected their own autonomous interests (Knorr, 1982). In any case, questions of the infinite and infinitesimals were on the minds of many philosophers and mathematicians in ancient Greece, as well as on those of some theologians in later centuries (Stump, 1982; Sylla, 1975).

It seems intuitively obvious that the number line is infinitely divisible. Given any two real numbers, no matter how close they are, one can find a number (their mean) between them. Inasmuch as the operation of finding a mean between two numbers can be iterated endlessly, it follows that between any two real numbers there are an infinity of real numbers. This leads to the mildly unsettling conclusion that all numbers lack nearest neighbors. If the number line, with these properties, is considered a legitimate analog of space and time—in the sense of distances between spaces or between times being faithfully represented by differences between numbers—then, as we shall see, many questions arise regarding the nature of motion and how it is possible.

☐ Infinite Series

Among the more powerful instruments in the applied mathematician's tool kit, particularly relative to the study and description of continuous change, are infinite series. Especially useful are series that, though infinite, converge on a finite value. Convergence here means that as the number of terms of the series increases, the value of the sum gets increasingly close to an ultimate value. Not all infinite series converge, and distinguishing between those that do and those that do not can be very difficult. Consider the two series,

$$\sum_{n=0}^{\infty} \frac{1}{2^n}$$

and

$$\sum_{n=0}^{\infty} \frac{1}{n+1}$$

TABLE 9.1. The (Approximate) Values of $\sum_{n=0}^{\infty}\dfrac{1}{2^n}$ **and** $\sum_{n=0}^{\infty}\dfrac{1}{n+1}$ **for the First Few Values of** *n*

	\multicolumn{7}{c}{*n*}
	0	1	2	3	4	5	6
$\dfrac{1}{2^n}$	1	0.5	0.250	0.1250	0.0625	0.03125	0.015625
$\sum_{n=0}^{\infty}\dfrac{1}{2^n}$	1	1.5	1.750	1.8750	1.9375	1.96875	1.984375
$\dfrac{1}{n+1}$	1	0.5	0.333	0.2500	0.2000	0.16667	0.142857
$\sum_{n=0}^{\infty}\dfrac{1}{n+1}$	1	1.5	1.833	2.0833	2.2833	2.45000	2.592857

The first of these series, which is a geometric series in that each term in the sum is the same multiple of the preceding term, converges to 2; the second series, generally known as the harmonic series, does not converge, but increases indefinitely as *n* increases. Table 9.1 compares the values of the terms and the sums for the first few values of *n*.

It is fairly obvious even from these few values that the first series is converging to 2. It is not clear from the table that the second series is not converging to any finite value; indeed, it is easy to believe that it is converging given that the difference between successive sums is decreasing. If it is diverging, it is doing so very slowly. In fact, the sum does not exceed 10 until one has added 12,367 terms, and to get to more than 100 requires 10^{43} terms. Nevertheless, if one could continue the series indefinitely, one would find that it does not converge. This series was proved to be divergent by Oresme in the 14th century; the proof, simple and clever, is reproduced by Maor (1987, p. 238). A different proof, published by Swiss mathematician Jakob Bernoulli and credited to his mathematician brother Johann in 1689, is reproduced by Dunham (1991, p. 196).

I have spoken of the number to which a converging series converges as the sum of the series, but does this make sense, given that the series never actually reaches the sum but only gets increasingly close to it? Sometimes use of the term *sum* is justified in the following way. Consider the geometric series $\sum_{n=1}^{\infty} \frac{1}{2^n} = 1$, which can be written as $S = \frac{1}{2} + \frac{1}{4} + \frac{1}{8} + \frac{1}{16} + \cdots = 1$. If we subtract from this $\frac{1}{2}S = \frac{1}{4} + \frac{1}{8} + \frac{1}{16} + \cdots$, we have left $\frac{1}{2}S = \frac{1}{2}$, from which it follows that $S = 1$.

Series with alternating positive and negative terms can converge as well as series with all positive terms. The series $1 - \frac{1}{2} + \frac{1}{3} - \frac{1}{4} + \frac{1}{5} - \ldots$ converges to ln 2, the natural logarithm of 2. Series with alternating positive and negative terms can be tricky, however. The terms of the preceding series can be reordered, for example, to produce a series that converges to $\frac{3}{2}$ ln 2 (see Byers, 2007, p. 142).

The comparison of the geometric and harmonic series illustrates that the appearance of convergence, as evidenced by a series' decreasing rate of growth, is not compelling evidence of actual convergence. To be sure that a series is converging, one needs a formal proof. An interesting aspect of the problem of proving convergence of a series is that it is possible to know that a series converges without being able to determine the value to which it converges. The series

$$\sum_{n=1}^{\infty} \frac{1}{n^2}$$

which we have already seen in Chapter 1, was shown by Jakob Bernoulli to converge to some number less than 2 several decades before Euler proved that it converges to $\frac{\pi^2}{6} = 1.6449\ldots$.

In Chapter 3, some series approximations to π and e were noted in the context of a discussion of numbers. Series approximations have been developed for a great variety of functions, including binomial, exponential, trigonometric, and logarithmic. The trigonometric sine function, for example, is approximated by the power series

$$\sin x = x - \frac{x^3}{3!} + \frac{x^5}{5!} - \frac{x^7}{7!} + \frac{x^9}{9!} - \ldots$$

Series approximations of numerous functions are readily available in engineering handbooks and books of standard mathematical tables.

Although few would question the usefulness of infinite series today, many difficulties were encountered by their early users. Kline (1980) describes those difficulties in some detail. "As Newton, Leibniz, the several Bernoullis, Euler, d'Alembert, Lagrange, and other 18th-century men struggled with the strange problem of infinite series and employed them in analysis, they perpetrated all sorts of blunders, made false proofs, and drew incorrect conclusions; they even gave arguments that now with hindsight we are obliged to call ludicrous" (p. 142). Of course, hindsight makes experts of us all, in our own eyes at least, but only with respect to intellectual struggles of the past; it is of no help while a struggle is going on, let alone about the future.

The kinds of problems encountered are illustrated by a series that was the subject of a correspondence between the Italian Jesuit priest and mathematician Luigi Grandi and Gottfried Leibniz, and became known as the Grandi series, or sometimes the Grandi paradox. Consider the ratio

$$\frac{1}{1+x} = 1 - x + x^2 - x^3 + x^4 - x^5 + \ \dots$$

From the left side of the equation, it is clear that if $x = 1$, the ratio equals 1/2. But the right side gives us

$$1 - 1 + 1 - 1 + 1 - 1 + \dots$$

which can be written as

$$(1 - 1) + (1 - 1) + (1 - 1) + \dots$$

the sum of which apparently is 0. But the original series can also be written as

$$1 - (1 - 1 + 1 - 1 + 1 - 1 + \dots)$$

and grouped as

$$1 - [(1 - 1) + (1 - 1) + (1 - 1) \ \dots]$$

the sum of which appears to be 1. Liebniz, among others, argued that the sum of this sequence should be taken to be 1/2, the mean of the results of the two groupings, and also the answer one would get from just the left side of the original equation. One also gets 1/2 as the answer if one writes the series as

$$\text{Sum} = 1 - 1 + 1 - 1 + 1 - 1 + \dots = 1 - (1 - 1 + 1 - 1 + 1 - 1 + \dots) = 1 - \text{Sum}$$

and Sum $= 1 -$ Sum only if Sum $= 1/2$. The resolution that is generally accepted today is to say that the sequence has no sum.

Kline notes that almost every mathematician of the 18th century made some effort to provide a logical foundation for the calculus, all to no avail. "The net effect of the century's efforts to rigorize the calculus, particularly those of giants such as Euler and Lagrange, was to confound and mislead their contemporaries and successors. They were, on the whole, so blatantly wrong that one could despair of mathematicians' ever clarifying the logic involved" (p. 151).

Kline's indictment of 17th- and 18th-century mathematicians' handling of the concepts of continuity and differentiability, which he calls the basic concepts in all of analysis, is severe: "One can only be shocked to learn how vague and uncertain mathematicians were about these concepts. The mistakes were so gross that they would be inexcusable in an undergraduate mathematics student today; yet they were made by the most famous men—Fourier, Cauchy, Galois, Legendre, Gauss—and also by a multitude of lesser lights who were, nevertheless, leading mathematicians of their times" (p. 161). Again, "from the standpoint of the logical development of mathematics, the principle of continuity was no more than a dogmatic ad hoc assertion intended to justify what the men of the time could not establish by purely deductive proofs. The principle was contrived and invoked to justify what visualization and intuition had adduced" (p. 164). Byers (2007) describes the idea of continuity as one of considerable complexity and one the understanding of which has evolved over many years: "The idea is not a single, well-defined object of thought but a whole process of successively deeper and deeper insights" (p. 239).

These observations prompt two thoughts. First, the rigor by which 19th-century mathematics was distinguished was perhaps, at least in part, a reaction against what was perceived to be the lack of rigor of previous times. Second, what is cognitively very difficult for one generation may be readily accepted, and found to be easy, by a subsequent one; intuitions are malleable and, while familiarity may sometimes breed contempt, it can also facilitate acceptance and assimilation.

It is important to bear in mind that the sum of a convergent infinite series gets increasingly close to its limit as terms are added to the series, but it never actually reaches the limit. For many purposes, it is convenient to treat the sum of a series as though it were equal to the limit, but my sense is that failure to bear in mind that the sum and limit are *not* the same is the basis of considerable confusion.

☐ Zeno Again

Zeno's paradoxes pose a compelling temptation for philosophers; few can resist the urge to comment upon them in some fashion. (Salmon, 2001, p. 42)

In trying to understand the struggles of 17th- and 18th-century mathematicians with concepts relating to the phenomenon of continuous change, it may help to remember that the famous paradoxes invented by Zeno of Elea had stood unexplained for over 2,000 years. (Nor have they yet been resolved to the satisfaction of everyone who has written or thought about them.) Most of what is known about Zeno's paradoxes is based on writings of philosophers (Plato, Aristotle, and others) other than Zeno himself. Of Zeno's own writings, very little survives.

Zeno proposed paradoxes of extension and of motion, both of which involve space and time. Grünbaum (1967) characterizes the paradoxes of extension this way:

> Zeno challenged geometry and chronometry to devise rules for *adding* lengths and durations which would allow an extended interval to consist of *unextended* elements. Specifically, Zeno challenged physical theory to devise additivity rules for length and duration which permit physical theory to assert *each* of the following assumptions without generating a paradox: (1) a line segment of physical space, whose length is positive, is a linear mathematical continuum of points, each of which is of length zero, (2) the time interval corresponding to a physical process of positive duration is a linear mathematical continuum of instants, each of which is of zero duration. (p. 3)

How, in short, can extended entities (space and time) be composed of entities of zero extension (points and instants)?

In his paradoxes of motion, Zeno argued that motion is impossible. In perhaps the best known of them, he contended that if, in a race between Achilles and a tortoise, Achilles gives the tortoise a head start, he can never overtake it, no matter how much faster he may be. Suppose Achilles is twice as fast as the tortoise and he gives the tortoise a head start of a specified distance. During the time it takes Achilles to cover the distance of the head start, the tortoise will have advanced by half that distance; during the time it takes Achilles to cover *that* distance, the tortoise will have advanced by half of *that* distance, and so on *ad infinitum*. So Achilles will not only fail to pass the tortoise, but he will never even catch up with it. (Falk [2009] engagingly draws a parallel between Achilles' race with the tortoise and the relationship between age and life expectancy as seen in life expectancy tables. As current age increases, remaining life expectancy decreases, but never actually goes to zero; no matter what the current age, there is a remaining life expectancy of some duration.)

Of course, everyone—presumably including philosophers—knows that Achilles will catch and pass the tortoise. And with knowledge of the speed with which both Achilles and the tortoise run and the head start that the tortoise is given, one can calculate precisely how long it will take him to do so. To use an example from Black (1950/2001), if Achilles runs

10 times as fast as the tortoise and the latter is given a 100-yard head start, Achilles will catch up with the tortoise when he (Achilles) has run $111\frac{1}{9}$ yards. If Achilles runs at a speed of 10 yards/second, it will take him $11\frac{1}{9}$ seconds to do it. Black notes that this type of calculation was seen as a resolution of the paradox by no less personages than Descartes, Peirce, and Whitehead, but he himself expresses doubt that it "goes to the heart of the matter. It tells us, correctly, when and where Achilles and the tortoise will meet, *if* they meet; but it fails to show that Zeno was wrong in claiming that they could not meet" (p. 70). So given that we know that Achilles will catch up to, and pass, the tortoise, what is the flaw in Zeno's argument that he cannot do so? This is one way of stating the challenge that has enticed numerous mathematicians and logicians and more than a few garden-variety folks over the two-and-a-half millennia or so since Zeno exposed his perplexing musings to his contemporaries and to posterity.

An argument closely related to that of the race between Achilles and the tortoise has it that in order to get from *A* to *B*, Achilles must first get halfway, then 3/4 of the way, then 7/8, then 15/16, and so on, always having to travel half of the remaining distance before completing the trip. In a variation on this theme, Rucker (1980) has mountain climbers climbing an infinitely high mountain composed of a series of ever-higher cliffs in two hours. The trick is to scale the first cliff in one hour, the second in half an hour, the third in a quarter hour, and to continue halving the time to scale each successive cliff, which will permit them to do an infinity of cliffs in two hours.

Grünbaum (1869/2001) distinguishes between a *legato* run, in which Achilles runs in an uninterrupted fashion toward his goal, and a *staccato* run, in which he runs for a quarter of a minute at twice the speed of his first half minute of the legato run, pauses for the same amount of time, runs for an eighth of a minute and pauses also for an eighth of a minute, and so on, halving the duration of the run and the pause, at each step. If we imagine Achilles Legato and Achilles Staccato racing each other, Achilles Staccato would be racing ahead of Achilles Legato at each segment and then resting while the latter catches up. If Achilles Staccato performs some task during each pause, he will have performed an infinity of them in a minute. As Moore (2001) points out, if he writes one of the digits of π during each pause, he will have done the complete expansion. "We are loath to admit this as a conceptual possibility," Moore notes, "although we seem bound to do so" (p. 4).

According to the *dichotomy*, another of Zeno's paradoxes of motion, in order to get from point *A* to point *B*, Achilles must first cover half the distance; in order to cover half the distance, he must first cover one-fourth of the total distance (half of the half); in order to cover one-fourth, he must first cover one-eighth; and so on. Inasmuch as covering each of these partial distances may be considered an act, in order to move any distance at all, Achilles must have already performed an infinite number of acts, which, the argument goes, is clearly impossible. The conclusion that appears to be forced by the dichotomy is that poor Achilles not only cannot catch the tortoise, but cannot even leave the starting line—of course, according to this argument, neither can the tortoise; racing, or any other act that requires motion, is impossible. The argument appears to rule out the possibility of anything at rest ever beginning to move, and it prompts the question: How, in fact, does something at rest begin to move. At one instant in time it is at rest and at a subsequent one (I did not say the immediately following one) it is moving; how did the transition from resting to in motion take place?

Still another of Zeno's paradoxes of motion involved the flight of an arrow. At any instant of time, the argument goes, the arrow occupies— that is, is stationary within—a region of space that is precisely equal to its length, and the same can be said of *every* instant of its flight. If, as claimed in this argument, it is stationary at every instant, its motion must be an illusion.

Many resolutions of one or another of Zeno's paradoxes have been proposed. Among them are the following:

- Space and time are not infinitely divisible. The universe is discrete. Both space and time have irreducible units—*hodons* and *chronons*, say— and these irreducible units have extent (are not points of zero extent or moments of zero duration). A piece of chocolate can be divided into parts only so long; eventually one comes to something (a molecule of chocolate) that, upon further division, is no longer chocolate. And upon further subdivision, one eventually comes to something (a quark?) which is not (yet) divisible. The idea that matter is composed of indivisible particles goes back at least to the classical Greek philosophers Leucippus of Miletus, Democritus of Abdera, and Epicurus of Samos, and it remains a viable possibility today.

  The application of mathematics to many problem areas in physics— celestial motion, electromagnetism, quantum theory—assumes the continuous nature of both space and time, but that space and time are continuous in reality is not beyond doubt (Casti, 2001; Smolin, 2001, 2004; Wheeler, 1968). Smolin has proposed a theory of quantum gravity that rests on the assumption that both space and time are quantal in nature. Space, according to this theory, comes in units of Planck length (about 10^{-33} cm), so the smallest admissible area is about 10^{-66} cm^2 and the smallest admissible volume is about 10^{-99} cm³. Time, according to Smolin's theory, moves by discrete jumps, each about the duration of Planck time, or about 10^{-43} s. One may wonder what is gained by hypothesizing that space (or time) is discrete, while allowing the unit that makes it discrete to have length, width and height (or duration), which presumably could be continuous entities. The important point in the present context is the idea that space and time are discrete for observational purposes; what goes on within a quantum being not determinable. In any case, Heisenberg's uncertainty principle rules out the simultaneous determination of an object's position and momentum with perfect accuracy. Zeno would have found all of this interesting.

According to Salmon (2001), William James, Alfred North Whitehead, and Henri Bergson all held that, while not proving the impossibility of motion, Zeno's paradoxes reveal the inadequacy of the mathematical account of continuity for the description of temporal processes. James and Whitehead, he argues, saw in these paradoxes a proof that temporal processes are discontinuous. It is one thing to say that the number line, which, after all, is a figment of the mathematician's imagination, has an infinity of points between any two points, and quite another to hold that physical space and time, each of which presumably exists independently of what the mathematician thinks, have an infinity of locations or moments between any two locations or moments.

- Motion should be conceptualized as a functional relationship. "A function is a pairing of elements of two (not necessarily distinct) classes, the domain of the function and its values. On the basis of this definition, if motion is a functional relation between time and position, then motion consists solely of the pairing of times with positions. Motion consists not of traversing an infinitesimal distance in an infinitesimal time; it consists of the occupation of a unique position at each given instant of time. This conception has been appropriately dubbed 'the at-at theory of motion.' The question, how does an object get from one position to another, does not arise" (Salmon, 2001, p. 23).

- Describing the movement of an object from one point to another as the completing of an infinite sequence of tasks is a misuse of language. Thomson (1954/2001b) makes an argument of this sort in addressing the question of whether it is possible to perform a "super-task" (e.g., complete an infinite number of tasks in a finite time). Thomson states this way the argument that it is not: "To complete any journey you must complete an infinite number of journeys. For to arrive from

A to B you must first go from A to A', the mid-point of A and B, and thence to A'', the mid-point of A' and B, and so on. But it is logically absurd that someone should have completed all of an infinite number of journeys, just as it is logically absurd that someone should have completed all of an infinite number of tasks. Therefore it is absurd to suppose that anyone has ever completed any journey" (p. 89). Noting that philosophers have differed with respect to whether the first or the second of the premises of this argument should be considered false, Thomson contends that the disagreement is moot because the argument is invalid in any case. It commits, he claims, the fallacy of equivocation, there being more than one connotation that can be given to the reference to completing an infinite number of journeys.

- The appearance of paradox is based on an unfounded assumption that properties of space and time are analogous in specific respects to those of the real number line. The real number line is infinitely divisible and consequently, as we have seen, no number has a nearest neighbor. If space and time are like the real number line in being infinitely divisible, then we may say that no point in space has a nearest neighboring point in space and no instant in time has a nearest neighboring instant in time, and that to get from one point to another, or from one instant to another, one must cross an infinity of intermediate points or instants. But are space and time like the number line in this respect? Why should we assume they are?

  According to Bergson (1911/2001), we should not. Supposing that what applies to the line of a movement—its divisibility into as many parts as we wish—applies also to the movement per se leads, he argues, to "a series of absurdities that all express the same fundamental absurdity" (p. 65). Black (1950/2001), if I understand him correctly, similarly holds that the paradox stems from misapplication of characteristics of the number line to space and time. "We can of course choose to say that we shall represent distance by a numerical interval, and that every part of that numerical interval shall also count as representing a distance; then it will be true a priori that there are infinitely many 'distances.' But the class of what will then be called 'distances' will be a series of pairs of numbers, not an infinite series of spatio-temporal things" (p. 80). Again, "Achilles is not called upon to do the logically impossible; the illusion that he must do so is created by our failure to hold separate the finite number of real things that the runner has to accomplish and the infinite series of numbers by which we describe what he actually does. We create the illusion of the infinite tasks by the kind of mathematics that we use to describe space, time, and motion" (p. 81). Wisdom (1951/2001) also distinguishes between mathematical distance and physical distance and objects to the use of the former to represent the latter. Unlike a mathematical point, a physical point, Wisdom argues, has some size, however small it may be. It follows that, unlike a mathematical distance, which can consist of an infinity of (mathematical) points, a physical distance can consist of only a finite number of (physical) points. So, in this view, Zeno's arguments apply to mathematical entities but not to physical ones.

- An object in motion is not equivalent to an object being at rest in a sequence of positions. Bergson (2001) dispenses with the paradox of the arrow this way: "The arrow never *is* in any point of its course. The most we can say is that it might be there, in this sense, that it passes there and might stop there. It is true that if it did stop there, it would be at rest there, and at this point it is no longer movement that we should have to do with" (p. 63). Movement, in Bergson's view, is not decomposable: "The arrow which goes from A to B displays with a single stroke, although over a certain extent of duration, its indivisible mobility" (p. 63). "A single movement is entirely, by the hypothesis, a movement between two stops; if there are intermediate stops, it is no longer a single movement" (p. 64).

  This argument is similar to one that might have been made by Aristotle, which, as paraphrased by Owen (1957/2001) in an imagined dialogue between Zeno and Aristotle, goes as follows: "If there is no time in a moment for the arrow to move there is no time for it to be stationary either. Movement involves having different positions at different moments, and accordingly rest involves having the same position at different moments. But we are considering only one moment, so neither of these ideas applies. In making either of them apply you [Zeno] treat the single moment as a period of time itself containing different moments" (p. 158). Owen argues that "talk of movement is appropriate only when we have in mind periods of time within which movements could be achieved … it is absurd either to say or to deny [that movements can be achieved *within* moments], for moments are not pieces of time such that within them any process can either take place or lack the time to take place. But this certainly does not show that the arrow is not moving *at* any moment. It is, of course: we have seen the sense in which it is. Whether it is, is a question of fact and not of logic" (p. 162).

- Rucker (1982) suggests a way out of the arrow paradox, in which the arrow is said not to be moving in any of the successive instants of time, which he believes not to have been published before. "According to Special Relativity, an arrow in motion experiences a relativistic length contraction proportional to its speed. So, in fact, the arrow's state of motion *is* instantaneously observable!" (p. 244). So has Zeno met his match in Dr. Einstein? The reader must judge for himself or herself; it is beyond me.

- Russell (1929/2001) argues that the arrow paradox rests on our strong tendency to assume that at any given instant when an arrow is in flight, there is a *next* position in which the arrow must be located in the next instant, and that the appearance of a paradox disappears when one realizes that there is neither a next position nor a next instant. It appears that Russell here tacitly accepts the infinitely divisible property of the number line as descriptive of physical space and time. Seeing the infinite divisibility of space and time as the resolution of Zeno's paradoxes is in stark contrast to the first view mentioned above, which sees the resolution in the assumption that space and time are not infinitely divisible.

- Some simply dismiss the paradoxes as nonsensical. French physicist Edme Mariotte (1678/1992), for example, makes the case that the answer to the argument that a man who runs twice as fast as another could never catch the slower runner if the latter had a head start of a league (about 5.56 kilometers) is the counterargument that if the faster one does a league in an hour, he will have covered three leagues in three hours, during which time the slower one will have covered only one and a half, which means the faster will overtake and pass the slower. The arguments on which the paradoxes are based are, in Mariotte's view, sophistical: Bodies obviously change positions (even if we do not understand how), and do so at different rates; to claim otherwise is nonsense.

- Zeno's own resolution of the paradox was that space, time, and motion are all illusory. This seems unlikely to appeal to many contemporary minds.

It will be clear to the reader that these "resolutions" are not all mutually consistent. I strongly doubt that any of them, or others that might have been included, will be compelling to everyone. What would constitute a compelling resolution is an interesting psychological question; the only safe conjecture, in my view, is that people who think about such things are likely to disagree on the matter for some time to come. As Salmon (2001) points out, "Each age, from Aristotle on down, seems to find in the paradoxes difficulties that are roughly commensurate with the mathematical, logical, and philosophical resources then available" (p. 44). Contemporary discussions of the paradoxes often raise the question of what is conceivable in view of the laws of physics as currently understood, and include references to such concepts as kinematics, dynamics, Newtonian mechanics, relativity theory, quantum mechanics, limitations imposed by the speed of light, minimal necessary conditions for emitting photons from a light bulb, and so on.

How should we view the paradoxes? Are they frivolous puzzles, mind games devoid of substance and unworthy of sober thought? Or do they present profound questions about the nature of such fundamental concepts as space, time, and motion? There are undoubtedly a range of views on this matter. I am inclined to agree with the assessment of Russell (1929/2001), who contends that although Zeno's paradoxes do not prove that motion and change are impossible, they are not "on any view, mere foolish quibbles: they are serious arguments, raising difficulties which it has taken two thousand years to answer, and which even now are fatal to the teachings of most philosophers" (p. 47). Salmon (2001) describes Zeno's paradoxes as having an onion-like quality: "As one peels away outer layers by disposing of the more superficial difficulties, new and more profound problems are revealed" (p. 43).

☐ The Dilemma of Divisibility

Questions of the nature of space, time, and motion—especially relating to whether they are to be considered discrete or continuous and infinitely divisible—have amused, bewildered, and tormented thinkers from Zeno's time to ours. Are things divisible indefinitely, or only within limits, beyond which no further division is possible? Miller (1982) refers to the question as the dilemma of divisibility and identifies its two horns as the *nihilistic horn*, which "starts from the proposition that magnitude is everywhere divisible and argues to the conclusion that the magnitude is thereby reduced to no extension or, more dramatically, to nothing at all," and the *atomistic horn*, which "starts from the premise that magnitude is not everywhere divisible, leading to the positing of extended but indivisible magnitudes" (p. 89).

Thomson (1954/2001b) makes a distinction between asserting that something is infinitely divisible—which means that "the operation of halving it or halving some part of it can be performed infinitely often"—and asserting "that the operation *can have been* performed infinitely often" (p. 91). He suggests that people have "confused saying (1) it is conceivable that each of an infinity of tasks be possible (practically possible) of performance, with saying (2) that it is conceivable that all of an infinite number of tasks should have been performed" (p. 92).

To clarify this distinction, Thomson describes a reading lamp that has an on-off button that if pushed when the lamp is off, turns it on, and if pushed when the lamp is on, turns it off. Suppose, he asks, that the lamp is off and the button is pushed once in a minute, again in the next half minute, once again in the next quarter minute, and so on, so that it is pushed an infinity of times within two minutes. What will be the state of the lamp at the end of two minutes, on or off? Thomson contrasts the question of what the consequence of the last push of the button would be and that of what the whole infinite sequence of button pushes would produce; the first question has no answer, he contends, because there is no last button push, but in his view the second question would seem to be a fair one.

Benacerraf (1962/2001), who sees the point of Thomson's lamp to be to demonstrate that the idea of completing a super-task is self-contradictory, contends that the argument is flawed. The flaw is that Thomson's description of the on-off states of the lamp applies only to instants of time *before* the two-minute mark; it says nothing about the state of the lamp *at* that time. "He does not show that to occupy all the points in an infinite convergent series of points *logically* entails occupying the limit point" (p. 120). Benacerraf's argument here has an analog in the observation that a convergent mathematical sequence never reaches the limit value. As already noted, to say that the series, $\sum_{n=1}^{\infty} \frac{1}{2^n}$, converges to 1 is not to say that it eventually actually reaches 1, but only that it gets ever closer to 1.

Benacerraf highlights the distinction between psychological and logical considerations by following the observation that it is not possible to imagine a circumstance in which one would be justified in saying that an infinite sequence of tasks had been completed with insistence that the inability to imagine something does not make it logically impossible. In a response to Benacerraf's critique of his argument, Thomson (2001a) acknowledges the validity of the critique and expresses an inclination "to think that there are no simple knock-down arguments to show that the notion of a completed ω-task [what Thomson earlier had referred to as a super-task] is self-contradictory" (p. 131).

The dilemma of divisibility, which has been a challenge to philosophers and mathematicians at least from the time of the ancient Greeks, remains a challenge to this day. Some of the associated problems have been discussed in the foregoing; there are many others. Here I will just mention a few of them:

- Imagine a cone being divided into two sections by a cut parallel to its base. How does the size of the bottom of the top section compare with the size of the top of the bottom one? If one says they are the same, this seems to imply that the shape is that of a cylinder, not a cone, because by slicing all the way up we find no instance in which the bottom of the top and the top of the bottom differ in size. If one says that the bottom of the top is smaller than the top of the bottom, this seems to require that a discrete jump has occurred from the one plane to the other, despite that there is no distance between them.

- Can a line be divided at every point? If points are not contiguous—if between any two points there are other points—how is it possible, even conceivable, to divide a line at any point? But suppose a line *can* be divided at any point. It would seem to follow that a line can be divided at *every* point. But if a line is divided at every point, what is left? Apparently nothing. If something is found to be left, this can only mean that the line was not divided at every point. But if division at every point leaves nothing, this seems to imply that a line—which is generally considered to be composed of points—is the sum of many nothings.

- Imagine a line segment bounded at one end by 0 and at the other by 10. Remove from it the subsegment bounded by 5 and 10. What is the largest value of the remaining subsegment bounded at the bottom by 0? If the line is infinitely divisible, the answer is that it has none. By removing a closed line subsegment from a closed segment in which it was contained, we have created a subsegment that has no maximum value. Is that not strange?

- What is a point? Does it exist physically? And if it does not, what constitutes the center of a circle or a sphere? And what do we call the place where a circle and a tangent to it touch?

- Can indivisible entities touch? By definition, indivisible entities have no parts (e.g., no edges); does it follow that to be juxtaposed means to be superposed or coincidental—or the same entity? If indivisibles cannot touch, can they lie in an ordered sequence with nothing between them?

- Does an instant of time have duration? If it does, then Aristotle's dictum that *a thing cannot be and not be at the same time* does not hold, because it could be during one part of an instant and not be during another part. But if an instant does not have duration, what is it? In what sense can it be said to exist?

- What is "now?" The past and future appear to abut. Does the present not exist, except perhaps as an illusion? Is time itself an illusion? If, as Aristotle argued, the past no longer exists, the future does not yet exist, and the present has no duration, what is left?

☐ The Continuity or Discontinuity
of the Number Line

The question of how to define the continuum of the number line has drawn the attention of some of the world's greatest mathematicians. The challenge has been to explain how the notion of a continuum can be made coherent. When a class of elements (points on a line, instants in time) has the property that any two elements, no matter how close they are, have elements between them, it is said to be dense with respect to the ordering relationship of betweenness.

Denseness is not to be confused, however, with continuity, or the absence of gaps: The rational numbers are dense, inasmuch as one can always find another rational number between any two of them; however, the rationals do not completely fill up the number line. Indeed, although both the rationals and the irrationals are infinite in number, the rationals, which are countable, are outnumbered by the irrationals (e, π, $\sqrt{2}$, $\sqrt{3}$, ...), which are not.

The question of whether the number line is continuous was addressed by Dedekind with his famous "cut" or "partition." Dedekind's cut divides the real numbers into two subsets, the members of one of which are all less than those of the other. A cut of the real number line need not occur at a rational number; if it occurs elsewhere than at a rational number, it can be used to define an irrational, a fact that Dedekind used to argue the continuity of the real number line. From the notion of a cut, it is a short step to the distinction between intervals that include one or both endpoints and those that do not. This allows an interval on the number line from, say 5 to 10 that includes the end points (5 and 10) to abut an interval from 10 to 15 that does not include the endpoint at 10. The latter interval is considered open at its lower end; it includes all values as close as one wishes to 10, but not 10 itself.

Lakoff and Núñez (2000) see Dedekind's work on continuity as illustrative of the *discretization* of mathematics (mentioned in Chapter 4). Dedekind's work, they argue, represented a major departure from the way continuity had been conceptualized for millennia, by introducing a metaphor that treated continuity as numerical completion rather than as a special concept. "Continuity no longer comes from motion but from the completeness of a number system. Since each number is discrete and each number is associated one-to-one with the points on the line, there is no longer any naturally continuous line independent of the set of points. The continuity of the line—and indeed of all space—is now to come from the completeness of the real-number system, independent of any geometry or purely spatial considerations at all" (p. 299).

The discretization illustrated in the work of Dedekind was continued by others, Lakoff and Núñez suggest, notably French mathematician Augustin Cauchy and German mathematician Karl Weierstrass, who further severed the conceptual dependence of arithmetic and calculus on geometry. The result was a discretized account of continuity. In Lakoff and Núñez's words: "Just numbers and logic. There is nothing here from geometry—no points, no lines, no planes, no secants or tangents. In place of the natural spatial continuity of the line, there is just the numerically gapless set of real numbers.... The function is not a curve in the Cartesian plane; it is just a set of ordered pairs of real numbers" (p. 313).

☐ The Continuity or Discontinuity of Space, Time, and Motion

The questions of the continuity (infinite divisibility) or discontinuity (discreteness) of the number line and that of the continuity or discontinuity of space, time, and motion are different, although they have been coupled tightly in treatments of both. Experiments have shown that many people do not make a clear distinction between mathematical and physical objects with respect to the question of their divisibility, believing either that division can be continued indefinitely in both cases or that there will come a point, again in both cases, at which further division will prove to be impossible (Stavy & Tirosh, 1993a, 1993b; Tirosh & Stavy, 1996).

The absence of nearest neighbors on the number line is the kernel of Zeno's dichotomy paradox: Inasmuch as to get started, Achilles must go first from his starting point to the next point, but there is no next point to which to go, so he is stuck on start. It is also the basis for numerous other paradoxes and puzzles. The following is from Normore (1982). It is now time t. At $t + 1$ second a light will be on, at $t + ½$ second it will be off, at $t + ¼$ second it will be on, and so on. Will the light be on or off *immediately* after t? Normore notes that an account by the medieval English logician Walter Burley has it that the statements "Immediately after t the light will be on" and "Immediately after t the light will be off" are both true. I leave it to the reader to consult Normore's chapter to see how this conclusion can be justified. (The reader will see a similarity between Nomore's question and that of Thomson, which was discussed earlier, but whereas Thomson asks what the state of the light will be after it has been turned on and off an infinity of times within a finite interval, Nomore asks how the state of the light can be changed at all.)

The dilemma represented by the notion of infinitely divisible space is illustrated by a contest in which the winner is the one who can stand closest to a building. Imagine (unrealistically) that it is possible to stand as close as one wishes. No matter how close one stands, it can be asked why one does not stand closer. The temporal analog is a contest in which the winner is the one who starts a process soonest following a specified instant, again supposing it is possible to start as soon as one wishes; in this case, no matter how soon after the instant one starts, it can be asked why one did not start sooner.

It was in part to answer questions like these that some of the ancient Greeks, notably Leucippus, Democritus, and Epicurus, argued that matter, time, and motion are discrete—composed of indivisible entities—atoms. But as Zeno demonstrated, atomism was not free of difficult questions. For example, if time is discrete—composed of a sequence of time atoms—how is motion, or change more generally, possible? If motion occurs, a thing in motion must be in different locations in space during different atoms of time. But if it is stationary during any given time atom, as the atomists held, how does it get from one space atom to another in successive time atoms? One proposed answer was that it does so by jerks or jumps. Another resolution of this problem was to deny that motion, as such, occurs: One does not say that something "is moving," only that something "has moved" (Diodorus, quoted in Sorabji, 1982, p. 61). "Diodorus accepts from Aristotle the idea that motion at indivisible times entails motion through indivisible places and that motion through indivisible places entails having moved without ever being in the process of moving" (Sorabji, 1982, p. 64).

Aristotle devoted much energy to refuting the idea, promoted by the atomists, that space, time, and motion are discrete. One of his several arguments against atomism was that indivisible entities could not be arranged so as to constitute a continuum, because contiguity or succession could not be achieved. A point in Aristotle's view has no extent; he defined it 2,500 years in anticipation of Dedekind, as a cut or division of the line. It marks the beginning or end of a line segment, but has no substance (Miller, 1982).

Aristotle's handling of the question of how change takes place, which—it appeared to some—involved the idea that something could be both x and *not-x* at the same instant, gave rise during the 14th century to what Kretzmann (1982a) refers to as "quasi-Aristotelianism." Kretzmann dismisses the proposal of quasi-Aristotelianism for dealing with Aristotle's problem as a nonsolution of a pseudoproblem and attributes it to a misreading of Aristotle. Spade (1982), who also notes problems with quasi-Aristotelianism, is less conclusively dismissive of it than is Kretzmann, arguing that it is interesting and not entirely outlandish, and that it deserves fuller investigation. For present purposes, the point is that Aristotle's views of the infinite persisted and motivated animated discussion and debate throughout the Middle Ages and continues to do so to this day.

Atomism had the challenge also of being clear about the nature of the atom. Is a point an atom (or a quark)? Is an instant? If an atom has extent or duration, however small, it would be divisible, at least conceptually, in which case, it does not really solve the various dilemmas it was invented to solve. And if it does not have extent or duration—is not divisible, even conceptually—the question is how to get something with extent or duration from extentless points or durationless instants—how to get something from nothing.

There is the question too of how to deal with the fact that objects move at different speeds. Suppose A travels twice as far as B during the same time. Does A jump twice as far as B during each instant? Or does A jump twice as frequently as does B, but covers the same distance on each jump? Miller (1982) puts the dilemma the atomist faces in dealing with travel at different speeds this way. "A pure atomist will hold that an atom A moves in an indivisible jerk over an indivisible magnitude in an indivisible time. If this atomist concedes that another atom B could move more slowly than A and agrees that a slower body covers a smaller magnitude in the same time, he will be driven to the conclusion that there is a smaller magnitude than 'the smallest magnitude.' The pure atomist can avoid self-contradiction only by refusing to concede that it is always possible to move faster or slower than any given moving body" (p. 110).

This atomistic dilemma is reminiscent of Aristotle's argument for the infinite divisibility of space and time. Aristotle distinguished between something that could be said to be (potentially) infinite by addition and something that could be said to be (potentially) infinite by division. His argument that space and time are infinite by division goes as follows:

> Consider two moving objects A and B, A being the faster. In the time t_1 that A moves a given distance d_1, B must move a shorter distance d_2. And in the shorter time t_2 that A moves the distance d_2, B must move a still shorter distance d_3. And so on *ad infinitum*. (Moore, 2001, p. 41)

Zeno, of course, held that his paradoxes prove that motion is impossible, but even if one does not accept his arguments to that effect, one might still see in them some relevance to the question of whether space, time, and motion are to be considered continuous or discrete. Although they prompt many questions that relate directly to this one, I do not see that they answer it. Perhaps space and time are not infinitely divisible. Perhaps they are quantal in nature and the appearance of continuity is due to the limited resolving power of our instruments of observation. If this is so, Zeno's paradoxes, resting as they do on the infinite divisibility of the number line, are not applicable to physical reality, because the number line is not descriptive of physical reality with respect to divisibility. It is conceivable that, at some time, it will be possible to determine that space and time are discrete, that they are both particulate. It is not clear, however, how it would ever be possible to determine that space and time are continuous. The most that could be determined is that they are effectively continuous within the limits of the resolving power of the instruments of observation and measurement at the time. That would not rule out the possibility of discreteness at a more precise level of observation. (The idea that space is quantal brings its own puzzles. Hermann Weyl (1949) points out, for example, that if we imagine a square area being composed of tiny indivisible square tiles, and if distance between two points is a function of the number of tiles between them, the diagonal of the area would be considered the same length as a side of it.)

It seems reasonably certain that questions of continuity and discreteness will fuel speculation for a long time to come. Dantzig sees the challenge as that of reconciling the "symphony of number," which plays in *staccato* with the "harmony of the universe," which knows only the *legato* form. But whether the universe really is best considered spatially and temporally continuous or discrete and which perspective—physical or mathematical—will have to adjust to effect a grand resolution appear to be open questions.

☐ Infinitesimals Persist

There are few ideas in the history of mathematics that have proved to be more controversial and, at the same time, useful than the concept of an infinitesimal. The idea is closely related to that of infinite divisibility. One conception of an infinitesimal is that of a quantity that is infinitely close to zero, but not equal to it. The elusiveness of the infinitesimal is seen in the common practice of treating this entity as though its value were either zero or not, depending on the demands of the occasion.

This dual-personality treatment of the infinitesimal was critical to the development and use of the calculus. It was also the focus of some of the harshest criticisms. Commenting on it, the noted Anglican cleric and philosopher Bishop George Berkeley asked, "And what are

these fluxions [of which mathematicians speak]? The velocities of evanescent increments. And what are these same evanescent increments?
They are neither finite quantities, nor quantities infinitely small, nor
yet nothing. May we not call them the ghosts of departed quantities...?"
(1734/1956, p. 292). The "ghosts of departed quantities" is a reference
to Newton's fluxions, today's differentials, and alludes to the practice
of mathematicians of giving them values only to take them away again
by assuming they go to zero. Davis and Hersh (1972) refer to Berkeley's
logic as unanswerable, but it did not prevent mathematicians from continuing to use infinitesimals to good effect.

Infinitesimals, as well as the allied concepts of the derivative and
the definite integral, remained very difficult and subject to severe criticism and even ridicule for a long time during the 17th and 18th centuries.
Like many other mathematical constructs, they were used effectively for
practical computational purposes long before anything approaching a
consensus as to what they meant was attained. That Newton and Leibniz
used infinitesimals does not mean that they were comfortable with the
concept, but only that they recognized its utility. Leibniz referred to
the infinitesimal as a *façon de parler*, and defended the use of it strictly
on practical grounds; even if one considers such a thing impossible, he
argued, it is an effective tool for calculation.

The concept of infinitesimals proved to be so problematic and resis-
tive to rigorous justification that many 19th-century mathematicians
refused to use it and made it superfluous with the development of the
theory of limits. "So great is the average person's fear of infinity that
to this day calculus all over the world is being taught as a study of *limit
processes* instead of what it really is: *infinitesimal analysis*" (Rucker, 1982,
p. 87). Ogilvy (1984) attributes early antagonism to the work of Newton
and Leibniz to the concept of the infinitesimal, which, he says, was exor-
cised during the 19th and 20th centuries. The exorcists, in this case, were
Cauchy and Weierstrass. Cauchy introduced the notion of a *limit* as the
value that a variable approaches and from which it eventually differs by
as small an amount as one wishes. Weierstrass formalized the idea by
defining the limit, L, of $f(x)$ as x approaches a, as

> For any $\varepsilon > 0$, there exists a $\delta > 0$ so that, if $0 < |x - a| < \delta$,
> then $|f(x) - L| < \varepsilon$

Something of the relief with which many mathematicians greeted the
introduction of the concept of a limit is captured in a comment by Kasner
and Newman (1940): "Because Weierstrass disposed of the infinitesimal
djinn, the calculus rests securely on the understandable and nonmetaphys-
ical foundations of limit, function, and limit of a function" (p. 332).

With respect to the claim of the subsequent exorcism of the very concept of infinitesimals, it undoubtedly is the case that, as currently taught, calculus makes much greater use of the concept of a limit than either that of an infinitesimal or of infinity. Moore (2001) describes the result: "What the calculus seems to do, once it has been suitably honed, is to enable mathematicians to proceed apace in just the sort of territory where the actual infinite might be expected to lurk, without having to worry about encountering it. They can uphold claims ostensibly about infinitesimals or about infinite additions, and they can even use the symbol '∞', knowing that they are only making disguised generalizations about what are in fact finite quantities. They still need not look at the actual infinity in the face" (p. 73).

On the other hand, it appears that the discreditation of infinitesimals has been less than complete and terminally effective; although many mathematicians avoid their use, infinitesimals continue to live and to be the focus of some attention. The angst about them has been felt primarily by pure mathematicians; physicists and engineers never stopped using them. Moreover, they have been returned to respectability among at least some mathematicians thanks to the work of American logician Abraham Robinson (1969) on nonstandard analysis and, in particular, his introduction of the concept of hyperreal numbers. Rucker (1982) argues that "Robinson's investigations of the hyperreal numbers have put infinitesimals on a logically unimpeachable basis, and here and there calculus texts based on infinitesimals have appeared" (p. 87). Explanations of Robinson's work (building on that of others, including logicians Thoralf Skolem, Anatoli Malcev, and Leon Henkin) and the significance of nonstandard analysis are provided by Davis and Hersh (1972) and Nelson (1977).

Nelson (1977) defines an infinitesimal as a number that lies between zero and every positive standard number, which is to say between zero and the smallest number one could conceive of writing. Doubt as to whether such entities exist does not preclude defining them. Infinitesimals, defined thus, are, as McLaughlin (1994) puts it, "truly elusive entities." They are to mathematics what quarks are to physics, only more so; quarks, although elusive in practice, are presumably observable in principle. Infinitesimals are, by definition, unobservable, immeasurable.

Their elusiveness rests on the mathematical fact that two concrete numbers—those having numerical content—cannot differ by an infinitesimal amount. The proof, by reductio ad absurdum, is easy: the arithmetic difference between two concrete numbers must be concrete (and hence, standard). If this difference were infinitesimal, the definition of an infinitesimal as less than all standard numbers would be violated. The consequence of this fact is that both end points of an infinitesimal interval cannot be labeled using concrete numbers. Therefore, an infinitesimal interval can never be captured through measurement: infinitesimals remain forever beyond the range of observation. (p. 87)

More simply, there is no smallest number greater than zero; name any number ever so small, and ever so close to zero, and one can always find another number closer to zero than that. Infinitesimals are not all clustered around zero; "every standard number can be viewed as having its own collection of nearby, nonstandard numbers, each one only an infinitesimal distance from the standard number" (p. 87).

Do infinitesimals exist? Does it matter whether they do? One position that some mathematicians have taken, either explicitly or implicitly by their effective use of them, is that the important thing is not whether infinitesimals exist but whether one can get correct solutions to mathematical problems by proceeding as though they do. Davis and Hersh (1972) contend that nonstandard analysis evades the question of whether infinitesimals really exist in some objective sense. "From the viewpoint of the working mathematician the important thing is that he regains certain methods of proof, certain lines of reasoning, that have been fruitful since before Archimedes" (p. 86).

Within the theory of measurement, which is a specialized subarea of abstract algebra, one issue of some importance is how to exclude infinitesimals from consideration. Typically, the algebraic structure is sufficiently strong so that one can define equally-spaced intervals. Any sequence of equally-spaced successive intervals is called a standard sequence. The structure is called Archimedean provided that every bounded standard sequence is finite, which property in effect rules out infinitesimals. See Luce and Narens (1992) for greater detail.

Despite their elusiveness—perhaps because of it—McLaughlin (1994) contends that infinitesimals, as made respectable by Robinson's introduction of hyperreals and subsequent developments—in particular a nonstandard analysis of Nelson, known as *internal set theory* (IST)— provide the basis for resolution of Zeno's paradoxes. The theory resolves the arrow paradox, in McLaughlin's view, by allowing the motion to occur inside infinitesimal segments, where it would be unobservable. The ineffability of such segments provides, he contends, a kind of screen or filter. Whether the motion inside an infinitesimal interval is uniform or discrete is indeterminate, because an infinitesimal is not observable. Mclaughlin contends that all of Zeno's paradoxes of motion are resolved as a consequence of basic features of IST, but one expects opinions to be divided as to whether the paradox has really been dispatched; unanimity among mathematicians on this point would be surprising.

Friedlander (1965) argues that the use of infinitesimals can involve mathematics in an uncertainty principle that is analogous to that of Heisenberg in physics. Consider the problem of finding a tangent to a curve:

> In plane geometry a tangent is, by definition, a straight line which has only one point in common with the curve in question, without intersecting the curve. As a tangent has direction—and it is the direction we are particularly interested in—one point is insufficient for its determination. Two points are necessary to fix a straight line in a certain direction. This is done in analytical geometry and its offshoot, calculus, by assuming two points, infinitesimally close to each other, but not contiguous to each other. Two geometrical points cannot be contiguous to each other, because they would coincide and would no longer be two points. ... If you are able to determine the direction of the tangent you cannot tell which one of the two points on the curve is the point of contact. If you are able to tell the point of contact you are unable to determine the direction as long as you don't employ the second point on the curve. The tangent presents a problem basically no different from the problem of Heisenberg's principle of uncertainty of Bohr's principle of complementarity. (p. 26)

Still tangents are used, and their slopes determined, to good effect.

Rucker (1982) notes that the *dt* of differential calculus, which is an infinitesimal, is considered close enough to zero to be ignored when added to a regular number, but sufficiently different from zero to be usable as the denominator of a fraction. As to the strange rules that this schizoid character follows: "Adding finitely many infinitesimals together just gives another infinitesimal. But adding infinitely many of them together can give either an ordinary number, or an infinitely large quantity" (p. 6). Noting also the apparently dual (or indeterminate) nature of *dt* in the development of the differential calculus, Moore (2001) describes the reasoning involved as fundamentally flawed and ultimately incoherent, resting as it does "on a certain notion of an infinitesimal difference (as not quite nothing, but not quite something either)" (p. 63). One can easily imagine what Bishop Berkeley might have said to all this.

Between the time of Zeno and that at which mathematicians began to attempt to develop useful approaches to dealing with (presumably) continuous variables, people moved around in the world and the phenomenon of one racer passing another was witnessed more than once; it was clear to perception that arrows do indeed fly through the air—motion manifestly occurs—but no widely accepted refutations of the paradoxes were forthcoming. Treating motion mathematically as a succession of states of rest, as suggested by Jean le Rond d'Alembert in the 18th century, does not quite satisfy the philosophical mind. As Dantzig (1930/2005) notes, "The identification of motion with a succession of contiguous states of rest, during which the moving body is in equilibrium, seems absurd on the face of it" (p. 132). No more absurd, but not less, Dantzig argues, "than length made up of extensionless points, or time made up of durationless instants" (p. 132).

Mathematicians made whatever concessions to absurdities were necessary to permit them to make progress on the solutions to mathematical and physical problems that captured their interests. Isaac Newton, working in the 17th century, ignored Zeno's paradoxes, if he ever heard of them, and he created the mathematics he needed to deal with continuous change. More generally, before the 19th-century emphasis on foundations and rigor, most notably by Cauchy, when the operations that were used by mathematicians produced presumably correct results, logical difficulties with them were of greater concern to philosophers than to the mathematicians who used them to advantage (Kaput, 1994). Bunch (1982) reflects the pragmatic attitude that most working mathematicians took with respect to the paradoxes that perplexed the philosophers. "What to do with a paradox? If you are sure that no contradiction results, incorporate the paradox into mathematics and declare it a paradox no longer" (p. 115).

Lakoff and Núñez (2000) see in the study of infinitesimals a lesson about mathematics that is deep and important—"namely, that ignoring certain differences is absolutely vital to mathematics" (p. 251). This lesson, they note, is contrary to the widely accepted view of mathematics as the science that is characterized by precision, and that never ignores differences, no matter how small. Calculus, they contend, is defined by ignoring infinitely small differences. More generally, "ignoring infinitesimal differences of the right kind in the right place is part of what makes mathematics what it is" (p. 253). What makes this acceptable without creating serious conceptual difficulties, in their view, is recognition of the metaphorical nature of the key concepts—especially infinity—that are involved.

☐ Psychology and the Paradoxes

Most of the discussion and debate about Zeno's paradoxes and their implications have centered on matters of philosophy and mathematics. However, the paradoxes are a rich source of grist for the psychologist's mill as well. The very fact of the persisting interest of thinkers in the paradoxes over two and a half millennia is an interesting phenomenon from a psychological point of view. What does the persistence of the paradoxes, despite countless attempts to dispatch them—and claims that they have been dispatched—tell us about the nature of human thought? What, from a psychological point of view, explains the amply demonstrated fact that some deep thinkers have found Zeno's arguments that motion cannot occur compelling, all the while moving about in the world like everyone else?

There are many psychological questions pertaining to specific aspects of the reasoning that the paradoxes prompt. Grünbaum (1955/2001a) asks, for example: "What is the basis for the view that the very meaning of temporal succession involves that events follow upon one another *seriatum*, like the consecutive beats of the heart, and not densely?" (p. 173). His speculation is that the answer is to be found in the way time-order is experienced in human consciousness:

> Since each act of thought takes a minimum positive amount of time rather than a mere instant of zero duration, it is inevitable that when we analyze the stream of consciousness into a succession of constituent moments or 'nows,' these elements are experienced as occurring in a discrete sequence. No wonder therefore that on such an intuitively grounded meaning of temporal succession, there is an ever present feeling that if physical events are to succeed one another in time, their order of occurrence must also be discrete, if it is to be a temporal order at all. (p. 173)

Grünbaum argues that refutation of Zeno's paradoxes of motion requires that this psychological understanding of temporal sequence be replaced by a conception based on a "strictly physical criterion of 'event B is later than event A' that does not entail a discrete temporal order, but allows a dense order instead" (p. 173).

The critical thing to note is that event *B* occurring later than event *A* does not entail that event *B* follows event *A* immediately (or that *any* event follows event *A* immediately). What is hard, but necessary, according to Grünbaum's conjecture, is to replace one's introspection-based belief that every event has an immediately successive event with a conception of temporal sequence for which that is not the case. "Upon freeing ourselves from the limitations of the psychological criterion of time-order by means of the constructive elaboration of an alternative, autonomous physical criterion, it becomes clear that the dense temporal ordering of the constituent point-events of a motion is no obstacle whatever to either its inception or its completion in a finite time" (p. 174). From here it is a short step to the conclusion that the inability to identify the location of Achilles the instant before he reaches his goal (in the dichotomy) or catches the tortoise (in the race) is no warrant for contending that it is impossible for him to do the one or the other. Grünbaum argues that the inability to identify a final subinterval in a progression during which a motion is completed does not preclude the existence of an instant that occurs after the motion has been completed.

In effect, Grünbaum dismisses both the psychological and the logical considerations as valid reasons for accepting Zeno's arguments. "In summary, Zeno would have us infer that the runner can *never* reach his destination, just because (1) in a finite time, we could not possibly contemplate one by one *all* the subintervals of the progression, and (2) for purely *logical* reasons, we could not possibly find the terminal instant of the motion in *any* of the $\aleph_0$ [Cantor's first-order infinity] subintervals of the progression, since the terminal instant is not a member of any of them. But it is altogether fallacious to infer Zeno's conclusion of infinite duration from these two premises" (p. 209).

Grünbaum (1969/2001b) argues that there is a threshold, or minimum, duration of time people can appreciate experientially, and that this minimum plays a role in several fallacies that people commit in reasoning about Zeno's paradoxes of motion. For example, because there is assumed to be a lower bound on the duration of any run that one can imagine, an infinity of runs, even of steadily decreasing lengths, would last forever.

Some of the difficulty people have with infinity, infinitesimals, and related concepts undoubtedly stems from the imprecision of natural language. Consider, for example, the word *more* in the context of the question of whether there are more points in a line segment extending from 0 to 2 than in one extending from 0 to 1. If B having more points than does A is taken to mean that B contains all the points in A and others besides, then the answer is yes, a line segment from 0 to 2 has more points than one from 0 to 1, inasmuch as the former has all the points of the latter and others as well. But if having more points is taken to mean having a greater number of points, the answer is no, because—it is generally agreed that—the number of points in any two line segments is the same (the same order of infinity) and independent of their length.

Another common word that requires careful definition in the context of discussion of Zeno's paradoxes is *task*. Several writers have raised the question of whether it is conceivable that an infinite number of tasks can be completed in a finite time (Grünbaum, 1967; Moore, 2001; Thomson, 1954/2001b). Benacerraf (1962/2001) notes that to show that the idea of an infinite number of tasks being performed in a finite time is self-contradictory, it would suffice to agree that, by definition, a *task* is something the performance of which takes some time and there is a lower bound on how little time a task can take.

A psychological question of considerable practical importance that is raised by the numerous treatments of Zeno's paradoxes in the literature is the following: How is it that some highly intelligent thinkers can consider claims to be compelling that others, equally intelligent, hold to be absurd? Discussions of the paradoxes are not the only context in which this question arises, of course, but it is one of them; *absurd* is a frequently used modifier in this literature. The literature is also replete with charges of inconsistency in argument and with claims and counterclaims of logical error and *non sequitor* reasoning. Why do writers, highly intelligent and presumably well versed in logic, often disagree about what does or does not follow from specific claims—about what does or does not constitute a valid argument when expressed in natural language (rather than in abstract form)?

There are, in short, many questions of psychological interest that are prompted by consideration of Zeno's paradoxes, and especially of the many efforts to resolve them, the critiques of those efforts and the countercritiques of the critiques. To date, the paradoxes have not received a lot of attention from psychologists, but there are many opportunities for exploration.

☐ A Personal Note

I will end this chapter with a personal anecdote. Consider the situation represented by Figure 9.1. The shortest path from *A* to *C* is the straight line, the length of which is $z = \sqrt{x^2 + y^2}$. An alternative, "city block" path, as shown in the leftmost diagram, would be from *A* to *C* by way of *B*. The length of this path is $x + y$, considerably longer than $\sqrt{x^2 + y^2}$. Or one could take the zigzag path shown in the next-to-leftmost diagram, going north from *A* halfway to *B*, then traveling the same distance east, before turning north again, and so on. It should be clear that the length of this zigzag path is exactly the same as the length of the city block path with only two legs, namely, $x + y$. As illustrated by the next two diagrams, one could repeat the process of halving the distance traveled north before turning east and halving the distance traveled east before turning north again, and in each

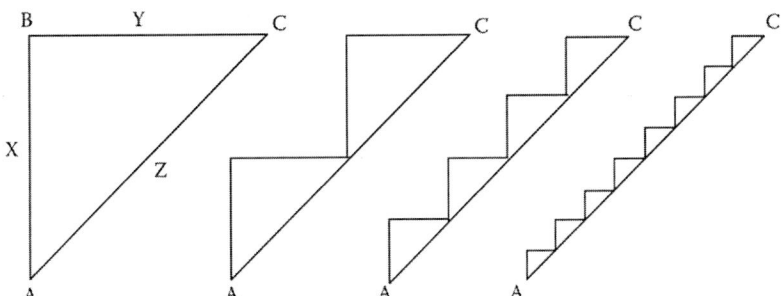

FIGURE 9.1 Illustrating the problem of the vanishing distance.

case one will have traveled the same distance, $x + y$, by the time one gets to C. No matter how many times this process is repeated, the length of the zigzag path to C remains $x + y$. But in the limit the zigzag path becomes the straight-line path from A to C, the length of which is $\sqrt{x^2 + y^2}$, does it not? What happened to the difference between $x + y$ and $\sqrt{x^2 + y^2}$? Note too that the area enclosed by the steps and the diagonal is reduced by half with each successive doubling of the number of steps; this value clearly gets ever closer to zero as the doubling of the number of steps is increased indefinitely.

This problem occurred to me on the occasion of driving from one point to another that is well represented by the two left-most diagrams. I was at A and the destination was C. I knew it was possible to go from A to B to C, but a passenger, who knew the area better than I, suggested that we take a shortcut by turning right on a street that was halfway between A and B, then left and right to arrive eventually at C. Of course, the "shortcut" saved no distance and introduced two unnecessary turns.

When it occurred to me that the shortcut maneuver could be repeated indefinitely without shortening the distance traveled, I believed for a time that I had stumbled onto a new mathematical puzzle. But alas, I eventually discovered that the problem, or at least the problem type, is well known. Friend (1954, p. 72) describes a version of it, without resolution, under the heading "The Field of Barley." Lakoff and Núñez (2000) describe a different version that they refer to as "a classic paradox of infinity." The latter version begins with a semicircle with diameter 1 drawn on a line from (0,0), to (1,0), as shown in Figure 9.2, top. The next step is to draw two semicircles, each with diameter 1/2 within the semicircle with diameter 1 (Figure 9.2, next to top). This process is repeated indefinitely, at each step drawing within each of the semicircles produced at the preceding step two semicircles with diameter reduced by half.

Consider now the semicircle with which the process begins. The length of its perimeter is $\pi/2$. Inasmuch as each of the semicircles drawn within the original one (Figure 9.2, next to top) has diameter 1/2, the length of its perimeter is $\pi/4$ and the sum of them is $\pi/2$. Every time the process of drawing within the existing semicircles two with diameter half that of those drawn on the preceding step, one produces a set of semicircles, the sum of the lengths of the perimeters of which is $\pi/2$. As the process is repeated indefinitely, the total area under the semicircles approaches 0, and the semicircles, in the aggregate, approach a straight line. Nevertheless, the sum of the lengths of their perimeters remains constant at $\pi/2$. How can this be?

Lakoff and Núñez argue that there are more possible resolutions than one of the semicircle paradox, and that what one accepts as a

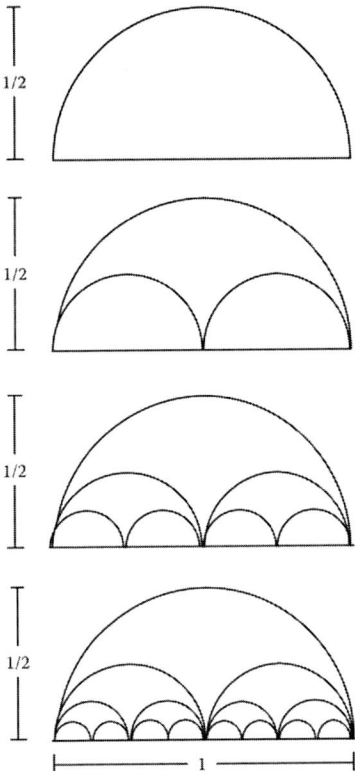

FIGURE 9.2 Illustrating again a vanishing area under a constant-length perimeter.

resolution is something of a matter of preference. Their own resolution (pp. 329–333) makes use of the concept of metaphor. The best I can do with respect to the city block paradox is to contend that the city block distance does not diminish, ever—that, conceptually, the doubling of the number of steps can be continued indefinitely and the zigzag route never does become a straight line. Similarly, neither do the semicircles become a straight line; they remain semicircles indefinitely, even though the area they encompass becomes vanishingly small. I leave it to the reader to decide whether these are acceptable resolutions.

That the area in both the city block and semicircle cases approaches 0 despite that the lengths of the perimeters of the figures remain constant should not be difficult to accept. It is obvious that figures with the same length perimeter can have different areas (a 2 × 2-ft square and a 3 × 1-ft rectangle each has an 8-ft perimeter, while the area of the first is 4 ft² and that of the second is 3 ft²). If one wants to decrease the area encompassed by a perimeter of given length, the most obvious way to do it is to increase the figure's length-to-width ratio while holding the length of the perimeter constant—to stretch the figure in one direction and compress it in the orthogonal one. But this is not what is happening with city block and semicircle figures. If we think of each of these figures as two-sided (the diagonal being one side of the city block figure and the steps the other; the diameter of the original semicircle being one side of the semicircle figures and the arc sides of the semicircles the other), as the areas within their perimeters are reduced with each iteration of the process, the ratio of the lengths of the two sides in each case is unchanging. Peculiar, but apparently not impossible.

10
CHAPTER

Predilections, Presumptions, and Personalities

> Mathematicians are finite, flawed beings who spend their lives trying to understand the infinite and perfect. (Schecter, 1998, p. 47)

Mathematicians represent as variable and colorful an assortment of personalities as do the people who comprise any disciplinary group. There is Archimedes (c. 287 BC–c. 212 BC), who, legend has it, could be so engrossed in a mathematical problem as to ignore the threat of a Roman soldier and to pay for the engrossment with his life. There is Girolamo Cardano (1501–1576), lawyer, physician, mathematician—self-proclaimed survivor of an attempted abortion. There is Blaise Pascal (1623–1662), who, after making seminal contributions to both mathematics and science, devoted the remainder of his short life to philosophy and religion. There is Isaac Newton (1643–1727), unsurpassed in his influence on mathematics and science, but with an ego bigger than life and a chronic need for stroking. There is Abraham de Moivre (1667–1754), who, upon noting that his sleeping time was increasing by about 15 minutes a night, predicted the day of his own death by calculating how long it would be before his sleeping time reached 24 hours, and then obligingly died on the predicted day. There is Marie-Sophie Germain (1776–1831), correspondent with Gauss and winner of a prize from the French Academy for work done under the pseudonym Antoine LeBlanc. There is political activist Evariste Galois (1811–1832), who lost his life at 20 in a duel of obscure instigation. There is

249

Bernhard Riemann (1826–1866), recognized as a genius but without a decent job and destitute for much of his adult life. There is Georg Cantor (1845–1918), who tamed infinities and succumbed to the frailty of his own mind. There is Srinivasa Ramanujan (1887–1920), who was unable to pass his school exams in India, but whose extraordinary mathematical intuition was recognized and cultivated by G. H. Hardy and J. E. Littlewod until his untimely death at 32. There is Norbert Wiener (1894–1964), a Harvard PhD at 18, widely recognized as the father of cybernetics, who worried in later years (Wiener, 1964) about some of the possible trajectories of machine intelligence. There is F. N. David (1909–1993), pioneering statistician and namesake of Florence Nightingale, who is generally recognized as the founder of the modern nursing profession and was herself a statistician of sufficient stature to be elected a Fellow of the Royal Statistical Society. There is Alan Turing (1912–1954), code breaker in World War II, computer visionary, and inventor of the universal Turing machine, dead by his own hand at 42. There is Paul Erdös (1913–1996), the peripatetic Hungarian mathematician whose long life appeared to be totally consumed by mathematics.

One or more book-length biographies have been written on most of the more famous mathematicians. Short biographical information is provided for many of them by several writers, including Turnbull (1929/1956), Bell (1937), Asimov (1972), Boyer and Merzbach (1991), Pappas (1997), and Fitzgerald and James (2007). Short biographies of notable female mathematicians have been provided by Henrion (1997). A few mathematicians have provided revealing and fascinating autobiographical accounts of what it is like to be a mathematician. Notable among these first-person accounts are G. H. Hardy's (1940/1989) *A Mathematician's Apology*, Norbert Wiener's (1953) *Ex-Prodigy: My Childhood and Youth* and (1956) *I Am a Mathematician: The Later Life of a Prodigy*, Stanislav Ulam's (1976) *Adventures of a Mathematician*, and Mark Kac's (1985) *Enigmas of Chance*. An index of biographies of essentially all notable mathematicians in history is given at http://www-groups.dcs. st-and.ac.uk/~history/BiogIndex.html. Generalizations are risky, but there are, I believe, some observations that can be made about attitudinal aspects of doing mathematics that are sufficiently in evidence among mathematicians to be viewed as characteristic of the field.

☐ Conservatism in Mathematics

> New ideas…are not more open-mindedly received by mathematicians than by any other group of people. (Kline, 1980, p. 194)

Mathematicians have sometimes been remarkably resistant to new ideas. Georg Cantor's work on the mathematics of infinity, for example, was attacked and ridiculed mercilessly by fellow mathematicians. Poincaré predicted that future generations would see it as a disease from which they happily had recovered. Kronecker was so upset by Cantor's work on the theory of sets that he prevented Cantor from getting any appointment in a German university and from publishing a memoir in any German journal.

The seminal work of Lobachevsky and Bolyai on non-Euclidean geometries was ignored by colleagues for three decades after it was first published. Gauss was reluctant to publish his work in this area, which predated that of Lobachevsky and Bolyai, because of concern for the ridicule it would evoke. Girolamo Saccheri narrowly missed inventing non-Euclidian geometry a century before the time of Lobachevsky and Bolyai because he could not accept his own results—which he published in *Euclid Vindicated From All Defects* in the year of his death—when they suggested the possibility of a geometry other than Euclid's. He, like countless others, put much effort into an attempt to prove Euclid's parallel postulate. His method led to the derivation of many theorems that were inconsistent with the parallel postulate and consistent with all the rest, but he could not bring himself to believe that a geometry could be constructed for which the parallel postulate did not hold. An amateur mathematician who apparently developed, but did not publish, a non-Euclidean geometry—which he referred to as astral geometry—before the work of Lobachevsky and Bolyai, was German professor of jurisprudence Ferdinand Schweikart, who corresponded with Gauss about his work.

French mathematician Gérard Desargues, who is remembered today as one of the founders of projective geometry, was ridiculed in his day and sufficiently discouraged by this treatment that he gave up his explorations in this undeveloped area of mathematics. Because every printed copy of his book published in 1639 was lost, much of what he did had to be rediscovered 200 years later.

The tenability of specific ideas often changes over time and differs from one culture to another. This is as true of mathematical ideas as of ideas in other domains of thought. Difficulties with such concepts as negative, irrational, and imaginary numbers stemmed at least in part from the assumption that the purpose of mathematics was to represent aspects of the physical world. When mathematics is seen as a body of rules for manipulating abstract symbols, of which numbers are examples, the same difficulties are less likely to be encountered; from this perspective the question of what $\sqrt{-1}$ "really means" does not arise. Surprisingly, however, even those extensions of the number system and other mathematical concepts that have seemed most abstract and philosophically problematic when first introduced have often found some physical interpretation in time and have proved to be useful for practical purposes. And they eventually have become sufficiently widely accepted to be taken for granted.

Is the resistance that mathematicians have shown to new ideas just another example of the general conservatism that human beings appear to have toward the unfamiliar, or does it rest on something distinctive about mathematics as a discipline? Whatever else it may signify, it points out the humanness of mathematicians. Kline (1953) describes this conservatism in less than complimentary terms: "Mathematicians, let it be known, are often no less illogical, no less closed-minded, and no less predatory than most men. Like other closed minds they shield their obtuseness behind the curtain of established ways of thinking while they hurl charges of madness against the men who would tear apart the fabric" (p. 397).

☐ Faith in Mathematics

One would normally define a "religion" as a system of ideas that contains statements that cannot be logically or observationally demonstrated. Rather, it rests either wholly or partially upon some articles of faith. Such a definition has the amusing consequence of including all the sciences and systems of thought that we know; Gödel's theorem not only demonstrates that mathematics is a religion, but shows that mathematics is the only religion that can prove itself to be one! (Barrow, 1990, p. 257)

Faith may seem a strange quality to associate with mathematics, but without it a mathematician would not get far. An obvious role that faith plays is that of permitting mathematicians to build on the work of others. "Mathematicians in every field rely on each others' work, quote each other; the mutual confidence which permits them to do this is based on confidence in the social system of which they are a part. They do not limit themselves to using results which they themselves are able to prove from first principles. If a theorem has been published in a respected journal, if the name of the author is familiar, if the theorem has been quoted and used by other mathematicians, then it is considered established. Anyone who has use for it will feel free to do so" (Davis & Hersh, 1981, p. 390).

Today teams of mathematicians may produce proofs that are sufficiently complex that no individual can vouch for the correctness of them in their entirety. When that is the case, each member of the team must rely on the competence and integrity of the others if they are to have any confidence in the aggregated results of their combined efforts.

There are roles that faith plays in mathematics that are more fundamental than those associated with the willingness of mathematicians to rely on the work of other mathematicians. Among the more forceful writers to make this point was Bishop Berkeley (1734/1956). His derisive commentary on the willingness of mathematicians to use what he considered nonsensical concepts in their development of the calculus were noted in the preceding chapter. His purpose was to show that religion is not alone in demanding faith on the part of its adherents, which he did by pointing out the ease with which some mathematicians accepted ideas that he saw as having less of a rational basis than some of the religious ideas that they just as easily dismissed. One might ask, Berkeley suggests, "whether mathematicians, who are so delicate in religious points, are strictly scrupulous in their own science? Whether they do not submit to authority, take things upon trust, and believe points inconceivable? Whether they have not *their* mysteries, and what's more, their repugnances and contradictions?" (p. 293).

It is much too easy with the benefit of hindsight to see Berkeley's objections to the calculus as the protestations of a man who had a vested interest in putting it down, but there is little reason to doubt the sincerity of his skepticism. Although the invention of Newton and Leibniz proved to be enormously useful, it had little in the way of a theoretical foundation until the concept of a limit was developed by Cauchy and Weirstrass nearly 200 years later. "Nobody could explain how those infinitesimals disappeared when squared; they just accepted the fact because making them vanish at the right time gave the correct answer. Nobody worried about dividing by zero when conveniently ignoring the rules of mathematics explained everything from the fall of an apple to the orbits of the planets in the sky. Though it gave the right answer, using calculus was as much an act of faith as declaring a belief in God" (Seife, 2000, p. 126).

We are unlikely to have trouble with the specific concepts Berkeley mentioned, and we may not wonder as he did about the calculus because familiarity with it as a practically useful tool has deadened our curiosity about its foundations or rational justification. There can be little doubt, however, that his general observations are true, if not of mathematicians, at least of those of us who have a passing acquaintance with some aspects of mathematics and use them to advantage on occasion. We accept that the product of two negative numbers is a positive number, that although it is all right to multiply by 0 it is forbidden to divide by it, that 0! is 1 as is 1!, and that we do so does not mean that we have a clear understanding of the bases of these rules. We use imaginary numbers to good effect, although we may not be able to imagine what they mean, and we readily include in our equations symbols that represent infinitely small or infinitely large magnitudes or quantities, although we cannot conceive of their referents.

Even our assumption that the real number system is appropriate to the description of natural phenomena is an article of faith. Penrose (1989) puts it this way: "The appropriateness of the real number system is not often questioned, in fact. Why is there so much confidence in these numbers for the accurate description of physics, when our initial experience of the relevance of such numbers lies in a comparatively limited range? This confidence—perhaps misplaced—must rest (although this fact is not often recognized) on the logical elegance, consistency, and mathematical power of the real number system, together with a belief in the profound mathematical harmony of Nature" (p. 87).

Barrow (1992) describes our acceptance of the whole enterprise of mathematics and its scientific applications in a similar fashion:

> We have found that at the roots of the scientific image of the world lies a mathematical foundation that is ultimately religious. All our surest statements about the nature of the world are mathematical statements, yet we do not know what mathematics "is"; we know neither why it works nor where it works; if it fails or how it fails. Our most satisfactory pictures of its nature and meaning force us to accept the existence of an immaterial reality with which some can commune by means that none can tell. There are some who would apprehend the truths of mathematics by visions; there are others who look to the liturgy of the formalists and the constructivists. We apply it to discuss even how the Universe came into being. Nothing is known to which it cannot be applied, although there may be little to be gained from many such applications. And so we find that we have adopted a religion that is strikingly similar to many traditional faiths. (p. 297)

There is a deeper sense still in which faith is necessary to mathematics and indeed to any activity in which reasoning plays a critical role. I am speaking of the need to accept the adequacy and unchanging nature of the rules of inference and to believe that the human mind is capable of comprehending and applying these rules. It is not clear that we could work on any other assumption, because if the rules of inference were to change from time to time, or if they were completely beyond our comprehension, we could have little hope of knowing anything. So the assumption is critical, but it is not demonstrably correct; it is an article of faith and can never be anything but. Mathematicians are constantly making claims about *all* numbers, *all* triangles, *all* circles, and other sets, the members of which are assumed to be infinite in number. Such claims cannot be verified empirically—we cannot check to see if what is claimed of the infinity of members of any set is true in every case. Nevertheless, we accept many such claims with complete confidence, and we can do so only because of our faith in rules of logic and in our ability to reason according to them.

☐ Passion in Mathematics

> Contrary to popular belief, mathematics is a passionate subject. Mathematicians are driven by creative passions that are difficult to describe, but are no less forceful than those that compel a musician to compose or an artist to paint. (Pappas, 1997, p. i)

To nonmathematicians, mathematics may seem to be the most dispassionate of activities; what could be less exciting than manipulating symbols on a piece of paper, or in one's head? In fact, it appears that at least the better mathematicians that the world has produced have found mathematics to be a totally absorbing activity and a source of intense pleasure, if, at times, also enormous frustration.

Biographers of great mathematicians have often described them as being obsessed with mathematics, as being unable sometimes *not* to think about mathematical problems, as becoming so engrossed in their thinking about mathematical matters as to be oblivious to what is going on around them. Asking how it was possible for one man to accomplish the colossal mass of highest order work that Gauss did, Bell (1937/1956) suggests that part of the answer was Gauss's *"involuntary* preoccupation with mathematical ideas." "As a young man," Bell writes, "Gauss would be 'seized' by mathematics. Conversing with friends he would suddenly go silent, overwhelmed by thoughts beyond his control, and stand staring rigidly oblivious to his surroundings" (p. 326). Tenaciousness was also one of Gauss's characteristics: Once engrossed in a problem, he stayed with it until he solved it. Gauss himself attributed his prodigious output to the constancy with which he thought about mathematics and suggested that if others reflected on mathematical truths as deeply and continuously as did he, they would make the same types of discoveries. Archimedes and Newton are both said to have often been so engrossed in a problem as to neglect to eat or sleep. Whether true or not, the well-known story of the death of Archimedes conveys a sense of how complete and uncompromising a mathematician's concentration on a problem of interest can be.

In his autobiography, Mark Kac (1985) describes being "stricken," at the age of 16, "by an acute attack of a disease which at irregular intervals afflicts all mathematicians ... I became obsessed by a problem" (p. 1). The problem that gripped him was the—self-imposed—challenge to derive the method for solving cubic equations that had been discovered, but not derived, by Girolamo Cardano in 1545. Kac reports having a number of such bouts with the "virus of obsession" during his life, but credits the first such experience, as a teenager in 1930, with establishing his lifelong commitment to mathematics. Here is how he describes this experience. "I rose early and, hardly taking time out for meals, I spent the day filling reams of paper with formulas before I collapsed into bed late at night. Conversation with me was useless since I replied only in monosyllabic grunts. I stopped seeing friends; I even gave up dating. Devoid of a strategy, I struck out in random directions, often repeating futile attempts and wedging myself into blind alleys" (p. 3). Kac's perseverance paid off and he was able to solve the problem on which he had been obsessing. "Then one morning—there they were! Cardano's formulas on the page in front of me." From that moment, Kac says, "having tasted the fruits of discovery," he wanted to do nothing but mathematics, and mathematics became his lifelong vocation.

Ulam (1976) stresses the importance to mathematical creativity of what he refers to as "'hormonal factors' or traits of character: stubbornness, physical ability, willingness to work, what some call 'passion'" (p. 289). In describing his own transformation from being an electrical engineer to being a mathematician, he notes that it was not so much that he found himself doing mathematics, but rather that mathematics had taken possession of him. He estimated that after starting to learn mathematics he spent on the average two to three hours a day thinking about mathematics and another two to three hours reading or conversing about it.

Keynes (1946/1956) suggests that Newton's peculiar gift was the ability to concentrate intently on a problem for hours, days, or weeks, if necessary, until he had solved it. Newton himself claimed that during the time of his greatest discoveries in mathematics and science (when he was in his early 20s) he thought constantly about the problems on which he was working.

Even if we discount somewhat the reports of extreme commitment to reflection by first-rank mathematicians, allowing for the possibility of some exaggeration in reports of the characteristics and behavior of individuals who have become legendary figures, we can hardly escape the conclusion that mathematics is unusual in its ability to capture the attention of certain minds and to hold it for long periods, sometimes a lifetime. This applies not only to the more glamorous aspects of mathematics, but sometimes to tedious ones as well. How else could we account for Napier's willingness to devote 20 years to calculating his table of logarithms, or for the expenditure of a similar amount of time by French astronomer Charles Delaunay to produce an equation that gives the exact position of the moon as a function of time, or for the dedication of German mathematician Ludolph van Ceulen of most of his life to the determination of the value of π to 35 places? This is not to suggest that mathematics is unique in this regard; some people devote their lives to thinking about music, philosophy, science, theology, medicine—or chess. It may be that in order to be first rate at anything, one must have the kind of commitment to it that leading mathematicians have appeared to have to mathematics.

☐ Capabilities, Personalities, and Work Habits of Mathematicians

> If today some earnest individual affecting spectacular clothes, long hair, a black sombrero, or any other mark of exhibitionism, assures you that he is a mathematician, you may safely wager that he is a psychologist turned numerologist. (Bell, 1937, p. 9)

Contrary to popular stereotypes, mathematicians span a very considerable range with respect to capabilities, personality characteristics, and work habits. Some, like Gauss, have remarkable computational ability; others, like Poincaré, have difficulty doing simple arithmetic. Some, like Gödel, are reclusive and taciturn; others, like American polymath John von Neumann, are sociable and gregarious. Some, like Erdös, work with numerous collaborators; others, like Hardy, work with only a very few; and still others generally work alone. British mathematician Arthur Cayley read the work of other mathematicians extensively; James Sylvester, also a British mathematician and Cayley's close collaborator, disliked learning what other mathematicians had done. Some mathematicians have had extraordinary memories—Ramanujan was a case in point—but unusual memory ability has not been considered a requirement for mathematical prowess.

Some mathematicians maintain highly regular work schedules, and rather regular daily routines more generally. Others lead much less highly structured lives. Some seem to be able to mix a certain amount of habitual routine with a good bit of less organized time. Von Neumann, for example, had the habit of spending some time writing before breakfast every day. A very sociable person, he enjoyed parties and informal gatherings, but he would sometimes drop out of conversations or withdraw from social affairs (even those he hosted) in order to work on a problem that had presented itself to his mind. Descartes, it is said, made a habit of staying in bed, thinking, every day until noon, a habit, if true, that was rudely ignored by Queen Christina of Sweden, who insisted, late in his life, on his appearance as her tutor in mathematics at 5:00 a.m. Poincaré appears to have done his mathematical problem solving in his head, while pacing, writing down what he had done only after having done it, and unlike most mathematicians, he remembered theorems and formulas primarily by ear. Believing that much mathematical problem solving occurred subconsciously during sleep, he also made it a point to get regularly a good night's rest. Mark Kac (1985) describes Stanislav Ulam's way of doing mathematics as "by talking ... wherever he found himself he talked mathematics day in and day out, throwing out ideas and generating conjectures at a fantastic rate" (p. xxi).

Some mathematicians have been immensely productive; the vast majority have been much less so. Bell (1937) puts Euler, Cauchy, and Cayley in a class by themselves in terms of sheer productivity, ranking Poincaré, who published almost 500 papers on new mathematics, and over 30 books, in addition to his popular essays and work on the philosophy of science, as a distant second. (Bell made this observation when Erdös was in his mid-20s, still early in his career as a publication machine.)

While mathematicians differ in the mentioned respects and others, there are characteristics that appear to be, if not essential, at least highly conducive to mathematical eminence. The ability to concentrate intensely is a case in point, and persistence appears to be an asset if not also a requirement for doing creative mathematics. Confidence in one's ability to see problems through to solution would seem to be necessary to keep one working when progress is proving to be very difficult. A certain intellectual independence, or toughness of mind, appears also to be an asset if not a requirement for work that departs from well-trodden paths and takes a field off in new directions. Perhaps it was the lack of this toughness of mind that cost Saccheri the lost opportunity, as noted in Chapter 4, to develop non-Euclidian geometry 100 years before others did so. The ability to invent new approaches to problems has been seen as one that distinguishes good from mediocre mathematicians. King (1992), for example, characterizes good mathematicians as those to whom the *problem* is paramount, who "once attracted to a problem of significance and elegance ... learn or create whatever mathematical methods are necessary to solve it." In contrast, mediocre mathematicians, in King's view, "are characterized by their tendency to use only the mathematics they already know and to search for problems that can be solved by these methods" (p. 35).

That mathematics can be, and often is, such an engrossing activity may be the basis of a common misconception about great mathematicians, which is that *all* they think about is mathematics. There may be mathematicians among the greats that fit this model, or come close to it, but there are also many who do not. Some of the more prominent mathematicians of history had exceptionally broad education and interests. Euler, arguably the most prodigious mathematician who ever lived, produced about 800 pages of mathematical output a year. His publications numbered almost 900 books and mathematical memoirs, in Latin, French, and German, many of which appeared only after his death. It is hard to imagine how he could have managed this if he spent much time thinking about anything other than mathematics. Nevertheless, in addition to mathematics, Euler studied theology, medicine, astronomy, physics, and oriental languages (Boyer & Merzbach, 1991). Euler was a devout Christian and family man, and a prolific correspondent; he is believed to have written some 4,000 letters, of which nearly three-quarters have been preserved (Beckman, 1971).

Newton's work in mathematics and science, which made him famous, occupied a relatively small part of his life, perhaps not more than about 10 years, and accounted for a relatively small percentage of the 1.3 million words that he wrote and left to posterity. At least as strong as his commitment to mathematical and scientific investigations was his lifelong interest in theology. Leibniz was a philosopher, theologian, linguist, and historian as well as a mathematician and logician, and was engaged for much of his life as a professional political advisor and diplomat. Pascal made lasting contributions to both mathematics and physics before devoting the last decade or so of his short life (dead at 39) to a study of scripture and vigorous defense of Christianity.

Although Gauss devoted a large percentage of his waking hours to mathematics, he also found time to master several languages (learning Russian by himself in two years, beginning at the age of 62) and to keep abreast of world affairs, to which he devoted an hour or so each day. His keen interest in languages and world affairs is especially noteworthy in view of the claim that the longest trip he is known to have taken was 27 miles from his home in Göttingen (Bell, 1946/1991).

The prototypical mathematician is undoubtedly a mythical character; as a group, mathematicians include as great a diversity of personalities as one can imagine. Some have led conservative—some might say boring—social lives; others have been very sociable party-loving folk; and some have been almost scandalously colorful. Some have been sickly; others have been noted for their physical vigor. Some have been political activists; others have showed little or no interest in the political world. And so on. The range of personalities, capabilities, and work habits adds considerable human interest to the story of mathematics.

☐ Extraordinary Numerical or Computational Skills

There are numerous reports in the literature of persons who are able to perform extraordinary computational feats. "Lightning calculators" or "calculating prodigies"—people who are able to produce almost instantly and without the use of paper and pencil answers to computational questions that even competent mathematicians would take considerable time to produce—have been of great interest to students of human cognition and the general public as well (Binet, 1894/1981; Hermelin & O'Connor, 1986, 1990; Hope, 1987; Hunter, 1962, 1977; Jensen, 1990; O'Connor & Hermelin, 1984; Pesenti, 2005; Rouse Ball, 1892; Smith, 1983; Tocquet, 1961; Treffert, 1988, 1989). Their feats include multiplying numbers with 10 or more digits, finding cube roots of very large numbers, solving cubic equations, finding logarithms, distinguishing prime numbers from composites, and identifying the day of the week on which a date in the distant past fell. Jensen (1990) describes the case of an otherwise intellectually unremarkable Indian woman, Shakuntala Devi, who, according to the *Guinness Book of World Records*, was able to find the product of two 13-digit numbers in 30 seconds. Devi (1977) has written an engaging book in which she describes many of the shortcut methods that can be used to facilitate the doing of mental arithmetic. In it she also reports having fallen in love with numbers at the age of three. "It was sheer ecstasy for me—to do sums and get the right answers. Numbers were toys with which I could play. In them I found emotional security; two plus two always made, and would always make, four—no matter how the world changed" (p. 9).

Although a few great mathematicians, including Ampère, Euler, Gauss, Hamilton, and von Neumann, have been extraordinarily good calculators, most have not. As a general rule, lightning calculators have not shown exceptional mental ability in other ways; many have had below-average intelligence; several have been autistic. The incongruity between the extraordinary ability of lightning calculators to calculate and their average or sometimes below-average abilities to perform other cognitively demanding tasks inspired the epithet *idiot savant*, a regrettable term often used in the past (much less often now) in reference to individuals who showed unusual abilities in specific other areas (notably music) as well.

One capability that many lightning calculators appear to have in common is an extraordinary short-term memory capacity, visual in most cases, auditory in some, and tactile in at least one documented case—the blind-from-birth Louis Fleury. Euler is said to have been able to do calculations mentally that required retaining in his head up to 50 places of accuracy. He appears to have had an extraordinary memory for all kinds of information, as evidenced by the claim that he memorized the entire text of Virgil's *Aeneid* as a boy and could still recite it 50 years later (Dunham, 1991).

Extraordinary memory for numbers is not a guarantee of similar capability for retention of nonnumerical information. Tocquet (1961) reports that a Mlle. Osaka—who could immediately give the 10th power of a two-digit number and the sixth root of an 18-digit number, and could repeat a string of 100 digits immediately after it was read to her, and again, both as given and in reverse order, when unexpectedly asked to do so after a 45-minute interval filled with conversation about other matters—was unable to learn the correct order of the letters of the alphabet. Others have presented evidence that lightning calculators are unlikely to have unusually large working memory capacities for nonnumerical information (Pesenti, Seron, Samson, & Duroux, 1999).

Among the more prodigious feats of memory for numbers is the widely publicized case of Rajan Mahadevan, who secured a place in the 1984 *Guinness Book of World Records* by memorizing π to 31,811 digits. Although commitment to such an amazingly long sequence to memory took considerable time and effort, Mahadevan has also demonstrated an ability to repeat a string of several dozen random digits upon hearing it once. Mahadevan's spectacular memory feats appear to rest on strategies he has developed specifically for recalling numbers, and his memory for nonnumeric information appears not to be extraordinary (Ericsson, Delaney, Weaver, & Mahadevan, 2004; Lewandowsky & Thomas, in press).

Tocquet (1961) gives several specific examples of problems solved by Jacques Inaudi, a computational prodigy who was studied by Alfred Binet, Paul Broca, and a committee of the French Académie des Sciences, whose members included Jean-Gaston Darboux and Henri Poincaré. Rouse Ball (1892) and Pesenti (2005), among others, also give numerous examples of the types of problems solved by lightning calculators, many of which are quite astounding. Besides having a prodigious memory, lightning calculators generally give evidence of their extraordinary ability relatively early in life, and many, though by no means all, tend to lose it later, sometimes after only relatively few years.

There have been attempts to attribute the extraordinary feats of lightning calculators to some combination of unusual memory, the previous commitment to memory of certain numerical facts, and learned or invented shortcut procedures for doing computations. The shortcut procedures usually have the effect of decreasing the amount of information that must be carried along in memory during the process of a lengthy computation.

There can be no doubt that it is possible to stage what appear to be spectacular computational feats that are in reality done with the help of scripted events, accomplices, and various forms of trickery. Moreover, there are facts about numbers that can be memorized and used to good effect in doing mental computations, and numerous learnable techniques (not normally taught in conventional mathematics courses) that can, once well learned, greatly simplify many computations. Tocquet (1961) describes several such techniques, for example, a shortcut memory-dependent method for mentally extracting cubic and higher roots from large numbers that are perfect powers. Is there something unusual left to explain when all such possible factors are taken into account? Tocquet contends that the answer to this question is yes. He points out, for example, that the methods he describes for extracting roots work only for certain roots (cubic, 5th, 9th, 13th, 17th, and 21st) and only when the number for which the root is wanted is a perfect power. Lightning calculators are not constrained to deal only with such cases. Tocquet favors the view that these people have somehow been able to use abilities that lie below the level of consciousness and that most of us have not been able to tap. "What seems essentially to characterize the lightning calculator is that to a greater degree than ordinary mortals he can use faculties which are in some way innate and which probably exist in a latent state in every human being" (Tocquet, 1961, p. 43).

Dehaene (1997) argues that the feats of lightning calculators can be attributed, at least for the most part, to an unusually large memory capacity (emphasized also by Binet), a passion for numbers, and great familiarity with many of them as a consequence of years of focusing on them—"Each calculating genius maintains a mental zoo peopled with a bestiary of familiar numbers" (p. 147)—and the learning of various shortcut mathematical procedures, some of which are described by Flansburg (1993), a proficient user of them. He rejects the idea that lightning calculation ability represents a mystery that defies a natural explanation. "A talent for calculation thus seems to arise more from precocious training, often accompanied by an exceptional or even pathological capacity to concentrate on the narrow domain of numbers, than from an innate gift" (p. 164). I think it fair to say that, to most of us who struggle to do two-digit multiplications without the crutch of paper and pencil, the documented feats of lightning calculators—whether mysterious or not—are impressive indeed.

☐ Mathematical Disabilities

Equally interesting as the question of the existence of extraordinary mathematical capabilities is that of whether there are specifically mathematical disabilities. The evidence is clear that people can lose mathematical ability as a consequence of brain injury, and losses of specific abilities can follow from lesions in specific cortical areas (Dehaene, 1997; Dehaene, Spelke, Pinel, Stanescu, & Tsivkin, 1999; Deloche & Seron, 1987; Gruber, Indefrey, & Kleinschmidt, 2001). This does not mean that any specific mathematical operation—counting, addition, multiplication, comparison—is accomplished by a specific area of the brain. It appears from results of cortical mapping studies that even the simplest of mathematical operations engage numerous cortical areas (Dehaene, 1996; Dehaene et al., 1996); nevertheless, that brain trauma can result in the loss of specific mathematical skills seems not in doubt.

Brain injury aside, is there such a thing as mathematical disability, as distinct from more general cognitive disability? Are there people who are perfectly capable of reasoning well in other contexts but lack the ability to learn to do mathematics? Belief that the answer is yes is sufficiently strong among some researchers and educators to have ensured a variety of names for such a disorder, including *dyscalculia, acalculia,* and *anarithmia* (Vandenbos, 2007). Estimates of the prevalence of school-age children with some form of mathematical disability vary, mostly from about 5% to about 8% (Badian, 1983; Fleishner, 1994; Kosc, 1974; Shalev, Auerbach, Manor, & Gross-Tsur, 2000).

Children can have difficulty learning basic mathematics for a variety of reasons other than an underlying cognitive deficit of some kind. Geary (1994) makes this point, and notes that in his own studies (e.g., Geary, 1990; Geary, Bow-Thomas, & Yao, 1992), only about half of the children who had been identified as having difficulty learning mathematics give evidence of a cognitive deficit. The goal of identifying learning disabilities in mathematics is complicated by poor achievement in mathematics having a variety of causes other than an actual cognitive disability. Geary and Hoard (2005) note that in the absence of measures specifically designed to diagnose mathematical disability, which have not yet been developed, what is commonly taken as evidence of such a disability is the combination of a score lower than the 25th or 30th percentile on a mathematics achievement test and a low-average or higher IQ. The intent of this combination criterion is to identify people whose mathematical disability is distinct from a general cognitive deficit, which might manifest itself in difficulty with mathematics but problems with other cognitive activities as well. Especially important is the dissociation of difficulty with mathematics from difficulty with reading, inasmuch as the two often occur together, and an inability to read comprehendingly would limit one's ability to deal with mathematical word problems (Aiken, 1972; Kail & Hall, 1999). For defenses of the position that the more general notion of *learning disability* is more of a political category than a scientific one, see Farnham-Diggory (1980) and Allardice and Ginsburg (1983).

Distinctions have been made both on the basis of the assumed origin of any particular mathematical deficit and on that of the nature of the disability. Disabilities that are assumed to be innate are sometimes referred to collectively as *developmental dyscalculia* (Butterworth, 2005; Kosc, 1974; Shalev & Gross-Tsur, 1993, 2001), and those that are known or believed to be the result of some type of trauma are covered by the term *acquired dyscalculia*. That mathematical disabilities differ in their nature is reflected in the variety of designations that one finds in the literature, which include *number-fact dyscalculia, procedural dyscalculia, spatial dyscalculia, alexia,* and *agraphia,* among others (Badian, 1983; Hécaen, 1962; McCloskey, Caramazza, & Basili, 1985; Temple, 1989, 1991). Overviews of these and other dysfunctions are provided by Geary (1994), Geary and Hoard (2005), and Macaruso and Sokol (1998). Several dissociations have been found in which one or more specific aspects of mathematical ability are impaired while others are not. In their review of work on developmental dyscalculia, Macaruso and Sokol (1998) mention the following:

> Dissociations between numeral processing and calculation (e.g., Grewel, 1969; McCloskey et al., 1985), between numerical comprehension and numeral production (e.g., Benson & Denckla, 1969; McCloskey et al., 1986; Singer & Low, 1933), and between Arabic and verbal numeral processing (e.g., Grafman, Kampen, Rosenberg, Salazar, & Boller, 1989; Macaruso et al., 1993, Noel & Seron, 1993). Within the domain of calculation, dissociations have been reported between operation symbol comprehension and other calculation abilities (Ferro & Botelho, 1980), between retrieval of arithmetic facts and execution of calculation procedures (e.g., Cohen & Dehaene, 1994; Sokol et al., 1991; Warrington, 1982), and between retrieval of arithmetic facts associated with different operations (e.g., Dagenbach & McCloskey, 1992; Lampl, Eshel, Gilad, & Sarova-Pinhas, 1994). (p. 208)

Dowker (1998), also with substantiating references, identifies some of the same dissociations as well as some not mentioned by Macaruso and Sokol, including between different arithmetical operations (Cipolotti & Delacycostello, 1995; Dagenbach & McCloskey, 1992; McNeil & Warrington, 1994), between oral and written presentation modes (Campbell, 1994; McNeil & Warrington, 1993), between conceptual and procedural knowledge (McCloskey, 1992; Warrington, 1982), and between calculation and estimation (Dehaene & Cohen, 1991; Warrington, 1982). The conclusion that such a diversity of dissociations seems to force is that mathematical competence is multifaceted and that different facets depend on the proper functioning of different neurological structures. Apparently, it is possible to lose (or fail to develop) certain capabilities while retaining (or developing) certain others in a remarkable array of combinations. This lends support to Dowker's (1998) contention that "there is no such thing as arithmetical ability; only arithmetical abilities. The corollary is that arithmetical development is not a single process, but several processes" (p. 275).

How much of the difficulty that many students have with beginning mathematics can be attributed legitimately to either genetic or environmental causes is a question of continuing interest and research. Twin and sibling studies suggest that siblings (especially twins) of children who have a math disability are more likely to have such a disability themselves than are siblings of children who do not have a math disability (Alarcon, Defries, Gillis Light, & Pennington, 1997; Shalev et al., 2001). One example of a genetic disorder that appears to result in difficulties in learning mathematics without necessarily affecting other cognitive abilities, such as reading and verbal reasoning, is Turner syndrome, which is caused by a chromosomal abnormality that occurs about once in 2,000 to 5,000 female births (Bender, Linden, & Robinson, 1993; Butterworth et al., 1999; Mazzocco, 1998; Mazzocco & McCloskey, 2005; Rovet, 1993; Rovet, Szekely, & Hockenberry, 1994; Temple & Marriott, 1998). A similar syndrome—the Martin-Bell, or fragile X, syndrome—also caused by a chromosomal abnormality, occurs with about the same relative frequency as Turner syndrome. It occurs in both males and females but tends to have more severe consequences in males. This too is associated with difficulties with math, but often with other cognitive problems as well (Mazzocco & McCloskey, 2005).

Spina bifida myelomeningocele, a spinal cord defect that occurs about once in 1,000 to 2,000 live births and affects both males and females, produces both physical and cognitive impairments, difficulty with mathematics among them (Barnes, Smith-Chant, & Landry, 2005; Dennis & Barnes, 2002; Friedrich, Lovejoy, Shaffer, Shurtleff, & Beilke, 1991; Wills, 1993; Wills, Holmbeck, Dillon, & McLone, 1990). Alzheimer's disease and other forms of dementia typically are accompanied by mathematical disability, some of which, perhaps, is specific to particular aspects of numerical or mathematical functions (Duverne & Lemaire, 2005; Kaufmann et al., 2002).

Although it has been possible to trace some incidences of mathematical disabilities to genetic origins, like those that produce Turner and fragile X syndromes, to date such traces account for a small percentage of children and adults who have difficulties with mathematics. According to one view, very little of the poor performance of many students in their beginning (or subsequent) courses in mathematics can be attributed to genetic and preschool environmental causes in combination. Allardice and Ginsburg (1983) express this view, for example, in noting that there are several reasons why studies attempting to demonstrate a neurological basis for difficulties in learning mathematics are inconclusive and in claiming that "the most obvious environmental cause of poor mathematics achievement is schooling that is especially inadequate in the case of mathematics" (p. 330).

Finally, the identification of specific types of mathematical disabilities is complicated by the unquestioned fact that many students experience anxiety in the study of mathematics, sometimes referred to, in extreme cases, as *mathophobia*. Such anxiety may, in some cases, be rooted in *bona fide* cognitive limitations peculiar to mathematics, but even when it is not—which could represent the large majority of cases—the fear of doing poorly can become a self-fulfilling prophecy, and early failures can reinforce the anxiety, more or less ensuring increasing difficulties at later stages of instruction.

☐ Collaboration and Recognition Among Mathematicians

> No matter how isolated and self-sufficient a mathematician may be, the source and verification of his work goes back to the community of mathematicians. (Hersh, 1997, p. 5)

A popular stereotype of mathematicians is that of somewhat reclusive individuals who spend most of their time secluded in the privacy of their own thoughts. Perhaps there have been, and are, mathematicians who fit this image; Cantor may have done so, and until he met British mathematician G. H. Hardy, the Indian genius Srinivasa Ramanujan appears to have worked pretty much by himself. Certainly much of the creative work that any mathematician does is intensely private. Ulam (1976) gives some credence to the stereotype of the mathematician as a withdrawn thinker by his observation that for some, mathematics can be an escape from the cares of the everyday world: "The mathematician finds his own monastic nitch and happiness in pursuits that are disconnected from external affairs. Some practice it as if using a drug" (p. 120).

On the other hand, collegial interactions and collaborations often play an indispensable role in creative mathematics. Ulam makes this point also and notes that much of the development of mathematics has taken place around small groups of mathematicians:

> Such a group possesses more than just a community of interests; it has a definite mood and character in both the choice of interests and the method of thought. Epistemologically this may appear strange, since mathematical achievement, whether a new definition or an involved proof of a problem, may seem to be an entirely individual effort, almost like a musical composition. However, the choice of certain areas of interest is frequently the result of a community of interest. Such choices are often influenced by the interplay of questions and answers, which evolves much more naturally from the interplay of several minds. (p. 38)

Ulam's own career bears testimony to the possibility of fruitful collaborations in mathematics, as do those of such different personalities as Hardy and Erdös. Hardy worked almost exclusively with two other mathematicians, John Littlewood and Ramanujan. Erdös is famous for the large number of people with whom he coauthored papers—507 according to Du Sautoy (2004). For many mathematicians, and not a few nonmathematicians, one's "Erdös number" has taken its place as a vital statistic along with one's age, height, weight, and, perhaps, IQ. One's Erdös number is the number of steps one must traverse in a coauthor chain to get from oneself to Erdös. Anyone who coauthored a paper with Erdös has an Erdös number of 1. Anyone who coauthored a paper with someone who coauthored a paper with Erdös has an Erdös number of 2, and so on. If you have coauthored a few papers, your Erdös number may be smaller than you realize. Du Sautoy (2004) has the number of mathematicians who have an Erdös number of 2 as more than 5,000. Hoffman (1998) and Schechter (1998) have provided very readable accounts of the life of this colorful man of numbers.

Other notable collaborations in the history of mathematics include those of Pascal and Fermat on the beginnings of probability theory, and of Cayley and Sylvester on the theory of invariants. Aczel (2000) argues that mathematical research is best done in a community, where ideas can be shared. "Working in isolation is hard and slow going, and there are many blind alleys into which a mathematician can stray when there is no possibility of sharing ideas with colleagues" (p. 99). Sharing ideas cannot only help people avoid blind alleys, but interactions among colleagues can spark insights that might not have occurred had the interactions not taken place. Collaboration, when effective, can also undoubtedly increase output. Erdös authored (or coauthored) over 1,500 papers in his lifetime, a number that puts him second in output only to Euler (Du Sautoy, 2004).

Collaboration among mathematicians, as judged by the authorship of their publications, has increased considerably in recent years. According to figures compiled by Jerrold Grossman and cited by Schechter (1998), "In 1940 about 90 percent of all mathematical papers were solo efforts; today that number has dropped to around 50 percent. Fifty years ago papers written with more than two people were almost unheard of, while today such multiple collaborations account for almost 10 percent of all published articles" (p. 182). Schoenfeld (1994b) argues that mathematics, like science, is a social rather than a solitary activity and that, as a consequence, the ability to communicate ideas with colleagues is an asset.

Because of the nature of some of the problems that mathematicians are working on today and because the only feasible approach to these problems involves breaking them into parts that can be dealt with simultaneously by different individuals or groups, usually with the help of considerable computing power, a great deal of collaborative mathematics is being done at the present.

With few, if any, exceptions, mathematicians who make major contributions to the advancement of the discipline build on the work of other mathematicians, living and dead. Andrew Wiles worked pretty much alone and somewhat secretly for several years on his proof of Fermat's last theorem, but he drew heavily from the work of predecessors and contemporaries in a variety of areas. He was the one who put it all together, which no one else had done, and therefore deserves the recognition he has received for his monumental feat, but without the work on which he drew, his proof would not have been possible. It is for this reason that Aczel (1996) could claim that the proof of Fermat's last theorem was the achievement of many mathematicians who lived between the time of Fermat and when the proof was finally accomplished.

☐ Competition, Rivalry, and Credit

> Despite the unworldly nature of mathematics, mathematicians still have egos that need massaging. Nothing acts as a better drive to the creative process than the thought of the immortality bestowed by having your name attached to a theorem. (du Sautoy, 2004, p. 171)

Du Sautoy (2004) describes mathematical research as "a complex balance between the need for collaboration in projects which can span centuries and the longing for immortality" (p. 171). It is easy to see how the longing for immortality can foster competition and rivalry and thereby be a hindrance to collaboration. Although I do not mean to equate either competition or rivalry with ill will, there are numerous examples in the history of mathematics of rivalries or disputes about precedence that became acrimonious. Flegg (1983) recounts one such dispute involving Girolamo Cardano, his pupil Ludovico Ferrari, and self-taught Italian mathematician Nicola Fontana (aka Tartaglia) that included accusations and counteraccusations of plagiarism and bad faith. "So unpleasant was the whole atmosphere generated by this dispute that Tartaglia was lucky not to have been murdered by Cardano's supporters" (p. 201). (Violence was not uncommon to the place and time. Tartaglia's father had been murdered in 1505, and he himself was maimed by a soldier's sword blow that slashed his jaw and palate when many citizens of Brescia were massacred by an invading army in 1512. Tartaglia's wound left him unable to speak normally, hence his nickname, Tartaglia, which means *stammerer*.)

The acrimonious dispute between Newton and Leibniz over the origination of the differential calculus, which involved many supporters of both men, is perhaps the best known of rivalries among the greats. (For details regarding claims and counterclaims, see Hall, 1980.) Historians of mathematics appear to be generally agreed that—despite the charges of plagiarism from both sides and considerable poetic license in retrospective accounts of relevant events—both Newton and Leibniz did original and immensely influential work in establishing the calculus as a major field of mathematics. Perhaps in part because of the prominence of the dispute over precedence between Newton and Leibniz, important work by others leading to the calculus may get less recognition than it deserves. Laplace (1814/1951) considered the true discoverer of the differential calculus to be Fermat, who had invented a method for finding minima and maxima before either Newton or Leibniz did their work. Dantzig (1930/2005) contends that had Fermat been more inclined to publish his work, he would have been remembered by posterity as the creator of both analytic geometry (for which we credit Descartes) and the calculus, and "the mathematical world would have been spared the humiliation of a century of nasty controversy" (p. 136). (As noted in Chapter 7, some credit for analytic geometry should also be given to Nicole Oresme, who used ordinate and abscissa to represent functions considerably before Descartes did so.) Unhappily, the dispute between two of the more fertile minds of all time had some negative effects on later mathematical progress, especially in England, for many decades.

Kronecker's incessant attacks on Cantor and his ideas are believed to have played some role in Cantor's recurring bouts of mental depression. How much of Kronecker's behavior in this regard can be attributed to a zealous desire to protect mathematics against Cantor's ideas, which he found profoundly disturbing, and how much can be attributed to some other type of motivation is not clear. That his attacks, including his blocking of Cantor's chance for a desired position at a Berlin university, had a devastating effect on Cantor personally appears clear.

The story of the long-lasting feud between British statisticians Ronald Fisher and Karl (né Carl) Pearson, as well as that between Fisher and British statistician Egon Pearson (Karl's son) and Polish statistician Jerzy Neyman, has been engagingly told by Salsburg (2001). Issues of precedence and claims of credit stealing tarnished even the reputation of the Bernoulli family, arguably one of the most mathematically gifted and productive families on record (paralleling in mathematics what the Bachs did in music).

French mathematician Gilles de Roberval was an unremitting critic of Descartes, and Descartes was not above ridiculing Roberval in personal terms, expressing to French monk-mathematician Marin Mersenne, for example, astonishment that he (Roberval) could pass among others as a rational animal (Watson, 2002). Descartes was quite capable of using flattery and expressions of adulation or insult and duplicity when it suited his purposes.

A very readable account of 10 of the major disputes between notable mathematicians, often involving many of their supporters, is provided by Hellman (2006). This and other accounts make it clear that many disputes have been motivated by personal concerns about recognition for work done. These give the lie to the romantic notion—if anyone has it—that great mathematicians do mathematics only for the satisfaction they derive from the activity. To be sure, they derive much satisfaction, but as a rule, they are not indifferent to recognition for their accomplishments. A possible exception was the prolific Euler, who is said to have sometimes withdrawn papers from publication in order to allow younger mathematicians to publish first. It seems also that some disputes have been motivated, at least in part, by concern about the implications that certain work could have for the future development of mathematics. This arguably could apply to contentions involving developments—such as Cantor's work on transfinite numbers and Russell's logicism—that had serious implications for the "foundations" of mathematics.

The question of the appropriateness of credit to mathematicians for specific discoveries is not always answered by accounts of explicit rivalries and attending evidence of the justification of claims of precedence. As is true also in science, the history of mathematics has its cases of misappropriated credit that have not been subjects of highly visible controversy. Dantzig (1930/2005) gives the example of the discovery of what he refers to as *Harriot's principle*, named for its discoverer, British astronomer-mathematician Thomas Harriot. The principle involves the procedure of writing polynomial equations, $P(x)$, in the form $P(x) = 0$, that is, with all nonzero terms of the equation on one side of the equals sign and 0 on the other. Given the equation $2x^2 + 5x = 12$, for example, Harriot's principle calls for transposing 12 from the right of the equals sign to the left of it so the equation becomes $2x^2 + 5x - 12 = 0$.

Dantzig describes this innovation, simple though it seems, as "epoch-breaking" because, among other things, it reduces the task of solving the equation to that of factoring a polynomial—in this case, $2x^2 + 5x - 12 = (2x - 3)(x + 4) = 0$, so $x = 1.5$ and -4. He notes, however, that because Descartes's book on analytic geometry, in which Descartes used Harriot's ideas without ascription, appeared soon after the publication of Harriot's *Praxis*, it was Descartes who was credited with the idea for nearly a century.

Correctly attributing credit is complicated in mathematics, as in science, by major developments being very seldom due exclusively to the work of a single individual, although they may become associated by posterity with a single name. There are many examples of people whose creative work in mathematics has been obscured by the glare of the light shone on the work of their eminent successors. Kline (1980) makes the point: "No major branch of mathematics or even a major specific result is the work of one man. At best some decisive step or assertion may be credited to an individual" (p. 84). As Boyer and Merzbach (1991) put it: "Great milestones do not appear suddenly but are merely the more clear-cut formulations along the thorny path of uneven development" (p. 332). The recorded history of mathematics, as the recorded history of science, is necessarily a gross simplification of the actual history. Such simplification is necessary, if for no other reason, to accommodate the limitations of our minds; a complete detailed account would be more than we could absorb.

☐ The Practice in Publishing Mathematics

> In developing and understanding a subject, axioms come late. Then in the formal presentations, they come early. (Hersh, 1997, p. 6)

Historically, publication practices have been strongly influenced by two factors. One was very practical. Before the days of tenure, contracts for university positions and renewals thereof often depended on winning public competitions in solving mathematical problems. The holder of a desirable position could be challenged to a contest at any time by another mathematician who coveted his chair. In such contests, both challenger and challenged could pose problems for the other to solve. Such a system provided strong motivation for a person who had developed an effective approach to the solution of some difficult problem to keep that knowledge to himself as a means of job security. If challenged by someone with an eye on his position, he could meet the challenger with a challenge of his own. The dispute already mentioned involving Cardano, his pupil Ferrari, and Tartaglia illustrates the seriousness attached to the guarding of mathematical secrets. The system of gaining or keeping academic positions by winning public competitions is a thing of the past. That does not mean that competition is no longer in play, but only that it operates in less explicit and visible ways.

A second factor that has greatly influenced publication by mathematicians is the strong preference among mathematicians to publish polished proofs but not the reasoning that produced them. Perhaps this is due in part to a sense of esthetics according to which well-constructed proofs are beautiful, whereas the often convoluted reasoning processes that lead to those constructions are not. Bell (1937) describes Gauss's attitude toward publishing this way: "Contemplating as a youth the close, unbreakable chains of synthetic proofs in which Archimedes and Newton had tamed their inspirations, Gauss resolved to follow their great example and leave after him only finished works of art, severely perfect, to which nothing could be added and from which nothing could be taken away without disfiguring the whole. The work itself must stand forth, complete, simple, and convincing, with no trace remaining of the labor by which it had been achieved" (p. 229). As a matter of principle, Gauss published only mathematical works that he considered perfect, and deliberately refrained from providing any clues to the thought processes that had brought him to the completed work. It has been said that his slowness to publish was problematic for his contemporaries because they could never be sure when they were about to go to print with a discovery that he had not already made it and simply failed yet to publish it. In Bell's view, Gauss's reluctance to expose work in progress may have delayed the development of mathematics by decades.

Having quoted Bell's mention of Archimedes, Newton, and Gauss in the same sentence, I should note that Bell considered these three to be the greatest mathematicians the world had yet produced. They, he says, "are in a class by themselves ... and it is not for ordinary mortals to attempt to range them in order of merit" (p. 218). It is of interest too to note the attitudes of these giants toward pure and applied mathematics. "All three started tidal waves in pure and applied mathematics: Archimedes esteemed his pure mathematics more highly than its applications; Newton appears to have found the chief justification for his mathematical inventions in the scientific uses to which he put them, while Gauss declared that it was all one to him whether he worked on the pure or the applied side" (Bell, 1937, p. 218). All three had their feet in both pure and applied worlds. If Archimedes esteemed pure mathematics more highly than its applications, this did not deter him from making some of the most heralded practical applications of all time, and if Newton was highly focused on applications, this did not get in the way of his making seminal contributions to the advancement of mathematical theory.

An aversion to exposing the psychology of mathematical thinking, as distinct from the results of such thinking, is seen in G. H. Hardy's (1940/1989) observation that the mathematician's function is to prove new theorems and not to talk about what he or other mathematicians have done. The distinction between doing mathematics and explaining how it is done was to Hardy a qualitative one; he saw the two kinds of activities as existing on different intellectual planes. In his view, the doing requires creativity; the describing or explaining requires something less. "There is no scorn more profound, or on the whole more justifiable, than that of the men who make for the men who explain. Exposition, criticism, appreciation, is work for second-rate minds" (Hardy 1940/1989, p. 61).

Hardy was apologetic—contrite—about the writing of his *Apology*: "If then I find myself writing, not mathematics, but 'about' mathematics, it is a confession of weakness, for which I may rightly be scorned or pitied by younger and more vigorous mathematicians. I write about mathematics because, like any other mathematician who has passed sixty, I have no longer the freshness of mind, the energy, or the patience to carry on effectively with my proper job" (p. 63).

Notable among the giants of mathematics for his somewhat unique attitude toward publishing his findings was Pierre de Fermat. Not a professional mathematician—he was a lawyer—Fermat nevertheless was a consummate doer of mathematics, and had an enormous impact on the development of the field, but published almost none of his findings in the conventional sense. It appears that he derived great pleasure from simply solving problems and felt no need to seek fame or recognition for his successes. On occasion he would let contemporaries know that he had solved a problem without revealing how, apparently as a challenge to them to do the same. Illustrative of this behavior was the announcement that he had proved that 26 is the only number that is sandwiched between a square (25) and a cube (27), without divulging the complicated proof.

What is known about Fermat's discoveries comes from his correspondence and notes in his belongings, many of which were in the margins of *Arithmetica*, a multivolume book written by Diophantus of Alexandria in the third century AD. Fermat had a copy of a 1621 Latin translation by Claude Gasper Bachet of a portion of *Arithmetica* (several volumes had been lost when much of the great library of Alexandria was destroyed in 389 and again in 642). This book, more than any other, stimulated Fermat's mathematical thinking, and many of the results of that thinking were recorded in notes written in the book's margins. Fortunately for posterity, Fermat's eldest son, Clément-Samuel, recognized his father's genius and saw to it that much of his work became widely available, in part by arranging the publication in 1670 of an edition of *Arithmetica* that contained Fermat's marginal notes, including the "last theorem."

Whether Fermat really had a proof of this theorem is not known. There are several possibilities. Perhaps he did have a proof. But if so, his proof certainly could not have been anything like the one for which Andrew Wiles became famous, which depended on several very complicated mathematical developments that occurred over the centuries since Fermat's time. The possibility that Fermat had a simpler proof will undoubtedly keep many people searching for it, or another relatively simple one, for a long time. Another possibility is that Fermat thought he had a proof, but that what he had would not have stood up under the scrutiny of other mathematicians. This possibility gains credence from many post-Fermat mathematicians, some eminent ones among them, having thought they had a proof, only to have its flaws pointed out by colleagues.

Is it conceivable that Fermat's claim to have a proof was a deliberate hoax? He appears to have gotten considerable enjoyment from tantalizing mathematicians with problems that he had been able to solve but the solutions of which he did not divulge. Singh (1997) says of Fermat's personal notes that they contained many theorems, but that they typically were not accompanied by proofs. "There were just enough tantalizing glimpses of logic to leave mathematicians in no doubt that Fermat had proofs, but filling in the details was left as a challenge for them to take up" (p. 63). In fact, most, if not all, of Fermat's theorems were proved by others over time. Did Fermat imagine that his notes in *Arithmetica* would be made public after his death? And is it possible that his marginal comment about his "last theorem" was a mischievous challenge to mathematicians to find a proof for it? It seems unlikely, but not entirely out of the question.

Borrowing a distinction made by Goffman (1963) in his observations of human behavior in social situations, Hersh (1997) distinguishes between "front" and "back" mathematics. "Front mathematics is formal, precise, ordered, and abstract. It's broken into definitions, theorems, and remarks. Every question either is answered or is labeled: 'open question.' At the beginning of each chapter, a goal is stated. At the end of the chapter, it's attained. Mathematics in back is fragmentary, informal, intuitive, tentative. We try this or that. We say 'maybe,' or 'it looks like'" (p. 36). The practice of publishing only front mathematics, Hersh suggests, is responsible for perpetuating several myths about mathematics. "Without it, the myths would lose their aura. If mathematics were presented in the style in which it's created, few would believe its universality, unity, certainty, or objectivity" (p. 38).

An interesting exception to the rule that mathematicians have tended to be reluctant to introspect, at least publicly, about their doing of mathematics is the legendary Archimedes. In *The Method*, the manuscript which was discovered only in 1906, he describes in detail a "mechanical" approach that he took in mathematical problem solving. It involved the imaginary balancing of lines as one might balance weights in mechanics. The approach led him to several beautiful mathematical discoveries regarding areas and volumes of curved planar figures and solids. Notable among more recent expositors of the type of thinking that often lies behind published mathematical proofs are George Polya (1954a, 1954b) and Imre Lakatos (1976).

Whatever their reasons for doing so, mathematicians' habit of publishing only the results of their thinking—the polished versions of proofs and not the unpolished methods by which they arrived at them—has had the unfortunate consequence of obscuring the nature of mathematical thought. And it has not prevented the publication of erroneous results. Court (1935/1961) notes that the list of names in Lecat's 1935 *Erreurs des Mathématiciens* looks pretty much like a *Who's Who in Mathematics*. The nonmathematician whose primary exposure to the discipline is via high school or college textbooks sees only a highly sanitized representation of what the thinking of mathematicians has produced, and gets little, if any, hint of the character of the thinking itself.

Although perhaps to a lesser degree than is true of mathematicians, scientists too tend to publish primarily the results of their thinking without revealing much about the processes that led to those results. Merton (1968) points this out: "The scientific paper or monograph presents an immaculate appearance which reproduces little or nothing of the intuitive leaps, false starts, mistakes, loose ends, and happy accidents that actually cluttered up the inquiry" (p. 4). Or, as Lindley (1993) puts it, scientists have a habit of covering their tracks when they publish. They do not often reveal all the false starts and sojourns down blind alleys that preceded the attainment of some publishable result. Even the reporting of "thought experiments" is less revealing than one might hope of scientific thinking in progress; Nersessian (1992) points out that by the time a thought experiment is presented, it always works, and the false starts that may have preceded the final result are seldom reported. This is fortunate in some respects—what are of lasting interest for most purposes are the reliable results that have been obtained and not the muddling that led to them, and publication of extensive records of the thought processes behind those results would, from one point of view, add an enormous amount of noise to the scientific literature. But the absence of such records makes the mathematical and scientific enterprises appear to be much more logically neat than they really are.

****

This chapter has focused on mathematicians as persons—their personalities, preferences, capabilities, work habits, sometimes quirks. Do mathematicians, as a group, share any characteristics that distinguish them from people in other professions? Clearly, to be a productive mathematician, one must be able to reason well, and for many areas of mathematics, to reason well about abstract entities. But the ability to reason well, and even to reason well abstractly, is called for by many other endeavors. Many of the other characteristics that are generally considered descriptive of especially productive mathematicians—the ability to focus intently on a problem, perseverance, commitment to one's work, satisfaction derived from solving problems—are not unique to mathematicians. Perhaps the most obvious thing that study of the lives and work of major mathematicians reveals is the great diversity of personalities, lifestyles, capabilities, and interests that this community contains. Whatever one's concept of the prototypical mathematician is, there are likely to be very few real live mathematicians who will fit it.

CHAPTER 11

Esthetics and the Joys
of Mathematics

> The mathematician's patterns, like the painter's or the poet's, must be *beautiful*; the ideas, like the colors or the words, must fit together in a harmonious way. Beauty is the first test: there is no permanent place in the world for ugly mathematics. (Hardy, 1940/1989, p. 85)

No one knows the extent to which the earliest attempts to count, to discover or construct patterns, to measure, and to compute were motivated by purely esthetic as opposed to practical interests. It seems not unreasonable to assume that both types of factors have been important determinants of mathematical thinking from the beginning, just as they are today. Tracing major developments in mathematics to their origins has proved to be very difficult, and in many cases impossible, because surviving evidence of the earliest efforts is sparse. Many of the written documents known to have been produced by the Egyptians, Mesopotamians, Greeks, Chinese, and other ancient cultures have not survived, so constructing a coherent representation of the course of development even during recorded history requires a considerable amount of conjecture, and there are many questions on which experts are not in agreement.

A few aspects of this history, however, are reasonably clear. There seems to be little doubt, for example, that esthetics, mysticism, and a general interest in matters philosophical energized much of the mathematical thinking of the ancient Greeks. In contrast, the Egyptians were motivated, in large part, by the mathematical demands of land measurement and pyramid building, and the study of algebra in Arabia may have been stimulated to some degree by the complicated nature of Arabian laws governing estate inheritance (Boyer & Merzbach, 1991).

According to Bell (1945/1992), the Greeks made a distinction between logistic and arithmetica, the former having to do with computation for practical purposes and the latter with the properties of numbers as such. Bell dismisses their accomplishments in logistic as "nothing that is not best forgotten as quickly as possible by a mathematician" (p. 50). The Greeks themselves appear to have held logistic in contempt, whereas they considered arithmetica to be worthy of the better minds among them.

Their work in mathematics, exemplified notably by that of Euclid and Pythagoras, put great emphasis on deductive proofs and treated mathematics as an abstract discipline that existed independently of the material world. The Pythagoreans considered mathematics to be more real, and more nearly perfect, than the world of the senses. This emphasis on deduction and abstraction was an immensely important contribution to the development of mathematics, but the downside of the Greek perspective was that it provided little incentive for finding practical applications of what they were developing, and perhaps as a consequence of their intense focus on abstract realities, they tended not to be keen observers of nature. They were nearly oblivious, for example, of the part played by curves of various types in the world about them. "Aesthetically one of the most gifted people of all times, the only curves that they found in the heavens and on the earth were combinations of circles and straight lines" (Boyer & Merzbach, 1991, p. 157).

An exception to the rule that the Greeks were not much interested in quantitative description was the great mathematician Archimedes, who applied his mathematics to the physical world and especially to practical engineering problems. He is reputed to have stalled the Roman's seige of Syracuse for two or three years by his inventions of various devices and instruments by means of which he twarted their efforts to take the city. It can be argued that Archimedes' work in physics was unrivaled until the time of Galileo.

☐ Number Mysticism

The origins of number mysticism, or numerology, are obscure. But if they did not originate it, the early Greeks, especially the Pythagoreans and Plato in his *Timaeus*, at least contributed substantially to it and helped ensure its long-term survival (Butler, 1970). To the Greeks, individual numbers, especially the first few integers, had great symbolic significance. Each was endowed with specific qualities, inferred sometimes from numerical or structural relationships that could be found within them. Six, for example, was associated with perfection because it was both the sum and the product of its divisors, which happened to be the first three integers: $1 + 2 + 3 = 6 = 1 \times 2 \times 3$. Four was associated with justice, because it could be factored into two equal numbers and equality was seen to be a defining property of justice. The same could be said of any square number, of course, but the Greeks tended to focus on the smallest number for which a property of interest held, and for some purposes 1 was not considered a number.

The fascination with numbers was deep, and the identification of their distinguishing properties was a serious undertaking by many ancient and medieval scholars. Publius Nigidius Figulus established a neo-Pythagorean school of philosophy in Rome that promoted numerology. Hopper (1938/2000) gives the following example of number fascination by Plutarch, whom he calls a superb demonstrator of the operations of Pythagorean mathematics: "He finds that 36 is the first number which is both quadrangular (6×6) and rectangular (9×4), that it is the multiple of the first square numbers, 4 and 9, and the sum of the first three cubes, 1, 8, 27. It is also a parallelogram (12×3 or 9×4) and is named 'agreement' because in it the first four odd numbers unite with the first four even $1 + 3 + 5 + 7 = 16$; $2 + 4 + 6 + 8 = 20$; $16 + 20 = 36$" (p. 45).

Hopper describes also the roles played by the Gnostics, philosophers originally in and around Alexandria, in perpetuating number mysticism through their efforts to integrate Greek philosophy, science, and Eastern religion, as well as those of the early Christian writers—notably St. Augustine—and medieval scholars in keeping number mysticism alive. The picture includes the development of *gematria*, in which letters of the alphabet are given numerical values and texts are searched for hidden meanings derived from those values, and *arithmology*, which involves the believed significance of, and powers ascribed to, certain integers. Gematria was central to the interpretation of Hebrew scriptures by the Kabbalists in medieval Europe (Aczel, 2000). The pervasiveness of the effect of number mysticism on the thinking of philosophers and scholars from the classical Greeks through the middle ages is easy for us today to fail to recognize; Hopper points out that no branch of medieval thought entirely escaped its influence.

Number mysticism was rooted in the idea that the world is understandable in quantitative terms because it was built according to mathematical principles. To the number mystic, that many aspects of nature could be described mathematically was less than surprising; it would have been surprising if this were not the case. Today the desire to describe the universe in mathematical terms appears to be at least as strong as it was in the days of the early Greeks, which is not to say that the motivation for doing so is necessarily the same.

We see vestiges of numerology today in various superstitions regarding numbers. The idea that individuals have lucky numbers is one such, and undoubtedly one that has been responsible for the loss of considerable money on lotteries and other games of chance. The widespread fear of certain numbers—13 in particular—is another case in point. The fear of 13, known clinically as triskaidekaphobia, has very real and substantial economic effects in terms of people absenting themselves from work, appointments, and business transactions on days of the month numbered 13. Ellis (1978) raises the question of how 13 came to be considered unlucky and, while not providing a definitive answer, discusses a variety of possible contributing factors. Whatever the answer, the superstition apparentle goes back at least to around 1780 BC; the Code of Hammurabi contains 281 numbered laws—1 through 12 and 14 through 282; there is no law numbered 13, presumably because 13 was considered an unlucky or evil number (http://leb.net/~farras/history/hammurabi.htm).

Although number mysticism is often the object of ridicule among mathematicians and scientists today, and much about it is very easy to criticize, it should not be assumed that it has ever been the province only of benighted cranks. As Brainerd (1979) puts it, "We would do well to remember that gematria, transcendental arithmetic, St. Isadore's dictionary, and the rest, no matter how far-fetched they may seem, were sober attempts by the leading scholars of the period to come to grips with what they believed to be fundamental problems. The mere fact that these efforts did not succeed or that they were predicated on a wildly improbable methodology does not make them any the less serious" (p. 16). Number mysticism has been taken very seriously, or at least some aspects of it have, by some of the most productive mathematicians and scientists the world has seen and by intelligentsia in other fields as well (Yates, 1964). Other equally productive and intelligent people have found it easy to dismiss it in its entirety. Still others may give it more credence than they realize, because they know it by other names.

Many of the properties of numbers discovered by ancient and medieval thinkers are not qualitatively different from the properties that get attention from number theorists today. What makes numerology different from number theory is the imputation by numerologists of mystic significance to numbers and the relationships among them. Number theorists find numbers and the relationships among them interesting in their own right, independently of either practical or metaphysical applications that might be made of their discoveries.

☐ The Pure Versus Applied Distinction

> The pure mathematician is much more of an artist than a scientist. He does not simply measure the world. He invents complex and playful patterns without the least regard for their practical applicability. (Watts, 1964, p. 37)

Pythagoras is often credited with originating the belief that mathematics is worth studying for its own sake, quite apart from any practical applications it might have, or, if not originating it, at least promoting it and giving it a form that would last until the present day. Euclid's contempt for the idea that the reason for pursuing mathematics is its practical value is seen in the often-told story of his response to a question from a student regarding what advantage he would gain by learning geometry. Euclid is said to have instructed his slave to "give him three pence, since he must make profit out of what he learns."

There is more than a hint of intellectual snobbery in the way the distinction between pure and applied mathematics has been articulated, at least by those who consider themselves on the pure side of this divide. The very use of the word *pure*, with its connotation of untainted and its moral overtones, to designate the doing of mathematics for its own sake reveals this prejudice. Plato deserves to share the credit with Pythagoras for the establishment of this view—a view that is not unique to mathematics, it must be said—because of his general association of usefulness with ignobility.

The disdain that some mathematicians have for the idea that mathematics gets its primary justification from its practical usefulness is seen also in a comment by Dantzig (1930/2005). "The mathematician may be compared to a designer of garments, who is utterly oblivious of the creatures whom his garments may fit. To be sure, his art originated in the necessity for clothing such creatures, but this was long ago; to this day a shape will occasionally appear which will fit into the garment as if the garment had been made for it. Then there is no end of surprise and of delight!" (p. 240). Court (1935/1961) makes a similar observation, and dismisses the idea that the solving of practical problems motivates most mathematicians. "He [the mathematician] likes to exercise his inventiveness and to display it before others, namely before those who can get as excited about it as he does himself. That is the best that can be said in defense of most of mathematics" (p. 216).

These sentiments are echoed in a comment by G. H. Hardy (1940/1989) in his *Apology* that is often quoted to illustrate the attitude that pure mathematicians sometimes convey about the possibility of using practical utility as an appropriate measure of the merit of their work:

> I have never done anything "useful." No discovery of mine has made, or is likely to make, directly or indirectly, for good or ill, the least difference to the amenity of the world. I have helped to train other mathematicians, but mathematicians of the same kind as myself, and their work has been, so far at any rate as I have helped them to do it, as useless as my own. Judged by all practical standards, the value of my mathematical life is nil; and outside mathematics it is trivial anyhow. (p. 150)

In an even more sweeping statement, Hardy gave essentially the same verdict with respect to mathematics generally: "If useful knowledge is, as we agreed provisionally to say, knowledge which is likely, now or in the comparatively near future, to contribute to the material comfort of mankind, so that mere intellectual satisfaction is irrelevant, then the great bulk of higher mathematics is useless" (p. 135).

It is difficult to know to what extent Hardy intended that these comments be taken at face value. He in fact did some mathematical work that was useful in the sense of being descriptive of real phenomena. A case in point is his work on population statistics, which led to the Hardy-Weinberg equilibrium, a law of genetics that relates gene frequencies across generations. It seems clear that Hardy was not apologizing, in the common sense of that word, for the way he had chosen to spend his life. He did not equate usefulness, as he used the term, with worth. He claimed of his own work that it was not useful. He did not claim that it lacked worth; indeed, quite the contrary. His judgment of his own life was that he had added something to knowledge, and that what he had added had a value similar to that of the creations of "any of the other artists, great or small, who have left some kind of memorial behind them" (p. 151).

Hardy contended that what distinguishes the best mathematics is its *seriousness*. As for what constitutes seriousness: "The 'seriousness' of a mathematical theorem lies, not in its practical consequences, which are usually negligible, but in the *significance* of the mathematical ideas which it connects. We may say, roughly, that a mathematical idea is 'significant' if it can be connected, in a natural and illuminating way, with a large complex of other mathematical ideas" (p. 89).

Dantzig (1930/2005) expresses a similar perspective in arguing that mathematical achievement is not to be measured by the scope of its applicability to physical reality but rather by standards that are peculiar to itself. "These standards are independent of the crude reality of our senses. They are: freedom from logical contradiction, the generality of the laws governing the created form, the kinship which exists between this new form and those that have preceded it" (p. 240).

Hardy acknowledged that it is difficult to be precise about what constitutes mathematical significance, but suggested that essential aspects include generality and depth. A general mathematical idea is one that figures in many mathematical constructs and is used in the proofs of different kinds of theorems. Hardy considered the concept of depth, in this context, to be very difficult to explain, but he believed it to be one that mathematicians would understand. Mathematical ideas, in his view, are arranged in strata representing different depths. The idea of an irrational number is deeper, for example, than that of an integer. Sometimes, though not always, in order to understand relationships at a given level, it is necessary to make use of concepts from a deeper level.

☐ The Joy of Discovery

Mathematicians do mathematics for a variety of reasons—to earn a living, to attain recognition, to contribute to the shared knowledge of the species. Not least among their motivations is the satisfaction they derive from working on and solving challenging problems. Byers (2007) expresses this reason somewhat rhapsodically: "Why do mathematicians work so hard to produce original mathematical results? Is it merely for fame and fortune? No, people do mathematics because they love it; they love the agony and the ecstasy. The ecstasy comes from accessing this realm of knowing, of certainty. Once you taste it, you can't but want more. Why? Because the creative experience is the most intense, most real experience that human beings are capable of" (p. 341).

The passion that mathematicians can have for their subject is seen in their descriptions of the sense of elation that solving a recalcitrant problem can bring. There are many accounts of this type of experience. The story of Archimedes running naked through the streets of Syracuse shouting "eureka" after solving the problem of determining whether the gold in the king's crown had been diluted with baser metal is presumably apocryphal, but it is a fine metaphor for the feeling that many mathematicians have described having upon finding the solution to a problem that had been consuming them.

Aczel (1996) reports Andrew Wiles's reaction upon finally seeing how to make his proof of Fermat's last theorem (that there exists no integral solution of $x^n + y^n = z^n$ for $n > 2$) work: "Finally, he understood what was wrong. 'It was the most important moment in my entire working life,' he later described the feeling. 'Suddenly, totally unexpectedly, I had this incredible revelation. Nothing I'll ever do again will ...' at that moment tears welled up and Wiles was choking with emotion. What Wiles realized at that fateful moment was 'so indescribably beautiful, it was so simple and so elegant ... and I just stared in disbelief'" (p. 132). Singh (1997) has chronicled Wiles's epic struggle—his obsession, as Wiles referred to it—with Fermat's last theorem, from his introduction to it as a 10-year-old boy through the 7 or 8 years of solitary concentration on this single problem as a professional mathematician, the exhilaration and acclaim that attended his announcement of a proof in 1993, the discouragement and depression that came with the discovery that the proof was flawed, and the elation that came again with the final repair of it and acceptance of its authenticity by the mathematical community. It is a gripping, emotionally charged story.

Du Sautoy (2004) quotes contemporary French mathematician Alain Connes describing his discovery of the joy of mathematics as a young boy—"I very clearly remember the intense pleasure that I had plunging into the special state of concentration that one needs in order to do mathematics"—and his comment as an adult that mathematics "affords—when one is fortunate enough to uncover the minutest portion of it—a sensation of extraordinary pleasure through the feeling of timelessness that it produces" (p. 305).

This is not to suggest, of course, that mathematicians live in a perpetual state of euphoria. Like everyone else, they have their ups and downs, but by their own testimony, many of them derive moments of extraordinary pleasure from working on, and especially from solving, challenging mathematical problems. It appears that, at least for some mathematicians, the joy of discovery is sufficiently strong to motivate long stretches of intense work between episodes of experiencing it. American mathematician Lipman Bers puts it this way:

> I think the thing which makes mathematics a pleasant occupation are those few minutes when suddenly something falls into place and you understand. Now a great mathematician may have such moments very often. Gauss, as his diaries show, had days when he had two or three important insights in the same day. Ordinary mortals have it very seldom. Some people experience it only once or twice in their lifetime. But the quality of this experience—those who have known it—is really joy comparable to no other joy. (Quoted in Hammond, 1978, p. 27)

Poincaré limited to a privileged few the full experience of the esthetic pleasure that mathematics can provide: "Adepts find in mathematics delights analogous to those that painting and music give. They admire the delicate harmony of numbers and of forms; they are amazed when a new discovery discloses to them an unlooked-for perspective, and the joy they thus experience has it not the esthetic character, although the senses take no part in it? Only the privileged few are called to enjoy it fully, but is it not so with all the noblest arts?" (quoted in Court, 1935/1961, p. 127). Bers surmises too that the enjoyment of mathematics experienced by mathematicians is largely unrealized by nonmathematicians, and expresses disappointment that mathematics does not compare favorably with music and art in this regard, which he assumes can be enjoyed by people other than musicians and artists. That this state of affairs is not of great concern to mathematicians generally is suggested in a responding comment by French mathematician Dennis Sullivan: "I don't particularly feel that it's important that mathematics be enjoyed by a lot of people; that would be very nice, but it's more important that mathematicians work on good problems and pursue mathematics" (quoted in Hammond, 1978, p. 33).

☐ Beauty and Elegance

Mathematics, rightly viewed, possesses not only truth but supreme beauty—a beauty cold and austere, like that of sculpture, without appeal to any part of our weaker nature, without the gorgeous trappings of painting or music, yet sublimely pure, and capable of a stern perfection such as only the greatest art can show. (Russell, 1910, p. 73)

Many mathematicians have, like Russell, lauded the beauty or elegance that is to be seen in mathematics, at least by the discerning eye, and have described mathematics as a form of art. Here are a few of the numerous examples of such expressions that can be found.

- "An elegantly executed proof is a poem in all but the form in which it is written" (Kline, 1953a, p. 470).

- "I am inclined to believe that one of the origins of mathematics is man's playful nature, and for this reason mathematics is not only a Science, but to at least some extent also an Art" (Péter, 1961/1976, p. 1).

- "The esthetic side of mathematics has been of overwhelming importance throughout its growth. It is not so much whether a theorem is useful that matters, but how elegant it is" (Ulam, 1976, p. 274).

- "The motivation and standards of creative mathematics are more like those of art than like those of science. Aesthetic judgments transcend both logic and applicability in the ranking of mathematical theorems: beauty and elegance have more to do with the value of a mathematical idea than does either strict truth or possible utility" (Steen, 1978, p 10).

- "Mathematics always follows where elegance leads" (Kaplan, 1999, p. 71).

- [Regarding a proposal to apply a particular set-theory concept to transfinite numbers] "The proposal is a piece of mathematical legislation, to be assessed, if at all, in terms of its power, elegance and beauty" (Moore, 2001, p. 151).

- "The poetry of science is in some sense embodied in its great equations" (Farmelo, 2003b, p. xi).

Kac (1985) describes his reaction upon reading a recommended book for the purpose of understanding Dedekind cuts. "As I read, the beauty of the concept hit me with a force that sent me into a state of euphoria. When, a few days later, I rhapsodized to Marceli [Marceli Stark, an older fellow student who had recommended the book] about Dedekind cuts—in fact, I acted as if I had discovered them—his only comment was that perhaps I had the makings of a mathematician after all" (p. 32).

The esthetic appeal of mathematics and the desire to produce mathematical results that will be seen as beautiful or elegant both by themselves and by others presumed to be qualified to have an opinion on the matter, appear to be very strong among mathematicians, especially those who have been most influential in shaping the field. Mathematicians often speak of beauty as a criterion by which the results of a mathematician's work should be judged, as in the comment of Hardy quoted at the beginning of this chapter. Du Sautoy (2004) sees an attraction to beauty as something innate to the mathematical mind. "The esthetic sensibilities of the mathematical mind are tuned to appreciate proofs that are beautiful compositions and shun proofs which are ugly" (p. 210). Hadamard (1945/1954) also expressed a similar sentiment in arguing that in mathematics the sense of beauty is not just *a* drive for discovery, but "almost the only useful one" (p 103); the idea of the *usefulness* of beauty in this context invites reflection.

Noting how remarkable it is that the mathematical world so often favors the most esthetic construction, Du Sautoy (2004) points out that Riemann's hypothesis (see p. 4), a proof of which has been sought by many top-grade mathematicians, "can be interpreted as an example of a general philosophy among mathematicians that, given a choice between an ugly world and an esthetic one, Nature always chooses the latter" (p. 55). The assumption appears to be that the beauty that is to be found in mathematics is a reflection of the beauty of physical reality. Many scientists have also taken the position that beauty should be a goal in the construction of scientific theories.

What is the basis of this interest in beauty? Is it the same in both mathematics and science? Is it rational, in either case, to expect or demand that the products of the discipline satisfy such a criterion? Is there an underlying assumption that the proper business of mathematics and science is to discover what can be discovered about reality and that truth—mathematical and physical—when seen as clearly as possible, must be beautiful? If the demand for beauty stems from some such assumption, is the assumption itself an article of blind faith? If such an assumption is not its basis, what is?

Whatever its basis, interest among mathematicians in beauty is undeniable and appears to be very strong. But what is beauty in this context? Indeed, what is beauty generally? The concept is exceedingly difficult to pin down with an objective definition. The dictionary and thesaurus are of remarkably little help. An observation by Mortimer Adler (1981), who includes beauty in his list of six great ideas, is telling in this regard. "The test of the intelligibility of any statement that overwhelms us with its air of profundity is its translatability into language that lacks the elevation and verve of the original statement but can pass muster as a simple and clear statement in ordinary, everyday speech. Most of what has been written about beauty will not survive this test. In the presence of many of the most eloquent statements about beauty, we are left speechless—speechless in the sense that we cannot find other words for expressing what we think or hope we understand" (p. 103).

There can be little doubt that beauty is subjective to a large extent: People differ greatly on what they consider beautiful or ugly. Arguably, though, there is likely to be greater agreement about what is beautiful in any particular context among people who are highly familiar with that context than among people who are not. Students of 17th-century Dutch paintings are more likely to agree on what constitutes a beautiful painting from this period and place than are people who have little knowledge of art; baseball fans are likely to be more consistent in distinguishing between beautiful and commonplace swings of a baseball bat than are people who have little knowledge of, or interest in, baseball. Similarly, one would expect to find greater agreement among mathematicians than among nonmathematicians regarding what should be considered beautiful in mathematics.

Adler (1981) makes a distinction between *enjoyable* beauty and *admirable* beauty. Enjoyable beauty is purely subjective, totally "in the eye of the beholder;" in contrast, admirable beauty is beauty that, by consensus of those presumably qualified to judge by virtue of their familiarity with the appropriate domain, meets some objective standards. The latter kind of beauty deserves to be appreciated; one should strive to appreciate it, and this may require acquiring some expertise in the domain. This line of thinking, if one accepts the distinction, prompts the question of what may constitute admirable beauty in mathematics.

Is it simplicitly? Regularity? Symmetry? Rigor? What is it about the perfect solids that makes them perfect in the eyes of those who gave them that label? That all the faces of a particular one are identical? That all the vertices of any given one will lie on a superscribed sphere? That the center of each one coincides with the center of its superscribed sphere? What prompted Moore (2001) to say of analytic geometry, which he describes as the casting of one whole body of mathematics in terms of another, that it is one of the greatest monuments to mathematical excellence because of its beauty, depth, and power? Can beauty be described or defined in a noncircular way? And in such a way that nonmathematicians will be able to recognize it when they see it?

What is it about the equation, attributed to Euler,

$$e^{i\pi} + 1 = 0$$

that many mathematicians find so beautiful? King (1992) points out that it satisfies what he refers to as the "aesthetic principle of minimal completeness" (p. 86). It contains the five most important constants of mathematics—e, i, π, 1, and 0—as well as the "paramount operations" of addition, multiplication, and exponentiation, and the "most vital relation" of equality, and nothing else. Some may find this to be an adequate explanation of why the equation appears to many to be beautiful; others may not. Another answer is that, to one who must ask, there is unlikely to be a persuasive answer. Wilczek (2003) more or less makes this point in contending that "it is difficult to make precise, and all but impossible to convey to a lay reader, the nature of mathematical beauty" (p. 158).

Although the foregoing equation is often presented as though it were invented, or discovered, by Euler in a vacuum, it follows from the relationship

$$e^{i\pi} = \cos \pi + i \sin \pi,$$

but it is nonetheless beautiful for that. For the reader who may be unfamiliar with the equivalence between the exponential function and the sum of two trigonometric functions, it is easily seen when each of the component expressions is represented in its power-series form:

$$\sin x = x - \frac{x^3}{3!} + \frac{x^5}{5!} - \frac{x^7}{7!} + \cdots$$

$$\cos x = 1 - \frac{x^2}{2!} + \frac{x^4}{4!} - \frac{x^6}{6!} + \cdots$$

$$e^x = 1 + x + \frac{x^2}{2!} + \frac{x^3}{3!} + \frac{x^4}{4!} + \frac{x^5}{5!} + \frac{x^6}{5!} + \frac{x^7}{7!} + \cdots.$$

Given that $i^2 = -1$, we can write

$$e^{ix} = 1 + ix + \frac{(ix)^2}{2!} + \frac{(ix)^3}{3!} + \frac{(ix)^4}{4!} + \frac{(ix)^5}{5!} + \frac{(ix)^6}{6!} + \frac{(ix)^7}{7!} + \cdots$$

$$= 1 + ix - \frac{x^2}{2!} - i\frac{x^3}{3!} + \frac{x^4}{4!} + i\frac{x^5}{5!} - \frac{x^6}{6!} - i\frac{x^7}{7!} + \cdots$$

$$= \left(1 - \frac{x^2}{2!} + \frac{x^4}{4!} - \frac{x^6}{6!} + \cdots \right) + i\left(x - \frac{x^3}{3!} + \frac{x^5}{5!} - \frac{x^7}{7!} + \cdots \right)$$

$$= \cos x + i \sin x.$$

When $x = \pi$, $\cos x = -1$ and $\sin x = 0$, so we can write $e^{i\pi} = \cos \pi + i \sin \pi = -1$, or equivalently, $e^{i\pi} + 1 = 0$.

Do mathematicians agree among themselves as to what is beautiful or elegant in mathematics and what is not? In fact, although it is claimed that mathematicians will generally agree in their judgments about which mathematical results deserve this description, there appears to be no well-supported account of what determines beauty or elegance in this context. (For examples of what many would consider unbeautiful equations that work, see du Sautoy, 2004, pp. 143, 200.)

King (1992) makes the interesting observation that although there exists a voluminous philosophical literature on esthetics, numerous works on the role of mathematics in art, and a few mathematical models of esthetics (a notable pioneering example of which is that of American mathematician George Birkhoff [1933]), very little attention has been given to mathematics *as* art—to the esthetics of mathematics per se. It appears, he suggests, that estheticians, like other educated people outside the mathematical aristocracy, have little concept of mathematical beauty or even awareness that such a thing exists. Mathematicians have no doubt of the beauty of their subject, but they have been content, for the most part, to enjoy it privately without taking the trouble—and trouble it would undoubtedly be—to shine the light in such a way that the rest of us could get a glimpse of it too. As King puts it, "Mathematicians have talked mathematics only to each other and about mathematics to no one at all" (p. 224).

King suggests that mathematicians judge the esthetic quality of mathematical ideas in terms of two principles—those of minimal completeness and maximal applicability:

> A mathematical notion N satisfies the principle of minimal completeness provided that N contains within itself all properties necessary to fulfill its mathematical mission, but N contains no extraneous properties.

> A mathematical notion N satisfies the principle of maximal applicability provided that N contains properties that are widely applicable to mathematical notions other than N (p. 181).

King sees these statements as somewhat more precise representations of the principles that Hardy was hinting at in his discussion of elegance in *A Mathematician's Apology*, in which he used such descriptors as seriousness, depth, economy, and generality. The principle of minimal completeness, which King associates with Occam's razor, requires that an elegant mathematical construct be complete but free of extraneous notions. For a proof to be elegant, for example, nothing essential can be missing, but nothing unessential can be contained. King's second principle articulates the idea that, other things being equal, the greater the range of applicability of the construct, the more elegant it is. Here elegance is more or less equated with conceptual power.

Weyl (1940/1956) was much impressed with symmetry as a mark of beauty: "Symmetry, as wide or as narrow as you may define its meaning, is one idea by which man through the ages has tried to comprehend and create order, beauty, and perfection" (p. 672). (For a very readable discussion of the concept of symmetry and its importance in mathematics and science, see Stewart and Golubitsky, 1992.) Consistent with this view is Weyl's description of the discoveries of the existence of the two most complex regular polyhedra—the dodecahedron (the 12-sided solid, each face of which is a regular pentagon) and the icosahedron (the 20-sided solid bounded by equilateral triangles)—as among the most beautiful in the history of mathematics.

Kline's (1956/1953) appraisal of projective geometry, which he sees as unique in many ways among the subareas of mathematics, references a number of characteristics that are at least suggestive of beauty: "No branch of mathematics competes with projective geometry in originality of ideas, coordination of intuition in discovery and rigor in proof, purity of thought, logical finish, elegance of proofs and comprehensiveness of concepts" (p. 641).

Among the more surprising searches for beauty in mathematics is that which involves prime numbers. To the nonmathematical eye the sequence of primes 2, 3, 5, 7, 11, 13, 17, 19, 23, 29, ... seems the epitome of haphazardness and unpredictability, hardly anything that could be described as elegant. But the attempt to "tame" the primes, to find a way to predict specific properties of the sequence, which was proved by Euclid to go on indefinitely, has engaged many of the greatest mathematicians, including Gauss, Riemann, Hilbert, and Hardy. There appears to be a deep belief among many mathematicians that there is a unique kind of beauty—du Sautoy (2004) calls it "the music of the primes"—in the prime sequence and it is just a matter of discovering precisely what it is.

Davis and Hersch (1981, p. 198) point to unification—"the establishment of a relationship between seemingly diverse objects"—as not only one of the great motivating forces in mathematics, but also one of the great sources of esthetic satisfaction in this field. The search for uniformity in what appear to be diverse phenomena—unity in diversity— has been a theme common to mathematics and science over the history of both.

Although there are some similarities in these references to mathematical elegance and beauty, and in others that could be cited, it is difficult to infer a definition to which all mathematicians would subscribe, King's thought-provoking proposal notwithstanding. One cannot expect to see beauty in mathematics in the absence of some minimal understanding of mathematics as a discipline. Mathematicians may differ among themselves as to precisely what determines whether a particular mathematical result should be considered beautiful, but my sense is that the more one understands of mathematics, the more likely one is to see beauty not only in many specific mathematical results, but also in the enterprise as a whole.

The following algebraic identity, which has been known at least since the publication of Fibonacci's *Liber abaci* in 1202, strikes me as beautiful, independently of any usefulness (unknown to me) that it may have:

$$(a^2 + b^2)(c^2 + d^2) = (ac \pm bd)^2 + (ad \quad bc)^2$$

where *a*, *b*, *c*, and *d* are any integers. What this identity says is that the product of two sums of two squares of integers is always expressible, in two different ways, as the sum of two squares of integers. The relationship may be written alternatively as

$$(a^2 + b^2)(c^2 + d^2) = u^2 + v^2 = x^2 + y^2$$

where

$$u = ac + bd,$$

$$v = ad - bc,$$

$$x = ac - bd,$$

and

$$y = ad + bc.$$

Try it with four random numbers, for *a*, *b*, *c*, and *d*. I just did it with 2, 34, 19, and 7.

$$(2^2 + 34^2)(19^2 + 7^2) = 475,600 = 276^2 + (-632)^2 = (-200)^2 + 660^2$$

Bell (1945/1992) claimed that "with the appropriate restrictions as to uniformity, continuity, and initial values when *a*, *b*, *c*, *d* are functions of one variable, the identity contains the whole of trigonometry" (p. 103). What he meant by that, I cannot say.

☐ Wonder

Dante Alighieri was fascinated by the fact that any angle inscribed inside a semicircle, no matter where on the semicircle its vertex lies, is a right angle; to him, this was wonder evoking. This relationship, illustrated in Figure 11.1, is known to us as the theorem of Thales, but the Babylonians were aware of it over a millennium before Thales of Miletus stated it.

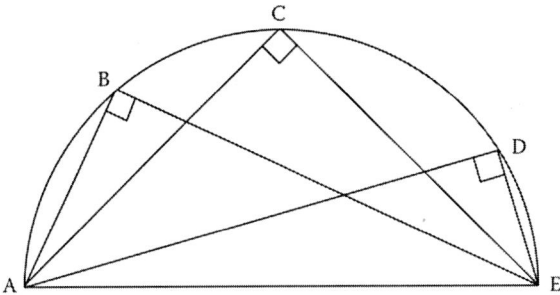

FIGURE 11.1 Illustrating the theorem of Thales, according to which any angle inscribed in a semicircle, such as ABE, ACE, and ADE, is a right angle.

Socrates, in his youth, wondered about why the sum and product of two 2s are the same. (Socrates did not give much thought to mathematics in his adult life, claiming to find the results of certain fundamental arithmetic operations incomprehensible.) The contemporary mathematician Richard Hamming (1980) confesses to being amazed by the possibility of abstracting numbers from the things numbered—that the universe is constructed in such a way that three things and four things always add to seven things, no matter what the things are that are counted. Peter Atkins (1994) calls mathematics "the profound language of nature, the apotheosis of abstraction and the archenabler of the applied" and argues that realization of this "should stir us into wondering whether herein lies the biggest hint of all about our origin and the origin of our understanding" (p. 95).

The childlike wonder that Einstein maintained about space and time is well documented, as is the fact that he himself attributed to it, at least in part, his formulation of the theory of relativity (Holton, 1973). Bertrand Russell (1955/1994) speaks of Einstein's expression of "'surprised thankfulness' that four equal rods can make a square, since, in most of the universes he could imagine, there would be no such things as squares" (p. 408).

What evokes wonder, or even awe, in one person may appear mundane to another. This could be because the person who wonders lacks the knowledge that reveals the mundaneness of the object of wonderment, or it could be because the one who does not wonder lacks the sensitivity or insightfulness to appreciate that a mystery is involved. A question that has not received much attention from researchers, to my knowledge, is whether a plausible objective basis for wonderment can be identified. What inspires wonder? Is it possible to specify the conditions under which wonder *should* be evoked?

These are important questions from an educational point of view. The teaching and learning of mathematics are discussed in later chapters of this book, but it seems appropriate to note here that wonderment is seldom mentioned in the literature on the teaching and learning of mathematics as a motivating factor in those contexts. This seems to me to indicate missed opportunities to provide young students with a sense of what a rewarding intellectual adventure the acquisition of mathematical knowledge can be. Whether wonder can be taught is perhaps itself a question for research. My suspicion is that, like other attitudes or perspectives, its most likely form of transmission is by contagion. Unhappily, there is little reason to doubt that a sense of wonder can be stifled if put down whenever it spontaneously appears.

☐ The Pythagorean Theorem

Every school child who has studied basic algebra or plane geometry has encountered the Pythagorean theorem, according to which the sum of the squares of the lengths of the two sides forming the right angle of a right triangle equals the square of the length of the side opposite that angle—the hypotenuse. That is, if X and Y are the two sides forming the right angle and Z is the hypotenuse,

$$X^2 + Y^2 = Z^2.$$

Although discovery of this relationship, which must be one of the best known of all mathematical theorems, is credited to the Pythagoreans, the relationship expressed by it was known by people in other cultures long before the time of Pythagoras. The Pythagoreans may have discovered it independently, however, and, in any case, they found the relationship to be awe inspiring and wondered much about it. Kepler referred to the Pythagorean theorem as one of "two great treasures" of geometry. (Kepler's other great treasure is discussed in the next section.)

In a series of articles in the *American Mathematical Monthly*, B. F. Yancey and J. A. Calderhead (1886, 1887, 1888, 1889) presented 100 proofs of the Pythagorean theorem. More recently, Elisha Loomis (1968) collected and classified 367 proofs of the theorem. A collection of 79 proofs, some with interactive illustrations, can be seen at http://www.cut-the-knot.org/pythagoras/index.shtml. It is doubtful if any other mathematical theorem has been proved in a greater number of ways than has the theorem of Pythagoras. Few, if any, other theorems have held the attention of so many for so long. While proving the theorem is not difficult, constructing a proof that differs from all those that have already been produced is a considerable challenge.

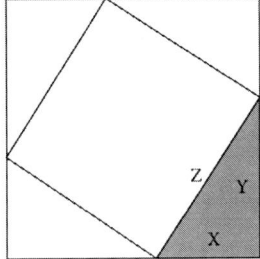 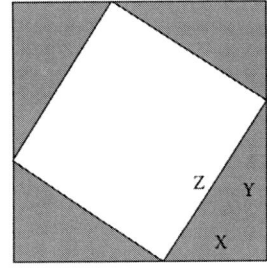

FIGURE 11.2 Geometric demonstration that a square erected on the hypotenuse of a right triangle equals the sum of the squares of the triangle's other two sides.

A particularly elegant proof, in my view, is represented in Figure 11.2. To prove: that given a right triangle with sides X, Y, and Z, Z the hypotenuse, $X^2 + Y^2 = Z^2$. Draw a square on Z and a circumscribing square with X and Y representing the corners of one side, as shown in Figure 11.2, left. It should be clear that the triangles around the inner square are all congruent, that is, have sides X, Y, and Z, as also represented in Figure 11.2, right. Inasmuch as the area of the inscribed square equals the area of the circumscribing square minus the areas of the four triangles, we have

$$Z^2 = (X + Y)^2 - 4\left(\frac{XY}{2}\right) = X^2 + 2XY + Y^2 - 2XY = X^2 + Y^2.$$

A similarly simple and elegant demonstration is shown in Figure 11.3, from which we get the equation

$$Z^2 = (Y - X)^2 + 4\left(\frac{XY}{2}\right) = Y^2 - 2XY + X^2 + 2XY = X^2 + Y^2.$$

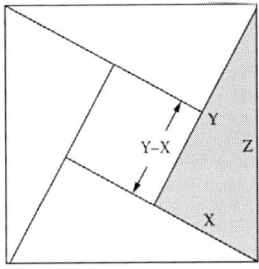

FIGURE 11.3 Another simple geometric demonstration that the square of the hypotenuse of a right triangle equals the sum of the squares of the other two sides.

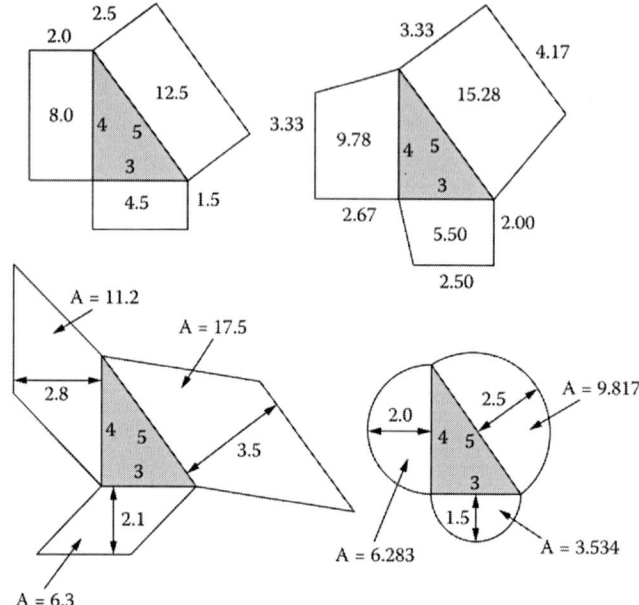

FIGURE 11.4 Illustrating with nonsquare rectangles (upper left), trapezoids (upper right), rhombuses (lower left), and semicircles (lower right) that when similar figures are erected on the sides of a right triangle, the area of the figure on the hypotenuse is equal to the sum of the areas of the figures erected on the other two sides. (The numbers in this figure and the following three are approximate.)

Although the Pythagorean theorem, as represented by the preceding equation, is well known, what is undoubtedly much less well known is that if any *similar* figures are erected on the three sides of a right triangle, the combined areas of the two smaller figures will equal the area of the largest one. Figure 11.4 illustrates this with rectangles, trapezoids, rhombuses, and semicircles. Figure 11.5 shows the same relationship with a variety of triangles. It even holds with circles, the diameters of which are equal to (or a fixed multiple of) the sides to which the circles are tangent, as illustrated in Figure 11.6.

These relationships may seem peculiar when first encountered, but they all follow from the simple facts that

1. if $X^2 + Y^2 = Z^2$, then $kX^2 + kY^2 = kZ^2$, where k is any constant;

2. the area of any regular planar figure, say W, can be expressed as a multiple of the area of a specified square; and

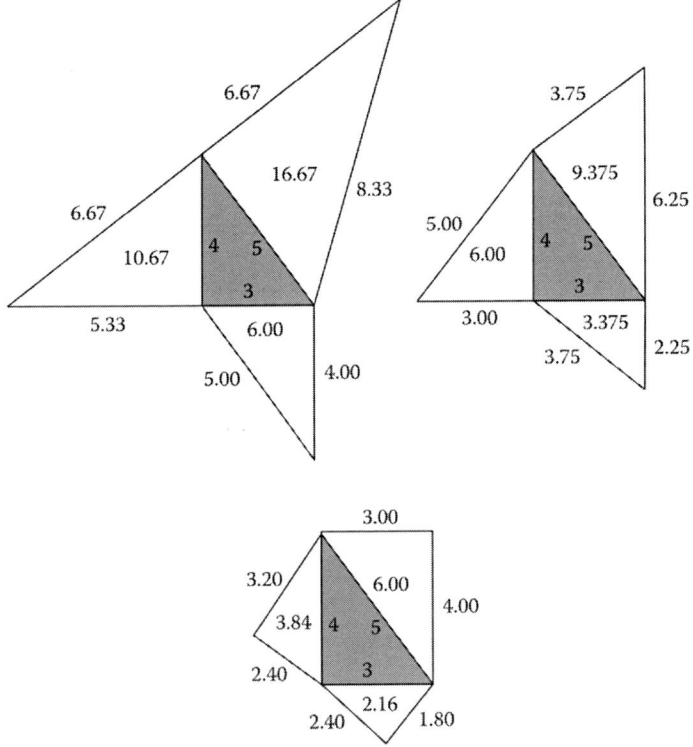

FIGURE 11.5 In each of these figures, the sides of the individual triangles are in the ratio of 3:4:5. In the upper left figure, the sides of the center triangle serve as the shorter of the sides forming the right angle of each of the abutting triangles. In the upper right figure, the sides of the center triangle serve as the longer of the sides forming the right angle of each of the abutting triangles. In the bottom figure, the sides of the center triangle serve as the hypotenuse of each of the abutting triangles. In all cases, the area of the largest abutting triangle is equal to the sum of the areas of the other abutting triangles.

3. if the figure W is erected on each side of the triangle such that the ratio of its area on any given side of the triangle to the area of the square on that side is the same for all three sides, the sum of the areas of W on the sides that form the right angle will be equal to the area of W erected on the hypotenuse.

In the illustrations in Figure 11.4, the values of k (the ratios of the areas of the figures to the areas of squares erected on the same sides) are 0.5, 0.611 (approximately), 0.393 (approximately), and 0.7 for the rectangles, trapezoids, semicircles, and rhombuses, respectively. For the

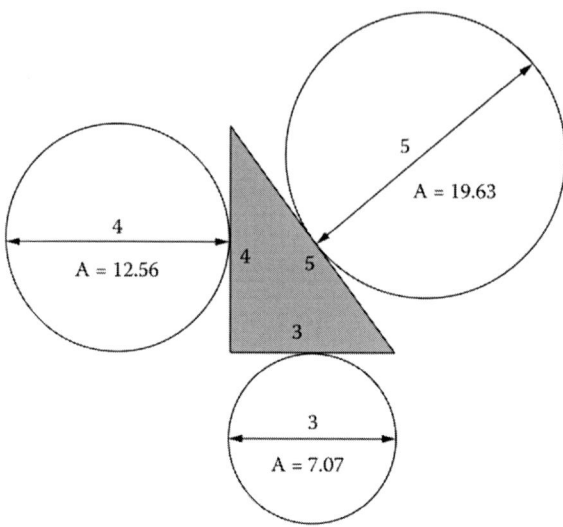

FIGURE 11.6 The diameter of each of the abutting circles is equal to the length of the side of the triangle that it abuts. The area of the largest circle is equal to the sum of the areas of the other two circles (area numbers are approximate).

triangles in Figure 11. 5, the values of k, clockwise from the upper left, are 0.667 (approximately), 0.375, and 0.24. For the circles in Figure 11.6, k = π/4 = 0.785 (approximately). In these illustrations, k < 1 in all cases, but that is not essential. If any regular polygon of more than four sides is erected on each side of a right triangle, using the side of the triangle as one of the sides of the polygon, the area of the polygon erected on the hypotenuse will equal the sum of the areas of those erected on the other sides. The values of k in each case will be greater than 1; for pentagons (see Figure 11.7) it will be 1.72; for hexagons, 2.60; for heptagons, 3.63 (all approximately); and so on. For convenience I have used three, four, and five triangles in all of the illustrations, but what is illustrated holds for all right triangles. Also, the figures in the illustrations are all simple geometric shapes, but this too is unnecessary. Imagine the outline of a bust of Pythagoras being drawn on the hypotenuse of a right triangle, and suppose that the ratio of the area enclosed by this figure to the area of a square erected on the hypotenuse to be k. If similar busts of Pythagoras are drawn on the other sides of the triangle such that the ratio of the areas enclosed to the areas of squares drawn on those sides is also k, then the sum of the areas within the bust outlines on the smaller sides will equal the area within the bust outline on the hypotenuse. One is led to wonder if the relationship represented by the Pythagorean theorem would be seen in any different light if the original discovery had been that the area

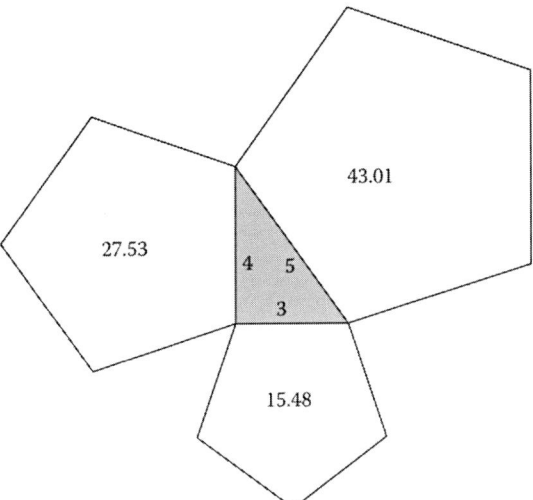

FIGURE 11.7 Illustrating that the area of a regular pentagon erected on the hypotenuse of a right triangle is equal to the sum of the areas of regular pentagons erected on the other two sides. The ratio of the areas of the pentagons to the areas of squares erected on the sides of the same triangle is approximately 1.72.

of an arbitrary figure erected on the hypotenuse of a right triangle would equal the sum of the areas of *similar* figures erected on the other two sides. Would realization of the generality of this relationship have astonished Kepler, or made the relationship with squares mundane?

☐ The Golden Ratio

The second great treasure of geometry, in Kepler's view, was what Euclid had referred to as the "division of a line into extreme and mean ratio"; it is known by various names, including golden ratio, golden section, and divine proportion. Consider a line segment of unit length. There is a unique point such that if the segment is divided at that point, the ratio of the larger segment—call it x—to the smaller is the same as the ratio of the whole segment to the larger segment, which is to say $\frac{x}{1-x} = \frac{1}{x}$. The value of this point, x, is approximately .618033988.... For this (and only this) value of x, not only is the ratio of the larger segment to its complement (the smaller segment) the same as the ratio of the whole to the larger segment, as the preceding equation indicates, but, as is easily verified, that the ratio of the smaller segment to the larger is equal to the larger,

$$\frac{1-x}{x} = x$$

the value of the larger segment plus 1 is equal to the reciprocal of the larger segment,

$$x + 1 = \frac{1}{x},$$

and the square of the larger segment is equal to its complement,

$$x^2 = 1 - x.$$

The ratio of the larger segment to the smaller, $\frac{x}{1-x}$, is conventionally represented by φ, has the value $\frac{\sqrt{5}+1}{2}$, or approximately 1.618, and is referred to as the golden ratio, golden section, golden number, golden mean, divine proportion, or, in Euclid's terms, extreme and mean ratio. *Golden ratio* is a misnomer, in a way, inasmuch as the ratio in this case is an irrational number, the value of which has been worked out to thousands of decimal places. Here 1.618 will suffice for our purposes. The ratio has a unique relationship, one might say, with itself; add 1 to it and you get its square,

$$\phi + 1 = \phi^2, \quad \text{or} \quad \phi = \phi^2 - 1,$$

subtract 1 from it and you get its reciprocal,

$$\phi - 1 = \frac{1}{\phi}, \quad \text{or} \quad \phi = 1 + \frac{1}{\phi}$$

and so on. These relationships provide a basis for some remarkable equations for the value of φ. The following examples are provided by Livio (2002). Consider

$$y = \sqrt{1 + \sqrt{1 + \sqrt{1 + \sqrt{1 + \dots}}}} .$$

Inasmuch as the square of the right-hand term is

$$1 + \sqrt{1 + \sqrt{1 + \sqrt{1 + \sqrt{1 + \dots}}}},$$

or $1 + y$, we have

$$y = y^2 - 1.$$

So the value of the golden ratio is the limit of this, some might say elegant, expression composed entirely of 1s:

$$\phi = \sqrt{1 + \sqrt{1 + \sqrt{1 + \sqrt{1 + \dots}}}}.$$

Or consider the continuing fraction

$$y = 1 + \cfrac{1}{1 + \cfrac{1}{1 + \cfrac{1}{1 + \cfrac{1}{1 + \dots}}}}$$

Inasmuch as the denominator of the second term on the right-hand side is y, we have

$$y = 1 + \frac{1}{y}.$$

So another elegant expression for ϕ, composed of nothing but 1s, is

$$\phi = 1 + \cfrac{1}{1 + \cfrac{1}{1 + \cfrac{1}{1 + \cfrac{1}{1 + \dots}}}}$$

The reader who finds such relationships fascinating will get much pleasure from Livio's exploration of this most unusual number.

To say that many people have found the golden ratio to be intriguing is an understatement. It has fascinated mathematicians at least since the days of the classical Greeks. It has been the subject of numerous books over the ages (the Italian mathematician Luca Pacioli published a book about it in 1509 that contains drawings by Leonardo da Vinci), and it continues to be a popular topic for writers on mathematics (e.g., Dunlap, 1997; Herz-Fischler, 1998; Huntley, 1970; Livio, 2002; Runion, 1990). Livio (2002) surmises that the golden ratio has probably inspired thinkers across disciplines more than any other number in the history of mathematics.

Basic to many of the fascinating properties of the golden ratio is that the line-sectioning process is self-perpetuating. If we take the longer of the two segments into which the original line segment was divided, which is to say the section of length x, and mark on it a point at a distance equal to the length of the smaller of the segments (i.e., $1 - x$) from one end, we will find that a division of the section of length x at this point will yield subsections with the same properties as those of the original line segment. The process can be continued indefinitely, and each sectioning will produce two segments such that the ratio of the shorter to the longer will be the same as the ratio of the longer to the whole.

A similar process of successive subdivisons can be applied in two dimensions. If we begin by drawing a rectangle, the longer and shorter sides of which are in the ratio 1.618 to 1, and then divide this figure into a square and a smaller rectangle by drawing a line segment 0.618 units from one of the shorter sides and parallel to it, we will find that the longer and shorter sides of the resulting smaller rectangle again have the ratio 1.618 to 1, or, if you prefer, 1 to 0.618. If we repeat this operation several times, each time dividing the small rectangle that resulted from the previous division into a square and a residual rectangle, the ratio of the longer and shorter sides of the resulting rectangles will remain constant at 1.618 to 1 (see Figure 11.8).

Given $x = 0.618 \ldots$, that is, $1/\phi$, the sequence x^0, x^1, x^2, x^3, $\ldots x^{n-1}$ has the property that every number except the first two is the difference

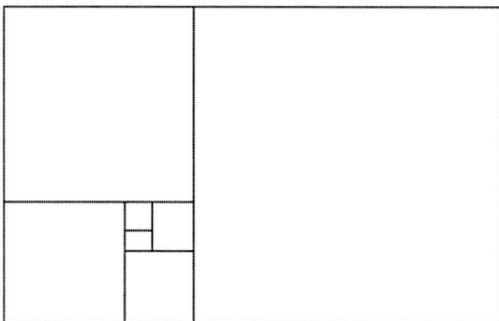

FIGURE 11.8 Showing the division of the golden rectangle within the golden rectangle into another square and smaller golden rectangle, and the result of repeating the process a few times.

between the two preceding numbers. Similarly, the sequence $1/x^0$, $1/x^1$, $1/x^2$, $1/x^3$, ... $1/x^{n-1}$ has the property that every number except the first two is the sum of the two preceding numbers. This is the defining property of the famous Fibonacci sequence 1, 1, 2, 3, 5, 8, 13, 21, 34, ..., x_n, ..., where $x_n = x_{n-1} + x_{n-2}$. The limit of the ratio of two adjacent numbers in this series as the length of the series increases, $\lim_{n \to \infty} x_n / x_{n-1}$, is the golden ratio, ϕ. What is perhaps more surprising is that this ratio is the limit of a sequence in which each number except the first two is the sum of the preceding two numbers, no matter what numbers are used to start the sequence. Suppose one starts, say, with the two numbers 17 and 342. This beginning produces the sequence 17, 342, 359, 701, 1,060, 1,761, 2,821, 4,582, 7,403, 11,985, ... The sequence of (approximate) ratios of each number with its immediate predecessor, beginning with 342/17, is 20.118, 1.050, 1.953, 1.512, 1.661, 1.602, 1.624, 1.616, 1.619. Successive ratios alternate between being larger and smaller than ϕ, and the magnitude of the difference gets progressively smaller.

The Fibonacci sequence is a very interesting phenomenon in its own right, and pops up unexpectedly in a great variety of contexts both in mathematics and in the physical (and biological) world (Garland, 2000; Livio, 2002; Stevens, 1974). The sequence is generated by adding numbers in Pascal's triangle (the coefficients of the expansion of the binomial $(a + b)^n$ in a certain order (Pappas, 1989, p. 41). Also, the sum of the squares of two successive Fibonacci numbers is always another Fibonacci number. The first few cases are shown in Table 11.1. The general formula

TABLE 11.1. Illustrating That the Sum of the Squares of Successive Fibonacci Numbers Is Another Fibonacci Number

Numbers	Sum of Squares
1, 2	5
2, 3	13
3, 5	34
5, 8	89
8, 13	233

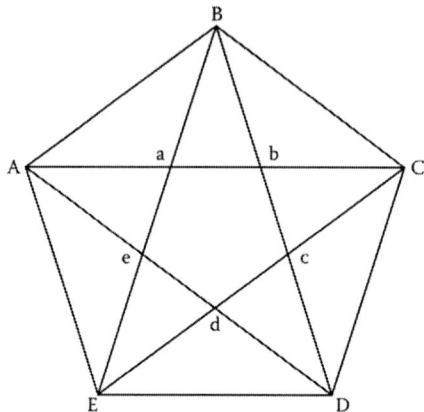

FIGURE 11.9 Showing the many instances of ϕ in a star inscribed in a pentagon.

is $(f_{n+1})^2 + (f_{n+2})^2 = f_{2n+3}$, where n is the ordinal position of the number in the conventional Fibonacci series, 1, 1, 2, 3, 5, 8, 13, ...

An aspect of the golden ratio that fascinated the ancient Greeks was its appearance in certain geometric patterns. If a five-pointed star is created by joining alternate vertices of a regular pentagon, as shown in Figure 11.9, the golden ratio is seen practically everywhere one looks. Considering only the area defined by the triangle ABC, the ratios AC/Ab, Ab/Aa, Aa/ab, AC/AB, and AB/Aa all are ϕ. By symmetry the same ratio is seen repeatedly also in triangles BCD, CDE, DEA, and EAB.

Note that within the star that sits in the pentagon there exists another pentagon, abcde, within which a star can be drawn by connecting its vertices. That star, too, has a pentagon within it, within which a still smaller star can be drawn, and so on *ad infinitum*. At each level, there occurs the same abundance of instances of ϕ as at the top level. Figure 11.10 shows the results of carrying the process to three levels.

Much has been written about the frequent appearance of dimensions of the golden ratio in ancient architectural structures and in classical paintings (Bergamini, 1963; Bouleau, 1963; Ghyka, 1927/1946). The idea is that many architects and artists constructed or painted structures the dimensions of which were related by the golden ratio, either because they intentionally made use of the ratio in proportioning, or because what they found to be pleasing to the eye turned out to have these proportions. The celebrated Swiss-French architect Charles-Édouard Jeanneret-Gris (aka Le Corbusier) based a theory of architecture on this idea.

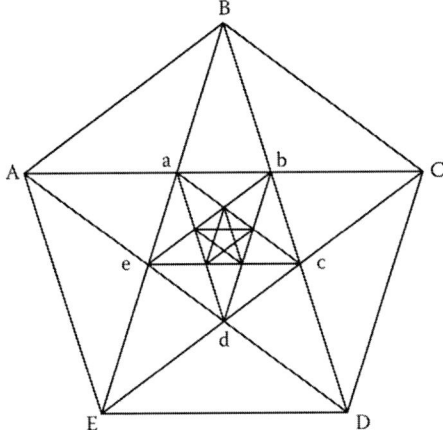

FIGURE 11.10 Nested stars and pentagons.

Numerous masterpieces have been analyzed with the hope of finding evidence of this notion. Livio (2002) contends that much of the evidence of the widespread influence of the golden ratio in architecture and painting is weak. Some exceptions involve a few relatively recent artists who have explicitly made use of the ratio in some abstract paintings. Livio also recounts numerous efforts to find evidence of the influence of the golden ratio in music and poetry and again concludes that much (though not all) of what has been seen has really been in the eye of the observer.

Among the earliest experiments performed by psychologists, notably German pioneer psychophysicist Gustav Fechner, were some designed to explore whether rectangles with sides in the golden ratio are esthetically distinct and preferred to those with other length-width ratios. Much of this work was flawed by a failure to consider the role that procedural artifacts could play in determining the experimental results. The judgmental context in which preferences were stated (e.g., the range of lengths and widths used and where the "golden" rectangle sat within that range) could affect the choices made; for example, and the analysis of aggregate data could lead to conclusions of a preference for golden rectangles, even if none of the individuals comprising the group showed such a preference (Godkewitsch, 1974). This work did not resolve the question one way or the other, but simply lost the attention of experimenters, who moved on to other things.

Hardly less interesting than the golden rectangle, though much less often discussed, is the golden triangle—a right triangle with sides in the

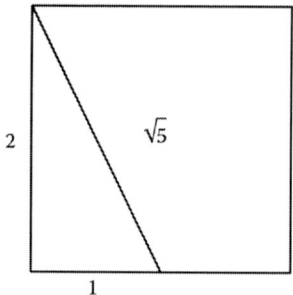

FIGURE 11.11 The golden triangle.

ratio $1:2:\sqrt{5}$. It is formed by dropping a line segment from an upper vertex of a 2×2 square to the center of the base, as shown in Figure 11.11.

If a sector of a circle with radius $\sqrt{5}$ is formed from the apex of the triangle to an extension of the square's base, the rectangle that is formed by extending the top of the square to match the extension of the base and connecting the two extensions, as shown in Figure 11.12, is the golden rectangle.

The rectangle to the left of the square is itself a golden rectangle, which can be divided into a square and a still smaller rectangle, as shown in Figure 11.13. The hypotenuse of the golden triangle constructed in the small square is the radius of a circle that is tangent to one side of the smaller golden rectangle. The smallest rectangle in this figure is also a golden rectangle and can be divided into a square and smaller rectangle still, and so on indefinitely, producing an endless series of golden rectangles.

Despite the attention that the golden ratio has received through the ages, it has not lost its charm or its ability to stimulate whimsical speculations

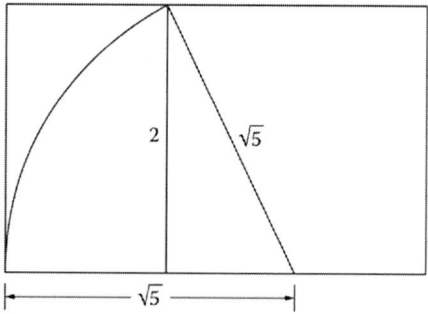

FIGURE 11.12 Illustrating the construction of a golden triangle from a square and an enclosed golden triangle.

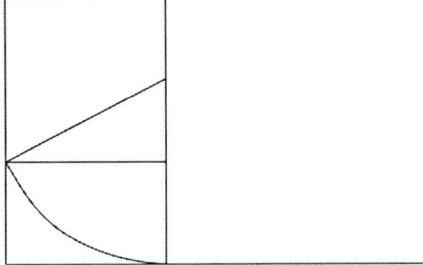

FIGURE 11.13 Showing the division of the golden rectangle within the golden rectangle into another square and smaller golden rectangle by use of the hypotenuse of a golden triangle.

about the numerical nature of reality. Without denying the possibility that unchecked, Pythagorean speculation can lead to nowhere, or worse, I confess to not being able to understand how anyone who stumbles onto this number can fail to wonder about it. Its geometric properties strike me as beautiful and fascinating; its penchant for appearing in unpredictable abstract places (the Fibonacci series being perhaps the best known among these) and its commonality (especially via the Fibonacci series) in nature are intriguing. (For an account of why the Fibonacci series and the golden ratio are encountered so frequently in botany, see Stewart, 1995b.)

☐ A Surprising Connection

There is nothing obvious about either the Pythagorean theorem or the Fibonacci series that would lead one to believe that one is related to the other. However, in 1948 Charles W. Raine pointed out that if for any four consecutive Fibonacci numbers the product of the outer two of these numbers and twice the product of the inner two are taken as two legs of a right triangle, the square of the hypotenuse of that triangle is also a Fibonacci number. Consider, for example, the four successive Fibonacci numbers 5, 8, 13, 21. According to Raine's rule, the lengths of the two legs of the triangle are $5 \times 21 = 105$ and $2 \times 8 \times 13 = 208$, and the hypotenuse is $\sqrt{105^2 + 208^2} = 233$, a Fibonacci number. The general expression for the relationship is $(a_n a_{n+3})^2 + (2a_{n+1}a_{n+2})^2 = (a_{2n+3})^2$, where a_n represents the nth number in the Fibonacci sequence. [This assumes determining n by counting from the first 1 in the series; if the first number in the series is considered to be 0, the correct formula is $(a_n a_{n+3})^2 + (2a_{n+1}a_{n+2})^2 = (a_{2n+2})^2$.]

As already noted, the ratio of the nth to the $(n - 1)$th Fibonacci number approaches the golden ratio with increasing n. In a discussion of

Raine's discovery, Boulger (1989) points out that as one goes farther in the Fibonacci sequence, forming triangles from four consecutive terms as described in the preceding paragraph, the golden ratio appears as the ratio of the sum of the shorter leg and the hypotenuse to the longer leg,

$$\lim_{n \to \infty} \frac{\text{shorter leg} + \text{hypotenuse}}{\text{longer leg}} = \phi.$$

☐ Recreational Mathematics

> Playing around with numbers has been a pastime for the past 3000 years and perhaps for much longer. (Flegg, 1983, p. 225)

To many people, the idea that mathematics can be fun would be curious if not incomprehensible. Such book titles as *Amusements in Mathematics* (Dudeney, 1958), *The Pleasures of Math* (Goodman, 1965), *Recreations in the Theory of Numbers* (Beiler, 1966), *The Joy of Mathematics* (Pappas, 1989), and *Trigonometric Delights* (Maor, 1998) would perhaps sound oxymoronic. But unquestionably some people derive great pleasure from mathematics—from discovering or simply contemplating mathematical relationships, from solving mathematical problems, from creating or experiencing mathematical patterns. What else but the unadulterated pleasure of solving problems or discovering relationships could motivate concerted efforts to find ever larger *perfect* numbers (numbers that equal the sum of their proper divisors; a proper divisor of *n* being any divisor of *n*, including 1, other than *n* itself) or *amicable* or *friendly* pairs of numbers (numbers each of which is equal to the sum of the proper divisors of the other).

The first few perfect numbers—6, 28, 496, 8128—were probably known to the ancients. Others have been discovered gradually over the centuries. The search has been aided greatly by Euclid's proof that if, for $p > 1$, $2^p - 1$ is a prime number (a Mersenne prime, named for Marin Mersenne, the French theologian-mathematician who later studied primes that are one less than a power of 2), then $2^{p-1}(2^p - 1)$ is a perfect number. Before the middle of the 20th century, the largest known prime was $2^{127} - 1$, a 39-digit number, which, according to Friend (1954), French mathematician Édouard Lucas spent 19 years checking; the perfect number that can be calculated from this prime, $2^{126}(2^{127} - 1)$, is 77 digits long. (Kasner and Newman [1940/1956] note that there is reason to believe that some 17th-century mathematicians had a way of recognizing primes that is not known to us. As evidence of this possibility they cite an occasion on which Fermat, when asked whether 100,895,598,169 was prime, was able to say straightaway that it was the product of two primes, 898,423 and 112,303.)

The appearance of high-speed digital computers on the scene accelerated the search for Mersenne primes and perfect numbers. The *Great Internet Mersenne Prime Search* (GIMPS) is an organized approach that taps the collective power of thousands of small computers to find ever larger Mersenne primes. As of October 2008, the largest known prime, found as a consequence of this effort, was $2^{43,112,609} - 1$. This means that the largest known perfect number as of the same date was $2^{43,112,608}(2^{43,112,609} - 1)$, which would require almost 26 million digits to write out. All known perfect numbers are even, but the nonexistence of odd ones has not been proved. A curious consequence of the search for perfect numbers is the finding of many numbers that are almost perfect in that their divisors add to one less than the number itself, and the failure to find any number that is almost perfect by virtue of its divisors adding to one more than itself, despite the lack of a proof of the nonexistence of the latter type of near miss.

Although the concept of amicable numbers was known to the ancient Greeks, they were perhaps aware of the existence of only the single pair 220 and 284. Fermat found a second pair—17,296 and 18,416—in 1636, and Descartes a third—9,363,584 and 9,437,056—two years later. Then, about the middle of the 18th century, Euler discovered a technique for generating such numbers and added 58 new pairs to the existing very short list (Dunham, 1991).

A 20th-century extension of the idea of amicable numbers is that of *sociable* numbers, defined as three or more numbers that, when treated as a loop, have the property that each number is the sum of the divisors of the preceding number. Singh (1996) gives, as an example of such a loop, 12,496; 14,288; 15,472; 14,536; 14,264. Each number is the sum of the divisors of the preceding number, and treating the set as a loop, the number that precedes 12,496 is 14,264.

Fascination with mathematical puzzles is as old as mathematics, and, as evidenced by the steady stream of books on the subject, it appears not to have abated over time. Claude Gaspar Bachet, a French nobleman and writer of books on mathematical puzzles, published in 1612 *Problèmes plaisants et délectables qui se font par les nombres*, which presented most of the more famous numerical or arithmetic puzzles known at the time. At least five subsequent, and greatly enlarged, editions of the book were published, one as late as the mid 20th century. Although the number of different puzzles that have been published is surely very large, many of the puzzles are variations on a few generic themes or scenarios, including weighing, river crossing, liquid pouring, dice tossing, and drawing colored balls from an urn.

One might be tempted to assume that puzzles are given little attention by serious mathematicians, but such an assumption would be wrong. Kasner and Newman (1940/1956) mention Kepler, Pascal, Fermat, Leibniz, Euler, Lagrange, Hamilton, and Cayley as among the many major mathematicians who have devoted themselves to puzzles. "Researches in recreational mathematics sprang from the same desire to know, were guided by the same principles, and required the exercise of the same faculties as the researches leading to the most profound discoveries in mathematics and mathematical physics. Accordingly, no branch of intellectual activity is a more appropriate subject for discussion than puzzles and paradoxes" (p. 2416).

Magic squares—square tables of numbers in which all rows, columns, and diagonals add to the same sum—go back to antiquity. Flegg (1983) gives as the oldest known magic square one that appears in a Chinese work on permutations that may have been written in the 12th century BC, but notes that, according to tradition, the idea of the magic square goes back to around 2200 BC. At one time, magic squares were believed by many people really to have magical powers; charms on which they were engraved were worn for protection against disease or other types of adversity. Flegg (p. 236) describes an interesting method of constructing magic squares of various orders, based on the way a knight moves on a chessboard.

Ellis (1978) also gives rules for constructing magic squares of different order and shows a 16×16 square, holding the numbers 1 through 256, reproduced here as Table 11.2, that he describes as "one of the most ingenious ever devised." The numbers in every row and every column add up to 2,056, as do those in every 4×4 block anywhere in the square; this includes *overlapping* blocks and blocks that wrap around either side-to-side or top-to-bottom. That is, a 4×4 block can be composed from cells in the right-most columns followed by cells in corresponding rows in the left-most columns, or from cells in the bottom rows followed by cells in the corresponding columns in the top rows. Every half column and every half row adds to 1,028, and every 2×4 and 4×2 block does so as well, again even when the square is treated as a wraparound device laterally and vertically. Ellis also points out regularities involving chevron patterns in the square. There may be a way of conceptualizing the production of such a square with the numbers 1 through 256 that makes it unremarkable, but, if so, I am unaware of it; it strikes me as a truly astounding accomplishment.

Sudoku is a puzzle with some similarities to magic squares that has become very popular, first in Japan and then elsewhere, including the United States, over the last 20 years or so. In the standard form, Sudoku is a 9×9 grid and the objective is to fill in the rows and columns in such a way that each of the numbers 1 through 9 appears once and only once in each row, in each column, and in each of the nine nonoverlapping 3×3 subgrids. The grid presented to the puzzle-doer has a few of the cells

TABLE 11.2. 16 × 16 Magic Square

200	217	232	249	8	25	40	57	72	89	104	121	136	153	168	185
58	39	26	7	250	231	218	199	186	167	154	135	122	103	90	71
198	219	230	251	6	27	38	59	70	91	102	123	134	155	166	187
60	37	28	5	252	229	220	197	188	165	156	133	124	101	92	69
201	216	233	248	9	24	41	56	73	88	105	120	137	152	169	184
55	42	23	10	247	224	215	202	183	170	151	138	119	106	87	74
203	214	235	246	11	22	43	54	75	86	107	118	139	159	171	182
53	44	21	12	245	236	213	204	181	172	149	140	117	108	85	76
205	212	237	244	13	20	45	52	77	84	109	116	141	148	173	180
51	46	19	14	243	238	211	206	179	174	147	142	115	110	83	78
207	210	239	242	15	18	47	50	79	82	111	114	143	146	175	178
49	48	17	16	241	240	209	208	177	176	145	144	113	112	81	80
196	221	228	253	4	29	36	61	68	93	100	125	132	157	164	189
62	35	30	3	254	227	222	195	190	163	158	131	126	99	94	67
194	223	226	255	2	31	34	63	66	95	99	127	130	159	162	191
64	33	32	1	256	225	224	193	192	161	160	129	128	97	96	65

Source: From Ellis (1978, p. 140).

already filled in. Although no mathematics is required to solve Sudoku puzzles—only logic is required—there are some interesting mathematical questions associated with the game. One question, not easy to answer, is how many ways can a Sudoku grid be filled in according to the rules of the game. Delahaye (2006) gives a number of about 6.7 × 10^{21} as an estimate produced with the help of a computer; he identifies a variety of strategies that people use to solve Sudoku puzzles and describes several variations on the Sudoku theme.

Mathematics that has been engaged strictly for its own sake—for fun, one may say—has frequently led to serious mathematical inquiry and notable discoveries with important practical applications. Kasner and Newman (1940) include the theory of equations, probability theory, calculus, the theory of point sets, and topology among the areas of mathematics that have grown out of problems first expressed in puzzle form.

However, that playing with mathematics could lead to such discoveries and applications is not necessary to justify the playing; the doing of mathematics is seen by many people—professionals and amateurs alike—as its own reward. Something of the pleasure that playing with mathematics can provide has been described by many writers, among them Kraitchik (1942), Friend (1954), Gardiner (1956, 1959, 1961), Dudeney (1958), Beiler (1964), Rouse Ball and Coxeter (1987), Pappas (1989, 1993), and Burger and Starbird (2005).

☐ Surprised by Simple Elegancies: Confessions of a Nonmathematician

> Few of us will ever scale the Himalayas of mathematics but we can all enjoy a stroll in the foothills where the dipping, twisting pathways of our basic number system lead to a bonanza of unexpected delights. (Ellis, 1978, p. 122)

The Pythagorean theorem has been mentioned several times in this book. That the square of the length of the hypotenuse of a right triangle is equal to the sum of the lengths of the squares of the other two sides is a useful bit of knowledge. The relationship appears also to be intrinsically fascinating to many people; how else are we to account for the extraordinarily large number of proofs of it that have been produced? One possible basis for the fascination is the surprising nature of the relationship. As Dunham (1991) puts it, "There is no intuitive reason that right triangles should have such an intimate connection to the sums of squares ... the Pythagorean theorem establishes a supremely odd fact, one whose oddness is unrecognized only because the result is so famous" (p. 53). Dunham quotes Richard Trudeau, who observes in his book *The Non-Euclidean Revolution*, "When the pall of familiarity lifts, as it occasionally does, and I see the Theorem of Pythagoras afresh, I am flabbergasted" (p. 53).

Another formula that is familiar to everyone who has had a high school algebra course, and one that we accept without surprise only because of its familiarity, is that which expresses the area of a circle as a function of its radius, $A = \pi r^2$. That π expresses the ratio of a circle's circumference to its diameter is unlikely to evoke wonder by anyone who knows anything about mathematics, because π is *defined* as that ratio. But how about the fact that the area of a circle is obtained by multiplying by π the area of a square whose side is equal to the circle's radius? This is not a matter of definition. Why should the constant that represents the ratio of a circle's circumference to its diameter turn out to represent also the ratio of the circle's area to a square erected on its radius? Archimedes constructed a proof that the area of a circle, A_{cir}, is equal to the area of a triangle, A_{tri}, whose base and height are equal, respectively, to the circle's circumference, c, and radius, r, that is, $A_{cir} = A_{tri} = \frac{1}{2}cr = \pi r^2$, and also described a geometric procedure for approximating the value of π, which bracketed the value between $3\frac{1}{7}$ and $3\frac{10}{71}$. So the relationship is established, but one is still left with a sense of wonder as to why it is what it is. One wonders too about the extent to which knowledge of this relationship may have encouraged the numerous people who have tried over the years to "square the circle" to believe that it should be doable.

The curve that describes the trajectory of a point on the edge of a disc as the disc is rolled along a straight line is known as a cycloid. The curve that describes the trajectory of fastest frictionless descent of an object from one point, A, to a lower one, B, is called a brachistochrone. What these two curves have to do with each other is anything but clear. The brachistochrone has to do with the time required to get from one point to another. The phenomenon the cycloid represents has nothing to do with time; it describes the path taken by a point on a rolling disc independently of how fast the disc is rolled. Nevertheless, the cycloid and the brachistochrone are, in fact, the same curve.

One would like to be able to explain coincidences like these in terms of some overarching theory from which the relationship in both contexts can be deduced. The situation is analogous to that faced by physicists attempting to make sense of the fact that gravitational mass (which involves a relationship of attraction between two bodies possibly separated by a great distance) and inertial mass (which involves acceleration of a body from the effect of force acting on it) should have the same value. Einstein, who called this coincidence astonishing, eventually deduced the *equivalence principle* from the non-Euclidean geometry of space assumed by his general theory of relativity. There are many coincidences in mathematics that await comparable enlightening.

I am not a mathematician, but even with my very limited mathematical knowledge, I find much to wonder about in the elegant mathematical relationships that one sees on every hand. Sometimes it is hard to tell whether a relationship is worthy of wonderment or not. Consider, for example, the way in which equal-radius circles pack. If I draw one such circle, and then surround it with as many abutting circles as I can, I find that six fit precisely. Having placed five, there is exactly room enough for the sixth. Similarly, in three-dimensional space, a sphere can be abutted by 12 other spheres of the same radius, and the fit is precise. Is this a mundane fact or an interesting one? I confess to being surprised by it and provoked to wonder why it should be so.

Here I want to focus on some very simple mathematical relationships that evoke wonder just because they *are* simple. One would not expect

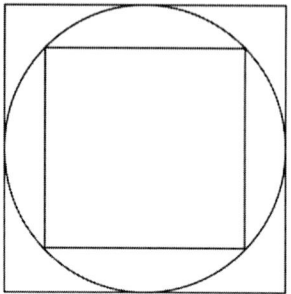

FIGURE 11.14 A square within a circle within a square.

them, I think, to be so, and one is surprised to discover that they are. Consider the situations shown in Figures 11.14 and 11.15. In the first case, a square circumscribes a circle, which circumscribes a square. In the second case, the roles of the square and circle are reversed. It is easy to show that, in both cases, the ratio of the areas of the outer and inner figures is precisely 2 to 1—not 1.98 to 1 or 2.02 to 1, but 2 to 1 exactly. Is this an interesting fact? Why should the ratios be integral? What does a circle have to do with a square that dictates such an elegantly simple relationship?

That the area of the inner square is half that of the outer square is readily seen by rotating the inner square 45 degrees and sectioning it with diameters of the circle as shown in Figure 11.16. It should be clear that each edge of the inner square divides each quadrant of the outer square into two areas of equal size.

That the same relationship holds when the roles of squares and circles are reversed is shown in Figure 11.17. Let R represent the radius of the outer circle and r the radius of the inner one. If r is set to 1, then, by the Pythagorean theorem, $R = \sqrt{2}$, from which it follows that the area of the inner circle is π and that of the outer one is 2π.

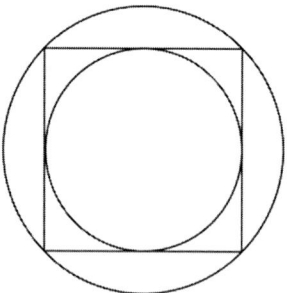

FIGURE 11.15 A circle within a square within a circle.

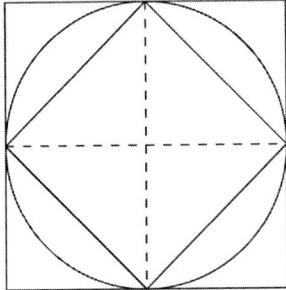

FIGURE 11.16 Showing that the area of the inner square is 1/2 the area of the outer square.

Figure 11.18 shows the comparable situations with the squares replaced by equilateral triangles. In the first instance, an equilateral triangle circumscribes a circle, which circumscribes an equilateral triangle. In the second, the roles of triangle and circle are reversed. As the reader may wish to verify, in each case the ratio of the area of the outer and inner figures is precisely 4 to 1.

Suppose I draw a right triangle with sides 3, 4, and 5 and an inscribed circle (see Figure 11.19). Is it not interesting that the circle has a radius of 1? Inasmuch as the circle's diameter is 2, the first five integers are represented in the basic measures of this figure. Six also is there as the area of the triangle, and with only a small stretch, one can see 7, 8, and 9 as the sums of the three possible pairings of the triangle's sides: 3 + 4, 3 + 5, and 4 + 5. π is there, too; do you see it? (If not, see the end of this chapter.)

I find these relationships surprising, although am not entirely sure that I should. What I find surprising about them is the integral relationships. If one inscribes an equilateral triangle within a circle, a circle within

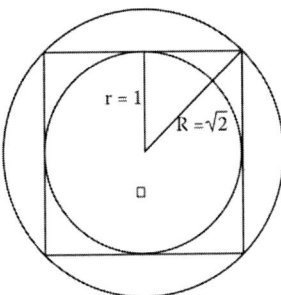

FIGURE 11.17 Showing that the area of the inner circle is 1/2 the area of the outer circle.

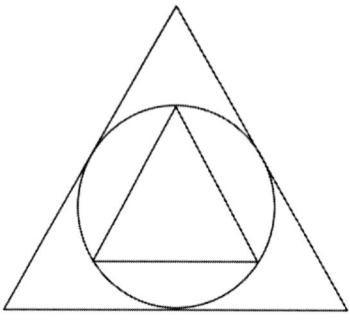

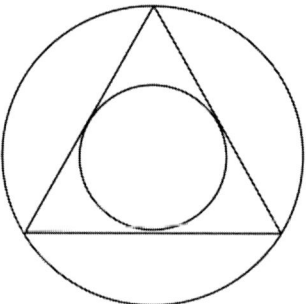

FIGURE 11.18 A triangle within a circle within a triangle, and a circle within a triangle within a circle. In each case the area of the innermost figure is 1/4 the area of the outer one.

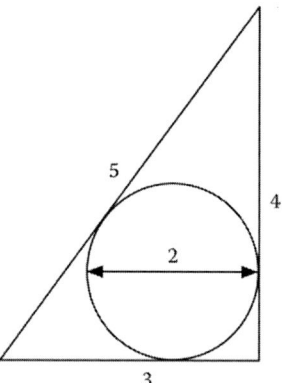

FIGURE 11.19 The radius and diameter of the circle are, respectively, 1 and 2; the sides of the right triangle are 3, 4, and 5.

that triangle, a square within that circle, a circle within the square, a pentagon within that circle, and continues in this fashion, inscribing in each successively smaller circle a polygon with one more side than the previous one, the radius of the circle one obtains in the limit is approximately 1/12 of the radius of the original circle (Kasner & Newman, 1940, p. 311). The important word here is *approximately*. This is what one would expect, in my view, when one inscribes regular geometric figures within other geometric figures; so I am surprised to find the area of a square inscribed in a circle that is inscribed within a square to be precisely 1/2 the area of the square within which the circle is inscribed.

Perhaps I am missing something that, if I could see it, would make the mundaneness of this and similar relationships obvious. But, in the absence of that insight, I am awestruck by their simplicity and elegance. Certainly such relationships are well known among mathematicians; nevertheless, I confess to a sense of joy in discovering them, as a non-mathematician, for myself. I have no expectations of ever discovering anything in mathematics that is not already well known, but this does not preclude my experiencing a feeling of delight upon occasionally discovering a relationship that is new to me. My experience leads me to give some credence to a claim made by Kaplan and Kaplan (2003): "Anyone who can read and speak (which are awesomely abstract undertakings) can come to delight in the works of mathematical art, which are among our kind's greatest glories" (p. 2).

Of course, upon becoming aware of these relationships, a real mathematician would want to know whether they generalize. For example, do the integral relationships between the areas of inscribed and circumscribed circles and triangles and squares generalize to all regular polygons? Alas, the answer is no. The area of a circle circumscribed around a regular pentagon is approximately 1.53 the size of a circle inscribed in the same pentagon. The same ratio holds between the area of a pentagon enclosing a circle and that of a pentagon within the same circle.

Nor do the integral relationships between the areas of inscribed and circumscribed circles and triangles and squares generalize to forms of higher dimension. The volume of a cube that circumscribes a sphere of radius r is

$$V_{CC} = (2r)^3$$

while the volume of a cube inscribed in a sphere of radius r is

$$V_{IC} = \left(\frac{2r}{\sqrt{3}}\right)^3.$$

So the ratio of the volume of a circumscribing cube to that of an inscribed cube is

$$\frac{V_{CC}}{V_{IC}} = \frac{(2r)^3}{\left(\dfrac{2r}{\sqrt{3}}\right)^3} = 3\sqrt{3}.$$

Given a sphere of radius 1, the volumes of the circumscribing and inscribed cubes are, respectively, 8 and $\frac{8\sqrt{3}}{3}$. Similarly, if we start with a sphere of radius r, inscribe within it a cube (a side of which would measure $\frac{2r}{\sqrt{3}}$), and then inscribe within that cube another sphere, the inner sphere will have a radius of $\frac{r}{\sqrt{3}}$, so the ratio of the volume of the circumscribing sphere to that of the inscribed sphere will be

$$\frac{V_{CS}}{V_{IS}} = \frac{\dfrac{4}{3}\pi r^3}{\dfrac{4}{3}\pi\left(\dfrac{r}{\sqrt{3}}\right)^3} = 3\sqrt{3}.$$

Again, given a sphere of radius 1 circumscribing a cube, which in turn circumscribes a sphere, the volumes of the circumscribing and inscribed spheres are, respectively, $\frac{4}{3}\pi$ and $\frac{4\pi}{9\sqrt{3}}$.

Despite their failure to extend to the areas of all regular polygons or to the volumes of three-dimensional shapes, I still find the integral relationships involving the areas of circles, triangles, and squares to be surprising and elegant. There are numerous relationships that are surprising, at least to some observers, because there is nothing that would lead one to expect the ratio between the two measures to be integral. Dunham (1991) notes several of them, all of which were proved by Archimedes. It is almost enough to make one a numerologist!

****

The examples of elegant or surprising mathematical relationships noted in this chapter are drawn primarily from basic algebra or geometry and are relatively simple, reflecting my own limited knowledge of mathematics. However, the testimony of mathematicians working in highly abstract and esoteric areas of mathematics makes it clear that there are elegancies, often surprising, to be found in their regions of operation as well, although appreciation of them may require levels of mathematical sophistication that few people attain. (π in Figure 11.19 is the area of the circle.)

The Usefulness of Mathematics

> There is one qualitative aspect of reality that sticks out from all others in both profundity and mystery. It is the consistent success of mathematics as a description of the workings of reality and the ability of the human mind to discover and invent mathematical truths. (Barrow, 1991, p. 173)

To say that mathematics is enormously useful is not to say that all of mathematics is equally useful, or even that *all* of mathematics is useful at all. To be sure, Florian Cajori (1893/1985) says, in his classic *The History of Mathematics*, that "hardly anything ever done in mathematics has proved to be useless" (p. 1). On the other hand, Davis and Hersch (1981) speculate that most of the millions of theorems contained in the mathematical literature are useless dead ends. And King (1992) makes the startling claim that "an ordinary mathematics research paper is read by almost no one except the author and the journal 'referee' who reviewed it prior to publication" (p. 38). Whatever the merit of any of these claims—it is easy to believe that the last one is true of most disciplines—they are of little moment to the pure mathematician, for whom neither the usefulness nor the popularity of theorems is the source of motivation for proving them. And practical minded souls who do see the value of mathematics in its applicability to real-world problems can comfort themselves with the thought that the practical value of that part of mathematics that *is* useful is sufficiently great to justify easily the entire enterprise on strictly pragmatic grounds.

Whatever the motivations that drive mathematicians to do what they do, there can be no doubt that mathematics, in the aggregate, has been essential to the development of civilization as we know it. Stewart (1995b) puts the matter this way: "If mathematics, including everything that rests on it, were somehow suddenly to be withdrawn from our world, human society would collapse in an instant" (p. 28). Even those areas of mathematics that have appeared most arcane or frivolous when first being explored have surprisingly often proved in time to have applications, sometimes in many domains. This seems like incredibly good fortune, especially in view of the fact that interest in applications is not what drove many of the most productive mathematicians to do their work. If, as we have seen that many modern mathematicians believe, mathematics can be pursued as an abstract discipline independently of its relationship to the physical world, why do esoteric developments in mathematics that are pursued for purely theoretical reasons so often turn out to have completely unanticipated practical applications? Indeed, why, if mathematics is an abstract discipline that is independent of the physical world, should it be useful in describing the world at all?

Although the usefulness of mathematics may be most obvious in science and technology, it extends to many other areas as well. In the ancient world the principal area of applications was probably trade. Over the centuries, mathematics has served, in addition to the purposes of science, technology, business, and commerce, those of art, war, mysticism, gambling, and religion. In short, while the role of mathematics in science has been well documented, "the Hand Maiden of the Sciences has lived a far more ravish and interesting life than her historians allow" (Davis & Hersh, 1981, p. 89).

Mathematics has played an especially important role in the arts through the centrality of such mathematical concepts as proportion, order, symmetry, harmony, and rhythm. This holds not only for music and architecture, where the influence may be most apparent, but in painting, sculpture, and poetry as well. Butler (1970) attributes great importance to Plato, Plotinus, Augustine, and Aquinas as purveyors of numerological ideas that had major influence on the art of the Renaissance. A recent and perhaps surprising application of mathematics to art is that of fractal geometry to an analysis of the paintings of Jackson Pollock (Taylor, 2002); objectives include those of determining the fractal dimensionality of his (and other abstract) designs, and of discriminating between genuine Pollocks and imitations.

Why mathematics is so useful is a mystery. Why should it be that, as Berlinski (1997) puts it, "By means of purely mathematical operations on purely mathematical objects—*numbers*, after all—the mathematician is able to say that *out there* this will happen or that will" (p. 97)? Stewart (1990) captures the uncertainty as to why mathematics is useful this way: "Perhaps mathematics is effective because it represents the under-lying language of the human brain. Perhaps the only patterns we can perceive are mathematical because mathematics is the instrument of our perception. Perhaps mathematics is effective in organizing physical exis-tence because it is inspired by physical existence. Perhaps its success is a cosmic delusion. Perhaps there are no real patterns, only those that we feeble-mindedly impose" (p. 7). Stewart notes that these are questions for philosophers and contends that "the pragmatic reality is that math-ematics is the most effective and trustworthy method that we know for understanding what we see around us" (p. 7).

☐ Utilitarian Interests

The belief that much of the most creative thinking in mathematics has been motivated by an interest in mathematics per se and not by any relationship it may bear to the physical world is held by many math-ematicians (Davis & Hersh, 1981; King, 1992). King argues, for example, that instances of new mathematics coming out of interest in real-world problems—of which the invention of the calculus is a prototypical case—are exceptions to the rule. Much more common, he suggests, is the case in which an area of mathematics that was originally developed for purely theoretical interest turns out to have unexpected practical applications. Even number theory, which is often considered the area of mathematics that is least motivated by possible applications, is being found useful in unanticipated ways, as, for example, in encryption for secure commu-nications. Moreover, work on number theory arguably had great impact on the development of both pure and applied mathematics indirectly. Richards (1978) makes the point:

> A good theorem will almost always have a wide-ranging influence on later mathematics, simply by virtue of the fact that it is *true*. Since it is true, it must be true for some reason; and if that reason lies deep, then the uncov-ering of it will usually require a deeper understanding of neighboring facts and principles. In this way number theory, "the Queen of Mathematics," has served as a touchstone against which many of the tools in other branches of mathematics have been tested. This, in fact, is the real way that number theory influences pure and applied mathematics. (p. 63)

There is also the view, however, that many areas of mathemat-ics have been developed by people who were keenly interested in ques-tions about physical reality, and that even those subjects that are usually considered pure mathematics were created, in many cases, in the study of real physical problems (Bell, 1946/1991; Kline, 1980). At least 1,000 years before Pythagoras, the Babylonians were considering questions involving the time required for money to double if invested at a specified annual rate of interest (Eves, 1964/1983). Trigonometry was created by the Alexandrians, notably the ancient Greek astronomers Hipparchus and Ptolemy, as a tool for enabling more precise predictions of the movements of the planets and other heavenly bodies. Having been developed primarily to serve the needs of astronomy, this area of mathematics, as in many other cases, proved to have applications in numerous other areas as well.

Ekeland (1993), who sees the development of mathematics as part of the general development of science and technology, attributes the growth of analysis to the interest of its developers in celestial mechanics and notes that the book in which Gauss established the foundations of geometry was also a treatise on geodesy. "Had historical circumstances been different, had there been different needs to satisfy, wouldn't mathematics have been different? If the Earth were the only planet around the Sun and if it had no satellite, we wouldn't have spent so many centuries accumulating observations and building systems to explain the strange movements of the planets among the stars, celestial mechanics wouldn't exist, and mathematics would be unrecognizable" (p. 55).

Others have argued the importance of observation of real-world phenomena as a source of mathematical ideas. British mathematician James Joseph Sylvester (1869/1956), for example, believed that most, if not all, of the great ideas of modern mathematics have had their origin in observation. Wilder (1952/1956) argues too that, although theoretically axioms need have no correspondence to anything, they usually are statements about concepts with which those who make them are already familiar. Psychologically, it is hard to imagine that it could be otherwise. As Wilder puts it, "We may say 'Let us take as undefined terms *aba* and *daba*, and set down some axioms in these and universal logical terms.' With no concept in mind, it is difficult to think of anything to say! That is, unless we first *give some meanings* to 'aba' and 'daba'—that is, introduce some *concept* to talk about—it is difficult to find anything to say at all" (p. 1660). Of course, interest in physical reality need not be understood as necessarily quite the same thing as the desire to have a practical impact.

☐ A Fuzzy Distinction

The line between pure and applied mathematics is a difficult one to draw in practice. This is especially so since the theorizing in some areas of science—quantum physics, for example—has become increasingly abstract and mathematical, and remote from the world of sense and perceptual experience. Bell (1946/1991) suggests that the line between "reputable and disreputable" (pure and applied) mathematics is drawn by the Pythagorean (pure) mathematician today "somewhere above electrical engineering and below the theory of relativity." He argues that despite the esteem in which pure mathematics has been held since the time of Pythagoras, "even a rudimentary knowledge of the history of mathematics suffices to teach anyone capable of learning anything that much of the most beautiful and least useful pure mathematics has developed directly from problems in applied mathematics" (p. 130). Probability theory, though not fitting Bell's criterion of "least useful pure mathematics," is an example of an important area of mathematics that grew out of attempts to solve applied problems, notably the collaborative attempt of Fermat and Pascal to specify the appropriate way to divide the stakes in a prematurely terminated game of chance.

On the other hand, one finds numerous examples in the history of mathematics of developments that have come out of work that appears to have been motivated totally by intellectual—what some might call idle—curiosity and that has then, surprisingly, turned out to be usefully applied to practical problems. A striking illustration of this fact is the centuries of work on the conic sections that eventually found applications in mechanics, astronomy, and numerous other areas of science.

One also finds many examples of the development of theoretical mathematics getting a push from the desire of people to work on real-world problems for which the then-current mathematics did not provide adequate tools. As just noted, an interest in solving wagering problems that began the development of probability theory is one case in point; the desire to work on problems of instantaneous change and continuity, which led to the development of the infinitesimal calculus, is another. Because science has become so dependent on mathematics, the recognition of a need for mathematical tools that do not exist can serve as a powerful motivation for the development of those tools. And the mathematical research that is done to fill the identified need may lead to developments that not only meet the need but have other unanticipated consequences as well.

In short, theoretical and applied interests appear to have coexisted throughout the history of mathematics. Work that has led to new developments has been motivated sometimes by the one and sometimes by the other; moreover, as Hersh (1997) points out, "Not only did the same great mathematicians do both pure and applied mathematics, their pure and applied work often fertilized each other" (p. 26). Whether either type of interest has been more important—more productive of new knowledge—than the other, it is probably not possible to say. For the most part, theoretical and applied interests have advanced together in a mutually beneficial way.

The Usefulness of Mathematics in Science

> The common theme that links Plato, Kepler, Einstein, the quantum theorists, and present-day string theorists is the belief that an understanding of the basic stuff of the universe will be found using mathematics. (Devlin, 2002, p. 68)

It is usefulness for the description of natural phenomena that is perhaps mathematics' most remarkable property. Mathematics has proved to be so useful in science that scientific progress has often been made on the basis of mathematical work done somewhat independently of empirical investigations. Mathematics is useful in science not only by reducing the mental effort required to solve certain types of problems, as Ernst Mach (1893/1960) once suggested that it should, but also by making the answering of many questions possible that otherwise would not be so. Kline (1953a) points out that the great scientific advances of the 16th and 17th centuries were in astronomy and in mechanics, and that, in both cases, they rested more on mathematical theorizing than on experimentation; Atkins (1994) argues that the Copernican revolution was less concerned with whether the sun encircled the earth, or vice versa, than with whether mathematical models are descriptive of reality. The application of geometry—Euclidean and non-Euclidean—to an understanding of space and time is beautifully described by Penrose (1978).

An especially useful 17th-century achievement was the development of the system of logarithms. Kasner and Newman (1940) hold that Napier's *Mirifici Logarithmorum Canonis Descriptio* (1614) is second in significance only to Newton's *Principia* in the history of British science. In providing a method for replacing multiplication and division with addition and subtraction, the raising to powers and the extraction of roots to multiplication and division, logarithms made otherwise inordinately tedious calculations relatively simple and straightforward.

The complementary and mutually reinforcing historical relationship between the study of nature and the development of mathematics has been captured nicely by Atkins (1994). "It should not be forgotten that mathematics and observation jointly squirm towards the truth. The process of discovery of the world is often a sequence of alternations between observations and mathematics in which the observations are stretched like a skin on to a kind of mathematical template. We refine and bootstrap ourselves into a mapping of the physical world by squirming forward, constantly comparing our expectations based on our current theory with observations they themselves suggest" (p. 105). Observation prompts the development of theory, and theory indicates what observations should be made in an endless spiral of looking, explaining, predicting, and looking again.

Smith (2003) argues that a mathematical model has two advantages over a verbal model. First, to write a mathematical model one has to be clear about what one is assuming. And if, in explicating such a model, one makes any assumptions of which one is not aware, others, in assessing the model, may see the need for the unstated assumptions and make them explicit. Expressing one's ideas in mathematical form is an effective way of clarifying them in one's own thinking. The second claimed advantage of mathematical models is that they make predictions that can be tested. This is not to deny that verbal models may make such predictions as well, but other things being equal, the mathematical model is the more likely of the two types to make predictions that are quantitatively precise, and thus easier to put to a rigorous test.

The degree to which an area of science is "mathematized" is sometimes taken as an indication of its maturity. As the Columbia Associates in Philosophy (1923) put it, "It does seem to be true that the more highly developed a science becomes, and the more knowledge we gain about the relations between its objects, the more its beliefs tend to fall into mathematical form, and to admit of treatment by purely mathematical methods" (p. 99). Today the term *mathematical* would undoubtedly be given a sufficiently broad connotation to include *computational*.

Some would argue that scientific explanations, or even descriptions, at their deepest level, must be mathematical; language without mathematics, in this view, is inadequate to the task. Peacocke (1993) makes this case: "There is a genuine limitation in the ability of our minds to depict the nature of matter at this fundamental level; there is a mystery about what matter is 'in itself,' for at the deepest level to which human beings can penetrate they are faced with a mode of existence describable only in terms of abstract, mathematical concepts that are products of human ratiocination" (p. 34). Again, "during the twentieth century we have been witnessing a process in which the previously absolute and distinct concepts of space, time, matter and energy have come to be seen as closely and mutually interlocked with each other—so much so that even the modification of our thinking to being prepared to envisage 'what is there' as consisting of matter-energy in space-time has to be superseded by more inclusive concepts of fields and other notions no longer picturable and expressible only mathematically" (p. 35).

It was noted in Chapter 4 that the history of mathematics has been characterized by increasing abstractness. Peacocke's comments suggest that the same trend is seen in science. Clearly the trend in the one domain is not independent of that in the other; the increasingly close coupling of mathematics and science ensures their correspondence in this regard.

☐ Surprised by Simplicity

One of the major characteristics of Renaissance science that distinguished it from the natural philosophy of the early Greeks was its emphasis on measurement and quantitative description. The classical Greek thinkers looked for explanations of physical phenomena, but the explanations they sought were of why things are the way they are, and they took no great pains to quantify nature or even always to make the observations necessary to check out their assumptions. Some of Aristotle's incorrect beliefs could easily have been corrected by observation.

For the first few centuries after the scientific revolution, mathematicians and scientists often were the same individuals. Their interest in discovering the nature of reality was intense. They looked for invariant relationships among the phenomena they studied, and when they found them, they attempted to express them as mathematical laws. A surprising result of this effort was the discovery that many aspects of the physical world and its behavior could be described by very simple mathematical equations. Consider, for example, Kepler's discovery of his three laws of planetary motion. The first of these laws states that the orbit of each planet around the sun is an ellipse with the sun at one of its foci. According to the second law, a straight line between the sun and a planet sweeps out equal areas in equal times as the planet proceeds on its elliptical orbit. The third law states that the cube of the mean distance of each planet from the sun is proportional to the square of the time taken by the planet to complete its orbit.

The elegant simplicity of these laws, in their mathematical formulation, is remarkable. It is easy to think of other equally simple laws that are descriptive of reality as we understand it. The various "inverse square" laws are cases in point. An inverse square law is a law that states that the strength (intensity, energy) at point B of some property that originates at point A is inversely proportional to the square of the distance between A and B. It describes the dissipation of any form of energy (gravity, electromagnetic radiation, electrostatic attraction, sound intensity) that propagates from a source equally in all directions, and follows from the fact that the surface area of a sphere increases by a factor of n^2 when its radius increases by a factor of n.

Barrow (1991) refers to the world being describable by mathematics as an enigma and to the simplicity of the mathematics involved as "a mystery within an enigma" (p. 2). Farmelo (2003b) expresses a similar sentiment: "Armies of thinkers have been defeated by the enigma of why most fundamental laws of nature can be written down so conveniently as equations. Why is it that so many laws can be expressed as an absolute imperative, that two apparently unrelated quantities (the equation's left and right sides) are exactly equal? Nor is it clear why fundamental laws exist at all" (p. xiii).

Kepler's discovery of the laws of planetary motion was a major achievement—some have held that it was the most impressive achievement in the history of science. That nature can actually be described in such simple terms was considered by Kepler, and by many others, as powerful evidence of design. He believed that in discovering these laws, he had been privileged to see the Creator's mark on nature more directly than it had been seen by anyone before him. Kepler (1619/1975) spoke of contemplating the beauty of his discovery "with incredible and ravishing delight" (p. 1009). He saw our ability to think quantitatively as a special endowment from the Creator, the purpose of which was to permit us to understand the mathematical harmonies of creation:

> God, who founded everything in the world according to the norm of quantity, also has endowed man with a mind which can comprehend these norms. For as the eye for color, the ear for musical sounds, so is the mind of man created for the perception not of any arbitrary entities, but rather of quantities; the mind comprehends a thing the more correctly the closer the thing approaches toward pure quantity as its origin. (Quoted in Holton, 1973, p. 84)

Kepler was not the last scientist or mathematician to have been awestruck by the ability of mathematics to describe nature—or, to put it the other way round, by nature's apparent obeisance to mathematics. Indian-American astrophysicist Subrahmanyan Chandrasekhar's reaction to Roy Kerr's finding of an exact solution to Einstein's general relativity equations that described a rotating black hole is another case in point:

> In my entire life, extending over forty-five years, the most shattering experience has been the realization that an exact solution of Einstein's equations of general relativity, discovered by the New Zealand mathematician Roy Kerr, provides the *absolutely exact representation* of untold numbers of massive black holes that populate the universe. This "shuddering before the beautiful," this incredible fact that a discovery motivated by a search

after the beautiful in mathematics should find its exact replica in Nature, persuades me to say that beauty is that to which the human mind responds at its deepest and most profound. (Quoted in Pagels, 1991, p. 71)

Does the fact that many aspects of nature can be described so effectively by simple mathematics reflect the basic simplicity of reality generally? Or might it be that simple mathematics works so well because it has been applied only to those aspects of nature that lend themselves to this kind of description, and that those aspects of nature constitute a tiny fraction of reality? If the latter is the case, the simplicity with which we are impressed may be the simplicity of our representations of nature and not of nature itself. Many writers have commented on the role that the successful application of analysis to problems of motion and time played in supporting the idea of a clockwork universe that was, in principle, completely describable in terms of differential equations (which represent the rates of change of the state variables of a system). However, some have also pointed out that, in reality, only relatively simple situations proved to be mathematically tractable. "Classical mathematics concentrated on linear equations for a sound pragmatic reason: it couldn't solve anything else.... Linearity is a trap. The behaviour of linear equations— like that of choirboys—is far from typical. But if you decide that only linear equations are worth thinking about, self-censorship sets in. Your textbooks fill with triumphs of linear analysis, its failures buried so deep that the graves go unmarked and the existence of the graves goes unremarked" (Stewart, 1990, p. 83). The answer to the question of whether nature is fundamentally simple is not known, and it is not clear whether it is knowable.

But simplicities are found even in irregular or "chaotic" systems. The finding that the ratio of the meandering lengths of rivers to their linear source-to-mouth distances tends, on average, to be approximately π is a case in point (Støllum, 1996). The constant $\delta \cong 4.669$ (Feigenbaum's number, named for its discoverer, contemporary mathematical physicist Mitchell Feigenbaum) is another. This constant is the factor by which one must decrease the flow of an output in order to double the period of a cascade (drips of a water faucet). "The Feigenbaum number δ is a quantitative signature for any period-doubling cascade, no matter how it is produced or how it is realized experimentally. That very same number shows up in experiments on liquid helium, water, electronic circuits, pendulums, magnets, and vibrating train wheels. It is a new universal pattern in nature, one that we can see through the eyes of chaos" (Stewart, 1995b, p. 122).

Since the appearance of computers on the scene, a form of mathematical description that has been used increasingly is that of numerical simulation. That it is possible to get very complicated structure and behavior from the iterative application of very simple mathematical rules has led to speculation that the complex structure and behavior that is seen in natural systems, including biological systems, may also rest on similarly simple mathematical rules. Validation of numerical models of natural systems is a nontrivial matter, however, perhaps impossible in some cases (Oreskes, Shrader-Frechette, & Belitz, 1994); and equivalence of results does not necessarily mean equivalence of method. Demonstration that process X yields product Y reveals one way of getting Y but does not prove that to be the only possible one.

Mathematical laws are not explanations of *why* the entities of interest behave as they do. That the gravitational attraction between two bodies is proportional to the product of their masses and inversely proportional to the square of the distance between them is a handy bit of knowledge, but the law of gravitation itself provides no clue as to why such a relationship should hold. It can be—has been—argued, however, that the kind of knowledge that is represented by the mathematical formulas that relate natural variables has been more responsible for the achievements of modern science than have causal explanations of the phenomena involved (Kline, 1953).

☐ Unanticipated Uses

> Mathematicians study their problems on account of their intrinsic interest, and develop their theories on account of their beauty. History shows that some of these mathematical theories which were developed without any chance of immediate use later on found very important applications. (Menger, 1937, p. 253)

When mathematical developments have grown out of work on practical problems, it is not surprising that what is developed is applicable to the problems of interest, inasmuch as they were developed with application to those problems in mind. However, many mathematical discoveries have been found to have unanticipated uses long after they were made, or applications to problems quite different from those that may have motivated their development. The connections between the discoveries and the applications often seem to be completely fortuitous. The conic sections (ellipses, parabolas, and hyperbolas) had no application beyond the amusement of mathematicians for 2,000 years after their discovery, and then they suddenly proved to be invaluable to the theory of projectile motion, the law of universal gravitation, and modern astronomy.

The ellipse—a conic section formed by cutting through a cone at an angle—is a plane curve defined as the set of all points such that the sum of the distances of each of the points from two fixed points—the foci—is constant. The Greeks contemplated this figure at least as early as the fourth century BC and discovered many of its properties, none of which was put immediately to any practical use. It is not difficult to imagine Kepler's delight when he found, 2,000 years later, that this simple figure described the orbits of the planets precisely, but it left unanswered the question of why it should be so.

British mathematician Arthur Cayley invented matrix algebra in the middle of the 19th century. It found little practical use for decades. In 1925, however, German physicist Werner Heisenberg saw in it the tool he needed to do his work in quantum mechanics. Complex numbers had no apparent practical applications when they were invented; now they are indispensable to electrical engineering and numerous other practical pursuits.

Another example of a mathematical construct that proved to have applications way beyond those that motivated its development is the Fourier series. Initially applied by French mathematician Joseph Fourier to the solution of problems of heat flow, the series became widely applied to the analysis of complex wave functions that are descriptive of numerous phenomena in acoustics, optics, seismology, and other fields.

In Chapter 8, in the context of a discussion of paradoxes of infinity, we noted such strange mathematical constructs as Sierpinski gaskets and Menger sponges—species of Cantor sets. Such figures are interesting to think about, but one may find it difficult to imagine much in the way of practical applications. In fact, it turns out that Cantor sets have proved to be descriptive of distributions in a variety of contexts. They have been used to represent the distribution of mass in the universe, as well as of "cars on a crowded highway, cotton price fluctuations since the 19th century, and the rising and falling of the River Nile over more than 2,000 years" (Peterson, 1988, p. 122).

In recent years the circulatory system has attracted attention as an example of a fractal in the biological world. The many-leveled branching structure accomplishes the remarkable feat of producing a system that takes up only about 5% of the body's volume and yet ensures that for the most part, no cell is more than three or four cells away from a blood vessel. Many biological structures are fractal-like in that they involve several levels of self-similar branchings or foldings, including the bile duct system, the urinary collecting tubes of the kidney, the brain, the lining of the bowel, neural networks, the placenta, and the heart (West & Goldberger, 1987).

Mathematicians and scientists have often been as surprised as everyone else when an esoteric area of mathematics turns out to be applicable to a problem area of practical importance. Group theory, which now has many applications both within mathematics and in areas of science such as crystallography, particle physics, and cryptography, was once judged to be useless by American mathematician Oswald Veblen; British mathematician–physicist Sir James Jeans, along with other experts, recommended in 1910 that it be removed from the mathematics curriculum at Princeton University for that reason (Davis & Hersh, 1981). Who could have anticipated that work on "Kepler's conjecture" regarding the optimal packing of constant-radius spheres would prove to be applicable to the current-day problem of devising error-detecting and error-correcting codes for communications systems?

The search for ever-larger prime numbers has been a diversion for some mathematicians for many years; every so often one sees an announcement in the popular press that someone has succeeded in finding a prime number larger than the largest one known before. It was hard to imagine that such discoveries had any practical use, until recently. Many modern encryption schemes used for purposes of security in computer networks are based on very large composite numbers being extremely difficult to factor. (Interestingly, it is very much easier to determine, with the help of a fast computer, whether a large number—say of several hundred digits—is prime than it is to factor the number if it is composite.) If I give you a number that happens to be the product of two very large primes (I know it to be because I produced it by multiplying the primes) you will have a very hard time finding the two prime factors. That I know the factors permits me to encode messages in a way that only someone who also knows the factors will be able to decode. There is an intense competition ongoing these days between those who wish to develop evermore secure codes and those who wish to break them; the challenge of the coders is to stay a jump ahead of the code breakers, and finding more efficient ways to factor ever-larger numbers that are the products of two primes is the basic goal of the latter. Although the competition is motivated by the practical concerns of network security, this work has led to the development of some deep and difficult mathematics.

I have noted already the enormous impact that the creation of non-Euclidean geometries had on mathematical thinking, and that this development forced a recognition of the possibility of constructing mathematical systems based on axioms that did not express self-evident properties of the physical world. Some of the axioms of the new geometries seemed to assert what obviously was not true of the perceived world.

It was surprising indeed, therefore, when even these renegade mathematical systems turned out to have profoundly significant applications, especially in Einstein's reconceptualization of the shape of space. Newman (1956b) contends that "the emancipation of geometry from purely physical considerations led to researches of the highest physical importance when Einstein formulated the theory of relativity" and describes this as "one of the agreeable paradoxes of the history of science" (p. 646). As an example of the surprising applicability of an aspect of mathematics that was originally developed for other than practical interests, however, it is only one among many.

☐ The Fading Distinction Between Science and Mathematics

As mathematics becomes increasingly abstract, one might expect its usefulness to science to decrease, because science is constrained by physical reality in a way that mathematics is not. Barrow's (1998) characterization of the difference between the study of formal systems (mathematics and logic) and physical science supports this expectation. "In mathematics and logic, we start by defining a system of axioms and laws of deduction. We might then try to show that the system is complete or incomplete, and deduce as many theorems as we can from the axioms. In science, we are not at liberty to pick any logical system of laws that we choose. We are trying to find the system of laws and axioms (assuming there is one—or more than one perhaps) that will give rise to the outcomes that we see" (p. 227). The appearance of the decoupling of mathematics from science has been the basis of some concern. Von Neumann (1947/1956), for example, worried about the danger "that the subject will develop along the line of least resistance, that the stream, so far from its source, will separate into a multitude of insignificant branches, and that the discipline [of mathematics] will become a disorganized mass of details and complexities" (p. 2063).

But here is an amazing thing, pointed out by Boyer and Merzbach (1991): Despite that most of the developments in mathematics since the middle of the 20th century have been motivated by problems in mathematics itself and have had little to do with the natural sciences, applications of mathematics to science have multiplied exceedingly during the same period. It seems that as mathematics has become increasingly abstract, it also has found increasingly powerful applications to the real world. Why is that? Is it because science itself is also becoming increasingly abstract? Is it the case, as Devlin (2000) claims, that although physicists' ultimate aim is to understand the physical world, they have been led into increasingly abstract mathematical universes?

Much of modern physics deals with aspects of the world that are beyond our powers of observation, even when aided by the most sophisticated instruments of technology. The primary tool for such work is mathematics. The indispensability of mathematics for some areas of science is illustrated by quantum theory.

> The history of quantum physics is the story of a continuing and highly successful search for simpler, deeper, more comprehensive, and more abstract unifying principles. These principles are not systematizations of experience. They are, as Einstein was fond of saying, 'free creations of the human mind.' Yet they capture, in the language of abstract mathematics, regularities that lie hidden deep beneath appearances. Why do these regularities have a mathematical form? Why are they accessible to human reason? These are the great mysteries at the heart of humankind's most sustained and successful rational enterprise. (Layzer, 1990, p. 14)

The recent interest among particle physicists in replacing the familiar concept of particles with the concept of "strings" was motivated by the mathematical intractability, because of the occurrence of infinities in them, of equations dealing with the force of gravity. These infinities occur when particles are treated as points, but not when they are treated as almost equally simple things, one-dimensional lines. Surprisingly, in addition to the simplification gained by doing away with the infinities, adoption of the string concept brought some other advantages as well (Greene, 2004; Gribbin & Rees, 1989). As of this writing, string theory lacks empirical support from the verification of predictions that it makes that simpler theories do not. Despite its esthetic appeal, it has been criticized as lacking testable predictions and therefore being inherently unfalsifiable (Woit, 2006). Nevertheless, it remains an active area of theoretical physics.

Science—at least physics—is so dependent on mathematics today that much of what has been discovered about the world cannot be communicated effectively apart from mathematics and cannot be understood in a more than superficial way by anyone who does not understand the mathematics in which it is represented. American physicist Richard Feynman (1965/1989), who had an unusual ability to make complicated ideas intelligible, held that mathematics is the element that is common to all of science and that connects its various parts. "The apparent enormous complexities of nature, with all its funny laws and rules ... are really very closely interwoven. However, if you do not appreciate the mathematics, you cannot see, among the great variety of facts, that logic permits you to go from one to the other" (p. 41). He took the position that there is a limit to how much can be explained without recourse to mathematical representations: "It is impossible to explain honestly the beauties of the laws of nature in a way that people can feel, without their having some deep understanding of mathematics. I am sorry, but this seems to be the case" (p. 40). Probably few observers would dispute these claims as they pertain to physics; whether they hold for chemistry, biology, economics, psychology, and the social sciences is more debatable.

The close coupling of physics and mathematics is seen very clearly in the work of Newton. Berlinski (2000) puts it this way. "Time, space, distance, velocity, acceleration, force, and mass are physical concepts. They attach themselves to a world of particles in motion. But the relationships among these concepts are mathematical. Some relationships involve little more than the elementary arithmetical operations; others, the machinery of the calculus. Without these specifically mathematical relationships, Newton's laws would remain unrevealing" (p. 186). Physics has even been defined as "the science devoted to discovering, developing and refining those aspects of reality that are amenable to mathematical analysis" (Ziman, 1978, p. 28).

The idea that mathematics is the language in which the book that we refer to as the universe is written goes back at least to Galileo: "Philosophy is written in this grand book, the universe, which stands continually open to our gaze. But the book cannot be understood unless one first learns to comprehend the language and to read the letters in which it is composed. It is written in the language of mathematics, and its characters are triangles, circles, and other geometric figures without which it is humanly impossible to understand a single word of it; without these, one wanders about in a dark labyrinth" (Galileo, quoted in Drake, 1957, p. 237). What has changed since the days of Galileo is that the language itself has evolved considerably; it now contains characters, constructs, and a level of descriptive power of which Galileo could hardly have dreamed. But the changes have served only to strengthen his point; the distinction between science and mathematics grows increasingly difficult to maintain as both areas become evermore abstract. Why should the book that is the universe be written in the language of mathematics? No one who has tried to read it doubts that it is, but neither has anyone given an answer to why it should be so that all, or even most, scientists or mathematicians find completely satisfactory.

The foregoing discussion focuses on the usefulness of mathematics in the physical sciences, and for good reason. Mathematics and science are so intertwined that "math and science," or "science and math," are often treated almost as a single word in the popular lexicon, especially in the context of discussions of education. But mathematics has many applications outside the physical sciences. To the extent that one wants to make a distinction between science and engineering, there can be no question of the importance of mathematics in the latter context as well as in the former. Projective geometry is used extensively in visual art. Trigonometry has, for centuries, been essential to navigation and wayfinding. The application of arithmetic to business and trade is probably as old as mathematics itself. Politics has its uses of math (Meyerson, 2002). The application of mathematics to ethics and governance was promoted by British utilitarian philosopher Jeremy Bentham (1789/1879) with his "hedonistic calculus"; variations on this theme have many proponents today. Recent and current applications of mathematics to problems of choice and decision making in essentially every conceivable context are sufficiently numerous to have motivated the establishment of several journals devoted solely to these topics. Mathematical modeling, so obviously effective in the physical sciences, is increasingly widely used for descriptive, predictive, and prescriptive purposes in the softer sciences as well.

In psychology, mathematics has been an essential ingredient since the time when controlled experimentation first became the method of choice for exploring sensation, perception, motivation, learning, and the countless other aspects of behavior and cognition of interest to psychologists. It has served the purposes of discovering and representing functional relationships in quantitative ways, of analyzing experimental data, and of constructing mathematical models intended to be either prescriptive for, or descriptive of, various aspects of behavior and mentation. There are journals devoted exclusively to the publication of mathematically oriented psychological work, notable examples of which are *Psychometrika* and the *Journal of Mathematical Psychology*. The early history of psychophysics has been told many times (e.g., Boring, 1957; Osgood, 1953; Woodworth & Schlosberg, 1954). Examples of classic work in mathematical psychology are provided in volumes edited by Luce (1960) and Luce, Bush, and Galanter (1963a, 1963b). Other notable examples of the application of mathematics in psychology include the theory of signal detection (Green & Swets, 1966; Swets, 1964), statistical decision theory (Arkes & Hammond, 1986; Edwards, Lindman, & Savage, 1963; Raiffa & Schlaifer, 1961; Von Winterfeldt & Edwards, 1986), and game theory (Luce & Raiffa, 1957; Rapoport, 1960). Laming (1973) gives an extensive and in-depth overview of the entire field as of the mid-1970s. Relatively recent overviews of mathematical psychology as a field are provided by Batchelder (2000) and Chechile (2005, 2006).

☐ Why Is Mathematics So Useful?

Why mathematics should turn out to be so immensely useful to our understanding of the world is seen by many to be a great puzzle. If mathematics has to do only with analytic truths—with tautological statements—how is it that it turns out to be so usefully applied to efforts to describe physical reality? There seems no obvious *a priori* reason why the world should be constructed so that its properties are describable in mathematical terms. Why, after all, should the gravitational attraction between two bodies vary inversely with the square of the distance between them? Why should a straight line between the sun and a planet sweep out equal areas in equal times as the planet orbits the sun? And, as if the ability of mathematics to describe aspects of the world that we can directly observe were not surprising enough, it appears to be descriptive also of phenomena governed by relativity and the quantum, which are outside our direct experience.

The puzzle that is represented by the usefulness of mathematics for the description of the natural world and the discovery of physical relationships has been noted by many mathematicians and scientists. Atkins (1994) refers to "the success of mathematics as a language for describing and discovering features of physical reality" as "one of the deepest problems of nature" (p. 99). Uspensky (1937) makes essentially the same point: "To be perfectly honest, we must admit that there is an unavoidable obscurity in the principles of all the sciences in which mathematical analysis is applied to reality" (p. 5). "Why," as Barrow (1992) puts the question, "does the world dance to a mathematical tune? Why do things keep following the path mapped out by a sequence of numbers that issue from an equation on a piece of paper? Is there some secret connection between them; is it just a coincidence; or is there just no other way that things could be?" (p. 4). "There is," he argues, "no explanation as to why the world of forms is stocked up with mathematical things rather than any other sort" (p. 271).

Even more puzzling is the observation by Bell (1946/1991) that "scanning each of several advanced treatises on the various divisions of classical physics—mechanics, heat, sound, light, electricity and magnetism—we note that two or more of them contain at least one pair of equations identically the same except possibly for the letters in which they are written" (p. 150). Why is it that an equation that was written to describe a specific phenomenon should serve equally well to describe another phenomenon of a qualitatively different type? What does a plane through a cone have to do with the trajectory of a projectile or the orbit of a planetary body? Why should the circles of Apollonius of Perga, discovered in the third century BC, turn out to describe so precisely the equipotentials of two parallel cylindrical electrical conductors (Beckman, 1971, p. 116)? And so on. These seem curious coincidences indeed.

Rucker (1982) takes "that *a priori* mathematical considerations can lead to empirically determined physical truths" as evidence that "the structure of the physical universe is deeply related to the structure of the mathematical universe" (p. 55). Polkinghorne (1998) goes a step farther and connects the structures of mathematics, the physical universe, and the human mind. He argues that the "use of abstract mathematics as a technique of physical discovery points to a very deep fact about the nature of the universe that we inhabit, and to the remarkable conformity of our human minds to its patterning" (p. 2). In his view, our ability to comprehend the microworld of quarks and gluons as well as the macro-world of big bang cosmology is also a mystery, which cannot reasonably be attributed to effects of the pressures for survival on the evolution of our intellectual capacity: "It beggars belief that this is simply a fortunate by-product of the struggle for life" (p. 3). "There is no a priori reason why beautiful equations should prove to be the clue to understanding nature; why fundamental physics should be possible; why our minds should have such ready access to the deep structure of the universe" (p. 4). Moreover, Polkinghorne (1991) argues, an explanation of the mathematical intel-ligibility of the physical world is not to be found in science, "for it is part of science's founding faith that this is so" (p. 76).

Wigner (1960/1980) sees the usefulness of mathematics—"unreason-able effectiveness of mathematics in the natural sciences"—as an enig-matic blessing. "The first point is that the enormous usefulness of mathematics in the natural sciences is something bordering on the mys-terious and that there is no rational explanation for it. ... The miracle of the appropriateness of the language of mathematics for the formulation of the laws of physics is a wonderful gift which we neither understand nor deserve" (pp. 2, 14). Daston (1988) speaks in similar terms: "For modern mathematicians, the very existence of a discipline of applied mathematics is a continuous miracle—a kind of prearranged harmony between the 'free creations of the mind' which constitute pure math-ematics and the external world" (p. 4). King (1992), who believes that the development of mathematics has been driven more by esthetic than by practical interests, contends that the miracle is really a "miracle of sec-ond order of magnitude." We are talking, he says, about the "paradox of the utility of beauty" (p. 121). Kline (1980) speaks of "a twofold mystery. Why does mathematics work even where, although the physical phe-nomena are understood in physical terms, hundreds of deductions from axioms prove to be as applicable as the axioms themselves? And why does it work in domains where we have only mere conjectures about the physical phenomena but depend almost entirely upon mathematics to describe these phenomena?" (p. 340).

This is a modern-day puzzle. To most mathematicians of a few centuries ago this question would not have arisen; the prevailing assumption was that the world was describable in mathematical terms because its Maker made it that way. Kline (1980) points out that the work of 16th-, 17th-, and most of the 18th-century mathematicians was a religious quest, a search for God's mathematical design of nature. "The search for the mathematical laws of nature was an act of devotion which would reveal the glory and grandeur of His handiwork" (p. 34). Kafatos and Nadeau (1990) put it this way: "This article of faith, that mathematical and geometrical ideas mirror precisely the essences of physical reality, was the basis for the first scientific revolution.... For Newton the language of physics and the language of biblical literature were equally valid sources of communion with the eternal and immutable truths existing in the mind of the one God.... The point is that during the first scientific revolution the marriage between mathematical ideal and physical reality, or between mind and nature via mathematical theory, was viewed as a sacred union" (p. 104).

Kline (1989), who does not believe, like King (1980), that mathematicians have been motivated more by esthetics than by practical interests, does not deny that mathematics is often applied in unanticipated ways, but contends that the development of the best mathematics has always been motivated by the quest to understand and describe physical reality, and he argues that this is why mathematics has proved to be so useful. Ziman (1978) makes a similar argument and likens the physicist who uses mathematics descriptively to a fisherman who concludes from his net catching only fish larger than the size of its mesh that as a "law of nature" all fish are larger than that size. Browder and Lane (1978) similarly reject Wigner's notion that mathematics is "unreasonably" effective in the physical sciences: "Because of its origins and its nature, mathematics is not unreasonably effective in the physical sciences, simply reasonably effective" (p. 345). But even if it were the case that mathematicians invariably attempted to develop mathematics that would be useful in describing the physical world, or that scientists focused only on those problems for which mathematics is useful, this would not explain why such a quest should prove to be successful. That the world is mathematically describable remains a mystery. It is a mystery that does not interfere with the further development and application of mathematics, but a mystery nonetheless.

While reflecting on the usefulness of mathematics, it is well to bear in mind that of all the work in mathematics that has been done to date, only a relatively small fraction has proven to be useful yet. Going a step further, Casti (1996) argues that scientists' insistence on making mathematics the universal language of science actually impedes progress on certain types of questions about the natural world. There is the danger, he contends, of finding answers to mathematical representations of questions about the world that are not answers to the questions themselves. He illustrates the point with reference to three well-known problems: the question of the stability of the solar system (the n-body problem of physics), the determination of how a string of amino acids comprising a protein will fold, and the question of financial market efficiency. The mathematical approach to such problems generally involves the construction of mathematical models. To be confident that one has obtained valid answers, one must either be sure the mathematical models used are faithful representations of the phenomena of interest—often they are gross oversimplifications—or abandon mathematics altogether. The point is amply illustrated also by the inability of economic models to predict the national and international economic turmoil of 2008 or to prescribe a clear path to resolution. Models of climate change, which also require dealing with many interacting variables, are proving to be very difficult to verify.

Lindley (1993) acknowledges both the attractiveness of neat mathematical descriptions of aspects of reality and the need to put them to empirical test. "The lure of mathematics is hard to resist. When by dint of great effort and ingenuity, a previously vague, ill-formed idea is encapsulated in a neat mathematical formulation, it is impossible to suppress the feeling that some profound truth has been discovered. Perhaps it has, but if science is to work properly the idea must be tested, and thrown away if it fails" (p. 13). Determining that a mathematical representation of a complex problem is veridical in all important respects generally is not an easy task. In practice what it boils down to is constructing arguments that most of the people who are presumed to be qualified to judge find compelling.

Bunch (1982) reminds us that the correspondence between mathematics and the real world is often not as precise as we may sometimes assume. Zeno, he of the paradoxes, Bunch argues, recognized that the mathematical and scientific ways of looking at the world could be contradictory. "As mathematics has grown independently (to some degree) over the centuries, it has been necessary again and again to change the rules slightly so that mathematical paradoxes become mere fallacies. But from the beginning, from Zeno's time, it was clear that mathematics does not correspond exactly to the real world" (p. 210). This is not to contend, Bunch notes, that mathematics is not useful in describing the real world. But there is no guarantee that what is discovered will fit with the mathematics in hand. "If you assume that an arrow behaves like a collection of mathematical points, you can use mathematics to describe its motion.

If you concentrate on the arrow being a finite collection of small packets of energy, none of which can be located at a particular mathematical point, then you are up the creek" (p. 210).

Dehaene (1997) argues that mathematics rarely agrees *exactly* with physical reality, and realization of this should make the mystery of its "unreasonable effectiveness" somewhat less mysterious. He suggests too that perhaps the effectiveness of the mathematics that has been applied to a description of the physical world is the result of a selection process that has ensured the development of mathematics in directions that are effectively applicable to real-world applications. "If today's mathematics is efficient, it is perhaps because yesterday's inefficient mathematics has been ruthlessly eliminated and replaced" (p. 251). The stated reservations notwithstanding, the applicability of mathematics—including simple mathematics—to the description of the physical world is a remarkable fact.

☐ Reality as a Consistency Check

We have already noted that Gödel (1930, 1931) demonstrated the unprovability of the consistency of any mathematical system sufficiently comprehensive to include all of arithmetic. In view of the desire of some mathematicians to treat mathematics as the epitome of pure abstract reasoning, unsullied by contact with the real world, it is ironic that the best, if not the only, test for the consistency of a set of axioms, postulates, or assumptions of a mathematical system that many mathematicians recognize is the test of a concrete representation or interpretation. "In general, a set of assumptions is said to be consistent if a single concrete representation of the assumptions can be given" (Veblen & Young, 1910/1956, p. 1698). "Representation" as used by Veblen and Young here is synonymous, I believe, with "interpretation" as used by Wilder (1952/1956): "If Σ is an axiom system, then an *interpretation* of Σ is the assignment of meanings to the undefined technical terms of Σ in such a way that the axioms become true statements for all values of the variables" (p. 1662).

Remember that in the view of many modern mathematicians, the terms of axioms of a mathematical system have no meaning; they are just abstract symbols. They may be *given* meanings when a mathematical system is applied to real-world problems, but those meanings are not deemed to be intrinsic to the mathematics. As an example of the assignment of an interpretation to a mathematical construct, Wilder points out that the expression $x^2 - y^2 = (x - y)(x + y)$ is meaningless—it cannot be said to be true or false—until some interpretation, such as "x and y are integers," is given to the variables. In Wilder's terminology, a system is said to be satisfiable if there exists an interpretation of it. And satisfiability, it is claimed, implies consistency. This is because, as Weyl (1940/1956) puts it, "inconsistency would *a priori* preclude the possibility of our ever coming across a fitting interpretation" (p. 1847). The logic, in other words, is: If a system is inconsistent, then an interpretation is impossible; therefore, if an interpretation is found, the system must be consistent.

There are two things to notice about the representation or interpretation test. It works only in one direction. If a concrete representation or interpretation of a system can be found, the system is said to be consistent. If a representation or interpretation has not been found, this is not compelling evidence that a system is inconsistent. This is not a serious limitation, because inconsistency is inherently easier to demonstrate than consistency in general. The difference is analogous to the difference between proving a true universal statement to be true and proving a false one to be false.

The second thing to notice is that the representation or interpretation test shifts the problem of demonstrating consistency from the domain of mathematics to that of the interpretation. If we are not sure that a system of axioms is consistent, why should we be confident that a concrete interpretation that is put on that system is consistent? The test involves the assumption that *nature is consistent*, or, as Court (1935/1961) puts it, "the fundamental belief that logical consistency is identical with natural consistency" (p. 27). This may be an assumption that most of us have little difficulty making, but it is an assumption nevertheless and should be recognized as such, and, again as Court points out, "it makes the consistency of nature to be one of the foundations, one of the cornerstones of the mathematical edifice" (p. 27).

Foundations and the "Stuff" of Mathematics

In spite, or because, of our deepened critical insight we are today less sure than at any previous time of the ultimate foundations on which mathematics rests. (Weyl, 1940/1956, p. 1849)

☐ The "Euclidean Ideal"

As already noted, for many centuries the prevailing view among mathematicians appears to have been consistent with what Hersh (1997) calls "the Euclidean ideal," according to which one starts with self-evident axioms and proceeds with infallible deductions. Plato, Aristotle, and other Greek philosophers of their era considered the axioms of mathematics to be self-evident truths. It was the beyond-doubt intuitive obviousness of certain assertions—two points determine a unique line; three points determine a unique plane—that qualified them to be used as axioms from which less intuitively apparent truths could then be deduced. Geometry (literally "earth measurement") was rooted in the properties of three-dimensional space.

Lakoff and Núñez (2000) associate the Euclidean view with a widely held folk "theory of essences," which they liken to Aristotle's classical theory of categories, according to which all members of a category were members by virtue of a shared essence. The essence of a member of category X was the set of properties that were necessary and sufficient to satisfy the criteria for membership. They contend that Euclid brought the folk theory of essences into mathematics by virtue of claiming that a few postulates characterized the essence of plane geometry. "He believed that from this essence all other geometric truths could be derived by deduction—by reason alone! From this came the idea that every subject matter in mathematics could be characterized in terms of an essence—a short list of axioms, taken as truths, from which all other truths about the subject matter could be deduced" (p. 109). In short, "the axiomatic method is the manifestation in Western mathematics of the folk theory of essences inherited from the Greeks" (p. 110).

In Lakoff and Núñez's view, the folk theory of essences, manifest in the axiomatic method of Western mathematics, is at the heart of much contemporary scientific practice, but not all. They argue that it serves the purposes of physics—which seeks the essential properties of physical things that make them the kind of things they are and provides a basis for predicting their behavior—but it is not useful in biology inasmuch as a species cannot be defined by necessary and sufficient conditions. They consider at least some areas of mathematics to be well served by the folk theory of essences. Algebra, the study of abstract form or structure, they claim, is about essence. "It makes use of the same metaphor for essence that Plato did—namely, Essence is Form" (p. 110). Lakoff and Núñez note that the axiomatic approach to mathematics was only one of many approaches that were taken (they point to the Mayan, Babylonian, and Indian approaches as examples of different ones), but it is the one that arose in Greek mathematics, became dominant in Europe beginning with Euclid, and shaped the subsequent development of mathematics in the West. They see the folk theory of essences, which was central to Greek philosophy, as key to this history.

The close coupling of mathematics with "obvious" truths about the physical world—as represented by the Euclidean ideal—prevailed for two millennia. The adequacy of this view began to be challenged with the appearance on the scene of such strange entities as imaginary or complex numbers and infinitesimals. As such concepts proved to be useful, even if difficult to take as representing anything real, mathematicians gradually became more accepting of entities on the basis of their utility even when the origin was their own imaginations unhampered by the need to identify real-world referents.

Kline (1980) credits the creation of strange geometries and algebras during the early 19th century with forcing mathematicians to realize that neither mathematics proper nor the mathematical laws of nature are truths. This realization, he argues, was only the first of the calamities to befall mathematics, and the soul searching it prompted led to another awakening, this one to the fact that mathematics did not have an adequate logical foundation. "In fact mathematics had developed illogically. Its illogical development contained not only false proofs, slips in reasoning, and inadvertent mistakes which with more care could have been avoided.... The illogical development also involved inadequate understanding of concepts, a failure to recognize all the principles of logic required, and an inadequate rigor of proof" (p. 5).

Kline refers to the perception of mathematics as a universally accepted, infallible body of reasoning, a view that was possible in 1800, as a grand illusion. He points out that even the rules of arithmetic are not immutable. Different arithmetics can be defined to serve different purposes. "The sad conclusion which mathematicians were obliged to draw is that there is no truth in mathematics, that is, truth in the sense of laws about the real world. The axioms of the basic structures of arithmetic and geometry are suggested by experience, and the structures as a consequence have a limited applicability. Just where they are applicable can be determined only by experience. The Greeks' attempt to guarantee the truth of mathematics by starting with self-evident truths and by using only deductive proof proved futile" (p. 95).

But even the view of mathematics as an infallible body of reasoning, which Kline considers possible in 1800, was based more on the practical utility of math than on the rigorous derivation of mathematical truths from indisputable first principles. Wallace (2003) describes what happened throughout much of the 1700s as something like a stock market bubble. Advance after advance yielded a situation, to use another of Wallace's similes, like a tree lush with branches but with no real roots. Davis (1978) argues that, despite that people have been computing for centuries, until the work of Alan Turing in the 1930s there was no satisfactory answer to the question of what constitutes a computation.

☐ Doubts and Emerging Perspectives Regarding Foundations

> Every philosophy of mathematics arises out of the sense that mathematics touches something that is profound yet difficult to make explicit. (Byers, 2007, p. 349)

The early part of the 20th century found mathematics, from a theoretical perspective, in an unhappy state. Rather than the rock-solid representation of indisputable eternal truth mathematics once was perceived to be, it had become a fractionated discipline with each of several groups of practitioners fully capable of pointing out the shortcomings of the views of each of the others. Kline summarizes the state of affairs as a consequence of the disagreement over foundations, and compares it with the perception of things at the beginning of the 19th century, this way: "The science which in 1800, despite the failings in its logical development, was hailed as the perfect science, the science which establishes its conclusions by infallible, unquestionable reasoning, the science whose conclusions are not only infallible but truths about our universe and, as some would maintain, truths in any possible universe, had not only lost its claim to truth but was now besmirched by the conflict of foundational schools and assertions about correct principles of reasoning. The pride of human reason was on the rack" (p. 257).

Attempts to deal with the question of the foundations of mathematics had produced several identifiable schools of thought by the early part of the 20th century. Dantzig (1930/2005) distinguishes two such schools, the proponents of which he refers to as *intuitionists* and *formalists*. What separates these schools, in Dantzig's view, are their perspectives regarding what constitutes a mathematical proof. "What is the nature of reasoning generally and mathematical reasoning in particular? What is meant by *mathematical existence*?" (p. 67).

Several other writers partition the major mathematical perspectives regarding foundations into more than two categories. Gellert, Küstner, Hellwich, and Kästner (1977, pp. 718–719) count three of them, represented by the views of *logicists*, *formalists*, and *intuitionists*. As the leaders of these schools, they identify, respectively, German mathematician-logician Gottlob Frege, German mathematician David Hilbert, and Dutch mathematician Luitzen Brouwer. Browder and Lane (1978) note the same three schools and the same leaders. To these three, Kline (1980) adds a fourth—*set theorists*, led by Ernst Zermelo. Casti (2001) also recognizes four main schools of thought, very similar, but not quite identical, to those listed by Kline. Casti's list contains: *formalism* (Hilbert), *logicism* (Bertrand Russell), *intuitionism* (Brouwer), and *Platonism* (Kurt Gödel, René Thom, and Roger Penrose). Hersh (1997) recognizes five schools: *logicism, formalism, intuitionism, empiricism,* and *conventionalism*. He also distinguishes what he refers to as three main philosophies: *constructivism, Platonism,* and *formalism*. It is not clear to me how schools and philosophies differ in this context, but inasmuch as Hersh identifies five schools and three philosophies, it appears that he sees some philosophical overlap among the schools. Barrow (1995) distinguishes *formalism, inventionism, Platonism, constructivism,* and *intuitionism*. What follows is a somewhat eclectic partitioning that appears to me to recognize the more important

TABLE 13.1. Major Schools of Mathematics, Their Kernel Ideas, and Notable Proponents

School	Kernel Ideas	Notable Proponents
Logicism	Math is reducible to logic	Frege, Russell, Whitehead
Intuitionism	Math is derived from human intuition; rejection of actual infinity	Brouwer, Poincaré, Kant
Constructivism	Close to intuitionism; limits math to objects that can be constructed; building blocks of math are natural numbers	Kroenecker, Lorenzen, Bishop
Set theory	Takes set (collection) as the foundational concept on which to build math	Zermelo, Fraenkel
Platonism	Sees math truths as real, abstract, and eternal	Gödel, Erdös, Penrose, Thom
Formalism	Math is rule-based symbol manipulation	Hilbert, Carnap, Tarski

kernel ideas promoted by one or another of the various schools, or philosophies, of mathematical thought. It will be clear that these ideas are not all independent of each other. The schools, kernel ideas, and notable proponents are listed in Table 13.1.

Logicism is the view that all mathematics is derivable from logic. Leibniz held this view before logicism was identified as a school. Notable among the founders of the school was Gottlob Frege, who, according to Moore (2001), was "by common consent, the greatest logician of all time" (p. 114). Frege's vision was that of being able to derive all mathematical theorems from a few foundational logical principles. But logicism got its strongest endorsement from the Herculean attempt by British philoshoper-mathematicians Alfred North Whitehead and Bertrand Russell to establish the logical foundation of mathematics in their frequently cited but seldom read, by some accounts, *Principia Mathematica*. Russell's identification with logicism is ironic inasmuch as it was his famous paradox involving sets of all sets that are not members of themselves that showed Frege's vision to be unattainable. (Are such sets members of themselves? If they are, they are not, and if they are not, they are.) Russell extricated himself from his paradox by defining a set as a different type of thing from its members, thus making the question of whether it could be a member of itself meaningless. The idea that mathematics could be reduced to logic has a twin in the idea that logic can be reduced to, or is a form of, mathematics.

Intuitionism, the founder of which is usually considered to be Dutch mathematician Luitzen Brouwer, sees the human mind as the ultimate sanction of any rules of thought. The foundation of mathematics is not logic, as the logicists contend, but psychology. As suggested by the subtitle of his book—*How the Mind Creates Mathematics*—Dehaene (1997) sees intuitionism as the most plausible of the various theories of the nature of mathematics that have been proposed, and argues that this view is supported by recent discoveries about the natural number sense that were not known to proponents of this theory, like Poincaré and Kant. "The foundations of any mathematical construction are grounded on fundamental intuitions such as notions of set, number, space, time, or logic. These are almost never questioned, so deeply do they belong to the irreducible representations concocted by our brain. Mathematics can be characterized as the progressive formalization of these intuitions. Its purpose is to make them more coherent, mutually compatible, and better adapted to our experience of the external world" (p. 246).

Constructivism, a close cousin to intuitionism, is described by Barrow (1995) as the mathematical version of operationalism and as a response to the problems created for formalism by the logical paradoxes of Russell and others. In the views of constructivists, mathematics is what can be constructed from certain undefined but intuitively compelling primitives. The natural numbers, according to this perspective, are the fundamental building blocks of mathematics; the concept of number neither requires nor can be reduced to a more basic notion, and it is that from which all meaningful mathematics must be constructed. Proponents include German mathematician-logician Leopold Kronecker, German philosopher-mathematician Paul Lorenzen, and American mathematician Errett Bishop.

Although, as already noted, Dehaene (1997) espouses intuitionism, he sees constructivism as an extreme form of intuitionism, as defended by Brouwer, and he rejects that. What Dehaene objects to is Brouwer's rejection of "certain logical principles that were frequently used in mathematical demonstrations but that he felt did not conform to any simple intuition" (p. 245). An example is the application of the principle of the law of the excluded middle to infinite sets. Constructivists are opposed to proofs by contradiction, which lead to such intuitive incongruities as the idea that most real numbers are transcendental despite the fact that relatively few examples of transcendental numbers have been identified (Byers, 2007). Byers distinguishes within constructivism a radical movement according to which it should not be claimed that any mathematical entities exist independently of what people do, which is to say that the only mathematical entities that exist are those that people have constructed.

Set theory attempts to build mathematics on the axiomatization of the fundamental concept of a set and of set relationships in order to avoid the categorical and self-referential paradoxes that Russell and others had discovered to be permitted by traditional (Aristotelian) logic: This statement is false; true or false? Every rule has an exception; including this one? Barbers in this town shave those and only those who do not shave themselves; do they shave themselves? Does the set of all sets that do not belong to themselves belong to itself? Russell's own approach to solving such paradoxes was his invention of the theory of types, which disallows the sorts of propositions that constituted the paradoxes, thereby—to use Wallace's (2003) term—effectively legislating them out of existence. Notable contributors to the axiomatization of set theory were the German mathematician Ernst Zermelo and the German-born Israeli mathematician Abraham Fraenkel, whose joint work, including refinements from others as well, is referred to as the Zermelo-Fraenkel (or simply ZF) set theory.

Platonism traces its roots to Plato's idealism, according to which real-world entities are mere shadows of ideals. (The term *absolutism* is also sometimes used to represent much the same view.) Platonists see mathematical concepts and relationships as having an existence outside of space and time that is more real than the tangibles of everyday experience. "The Platonist regards mathematical objects as already existing, once and for all, in some ideal and timeless (or tenseless) sense. We don't create, we discover what's already there, including infinites of a complexity yet to be conceived by mind of mathematician" (Hersh, 1997, p. 63). Platonists, to whom mathematics is a reflection of reality and mathematical truths are discoveries, are not surprised by the usefulness of mathematics in describing real-world relationships and solving real-world problems. Proponents include Kurt Gödel, Paul Erdös, Roger Penrose, and René Thom.

Formalism, championed first by German mathematician David Hilbert and later by German philosopher Rudolph Carnap and Polish logician-mathematician Alfred Tarski, sees any system of mathematics as a creation of the human mind. The doing of mathematics is the manipulation, according to arbitrary rules, of abstract symbols devoid of meaning and independent of physical reality. The way to play the game is to define terms, state axioms (givens), set operational rules for manipulating symbols—transforming statements from one form to another (making inferences)—and see where it all leads. One accepts as part of the system any statement that can be derived from the original statements and other statements already derived from the original ones, in accordance with the symbol manipulation rules. The only requirement is internal consistency; everything else is arbitrary. One need not worry about what the system's elements "really are," and whether the axioms are true empirically is irrelevant. Nor is it necessary to attach some physical meaning to the theorems that are derived from them. Nothing need correspond to anything that is believed to be true of the physical world. Correspondence is not prohibited, just irrelevant. If the axioms happen to be empirically true, and the proofs of the theorems are valid, the theorems can be assumed to be empirically true as well, but this would be an extra benefit. And mathematics does not provide the wherewithal to determine whether the axioms are true. It should be clear that this conception allows the existence of many mathematical systems—it is not required that one system be consistent with another. Formalism is the most abstract of the various schools.

Brainerd (1979) also describes formalism as the view that mathematics is a game played on sheets of paper with meaningless symbols. Mathematical formulas may be applied to real-world problems, and when they are, they acquire meaning and can be said to be true or false. "But the truth or falsity refers only to the physical interpretation. As a mathematical formula apart from any interpretation, it has no meaning and can be neither true nor false" (p. 139). Bell similarly (1946/1991) describes the formalist's conception of mathematics as that of "a meaningless game played with meaningless marks or counters according to humanly prescribed rules—the humanly invented rules of deductive logic" (p. 339).

Byers (2007) describes formalism as an ambitious—but failed—attempt to remove subjectivity from mathematics. The attempt failed, he contends, because mathematics is infinitely richer and more interesting than is apparent in the picture presented by formalism. Moreover, "the attempt to access objective truth and certainty by means of logic is fundamentally flawed, and not only because of the implications of Gödel-like theorems. Logic does not provide an escape from subjectivity. After all, what is logic, in what domain does it reside? Surely logic represents a certain way of using the human mind. Logic is not embedded in the natural world; it is essentially a subjective phenomenon" (p. 354). It should be apparent, Byers argues, that formalism misses most of mathematics. "Where do the axioms come from that form the foundations of a formal system?" Clearly they come from human thought. And the axioms that people propose are not arbitrary, although there is nothing in the perspective of formalism to prevent them from being so. Most axioms that might conceivably be proposed would not be considered by most mathematicians to be worth exploring, and decisions about which possibilities are worth exploring and which are not must come from outside the tenets of formalism and are subjective through and through.

The foregoing is a rough partitioning of the major schools of thought regarding the foundations of mathematics. I make no claim to it being the best partitioning that can be done. Missing from it are several other perspectives that are mentioned in the literature, including empiricism (mathematics is discovered by empirical research), conventionalism (the "truths" of mathematics are true only in the sense of being agreed to by society), fictionalism (mathematical "truths" are fictional, though useful), and inventionism (mathematics is what mathematicians do, period). For present purposes, the main point is that mathematics is not seen by all mathematicians through the same lens, and I doubt the possibility of classifying the various perspectives in a way with which all mathematicians will agree.

How the prevailing view of the nature of mathematics has changed over the centuries is described by Maor (1987) this way:

> Ever since the time of Thales and Pythagoras, mathematics has been hailed as the science of absolute and unfailing truth; its dictums were revered as the model of authority, its results trusted with absolute confidence. "In mathematics, an answer must be either true or false" is an age-old saying, and it reflects the high esteem which layman and professional alike have had for this discipline. The 19th century has put an end to this myth. As Gauss, Lobachevsky, and Bolyai have shown, there exist several different geometries, each of which is equally "true" from a logical standpoint. Which of these geometries we accept is a matter of choice, and depends solely on the premises (axioms) we agree upon. In our own century Gödel and Cohen showed that the same is true of set theory. Since most mathematicians agree that set theory is the foundation upon which the entire structure of mathematics must be erected, the new discoveries amount to the realization that there is not just one, but several different mathematics, perhaps justifying the plural "s" with which the word has been used for centuries. (p. 258)

Hersh (1997) contends that, despite that formalism had become the predominant position in textbooks and other official writing on mathematics by the mid-20th century, nearly all mathematicians were, and are, closet Platonists. "Platonism is dominant, but it's hard to talk about in public. Formalism feels more respectable philosophically, but it's almost impossible for a working mathematician to really believe it" (p. 7). Or as Davis and Hersh (1981) put it, "Most writers on the subject seem to agree that the typical working mathematician is a Platonist on weekdays and a formalist on Sundays" (p. 321).

Formalists and Platonists can coexist harmoniously because they do not differ on the matter of how to go about proving theorems; they differ only on the philosophical question of whether mathematics exists independently of mathematicians, and this is a difference that need not get in the way of doing mathematics. In contrast, the position of the constructivists, of whom there were relatively few by mid-century, illegitimizes any areas of mathematics (e.g., those involving infinities) that cannot be constructed from the natural numbers.

Dehaene (1997) criticizes formalism on the grounds that it does not provide an adequate explanation of the origins of mathematics. And clearly its content is anything but arbitrary. "If mathematics is nothing more than a formal game, how is it that it focuses on specific and universal categories of the human mind such as numbers, sets, and continuous quantities? Why do mathematicians judge the laws of arithmetic to be more fundamental than the rules of chess? ... And, above all, why does mathematics apply so tightly to the modeling of the physical world?" (p. 243). Barrow (1995) also argues that formalists, like inventionists, have a difficult time accounting for the practical usefulness of mathematics.

☐ How Serious Is the Problem of Foundations

> Concern about foundations should come, if at all, after one has a firm intuitive grasp of the subject. (Kac, 1985, p. 111)

The inadequacy of mathematics' logical foundation, when it first came to light, was perceived by many as a fixable problem, and it was the various efforts to fix it that eventuated in the several different schools of mathematical thought, distinguished by what the fixers believed the foundation should be. Undoubtedly the best known, and probably the most ambitious, of these efforts was Whitehead and Russell's (1910–1913) *Principia Mathematica*. The laboriousness of this task and the lack of impact of the result on practicing mathematicians are captured in a comment by Hersh (1997): "I'm told that finally on page 180 or so they prove 1 is different from 0" (p. 28).

The debate on the issue of what would constitute an adequate foundation was ongoing when Kurt Gödel, in his 1931 paper "On Formally Undecidable Propositions of *Principia Mathematica* and Related Systems," demonstrated that none of the collections of principles adopted by any of the schools was adequate to prove the consistency of any mathematical system sufficiently complicated to include arithmetic. Gödel himself did not subscribe to the formalist's idea that mathematics has no meaning; generally considered a Platonist, he believed that the axioms of set theory "force themselves upon us as being true" and held that mathematical intuition is no less reliable than sense perception (Gödel, 1947/1964, p. 272).

Dyson (1995) credits Gödel's theorem with showing that pure reductionism does not work in mathematics: "To decide whether a mathematical statement is true, it is not sufficient to reduce the statement to marks on paper and to study the behavior of the marks. Except in trivial cases, you can decide the truth of a statement only by studying its meaning and its context in the larger world of mathematical ideas" (p. 6). Moore (2001) describes the consequences of Gödel's work for the foundations of mathematics this way: "What Gödel showed was that no such system [set theory] would ever be strong enough to enable us to prove every truth about sets—unless it was inconsistent, in which case it would enable us to 'prove' anything whatsoever, true or false" (p. 130). Rucker (1982) likens Gödel's demonstration of the incompleteness of mathematics to the Pythagoreans' discovery of the irrationality of $\sqrt{2}$. What the incompleteness theorem makes clear, he argues, is the distinction between truth and provability. "If we have correct axioms, the provable statements will all be true, but not all the true statements will be provable" (p. 207). But this does not mean that provability is a useless concept; we cannot do away with it in favor of truth, "because we have no finite definition of what 'truth' means" (p. 207).

So we see that there are many schools of thought regarding the foundations of mathematics. Precisely how many there are is a matter of opinion. I have noted seven kernel ideas, but I am not contending that each of them represents a unique school of thought. However the main schools are defined, boundaries between them are likely to be fuzzy. As we have seen, Barrow (1995) identifies five schools, including constructivism and intuitionism, but he equates these two schools by noting that, because of its appeal to intuition as the justification of the basic building blocks, constructivism became known as intuitionism. I have noted too that Dehaene (1997) sees constructivism as an extreme form of intuitionism, but does not equate the terms, because he accepts intuitionism as the most plausible of the various schools, while rejecting what he considers to be its most extreme form. There are many other instances of imprecise boundaries between, or overlap among, the schools as conceptualized by different writers.

The situation has not improved in recent years. Each of the major schools, however one identifies them, developed various factions within them. The foundations remain as much a matter of dispute as ever. "The claim therefore to impeccable reasoning must be abandoned.... No school has the right to claim that it represents mathematics. And unfortunately, as Arend Heyting remarked in 1960, since 1930 the spirit of friendly cooperation has been replaced by a spirit of implacable contention" (Kline, 1980, p. 276). The spirit of contention notwithstanding, Hersh (1997) claims that one view—logico-set theoreticism—dominates the philosophy of mathematics today, but that it is a tenuous domination, "not clear and indubitable like elementary logic, but unclear and dubitable" (p. 149). Even what constitutes a set is a matter of some dispute. The concept and its ramifications for mathematics appear to be continuing to evolve. "The universe of set theory is infinitely elusive; if there is one thing we can be sure of, it is that the set theories of the future will be vastly more inclusive than anything we have ever dreamed of" (Rucker, 1982). The issue of foundations is far from settled.

In view of the lack of agreement among mathematicians regarding the foundations of the discipline and of the many shortcomings that have been identified in the original developments of geometry, arithmetic, algebra, the calculus, and other areas of mathematics, it is remarkable that mathematics remains the robust and immensely useful undertaking that it is. Why, especially given its "independence" from physical truth, should it be so useful in describing, and in facilitating control of, the physical world? This question was the focus of the preceding chapter. Here I wish to point out four facts that illustrate the intuitive basis of mathematics and of reasoning more generally.

First, concerted efforts were made during the 19th century by George Boole, Ernst Schröder, Charles Peirce, Gottlob Frege, Guiseppe Peano, and others to "rigorize" both logic and mathematics. Kline (1980), who describes these efforts, notes that they revealed something about the development of mathematics, because the guarantee of soundness they were intended to provide turned out to be largely gratuitous. "Not a theorem of arithmetic, algebra, or Euclidean geometry was changed as a consequence, and the theorems of analysis had only to be more carefully formulated.... In fact, all that the new axiomatic structures and rigor did was to substantiate what mathematicians knew had to be the case. Indeed, the axioms had to yield the existing theorems rather than different ones because the theorems were on the whole correct. All of which means that mathematics rests not on logic but on sound intuitions" (p. 194). I would only add that logic too must rest on sound intuitions, inasmuch as there is nothing else from which it can get its sanction.

Second, mathematicians, logicians, and philosophers argue about the logical foundations of mathematics or the mathematical foundations of logic, and they disagree among themselves about many basic points relating to rules of reasoning. In view of the existence of several different schools of thought regarding the assumptions on which mathematical and logical reasoning rests, it is remarkable, perhaps ironic, that arguments *about* these issues are not impeded by the unresolved question of the nature of argumentation. People from different schools of thought seem to be able to engage in these arguments without reserve; their different perspectives about the basics do not seem to get in the way at all. This is surprising. One might think that unless there could be agreement on the rules, there could not really be a dispute. Apparently, at least for practical purposes, there is a greater degree of tacit agreement regarding the rules of argument than the disputes about foundations might lead one to believe.

Third, disputes about foundations seem not to have interfered much, if at all, with the application of mathematics to the solution of practical problems of a wide variety of types. For the most part, applied mathematicians have not only continued using mathematics to great advantage despite the arguments about how the entire enterprise is to be justified, but they have essentially ignored the fact that the arguments were going on. In many cases, concepts that have frustrated mathematicians in recent years were simply applied to problems by their first users without much concern for philosophical or metaphysical significance. The differential calculus had been used to good effect for 200 years following its development by Newton and Leibniz before there existed a well-founded explanation of why it worked; as Stewart (1995b) points out, while mathematicians were worrying about how to make the calculus sound and philosophers were declaring it nonsensical, physicists were using it successfully to understand nature and to predict its behavior. The enigma of the usefulness of mathematics was deepened by the formalist's view that mathematics is strictly symbol manipulation and bears no necessary relationship to the physical world. It was deepened again by the demonstration that the people who best understood the symbol manipulation that constitutes mathematics could not agree on the nature or the adequacy of the foundation on which it rests. But enigma or no, mathematics continues to work; eclipses are predicted, bridges are built, satellites are launched, and new applications are found continually.

Fourth, suppose that mathematicians *did* agree with respect to what constitute the foundations of mathematics; how would we know whether, in their aggregate wisdom, they were right? The history of mathematics is full of examples of beliefs held by one generation of mathematicians that were considered untenable by subsequent generations. Again, the same observation can be made with respect to the foundations of logic. The mathematicians and the logicians may or may not come to agree among themselves on these matters, but you and I have to decide whether to accept what they say, and to do that, unless we are willing to act on blind faith, we have to appeal to our intuitions—to accept what seems intuitively to us to be right and to reject what does not.

☐ Discoveries or Inventions?

> How "real" are the objects of the mathematician's world? From one point of view it seems that there can be nothing real about them at all. Mathematical objects are just concepts; they are the mental idealizations that mathematicians make, often stimulated by the appearance and seeming order of aspects of the world about us, but mental idealizations nevertheless. Can they be other than mere arbitrary constructions of the human mind? At the same time there often does appear to be some profound reality about these mathematical concepts, going quite beyond the mental deliberations of any particular mathematician. It is as though human thought is, instead, being guided towards some eternal external truth—a truth which has a reality of its own, and which is revealed only partially to any of us. (Penrose, 1989, p. 95)

Does mathematics exist independently of human minds; are its concepts and relationships there to be discovered, or are they entirely human inventions? Do the concepts with which mathematics deals represent things that exist in the real world, outside the heads of mathematicians? Is there really such a thing as an infinitesimal? A circle? A number? Where does one look in the real world to find an infinity? Are these constructs only computational conveniences, invented to facilitate the solving of certain types of problems, but having no real referents? Is it the case, as Carl Hempel (1945/1956b) argues, that "the propositions of mathematics are devoid of all factual content" and "convey no information whatever on any empirical subject matter" (p. 1631)? These are not new questions; they have been asked many times, and certainly not for the last time. Given the still unsettled question of the foundations of mathematics, and the existence of several schools of thought as to what the essence of mathematics is, it should not be surprising that these questions are also unsettled.

When mathematicians prove new theorems or develop new areas of mathematics are they engaged in a process of discovery or one of invention? Are they learning something about reality—mathematical reality—or are they creating symbolic systems that are completely arbitrary except for the requirement to be internally consistent as judged by some prescribed logic. As we have seen, mathematicians themselves have not been of one mind on this question.

Hardy (1940/1989) considered mathematical theorems to be discoveries, not creations. Mathematical reality, in his view, exists outside us, and the mathematician's function is to observe that reality. He considered the reality of mathematics to be more real, in some fundamental sense, than the subject matter of physics. We cannot know what the subject matter of physics really is or really is like, but only what it seems to be. Mathematical objects in contrast *are* what they *seem* to be: "317 is a prime, not because we think so, or because our minds are shaped in one way rather than another, but *because it is so*, because mathematical reality is built that way" (p. 130).

A similar view is expressed by Polkinghorne (2006). "It is difficult to believe that they [the truths of mathematics] come into being with the action of the human mind that first thinks them. Rather their nature seems to be that of ever-existing realities which are discovered, but not constructed, by the explorations of the human mind" (p. 90).

Penrose (1989) takes essentially the same position. "The Mandelbrot set is not an invention of the human mind: it was a discovery. Like Mount Everest, the Mandelbrot set is just *there*! Likewise, the very system of complex numbers has a profound and timeless reality which goes quite beyond the mental constructions of any particular mathematician" (p. 95). The history of complex numbers is instructive in this regard, Penrose argues. Although this construct was introduced originally for the specific purpose of enabling the taking of square roots of negative numbers, its use brought additional unanticipated benefits. But the properties that yielded these benefits were *there* before they were discovered. "There is something absolute and 'God-given' about mathematical truth," Penrose (1989) contends. "Real mathematical truth goes beyond man-made constructions" (p. 112).

Penrose qualifies this view to the extent of allowing that sometimes mathematicians do invent constructs in order to achieve specific goals, such as the proof of a recalcitrant theorem. What distinguishes true discoveries from such inventions, in his view, is that more comes out of the structures that result from discoveries than is put into them in the first place, whereas this is not true in the case of inventions. Not surprisingly, Penrose holds mathematical discoveries in higher regard than mathematical inventions, as he distinguishes them.

Beyond perceiving mathematics to be real, independently of the discovery of its truths, some see mathematics as more basic than physical reality in the sense of dictating what is possible in any world. "Far from being an arbitrary creation of the human mind, such mathematical facts [of logic, arithmetic, geometry, and probability] have (in my view) universally held before the emergence of life, constraining what is possible in any world. Indeed, abstract mathematical constraints may have determined not only the form of the universe and its physical laws (as some theoretical physicists now suggest) but also the forms of evolutionarily stable strategies, of sustainable social practices, and of the laws of individual thought, whenever and wherever life emerged. Leibniz's claim that this is the best of all possible worlds may have been correct only in that, at the level of abstract principles anyway, this is the only possible world" (Shepard, 1995, p. 51).

Kline (1953a) sees things differently: "Mathematics does appear to be the product of human, fallible minds rather than the everlasting substance of a world independent of man. It is not a structure of steel resting on the bedrock of objective reality but gossamer floating with other speculations in the partially explored regions of the human mind" (p. 430). In a more recent book, Kline (1980) makes the argument that mathematics, or at least the best mathematics, has been developed by people whose main interest in the subject was what it could help reveal about the nature of the physical world; nevertheless, mathematics, per se, he sees as "a human activity and subject to all the foibles and frailties of humans. Any formal, logical account is a pseudo-mathematics, a fiction, even a legend, despite the element of reason" (p. 331).

Over half a century ago, Kasner and Newman (1940) dismissed Platonism as a view of the past: "We have overcome the notion that mathematical truths have an existence independent and apart from our own minds. It is even strange to us that such a notion could ever have existed.... Today mathematics is unbound; it has cast off its chains" (p. 359). More recently, Dehaene (1997) takes a similar position in contending that Platonism "leaves in the dark how a mathematician in the flesh could ever explore the abstract realm of mathematical objects. If these objects are real but immaterial, in what extrasensory ways does a mathematician perceive them?" (p. 242). The feeling that Platonists have that they are studying real objects that exist independently of the human mind is, in Dehaene's view, an illusion. I think it fair to say, however, that Platonism, while perhaps not the prevailing view, is far from dead in the 21st century, and its persistence proves the point that one person's illusion is another's vision.

Lakoff and Núñez (2000) argue that if transcendental Platonic mathematics exists, human beings can have no access to it. Human understanding of mathematics is limited by the affordances and constraints of the human brain and mind, and there is no way of knowing whether proved theorems have any objective truth external to human beings. Belief in Platonic mathematics, in their view, is a matter of faith, about which there can be no scientific evidence, for or against. They reject the idea—which they see as part of a myth that they call "the romance of mathematics"—that "mathematics is an objective feature of the universe; mathematical objects are real; mathematical truth is universal, absolute, and certain" (p. 339).

Lakoff and Núñez (2000) make the telling point that "there is no way to know whether theorems proved by human mathematicians have any objective truth, external to human beings or any other beings," and argue that "all that is possible for human beings is an understanding of mathematics in terms of what the human brain and mind afford" (p. 2). So much seems obvious, but of course this does not make mathematics unique; human understanding generally must be limited by the conceptual affordances and constraints of the human brain and mind.

Giving the perspective of a cultural anthropologist, White (1947/1956) defends the contention that the opposing propositions—(1) that mathematical truths exist independently of the human mind and (2) that mathematical truths have no existence apart from the human mind—are both valid. His acceptance of such apparently contradictory assertions rests on the idea that "the human mind" has different referents in the two assertions. In the first, it refers to the mind of an individual person; in the second, it refers to cultural tradition of mankind as a whole. "Mathematical truths exist in the cultural tradition into which the individual is born, and so enter his mind from the outside. But apart from cultural tradition, mathematical concepts have neither existence nor meaning, and of course, cultural tradition has no existence apart from the human species. Mathematical realities thus have an existence independent of the individual mind, but are wholly dependent upon the mind of the species" (p. 2350). Are the truths of mathematics discovered, or are they man-made? They are both, White contends: "They are the product of the mind of the human species. But they are encountered or discovered by each individual in the mathematical culture in which he grows up" (p. 2357).

The numerous occurrences of simultaneous "discoveries," White argues, demonstrates the importance of the accumulated store of mathematical knowledge in producing them; the role of the individual human brain is only that of a "catalytic agent" in the cultural process. "In the process of cultural growth, through invention or discovery, the individual is merely the neural medium in which the 'culture' of ideas grows" (p. 2358). The quality of the individual human brain has not changed over recorded time, White argues, but recent mathematical discoveries or inventions could not have been made before the prerequisite mathematics had been developed; but "when the cultural elements are present, the discovery or invention becomes so inevitable that it takes place independently in two or three nervous systems at once" (p. 2359).

Byers (2007) also argues that whether mathematics is considered invented or discovered is a matter of perspective and that both perspectives are legitimate, despite the fact that they conflict. "'Discovery' and 'invention' evoke equally valid, consistent frames of reference that are clearly in conflict with one another" (p. 360). "Mathematics is that one unified activity that looks like discovery when you think of it from one point of view and appears to be invention when regarded from another" (p. 361). Byers makes the perspective a matter of which side of some mathematical truth one is on. Before the establishment of some relationship, the work looks like what is needed is some creativity; after the fact it looks like something has been discovered—"there is the sense that it was there all the time, waiting for us" (p. 362). Invention emphasizes the subjective aspect of mathematics; discovery emphasizes the objective nature of relationships that are found. These ideas are illustrative of Byers's emphasis on the role of ambiguity as a major source of mathematical creativity.

It is clear from all this that there is a profound difference of opinion among mathematicians as to whether the business of mathematics is discovery or invention. Bell (1946/1991), who discusses this difference at length, suggests that it is an irreconcilable one and one that has significant implications for other views that one may hold. "Neither the necessity nor the universality [of mathematics] is taken for more than a temporary appearance by those who believe mathematics and logic to be of purely human origin. Others, including many who believe that numbers were discovered rather than invented, find in mathematics irrefutable proof of the existence of a supreme and eternal intelligence pervading the universe. The former regard mathematics as variable and subject to change without warning; the latter see mathematics as a revelation of permanence throughout eternity, marred only by such imperfections as are contributed by the inadequacies of human understanding" (p. 60).

The development of non-Euclidian geometries, which is usually taken to be the event that, above all others, destroyed the idea that the axioms of geometry represent self-evident truths about the physical world, also prompted the distinction between mathematical and physical geometries, and the attendant assumption that only the latter correspond necessarily to the characteristics of physical reality. With respect to the question of which, if any, of the various geometries that have been developed is true, Bell (1946/1991) points out that each, including Euclid's, "when obvious blemishes are removed," is self-consistent and therefore true in a mathematical sense. With respect to which is physically true, each can be usefully applied to physical world problems and therefore can be considered true in the physical sense for the range of phenomena for which it is appropriate, but inasmuch as these geometries are mutually incompatible, no two are both factually true for the same range. The freedom that geometers realized, following the work of Bolyai, Gauss, Lobachevski, and Riemann, to invent new geometries without thought of the "obvious truth" of their axioms was soon claimed also by algebraists who began to invent algebras whose axioms needed no justification beyond internal consistency. However, the freedom referred to here is highly constrained. The mathematician who would invent a new area of mathematics is free to select the axioms that define the area with no requirement that the system bear any relationship to the physical world, but beyond the "givens," it is anything but arbitrary. And determining what the givens imply is very much a matter of discovery.

Jourdain (1956) distinguishes between Mathematics (capital M), "a collection of truths of which we know something," and mathematics (small m), "our knowledge of mathematics." In his view, Mathematics (M) is eternal and unchanging; mathematics (m) changes over time. Mathematics (M) is truth; mathematics (m) represents what has been discovered about that truth at any point in time. Jourdain's distinction provides a framework in which it is possible to fit other views. Borrowing his terms, one would say that Hardy and Polkinghorne had in mind Mathematics (M), whereas Kline was speaking of mathematics (m). They, of course, might not agree with having their views pigeonholed in this way.

Perhaps on the question of discovery versus invention, we can do no better than admit that, to a large extent, the answer is a matter of perspective, and that it can change from time to time. Mazur (2003) puts the difficulty of getting an either–or answer to the question that will remain stable this way: "On the days when the world of mathematics seems unpermissive with its gem-hard exigencies, we all become fervid Platonists (mathematical objects are 'out there,' waiting to be discovered—or not) and mathematics is all *discovery*. And on days when we see someone who, Viète-like, seemingly by will power alone, extends the range of our mathematical intuition, the freeness and open permissiveness of mathematical invention dazzle us, and mathematics is all *invention*" (p. 70).

Or, one might accept Hersh's (1997) contention that mathematics involves *both* discovery and invention, that the two processes distinguish two kinds of mathematical advance. "When several mathematicians solve a well-stated problem, their answers are identical. They all *discover* that answer. But when they create theories to fulfill some need, their theories aren't identical. They *create* different theories.... Discovering seems to be completely determined. Inventing seems to come from an idea that just wasn't there before its inventor thought of it. But then, after you *invent* a new theory, you must *discover* its properties, by solving precisely formulated mathematical questions. So, inventing leads to discovering" (p. 74).

The disagreement among mathematicians on the question of whether their work has to do primarily with discovery or invention, or both, appears to be a deep and philosophically grounded one. It is an old disagreement that persists, and it is not likely to be resolved to everyone's satisfaction anytime soon, if ever. Mathematics itself is not going to provide the resolution. How one thinks about this question will depend, to no small degree, on other beliefs that one holds about the nature of reality and how one accounts for structure and regularity in the universe. One cannot think about the nature of mathematics at a very deep level without encountering philosophical questions that do not admit of mathematical answers.

Throughout this book I have referred to mathematical advances sometimes as inventions and sometimes as discoveries, using whichever term seemed more natural in the context, but I have not made a sharp distinction between the two. I believe that in many cases, if not most, which term one uses is a matter of personal preference. If pressed to make a distinction, however, I would say that I find it natural to consider the new realization of a mathematical relationship as a discovery and the construction of a proof of the relationship as an invention. Realizing that if the lengths of the two shorter sides of a right triangle are x and y, the length of the remaining side is $\sqrt{x^2 + y^2}$ seems to me to be appropriately considered a discovery, whereas the many proofs that have been offered of this relationship seem to me to be better seen as inventions. More generally, theorems, in my view, are better considered to be discoveries, and proofs to be inventions (although I find it easier to consider the latter to be discoveries than to consider the former to be inventions). Other examples of what seem to me to be appropriately considered inventions are numerals, symbols, notational conventions, axioms, and algorithms. Discoveries include conjectures (which sometimes morph to proofs), relationships (that the area of a rectangle is the product of its length and width, that the circumference of a circle is the product of π and its diameter), and applications (discoveries that certain types of mathematics can be usefully applied to specific practical purposes). In this view, the answer to the question of whether mathematicians are in the business of discovering or inventing is that they are in the business of doing both.

☐ More on Mathematics and Logic

Mathematics is the science of the logically possible. (Le Corbeiller, 1943/1956, p. 876)

Given the deductive nature of theorem proving, and much of mathematical reasoning more generally, it would be surprising if mathematics bore no relationship to logic. And indeed, no one, to my knowledge, argues that mathematics and logic are independent. As we have seen, different opinions have been expressed regarding exactly what the nature of the relationship is. Some consider mathematics to be founded on logic; others would have it the other way around; still others see them as equivalent.

Carl Hempel is in the first category. "Mathematics is a branch of logic. It can be derived from logic in the following sense: (a) All the concepts of mathematics ... can be defined in terms of four concepts of pure logic. (b) All the theorems of mathematics can be deduced from those definitions by means of the principles of logic" (1945/1956b, p. 1630). Others who share Hempel's view that mathematics can be derived from logic include Lewis and Langford (1932/1956), Nagel (1936/1956), and Morris (1987), the last of whom defines mathematics as "the science of finding logical consequences" (p. 179).

American philosopher and polymath Charles Sanders Peirce (1839–1914) is in the second category. "It does not seem to me that mathematics depends in any way upon logic. It reasons, of course. But if the mathematician ever hesitates or errs in his reasoning, logic cannot come to his aid. He would be far more liable to commit similar as well as other errors there. On the contrary, I am persuaded that logic cannot possibly attain the solution of its problems without great use of mathematics. Indeed all formal logic is merely mathematics applied to logic" (1902/1956a, p. 1773).

Bertrand Russell (1901/1956a) represents the third position. Commenting on British philosopher-mathematician George Boole's *Laws of Thought*, he says it "was in fact concerned with formal logic, and this is the same thing as mathematics" (p. 1576). Others, however, have taken the monumental work of Whitehead and Russell (1910–1913), as represented in the *Principia Mathematica*, as establishing the primacy of logic and the dependence of mathematics on it: Lewis and Langford (1932/1956) point to the *Principia* as the warrant for their claim: "It [mathematics] follows logically from the truths of logic alone, and has whatever characters belong to logic itself" (p. 1875).

There is general agreement that the relationship between mathematics and logic is very close, but not on the question of the precise nature of that relationship. As the cited comments show, opinions differ, in particular, regarding which should be considered basic and which derivative. Whichever way one jumps on this issue, one is assured of landing on something less than infinitely firm ground. If mathematics underlies logic, its own foundations are very uncertain, and if logic is the more basic, where does it get its warrant? As Austrian economist-philosopher Ludwig von Mises points out, "There is by no means an eternally valid agreement about the admissible methods of logical deduction" (1951/1956, p. 1733).

The distinction between mathematics and logic has been blurred considerably in recent years by the appearance of the digital computer on the scene. Especially is this true with respect to computation, admittedly only one aspect of mathematics, but an important one. It is very difficult to tell, when one looks at the operation of a digital computer at the level of the most basic operations, what to consider logic and what to consider mathematics. At one level of description, a computer is a device that operates on binary variables, changing their values from 0 to 1, or from 1 to 0, or leaving them alone.

Inasmuch as each binary variable can have either of two values, the values of n binary variables can be combined in 2^n ways. Table 13.2 shows all possible combinations of the values of three binary variables.

TABLE 13.2. All Possible Combinations of Three Binary Variables

A	B	C
0	0	0
0	0	1
0	1	0
1	0	0
0	1	1
1	0	1
1	1	0
1	1	1

TABLE 13.3. Truth Table for the Propositional Connective *and*

A	B	A and B
0	0	0
0	1	0
1	0	0
1	1	1

A *function* of binary variables is a binary variable whose value depends on the value or values of one or more other binary variables. Thus, if A and B are binary variables, and the value of B is invariably opposite that of A (if $A = 0$, $B = 1$, and if $A = 1$, $B = 0$), B may be said to be a function of A. Or if A, B, C, and D are all binary variables, and the value of D is 1 if and only if (iff) the values of A, B, and C are 0, 0, and 1, respectively, we may say that D is a function of A, B, and C. A, B, and C are said to be independent variables, whereas D is referred to as a dependent variable, reflecting that the value of D depends on the values of A, B, and C.

Functional relationships among binary variables can be represented in a variety of ways, one of which is the truth table. In the logic of propositions, truth tables are used to show the dependency of a compound statement on the truth or falsity of its components. For example, if we let A represent the statement "Today is Friday" and B the statement "Today is the 13th," then Table 13.3 shows the dependency of the truth (represented by 1) or falsity (0) of the compound statement "Today is Friday (and today is) the 13th" on the truth (or falsity) of each of the statement's components. As indicated in the table, given the connective *and*, the compound statement is true if and only if both of its component statements are true.

The truth table for a two-variable function (as shown in Table 13.3) has four rows (to accommodate the four possible combinations of the values of the two variables). A function of the two variables (represented in the table by *A and B*) can have either 0 or 1 associated with each row of the table, so the total number of functions that can be defined on two binary variables is 2^4 or 16; more generally, given n binary variables, we are able to define 2^{2^n} functions. Table 13.4 shows all 16 functions of two binary variables, and Figure 13.1 shows the same functions as Venn diagrams.

TABLE 13.4. Sixteen Functions of Two Binary Variables

A	1	1	0	0	
B	1	0	1	0	
F_1	0	0	0	0	*not-(A or not-A)*
F_2	1	0	0	0	*A and B*
F_3	0	1	0	0	*A and not-B*
F_4	0	0	1	0	*not-A and B*
F_5	0	0	0	1	*not-(A or B)*
F_6	1	1	0	0	*A*
F_7	1	0	1	0	*B*
F_8	1	0	0	1	*(A and B) or (not-A and not-B)*
F_9	0	1	1	0	*(A and not-B) or (not-A and B)*
F_{10}	0	1	0	1	*not-B*
F_{11}	0	0	1	1	*not-A*
F_{12}	1	1	1	0	*A or B*
F_{13}	1	1	0	1	*A or not-B*
F_{14}	1	0	1	1	*not-A or B*
F_{15}	0	1	1	1	*not-(A and B)*
F_{16}	1	1	1	1	*A or not-A*

If one looks carefully at the table, one sees that the rows representing functions can be divided into subsets on the basis of the number of 1s each contains. There is one row with no 1s, four with one, six with two, four with three, and one with four, which is to say that there is one function (F_1) that has the value 1 for none of the possible combinations of A and B, there are four functions ($F_2 - F_5$) that have the value 1 for

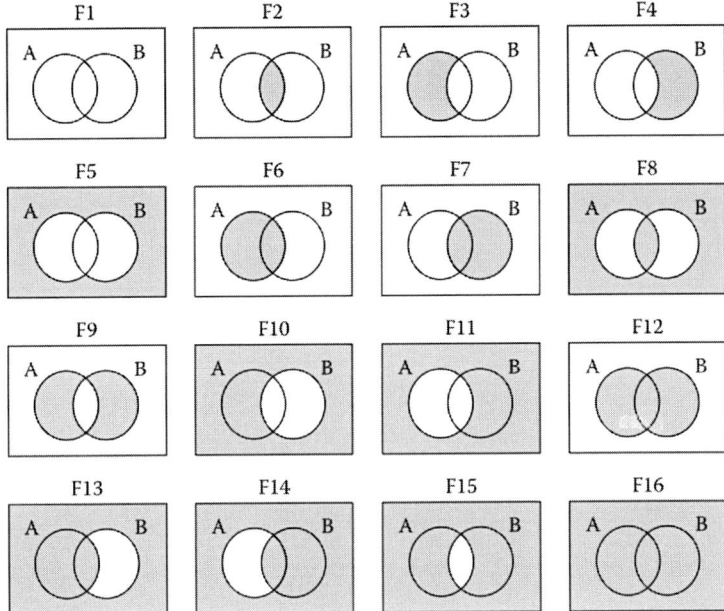

FIGURE 13.1 Venn diagrams of the 16 functions of two binary variables. The filled portion of each diagram shows the conditions under which the function is true.

a single combination of A and B, there are six ($F_6 - F_{11}$) that have the value 1 for two combinations, and so on. The numbers 1, 4, 6, 4, 1 are the coefficients of the successive terms of the expansion of a binomial raised to the fourth power, $(a + b)^4$. Any particular binomial coefficient, $\frac{m!}{k!(m-k)!}$, represents the number of combinations that can be made from m things taken k at a time. To generate all possible functions of two binary variables, we have combined the basic combinations (the four possible combinations of A and B) in all possible ways, taking zero at a time, one at a time, two at a time, and so on. Given that the number of functions of n binary variables grows as 2^{2^n}, the number of functions increases with n very rapidly. With three binary variables, the number of functions is 256; with four, it is 65,536; and with five, it is over 4 billion.

Some of the functions in Table 13.4 are familiar from common parlance. F_2, as already noted, is *and*; F_9 and F_{12} are, respectively, *exclusive or* (A or B but not both), and *inclusive or* (A or B or both). F_6 is A, F_7 is B, F_{11} is *not-A*, and F_{10} is *not-B*. F_{14} deserves special attention. This function represents the truth functional interpretation of *If A then B*, known generally in treatments of conditional reasoning as the *material conditional*. *If A then B* is considered

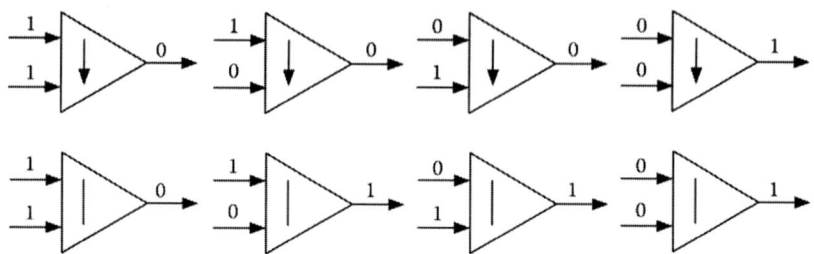

FIGURE 13.2 Logic gates for *nor* (top) and *nand* (bottom) functions.

to be false only in the case that *A* is true and *B* is false. The same function may be described also as *not-A or B*. (Generally when *or* is used without a qualifier, it is intended to represent *inclusive or*, as it does here.) *If … then … * statements in everyday discourse frequently—perhaps more often than not—have an interpretation other than the material conditional.

Two other functions in Table 13.4 that deserve special mention are F_5 and F_{15}. The first, *not-(A or B)*, is referred to as *nor* (short for *neither … nor …*), and the second, *not-(A and B)*, is referred to as *nand* (short for *not and*). The *nor* function is conventionally represented by a down-pointing arrow—$A{\downarrow}B$—which is sometimes referred to as the Peirce arrow, after American logician-philosopher Charles Peirce. The *nand* function is typically represented by a vertical line—$A|B$—which is generally referred to as the Sheffer stroke, after American logician Henry Sheffer. *Nor* and *nand* are shown as logic gates in Figure 13.2, top and bottom, respectively.

Each of the gates in Figure 13.2 shows the output of a gate for a particular combination of inputs. For what follows it will be convenient to show a string of outputs for strings of inputs, as represented in Figure 13.3.

What makes the *nor* and *nand* functions special is that each is *functionally complete*, which is to say that a device that is capable of performing any logical or mathematical function can be constructed from (multiple copies of) either one of them alone, and without the use of any

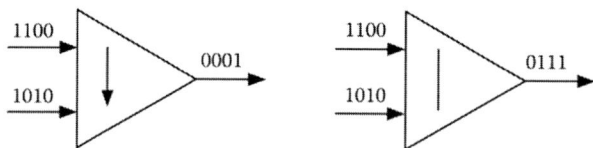

FIGURE 13.3 Showing the string of outputs yielded by a *nor* (left) and a *nand* (right) gate, given the indicated strings of inputs.

TABLE 13.5. *And, or,* and *not* Functions Constructed From Only *nor* or Only *nand* Functions

	Using Only *nor* ($\downarrow$)	Using Only *nand* ($\mid$)
A and B	$(A \downarrow A) \downarrow (B \downarrow B)$	$(A \mid B) \mid (A \mid B)$
A or B	$(A \downarrow B) \downarrow (A \downarrow B)$	$(A \mid A) \mid (B \mid B)$
Not A	$(A \downarrow A)$	$(A \mid A)$

other functions. The equations in Table 13.5 and the logic diagrams in Figure 13.4 show how to build *and, or,* and *not* functions from *nor* or *nand* functions. The "functional completeness" of *nor* and *nand* follows from the interesting fact that one can get truth from falsity ("It is false that it is false" is equivalent to "It is true") but not falsity from truth ("It is true that it is true" is not equivalent to "It is false").

Logic circuitry in electronic computers is typically built from components representing the functions *and,* (inclusive) *or,* and *not.* And the same functions are used to build circuits that can perform addition and

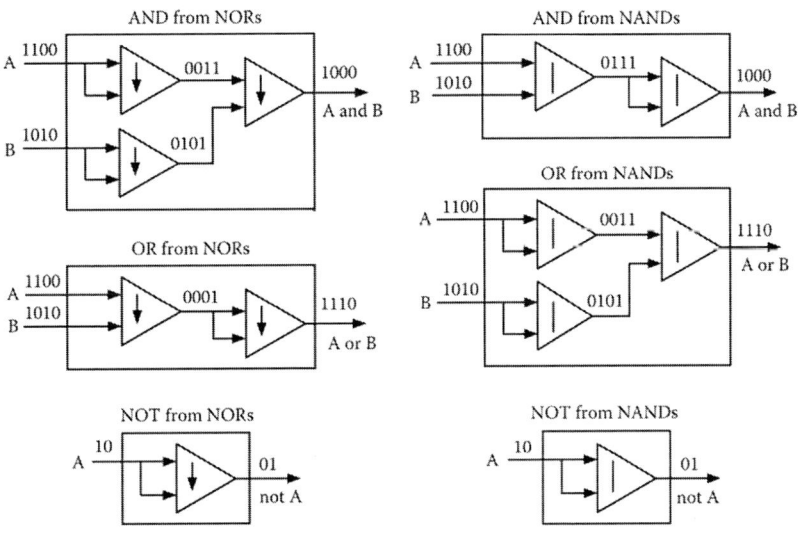

FIGURE 13.4 *And, or,* and *not* logic circuits built from *nor* (left) and *nand* (right) gates.

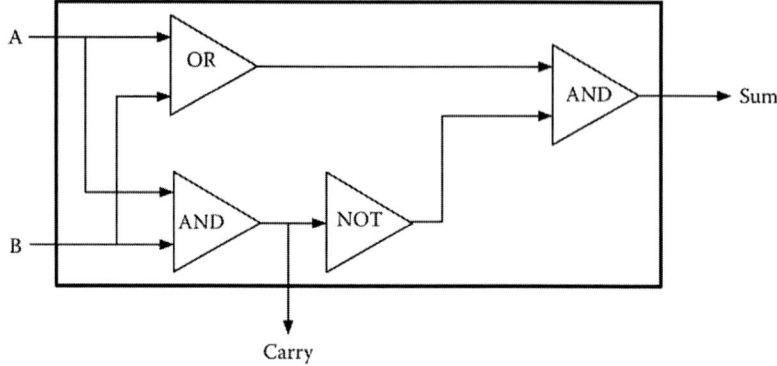

FIGURE 13.5 Half adder built from logic gates for *and*, *or*, and *not*.

other basic mathematical operations. Figure 13.5 shows, for example, how a circuit to add binary numbers can be built from logic gates for the functions *and*, *or*, and *not*. This circuit is a *half adder*, which means that, given two binary digits as input, it will produce the sum (1 if the inputs are 0 and 1; 0 if they are both 0 or both 1) and a carry (1 if both inputs are 1; 0 otherwise). To make a *full adder*, which will accept three inputs, one representing the carry from the preceding addition of two digits, one would combine two half adders in the appropriate way, still using only *and*, *or*, and *not* components.

Figure 13.6 shows how a full adder can be constructed from only *nor* gates. (The reader may wish to try to sketch one that uses only *nand* gates.) It has always struck me as surprising and more than a little interesting that a single function (either *nor* or *nand*) is all that is needed to produce any desired logical or mathematical operation. Real computers are not built using only a single type of logic gate, because it is more efficient to use several types of gates, but in principle they could be.

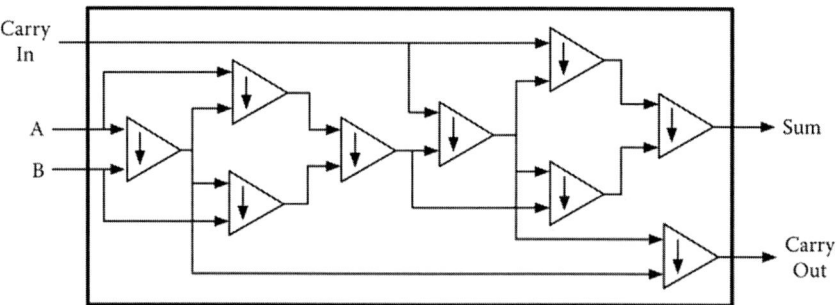

FIGURE 13.6 A full adder made from only *nor* gates.

The *nor* and *nand* functions are a bit of a mystery from a psychological point of view. It is curious that, despite their versatility, they are not much used in common discourse, as compared, say, to the (inclusive) *or* and *and* functions, and they have not been of much interest to either linguists or cognitive psychologists. Presumably the lack of focus on them by researchers is a consequence of their very little use in everyday language and reasoning. But why we, as a species, appear not to have developed much use for these remarkable logical elements is an interesting question.

The purpose of this little excursion into binary logic is to note the possibility of effecting any computation with a device built of simple logic circuitry. Does this demonstrate that mathematics rests on logic? Only if mathematics is taken to be synonymous with computation, which it clearly is not. However, computation is an important part of mathematics, and that it can be accomplished with simple logic gates is significant. Moreover, as has been becoming increasingly clear, those same logic gates can, when organized in sufficiently complex ways, yield performance of other types (pattern recognition, problem solving) that are also important aspects of mathematics. What the limits are of what can be done with these gates remains to be seen.

Preschool Development of Numerical and Mathematical Skills

What are the numerical and mathematical capabilities of children and how do they change over the first few years of life? How can children's understanding of mathematics and their competence to solve mathematical problems—to *do* mathematics—be enhanced by formal education? Much research has been addressed to these and closely related questions, especially during the last few decades.

This chapter and the following two focus on the acquisition of numerical and mathematical competence. This chapter deals primarily with the preschool years. The next one focuses on the learning and teaching of elementary mathematics during the early school years. The chapter following that one relates to the acquisition of skill in mathematical problem solving. While the distinction between the ability to perform mathematical operations, such as those of basic arithmetic, and mathematical problem solving is a common one (Dye & Very, 1968; Geary, 1994), I make it here largely as a matter of organizational convenience; the distinction becomes fuzzy when one considers that what constitutes a problem for one person may not be a problem for another whose mathematical competence is at a different level.

Efforts have been made to identify distinct capabilities that are essential to the development of mathematical competence. Spelke (2005), for example, suggests that research has provided evidence of five cognitive systems that are at the core of mathematical thinking by adults and that serve the following purposes: representation of small exact numbers (one, two, three) of objects; representation (approximate) of large numerical magnitudes; number words and counting; environmental geometry; and landmarks. Each of these component systems, she suggests, is active relatively early in life. While the following discussion is not organized in terms of these hypothesized systems, or any other taxonomy, it supports the idea that mathematical reasoning, even as seen in very young children, is multifaceted and complex.

☐ Numerosity Perception

According to the National Research Council's Mathematics Learning Study Committee, a set of concepts associated with number is at the heart of preschool, elementary school, and middle school mathematics, and most of the debate about how and what mathematics should be taught at these levels revolves around number (Kilpatrick, Swafford, & Findell, 2001). The National Mathematics Advisory Panel (2008) distinguishes between competencies that comprise the number sense that most children acquire before formal schooling ("an ability to immediately identify the numerical value associated with small quantities (e.g., 3 pennies), a facility with basic counting skills, and a proficiency in approximating the magnitudes of small numbers of objects and simple numerical operations") and the "more advanced type of number sense" that schooling must provide ("a principled understanding of place value, of how whole numbers can be composed and decomposed, and of the meaning of the basic arithmetic operations of addition, subtraction, multiplication, and division" as well as an ability to deal with numbers "written in fraction, decimal, percent, and exponential forms") (p. 27).

A rudimentary sensitivity to numerosity appears to be present very early in children, even, according to some observers, during the first few weeks, if not days, of life (Antell & Keating, 1983; Strauss & Curtis, 1981; van Loosbroek & Smitsman, 1990). Evidence has been reported of the beginnings of counting-like behavior, such as the ability to distinguish sets with different (small) numbers of members, before the end of the first year (Durkin, Shire, Riem, Crowther, & Rutter, 1986; Feigenson & Carey, 2003; Haith & Benson, 1998; Mix, Huttenlocher, & Levine, 2002a, 2002b; Starkey & Cooper, 1980; Wynn, 1995; Xu, 2003; Xu & Spelke, 2000). Studies suggest that infants less than one year of age are likely to notice differences between 2 and 3 and, in some instances, perhaps between 3 and 4, but not between 4 and 5 (Starkey & Cooper, 1980; Strauss & Curtis, 1981). Six-month-olds can also distinguish between sets of relatively large numbers of members provided the numerical differences are sufficiently large (Brannon, 2002; Lipton & Spelke, 2003).

The behavioral evidence of this type of discrimination in most studies of infants' grasp of numerosity is the time spent looking at a display; an assumption underlying interpretation of their behavior is that infants tend to look longer at novel scenes than at scenes with which they have become familiar—to which they have become habituated. When, following habituation with a display of two objects, an infant looks longer at a display of three objects than at one of two objects, this is taken as evidence of recognition of the difference between two and three objects.

In a different scenario, infants watch a number of objects being placed in a box into which they can reach but cannot see. If they watch n objects being placed in the box and then reach in n (and only n) times to retrieve the objects one at a time, this is taken as evidence that they are aware of the number of objects that were placed in the box. Use of this technique has shown that infants of approximately one year of age can keep track of up to three objects, but rarely up to four (Feigenson & Carey, 2003).

The beginnings of a sensitivity to changes in number are also seen during the first year of life. Infants give evidence of being surprised when adding an item to another item appears to result in one rather than two items, when adding one item to two items appears to result in two items rather than three, or when removing one of two items appears to leave two items rather than one (Koechlin, Dehaene, & Mehler, 1997; Simon, Hespos, & Rochat, 1995; Wynn, 1992a).

The ability to distinguish numerosity is not limited to the distinction of numbers of objects. Wynn (1996) showed that six-month-old infants can distinguish between a puppet jumping three times and the same puppet jumping twice. Nor is the ability limited to visual stimuli: Six-month-old infants are also able to distinguish between a sound that occurs twice and one that occurs three times (Starkey, Spelke, & Gelman, 1990).

The extent to which such distinctions are made on the basis of numerosity per se, rather than correlates of numerosity (e.g., surface area covered by visual objects or the density of objects in a fixed space; the duration of an auditory sequence as distinct from the number of sounds in it), has been a matter of debate, as has the question of whether numerosity information is stored in short-term memory in analog form (as magnitudes) or as discrete representations of the perceived objects (Carey, 2001, 2004; Dehaene, 1997; Feigenson, Carey, & Hauser, 2002; Feigenson, Carey, & Spelke, 2002; Gallistel & Gelman, 1992; Pylyshyn, 2001; Wynn, 1992b; Xu, 2003). That numerosity per se can be discriminated gets strong support from the finding that when shown side-by-side visual displays, one with two common objects and one with three, six- to eight-month-old children tend to fixate longer on the display on which the number of objects matches the number of sounds they hear from a loudspeaker positioned between the displays (Starkey, Spelke, & Gelman, 1983, 1990). Conceivably infants discriminate numerosity sometimes independently of other factors that typically correlate with it, and sometimes on the basis of the correlates.

Some investigators of quantity perception in children contend that children (perhaps animals as well) have two systems for distinguishing quantities, one of which works primarily with very small quantities (one to three or four) and the other of which deals with anything larger (Brannon & Roitman, 2003; Carey, 2004; Mix et al., 2002a). The first of these systems represents small quantities discretely; the second one represents larger quantities in an analog fashion. Carey (2004) argues that before they acquire language, infants develop several different types of representations of numerical quantities, at least two of which are developed also by other species. One such representation codes numerical quantities as analog magnitudes, the magnitude symbol (e.g., line length) increasing with the number of items in the set being represented. This type of representation serves to make quantitative comparisons (which set has the larger number of members) and perhaps some elementary arithmetic operations, such as addition and subtraction. The second type of representation is discrete. In it there is a one-to-one correspondence between the items being represented and the tokens in the representation; thus, a set of three items might be represented by a mental image of a set of three boxes. Neither of these two systems includes symbols for numbers per se.

By 30 months or so, many children appear to realize that a set of four items is larger than a set of three items (knowledge of ordinality) even if they are unable to label correctly which set contains four and which contains three (knowledge of cardinality) (Bullock & Gelman, 1977). Some investigators contend that even at 18 months some children show a rudimentary awareness of ordinality—for example, realization, in some sense, that three items are more than, and not just different from, two (Cooper, 1984; Strauss & Curtis, 1984).

☐ Number Naming

Many children acquire the ability before attending school to say the number words, *one, two, three, ...,* in the correct order at least to *ten*, and to recognize the numerals on sight (Fuson, 1991; Gelman & Gallistel, 1978;

Sarnecka & Gelman, 2004; Siegler & Robinson, 1982). The ability to say the number words in sequence up to about 100 is acquired gradually by children from the time they are two years old until about the age of six; however, numerous factors contribute to individual differences in this regard (Fuson & Hall, 1983).

The ability to say which of two single-digit numbers is the larger is not likely to be acquired before the age of four or five (Schaeffer, Eggleston, & Scott, 1974; Sekuler & Mierkiewicz, 1977), at which age children may also be able to judge whether a collection has more or fewer objects than a specified (relatively small) number (Baroody & Gatzke, 1991) and which of two collections is closer to a specified number (Sowder, 1989). It appears that generally, though not invariably, ordinal concepts are learned before cardinal concepts (Brainerd, 1979); however, less research has been addressed to the question of how children learn to use ordinal number words and acquire an understanding of ordinal relationships than has been done on the acquisition of competence with cardinal numbers (Fuson & Hall, 1983).

The verbal names of numbers are typically learned before their visual representations, just as spoken words are learned before their visual representations. That brain injuries can result in loss of the ability to recognize spoken numbers while leaving the ability to recognize printed numbers intact, or conversely, suggests that the two abilities are independently encoded neurologically (Anderson, Damasio, & Damasio, 1990; Cipolotti, 1995; Cohen & Dehaene, 2000). Just as children generally can recognize letters and words in print before they can write them, they generally can recognize one- and two-digit numerals before they can write them (Baroody, Gannon, Berent, & Ginsburg, 1983).

Differences in the principles that guide number naming in different languages appear to have implications for the ease with which the names of numbers greater than 10 are learned (Fuson, Richards, & Briars, 1982; Miller & Paredes, 1996). Asian children tend to master the number names for numbers above 10 more quickly than do English speakers, possibly because of the greater regularity of number names in Asian languages (Miller, Smith, Zhu, & Zhang, 1995; Miller & Stigler, 1987; Miura, Okamoto, Kim, Steere, & Fayol, 1993; Miura et al., 1994). That in English the names for 10, 11, and 12 are *ten*, *eleven*, and *twelve*, instead of, say, *ten*, *ten-one*, and *ten-two*, obscures the base 10 structure of the system and complicates the English-speaking child's learning that *thirty* is followed by *thirty-one* and *thirty-two* rather than by *twenty-eleven* and *twenty-twelve*. This feature is shared by most European languages. In contrast, number names in Asian languages, such as Chinese and Japanese, typically make the base 10 nature of the system explicit in the use of names that are equivalent, in the specific languages, to *ten-one*, *ten-two*, and so on.

Explicit representation of base 10 structure in number names is found also in parts of Africa (Posner, 1982). (The obscuration of the base 10 structure in 11 and 12 may be the result of linguistic contraction: the German *elf* and *zwölf* may be contractions of *ein-lif* and *zwo-lif*—*one-ten* and *two-ten* in old German [Dantzig, 1930/2005, p. 12].)

A similar distinction holds for ordinal as well as for cardinal numbers. In Chinese, for example, ordinals are formed by adding an ordinal prefix to cardinal number names, which makes the correspondence between the first few ordinals and cardinals apparent, while there is little in *first, second*, and *third* to remind one of *one, two*, and *three* (Miller, Kelly, & Zhou, 2005). An additional confusing infelicity in English number names is that, unlike the printed numbers in which the 10s digit always precedes the 1s digit, in verbal number names this principle is followed with 20 and larger, but with the teens the 10s digit is represented after the 1s digit.

Munn (1998) argues that learning number symbols (visual representations of numerals) is more problematic than learning other aspects of number and that, for this reason, some children fail to see the correspondence between quantities and their symbols. She speculates that children may create three different cognitive models regarding numbers: "one ... around the verbal number system, another around objects used in counting and calculating, and yet another around number symbols" (p. 57). She contends that unless "adults deliberately foster a cognitive model that links objects with symbol systems, children will find it hard to see how the logical structure of concrete objects maps onto that of numeric symbol systems" (p. 57). Miller and Paredes (1996) note that the variety of ways in which systems for representing numbers differ affect the course of children's mathematical development. Differences in how systems reflect the base 10 principle, for example, can affect the time it takes children to learn a system and the types of difficulties they encounter in doing so.

Fuson and colleagues have studied the effects of differences in number names across languages, especially differences between Asian and European languages, not only on the ease with which children acquire the ability to do basic arithmetic but also on how readily they come to understand the principles of base 10 notation (Fuson & Kwon, 1992a, 1992b; Fuson, Stigler, & Bartsch, 1988). Miura et al. (1993) have done similar studies with French and Swedish. Geary (1994) sees the differences between Asian and English or European number names to be responsible, to some degree, for certain specific difficulties (e.g., with the concept of borrowing in doing subtraction with multidigit numbers) that are commonly seen with American children (VanLehn, 1990; Young & O'Shea, 1981) but less so with East Asian (Korean) children (Fuson &

Kwon, 1992b). More generally, poor understanding of the base 10 system is seen as the basis for problems some students have in acquiring skill in multidigit computation (Cauley, 1988; Hiebert & Wearne, 1996).

☐ Naming Quantities

That competence in the production and use of numerals in counting or recording the results of counting typically increases gradually during preschool years is well documented and not surprising (Bialystok & Codd, 1996; Hughes, 1986, 1991; Sinclair, 1991; Sinclair & Sinclair, 1984). Still, many children enter school with a shaky grasp of the process of counting and the use of numerals. Once learned, the process of counting seems quite straightforward, but many studies of how children learn to count and how their skill in this regard increases during the first few years of life have revealed that acquisition of this ability is a more complex process than is apparent to the casual observer.

The abilities to say number words in sequence and to recognize numerals on sight are not, by themselves, compelling evidence of an ability to count in the sense of determining the number of items in a set (Hughes, 1986, 1991; Sinclair, 1991; Sinclair, Siegrest, & Sinclair, 1982). Among the principles that must be learned before a child can be said to have a good grasp of what it means to count the items in a set is that of one-to-one correspondence between the numbers used and the items in the set to be counted; otherwise, there is nothing to prohibit counting the same item more than once (Gelman & Gallistel, 1978; Montague-Smith, 1997; Starkey, 1992). One must come to understand, too, that when putting number names into one-to-one correspondence with the items in the set, the last number named is the number of items in the set. It is possible to be able to count in some sense without realizing that the purpose of counting is to determine the number of items in a set, and this realization may not come until about the fourth year (Wynn, 1990, 1992b).

One must also appreciate what Gelman and Gallistel (1978) call the *order irrelevance principle*, according to which one gets the same number independently of the order in which the items are counted. Briars and Siegler (1984) found that four- to five-year-old preschoolers typically recognized the principle of word-to-object correspondence as essential to counting before they knew that certain other aspects of observed counting episodes (counting adjacent objects sequentially, counting objects left to right) are not essential. Steffe and colleagues (Steffe, Thompson, & Richards, 1982; Steffe, von Glasersfeld, Richards, & Cobb, 1984) argue that counting skills emerge as a progression, starting with the ability to count only tangible objects and moving through several levels of increasing abstraction to the eventual ability to count imagined entities. Sophian (1988, 1995, 1998) presents evidence from several studies that suggests that although children's early counting conforms to principles such as those identified by Gelman and Gallistel, their understanding of the significance of those principles is acquired only gradually over several years. Carey (2004) argues that children first become *one-knowers* (learn to distinguish one from many), then *two-knowers* (distinguishing one and two from more than two), then *three-knowers* (able to count to three), and eventually to understand how counting more generally works, each advance in the process taking several months. Going from the ability to distinguish between one and more than one and being able to count beyond three may take a couple of years.

Children commonly point to objects in the process of counting them. This involves the use of pairing in three ways: between a number word and a pointing action, between a pointing action and an item in the set being counted, and between the number word and the item (Fuson & Hall, 1983). Children find it easier to count (a small number of) objects if they can physically move the objects as they do the count (say the number names) than if the objects cannot be moved (as spots on a piece of paper), and they find it easier to count all the items in a set than to "count out" a subset of a specified number of the items (Resnick & Ford, 1981; Wang, 1973; Wang, Resnick, & Boozer, 1971).

Citing numerous studies (Baroody, 1992; Frydman & Bryant, 1988; Frye, Braisby, Lowe, Maroudas, & Nicholls, 1989; Fuson, 1988, 1992a; Sophian, 1992; Wynn, 1990, 1992b), Bialystok and Codd (1996) contend that the evidence is compelling that below the age of five or six, children generally do not comprehend very well the principle of cardinality—the relation between the numbers in the counting sequence and the abstract notion of quantity—and their ability to produce numerical notation is no guarantee that they understand what the symbols represent. They argue that the ability to count, coupled with the ability to identify numeric notation, is not compelling evidence of an understanding that the two processes are linked. In a seven-month study of two- and three-year-olds, Wynn (1992b) showed that children come to realize that each of the counting words refers to a specific numerosity before they realize to which numerosity it refers. Thompson (1994b) contends that while concepts of quantity and concepts of number are closely related, they are not the same—one evidence of the difference being that number concepts readily become confounded with matters of notation and language, whereas quantity concepts tend not to do so.

In brief, learning to count involves the acquisition of the ability to make several kinds of distinctions, including, but not limited to, those

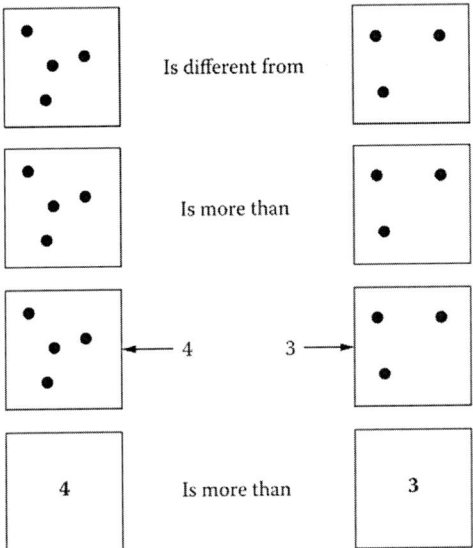

FIGURE 14.1 Some of the distinctions involved in learning to count.

identified in Figure 14.1: recognizing that two collections are different in number, recognizing which of the two contains the larger number, putting a numerical label on a collection, and recognizing which of two numerals represents the larger quantity.

An important aspect of the understanding of number is the realization that a set of objects can be thought of as composed of various combinations of subsets: that a set of six items can be composed of subsets of four and two items, of three and three items, or of one, two, and three items. Subset-set relationships of this sort are sometimes referred to as *part-part-whole* relationships (Fischer, 1990), and an understanding of the principle has been called a major breakthrough in children's mathematical development (Van de Walle & Watkins, 1993).

☐ Grouping and Unitizing

When there are more than a few objects to be counted, errors are easily made. One of the tricks that children appear to learn more or less spontaneously is that of grouping objects into clusters of modest size, counting the number in each cluster, and adding up the results (Beckwith & Restle, 1966; Carpenter & Moser, 1983; Steffe & Cobb, 1988). If each cluster is made to contain the same number of items, adding up the amounts in the clusters would be the equivalent of multiplying the number of items per cluster by the number of clusters.

The process of constructing a reference unit—sometimes referred to as *unitizing*—is seen as instrumental to the development of skill in multiplication, division, and related operations (Lamon, 1994). Unitizing provides a unit, other than 1, in terms of which situations can be conceptualized. This process is also referred to as *norming* (Freudenthal, 1983; Lamon, 1994). Once established, a unit can be the basis of a hierarchy of clusters—a unit, a unit of units, a unit of units of units, and so on—a principle that is seen in a place notation number system such as the Hindu-Arabic system, which organizes quantities in terms of units of 10, units of units of 10 (hundreds), units of units of units of 10 (thousands), and so on. Many students have difficulty with early arithmetic because they do not have a good grasp of the principles of place value notation (Payne & Huinker, 1993); even many 13-year-olds lack a good understanding of place notation (Baroody, 1990; Kamii, 1986; Kouba & Wearne, 2000). Conversely, a good understanding of the principles of place notation facilitates the acquisition of computational skills (Fuson & Briars, 1990; Hiebert & Wearne, 1996; Resnick & Omanson, 1987). The learning of this representational system has been likened to the learning of a foreign language (Kilpatrick et al., 2001). Behr, Harel, Post, and Lesh (1994) describe an instructional approach that is built on the notion of units of quantity that they expect will help children extend knowledge about addition and subtraction to the domain of multiplication and division.

☐ Nature and Nurture

Differences in arithmetic ability are evident well before formal schooling (Geary, Bow-Thomas, Fan, & Siegler, 1993; Geary, Bow-Thomas, Liu, & Siegler, 1996), and the magnitude of the differences increases during the first few years of school (Stevenson, Chen, & Lee, 1993; Stevenson, Lee, & Stigler, 1986). What are the origins of these differences? What determines mathematical potential and the ease with which that potential can be developed? How much of a role does genetics play? Are people with different ethnic heritages likely to differ in their potential to acquire mathematical skill? Do males and females differ in this respect? To what extent does the culture in which one is raised influence the development of whatever potential one has?

No one seems to doubt that both innate and learned factors are involved; the question is one of relative importance. Studies of homozygotic and heterozygotic twins have produced results that have been interpreted as indicating that nature and nurture are roughly equally important (Vandenberg, 1966), but this interpretation has not been universally accepted. There is a lively debate about the question and the matter is far from settled (Geary, 1994). Specific theoretical positions have been defended by Gelman and Gallistel (1978), Briars and Siegler (1984), Fuson (1988), Siegler and Jenkins (1989), and Sophian (1998), among others. Whatever the facts of the matter, beliefs about the relative importance of nature and nurture appear to differ across cultures: According to Stevenson and Stigler (1992), for example, Japanese parents are inclined to attach more importance to effort and quality of teaching as determinants of children's acquisition of mathematical competence, whereas American parents tend to attribute differences in performance to differences in innate talent. Hess, Chang, and McDevitt (1987) report a similar difference between the beliefs of Chinese and American mothers.

The idea that the capacity to develop numerical ability is innate and universal gets support from the claim that the numerical and informal mathematical skills that children acquire by the age of about seven are very similar across many cultural and social groups (Ginsburg, 1982; Ginsburg, Posner, & Russell, 1981a, 1981b; Ginsburg & Russell, 1981; Klein & Starkey, 1988; Petitto & Ginsburg, 1982). Ginsburg and Baron (1993) summarize evidence on the point from cross-cultural research this way: "The general finding is that children from various cultures, literate and preliterate, rich and poor, of various racial backgrounds, all display a similar development of informal addition (and other aspects of informal mathematics, like systems for counting and enumeration)" (p. 8). (Barrow, 1992, in a footnote on p. 34, refers to Australian friends who claim that children from aboriginal groups that do not use counting have no unusual difficulty learning math when placed in modern educational situations.)

Ginsburg and Baron (1993) caution that although the research shows similarities across cultures, races, and classes in the general course of development of informal mathematical abilities, it does not show that children from the various groups are identical in their mathematical thinking. Moreover, no one denies that there are cultural and social influences on the ways in which the skills are acquired (Geary, Fan, & Bow-Thomas, 1992; Saxe, 1991; Saxe, Guberman, & Gearhart, 1987). There is a school of thought that holds that knowledge of arithmetic should grow out of social contexts and that the practice of teaching arithmetic divorced from social contexts and only later making practical applications through the teaching of problem solving is not the way it should be done (Behr et al., 1994).

Geary (1996a) contends that there exists "a biologically primary numerical domain which consists of at least four primary numerical abilities: numerosity (or subitizing), ordinality, counting, and simple arithmetic" (p. 152). He argues, however, that primary abilities are likely to develop into mathematical competence only with the help of appropriate instruction. "Most of children's knowledge of complex arithmetic and complex mathematics emerges in formal school settings (Ginsburg et al., 1981) and only as a result of teaching practices that are explicitly designed to impart this knowledge" (p. 155).

Much of the work aimed at identifying mathematical potential has been done from a componential perspective that assumes mathematical ability is the result of some combination of underlying component abilities and that the goal is to determine what those underlying abilities are. Not surprisingly, for many studies with this perspective, beginning with the work of Spearman (1904, 1923) and Thurstone (1938; Thurstone & Thurstone, 1941), factor analysis has been a technique of choice (Dye & Very, 1968; Ekstrom, French, & Harman, 1979; Goodman, 1943; Meyer, 1980; Osborne & Lindsey, 1967). Most of the factor analytic studies that have been done have not been motivated by the desire to identify components of mathematical ability in particular—as distinct from other types of cognitive ability—but they have often resulted in the identification of one or more factors that would be considered important for mathematical competence. Two that have been identified relatively consistently are numerical facility and mathematical reasoning; a few others have been identified in some studies, but much less consistently (Geary, 1994). Geary refers to numerical facility as "among the clearest and most stable factors identified across decades of psychometric research" (p. 139) and sees the stability of this factor as strongly supportive of the conclusion that "arithmetic involves a fundamental domain of human ability" (p. 140). He also points out a lack of consensus regarding whether, in the absence of experience in solving mathematical problems, mathematical reasoning ability really differs substantively from reasoning ability more generally.

Undoubtedly among the best known and most influential studies of children's acquisition of numerical concepts are those conducted by Piaget and his associates (Inhelder & Piaget, 1958, 1964; Piaget, 1942, 1952; Piaget & Inhelder, 1969). For present purposes it suffices to note that Piaget's theory of stages of logical development is highly controversial, as is the question of whether it illuminates the acquisition of number concepts. Brainerd (1979) summarizes a review of several studies inspired by Piaget's theory by saying that while "there may be a positive statistical relationship between children's arithmetic competence and their grasp of ordination, cardination, or both ... we still are completely in the dark about whether there is a developmental relationship between these variables" (p. 120). More generally, Piaget's work raised a host of questions about the acquisition of numerical and other mathematical concepts and inspired a great deal of experimentation, but did not settle many of the questions it raised.

☐ Beginning Mathematics

One reason for studying mathematical abilities of preschool children is the belief that for instruction to be maximally effective, it must take into account what children do and do not already know when they arrive at school (Fuson et al., 2000; Ginsburg, Klein, & Starkey, 1998; Seo & Ginsburg, 2003). As we have noted, long before they encounter efforts by adults to teach them elementary mathematics, children give evidence of innate appreciation of some of the rudiments of arithmetic. They appear to notice, for example, when the number of items in a set that they have been observing has changed as a consequence of simple arithmetic transformations (addition or subtraction of an item or items) (Gelman, 1972; Gelman & Gallistel, 1978; Huttenlocher, Jordan, & Levine, 1994; Levine, Jordan, & Huttenlocher, 1992; Sophian & Adams, 1987; Wynn, 1992a). Not everyone agrees on how such findings are best interpreted (Bisanz, Sherman, Rasmussen, & Ho, 2005).

Citing numerous sources, among them Booth (1981), Erlwanger (1973), and Ginsburg (1977), Steffe (1994) argues that when children are faced with their first arithmetical problems, they attempt to solve them using whatever mathematical schemes they already have, and they persist in using those schemes—in preference to others they are being taught—so long as they prove to produce answers that teachers accept as correct. He contends further that teachers often remain unaware that students are using their own methods. Rather than discouraging children from using their own methods, as much conventional teaching practice does, Steffe argues that teachers should try to understand those methods and build on them, and that ignoring child-generated algorithms in teaching basic arithmetic is a serious mistake. Much of the research on preschool mathematics is aimed at identifyng methods and algorithms that children discover on their own.

Not surprisingly, many of the techniques that preschool children invent for accomplishing elementary mathematical operations are based on counting in one way or another (Carpenter, 1985; Carpenter & Moser, 1982, 1983; Fuson, 1992b; Ginsburg, 1989; Groen & Resnick, 1977; Kamii, 1985; Kaye, 1986; Maxim, 1989; Reed & Lave, 1979; Resnick, 1983). It is not unusual for three-year-olds to be able to do some simple addition and subtraction, or at least to understand (not yet having a concept of negative numbers) that addition increases the numerosity of a set and subtraction decreases it (Starkey, 1992; Starkey & Gelman, 1982). (Of course, later they have to unlearn this principle when they begin adding and subtracting negative numbers.) The fuzziness of the line between counting and computing is seen in the common use of counting, often with the aid of fingers, by preschool children to do simple addition and subtraction, which most of them can do by the age of five or six (Geary et al., 1992; Siegler & Shrager, 1984; Starkey & Gelman, 1982). Many children are also capable of inventing procedures for carrying out multidigit operations (Carpenter, Franke, Jacobs, Fennema, & Empson, 1998; Hiebert & Wearne, 1996) and operations with fractions (Huinker, 1998; Lappan, Fey, Fitzgerald, Friel, & Phillips, 1996; Mix, Levine, & Huttenlocher, 1999).

More will be said about the methods and strategies that children use when beginning to do arithmetic. This topic has motivated a great deal of research. Often it is not clear whether the methods and strategies that have been observed were developed during preschool years or only after the beginning of formal training. *Preschool* is a somewhat imprecise status inasmuch as many children whose mathematical performance has been studied have been exposed to some formal instruction by virtue of participation in prekindergarten programs that might be considered schooling by some definitions.

☐ Preschool Facilitation of Mathematical Development

Considerable attention is being given to the question of what should be done to facilitate the development of mathematical and premathematical abilities in preschool children, to enhance and sharpen the informal mathematical skills they may have acquired naturally well before they begin their formal schooling (Baroody, 1992, 2000; Ginsburg & Baron, 1993; Payne, 1990). Does it make sense to provide mathematics instruction to preschoolers? Baroody (2000) asks this question and answers it in the affirmative. He argues that preschoolers have impressive informal mathematical strengths as well as a natural inclination for numerical reasoning, and that being so, it makes sense to involve them in "engaging, appropriate, and challenging mathematical activities" (p. 64). Regarding the question of how preschoolers should be taught mathematics, he advocates engaging them in "purposeful, meaningful, and inquiry-based instruction" (p. 66). That there is much to be gained by exposing preschool children to aspects of mathematics, in both preschool programs (Clements, 2001) and the home (Starkey & Klein, 2000), seems likely, although there are differences of opinion regarding precisely what that exposure should entail.

Standards proposed by the National Council of Teachers of Mathematics (1989, 2000), and adopted or adapted by most of the states (Blank, Manise, & Braithwaite, 2000), emphasize the role of play in developing mathematical concepts in young children. Games and activities designed to facilitate the development of mathematical capabilities in preschoolers have been addressed to a variety of types of skills, including classification and set creation, matching (one-to-one correspondence), ordering and seriation, quantitative comparing (more or less), and counting (Kaplan, Yamamoto, & Ginsburg, 1989; Moomaw & Hieronymus, 1995; Van De Walle, 1990).

Teachers and caretakers of preschoolers may be aware of the importance of play in developing mathematical skills without knowing how to use it effectively for this purpose. In the absence of formal assessment techniques, they may find it difficult to assess the ability of young children to use mathematical concepts in their play. Kirova and Bhargava (2002), who make this point, provide checklists intended for use in assessing children's ability to do matching (one-to-one correspondence), classification, and seriation. Brainerd (1979) notes the trickiness of determining whether a child really uses one-to-one correspondence in determining whether two sets are of equal number, arguing that one generally has at least one other cue (relative lengths of linear sequences, relative density of two-dimensional groups) in addition to correspondence on which the judgment of relative manyness could be made. He contends that when trying to determine whether a child understands cardination (the logical connection between correspondence and manyness), judgments of classes containing the same number of items are more revealing than judgments of classes containing an unequal number of items. "When a child correctly judges that two classes contain unequally many terms, we cannot be certain that correspondence was the basis for the judgment. In the case of correct judgments of equal manyness, however, correspondence is the only basis for such judgments of which we are aware" (p. 132). Brainerd presents data from studies of his own (and a replication by Gonchar, 1975) that he interprets as strong evidence in favor of the hypothesis that development of an understanding of ordination generally precedes that of cardination, and that understanding of ordination is more important than understanding of cardination for success in beginning arithmetic. "In the absence of either contradictory data or logically compelling counterarguments, one can therefore provisionally conclude that the human number concept, as indexed by arithmetic competence, initially evolves from a prior understanding of ordinal number and not from prior understanding of cardinal number" (p. 205).

As already noted, children generally learn to count—by some definitions—long before entering school. Many books have been written for use with children who are just beginning to learn to read that feature numerals or number names. Flegg (1983) expresses concern about children's books that associate numerals with pictures in the same way as they associate letters or words with pictures. He contends that the presentation of the symbols 1, 2, 3, ..., as if they were equivalent to letters of the alphabet, overlooks a fundamental difference between letters and numerals. "A picture of a cat, together with the letter 'c' or the word 'cat' is not conveying the same sort of information as a picture of two cats together with the numeral '2' or the word 'two'. The letter 'c' stands for a sound—the sound with which the spoken name of the animal pictured begins. The word 'cat' is the written form of that name. The numeral '2', on the other hand, is not the name of the animals pictured, spoken or written, nor is it related directly to that name. It is a property in some way possessed by the two cats by virtue of their being two of them" (p. 281). Flegg questions whether numerals should be introduced before they are needed to facilitate written arithmetic operations and whether it might not be better to concentrate initially on number words.

Noting that the concept of numbers in the abstract emerged relatively late in the history of mathematics, Flegg argues that it probably should be deferred pedagogically as well. "The concrete should always precede the abstract—abstract concepts are very difficult to assimilate unless there have been plenty of concrete examples with which the pupil has become familiar" (p. 282). He speculates that irreparable damage may be done by having children do calculations before they have had adequate time to explore number concepts at the level of number words and symbols. Flegg's speculations cannot be said to rest on solid empirical evidence, but they raise some questions that deserve investigation. Relatively little is known about the effects on eventual mathematical competence of the ways in which young children are introduced to number concepts in their preschool years.

A point that Flegg makes that I find especially thought-provoking is the importance, from the earliest encounters with mathematics, of keeping an impetus on excitement and discovery. "It all begins with numbers—if children come to fear them or to be bored with them, they will eventually join the ranks of the present majority for whom the word 'mathematics' is guaranteed to bring social conversation to an immediate halt. If, on the other hand, numbers are made a genuine source of adventure and exploration from the beginning, there is a good chance that the level of numeracy in society can be raised significantly" (p. 290). It would be good to know if that is the case.

There is a need too for research on the assessment of mathematical potential. Allardice and Ginsburg (1983) note that as of the time of their writing almost no effort had been made to assess children's learning potential, and they put a focus on this topic at the top of their short list of needs for future research. The standard IQ score, especially the g component, which is sometimes equated with fluid intelligence, is a reasonably good indicant of mathematical potential and predictor of how well students who are given special mathematical training are likely to do (Carroll, 1996; Lubinski & Benbow, 1994; Lubinski & Humphreys, 1990; Stanley, 1974), but assessing the potential to become proficient at higher mathematics—as distinct from doing well on cognitively demanding tasks generally—remains a challenge.

Mathematics in School

In most developed countries, during the earliest years of formal schooling much emphasis is put on the explicit teaching of mathematical skills and skills that are assumed to be prerequisite to the acquisition of mathematical competence. It is generally recognized that in the absence of such schooling, children are unlikely to acquire mathematical knowledge beyond a rudimentary level. As one of the "three Rs," arithmetic has been a staple of elementary education since the beginning of universal public education in the United States and was prominent in schooling long before then.

☐ The Current Situation

The foregoing chapters have dealt, for the most part, with mathematical reasoning independently of where it occurs. For the immediately following comments, I ask the reader's indulgence to focus on the current status of mathematical education in the United States. Concern in the United States about elementary and secondary mathematics education has been fueled by the results of numerous studies showing that U.S. school children tend to do poorly on international tests of mathematical ability (Byrne, 1989; Carpenter, Corbitt, Kepner, Lindquist, & Reys, 1980; Crosswhite, Dossey, Swafford, McKnight, & Cooney, 1985; Dossey, Mullis, Lindquist, & Chambers, 1988; Husén, 1967; Lapointe, Mead, & Askew, 1992;

McKnight et al., 1987; Mullis, Dossey, Owen, & Phillips, 1991; Schmidt, McKnight, Cogan, Jakwerth, & Houang, 1999). The performance of U.S. students in mathematics compares unfavorably to that of children of the same age or school grade in several other countries, especially in East Asia, including Japan, South Korea, Taiwan, Hong Kong, Singapore, and mainland China (Geary, 1996; Peak, 1996, 1997; Song & Ginsburg, 1987; Stevenson, Chen, & Lee, 1993; Stevenson & Lee, 1998; Stevenson, Lee, & Stigler, 1986; Stevenson et al., 1990; Stevenson & Sigler, 1992; Towse & Saxton, 1998). U.S. students also score below the average of students in the 30 member countries (industrialized democracies) of the Organization for Economic Cooperation and Development (Program for International Student Assessment, 2006).

Although interpretation of cross-national comparisons is tricky (Brown, 1996; Reynolds & Farrell, 1996), that U.S. children—as well as those in much of Europe—tend to do poorly on tests designed for such comparisons, or at least that they have indeed done poorly on several assessments in recent years, is not in question. The National Mathematics Advisory Panel (2008) was able to note some positive trends in the scores of fourth and eighth graders in recent national assessments, but it reported broad agreement among its diverse membership on the point that "the delivery system in mathematics education [in the United States] is broken and must be fixed" (p. xiii).

In addition to test results, there are other indications that elementary mathematics education is in trouble in the United States. The difficulty level of the types of problems that comprise basic arithmetic can be from one to three years behind that used in comparable grades in Asian countries (Fuson, Stigler, & Bartsch, 1988; Stigler, Fuson, Ham, & Kim, 1986). U.S. textbooks for elementary mathematics may cover a broader range of topics than do those of several other countries, but the coverage is claimed to be less substantive (Mayer, Sims, & Tajika, 1995; Schmidt et al., 1999; Schmidt, McKnight, & Raizen, 1997).

Geary (1994) draws from some of these and similar findings the "very clear" conclusion that among industrialized countries, American children are among the more poorly educated in mathematics. By international standards, he contends, the mathematics curriculum in the United States is developmentally delayed. On the basis of their review of work on conceptual and procedural knowledge acquisition, Rittle-Johnson and Siegler (1998) conclude that the results, in the aggregate, suggest that conceptual and procedural knowledge are related, inasmuch as Asian children have both and American children have neither.

A similarly negative view of mathematical education in American schools has been expressed by Dreyfus and Eisenberg (1996). "Lists of 'simple' problems that no one seems to be able to do are becoming so long that it is embarrassing to continue the practice of making them…. So many students are deficient in so many simple skills that we are in the midst of an epidemic of ignorance running wild" (p. 256). In its Project 2061 report, *Science for All Americans*, the American Association for the Advancement of Science (1989) characterized the problem in equally dire terms: "A cascade of recent studies has made it abundantly clear that by both national standards and world norms, U.S. education is failing to educate too many students—and hence failing the nation. By all accounts, America has no more urgent priority than the reform of education in science, mathematics, and technology" (p. 3).

Concern about the teaching of mathematics is not limited to elementary and secondary schools. Here is one assessment of the situation at the college level. "Mathematics education at the college level is in a sorry state. Students are turning away from mathematics, and those who do stay do not seem to learn very much. Our students do very poorly on national and international assessments. Our school teachers seem almost to be afraid of the subject. Industry complains and does its own teaching to make up for employees' mathematical deficiencies" (Dubinsky, 1994a). Selden, Selden, and Mason (1994) contend that even good calculus students cannot solve nonroutine problems.

Determining the basis of this poor showing is understandably deemed to be very important by U.S. educators, because what they should do if they wish to close the achievement gap depends on whether the difference proves to be the result of genetics (Does the genetic makeup of East Asians give them an edge in mathematics?), cultural differences (Is the development of mathematical competence more highly valued by East Asian families than by American families?), attitudes toward education generally (Do Asian parents have greater expectations and ambitions for educational achievement by their children?), beliefs (Are Asians more likely to believe that academic success depends largely on effort, while Americans are more likely to attribute academic success or failure to possession or lack of native ability?), instructional approaches (Has the East Asian educational system developed more effective ways of teaching mathematics?), opportunity to learn (Do East Asian classrooms simply devote more time to the teaching of mathematics?), or some combination of these and perhaps other variables as well.

All of these possibilities have advocates, and it is not clear that any of them has been conclusively ruled out. However, following a review of the results of several studies that focused on the question of why East Asian students consistently outperform U.S. students on tests of mathematical ability, Geary (1996) concludes that the performance difference is largely attributable to differences in schooling. Asian students, he notes, get more classroom exposure to mathematics and do more homework than their U.S. cohorts, and mathematical competence appears to have a higher value within East Asian cultures than in the United States. "The bottom line is that U.S. children lag behind their international peers in the development of secondary mathematical abilities because U.S. culture does not value mathematical achievement. East Asian children, in contrast, are among the best in the world in secondary mathematical domains, because Asian culture values and rewards mathematical achievement" (p. 166).

Citing Song and Ginsburg (1987) and Stevenson and Lee (1990), Ginsburg and Baron (1993) contend that American and Asian children perform at approximately the same level during preschool and kindergarten years and that the differences emerge only after a year or two of schooling. These differences, they argue, are primarily the result of differences in motivation and teaching. "Asian children are taught that by being diligent and working hard they will be able to master even those areas of learning that they find very difficult. Success is attributed to hard work more than to innate ability. All children are expected to work hard and all children are expected to succeed. And generally they do" (p. 18).

The generalization that American and Asian children perform at about the same level before they begin formal schooling has been challenged by data obtained by Siegler and Mu (2008) comparing the performance of Chinese and American preschoolers on number-line and addition tasks. The number-line task required the children to locate each of 26 numbers (between 3 and 96) on a number line; the addition task required performance of 70 addition problems with sums between 2 and 10. The Chinese children did significantly better than their American peers on both tasks.

That Asian immigrants to the United States—whether they received their elementary mathematics training in the United States or in Asia—tend to do better at mathematics than do American-born students has been attributed to higher-ability Asian parents being more likely than lower-ability Asian parents to immigrate to the United States and to the relatively high value that Asian parents place on academic achievement, which translates into relatively more time spent on homework (Caplan, Choy, & Whitmore, 1992; Tsang, 1984, 1988).

The possibility of attributing the poor showing of American students in mathematics to overcrowded classrooms or inadequate physical infrastructure appears to be ruled out by the fact that classrooms in Asia often include 40 to 50 students, whereas in the United States the average tends to be between 20 and 30, and the physical infrastructure of American schools is among the best anywhere (Lapointe et al., 1992).

An extensive review of the effects of class size on student achievement in the United States showed the benefit of smaller classes to be greater for lower grades (1–3) than for higher ones, but relatively small in all cases (Ehrenberg, Brewer, Gamoran, & Willms, 2001). For an example of a study finding a relatively large effect, see Kruger (1999).

Attributing the poor mathematical performance of American students to inadequate funding of U.S. education is challenged by the fact that as of the time of the first IEA assessment (Husén, 1967) the United States was spending roughly 6.5 times as much per student on education as was Japan, and despite that the total U.S. expenditure on education grew from 4.5% of the gross national product to 7.5% between the late 1960s and the early 1990s, SAT scores were lower in 1991 than in 1964 (Geary, 1994). Particularly disheartening is the finding that the expressed degree to which U.S. students like mathematics decreases substantially over the middle school through high school years (grades 6 through 12) (Brush, 1985).

It is, of course, one thing to note the disheartening state of mathematics education in the United States and quite another to articulate a workable route to significant improvement. I have no illusions about being able to do the latter, but, clearly, improving the current situation is a challenge of great national concern. My hope is that this book may contribute in a small way to a better understanding of the problem and perhaps prompt some useful thoughts about possible approaches to specific aspects of it.

☐ Goals of Teaching Mathematics

Concern has been expressed not only about the need for more effective methods for teaching mathematics but also about the need for a clearer articulation of what the goals of mathematical instruction should be. "Perhaps the most serious impediment to the design of more effective learning environments or instructional materials is the fact that, in general, we lack principled descriptions of the forms of understanding we seek to develop in students" (Wenger, 1994, p. 245).

In 1991 the U.S. Department of Education set as one of six goals to be realized by 2000: "U.S. students will be first in the world in science and mathematics achievement" (U.S. Department of Education, 1991, p. 3). In retrospect, this may have been an unrealistic goal, given how poorly U.S. students were doing on international assessments at the time it was set. In any case, the goal did not come close to being met.

Here I wish to make two distinctions relating to the question of what the goals of teaching mathematics in public school systems should be. The first contrasts two possibilities, each of which is sometimes stated or assumed to be in effect:

- Increasing the proportion of students of a given age who meet specified minimum standards

- Giving all students a better opportunity to realize more of their potential

The first of these possible goals focuses on the lower end of the performance distribution. The main objective is to bring the poorer performers up to the standards that have been set. Much has been written about children differing greatly with respect to the mathematical knowledge and skills that they acquire before school. Some of the disparity has been associated with socioeconomic status. As the National Mathematics Advisory Panel (2008) puts it: Most children from low-income backgrounds enter school with far less knowledge than peers from middle-income backgrounds, and the achievement gap in mathematical knowledge progressively widens throughout their PreK-12 years" (p. xviii). Adey and Shayer (1994, p. 171) argue that a reasonable goal of an effort to enhance thinking generally—assuming an effective combining of programs to develop thinking capabilities with improvements in instruction itself—is to bring the mental development of all children to the range that currently encompasses the top 30% of students. According to this view, the goal should be to decrease the spread of the ability range by moving those students who are currently at the bottom of it closer to the top.

One way to accomplish the alternative goal of giving all students a better opportunity of reaching their full potential is that of using teaching techniques that are adaptive to the aptitudes of individual students. If successful, this approach would increase the average performance level across the board, and it should have a salutary effect of increasing the population of people who are qualified to fill jobs that require high levels of mathematical competence. There is also the possibility that it would increase the spread of the performance continuum—amplify individual differences—by helping the more capable students more than the less capable ones. This possibility was noted decades ago by Carroll (1967). Recently, Gottfredson (2005) has made the same point and surmised that the result would not be welcome universally. "Targeting instruction to students' individual cognitive needs would likely improve achievement among all, but it would not cause the slow learners to catch up with the fast. The fast learners would improve more than the slow ones, further widening the learning gap between them and seeming to make the 'rich richer.' This is currently politically unacceptable" (Gottfredson, 2005, p. 168). Notably missing from the recommendations of the National Mathematics Advisory Panel (2008) is much attention to the question of how to give mathematically gifted students the best opportunities to realize their potential. The single reference to gifted students in the 45 recommendations in the executive summary notes that "with sufficient motivation [they] appear to be able to learn mathematics much faster than students proceeding through the curriculum at a normal pace, with no harm to their learning, and should be allowed to do so" (p. xxiv).

This is an extraordinarily important issue. It may well be the case generally that when powerful tools—including effective teaching and learning techniques—become available to everyone, those people who are more capable (interested, motivated) tend to use them to better advantage than those who are less so. It may be possible to define approaches to the teaching of mathematics that will help less capable students to learn faster and better than they otherwise would, but the idea that it would be good to minimize, or even reduce the range of, individual differences as they are expressed in mathematics is neither realistic nor desirable, in my view.

The second distinction relating to goals that I want to note is made by Papert (1972). This is the distinction between the goal of teaching children about mathematics and that of teaching them to be mathematicians. Some feeling for the goal of teaching students to be mathematicians is captured in Schoenfeld's (1987b) description of his experience of teaching a college course in mathematical problem solving for many years. "With hindsight, I realize that what I succeeded in doing in the most recent versions of my problem-solving course was to create a microcosm of mathematical culture. Mathematics was the medium of exchange. We talked about mathematics, explained it to each other, shared the false starts, enjoyed the interaction of personalities. In short, we became mathematical people. It was fun, but it was also natural and felt right. By virtue of this cultural immersion, the students experienced mathematics in a way that made sense, in a way similar to the way mathematicians live it. For that reason, the course has a much greater chance of having a lasting effect" (p. 213). Schoenfeld's classroom approach is described in his 1985 *Mathematical Problem Solving* and in numerous other publications. One surmises that the effectiveness of this approach depends, to no small degree, on an extraordinarily knowledgeable and committed teacher.

☐ Drill and Practice Learning

> Visions of draconian teachers demanding insane memorization of mean-
> ingless mumbo-jumbo prevent a large number of people from reacting
> normally to the opportunities offered by contemporary mathematics.
> (Steen, 1978, p. 2)

Interest among psychologists in the psychology of mathematics and
its implications for the teaching of arithmetic to children goes back at
least to American psychologist Edward L. Thorndike. His *The Psychology
of Arithmetic*, which appeared in 1922, gives a prescription for teaching
arithmetic to children based on his "law of effect" (Thorndike, 1913),
and the idea that mathematical skill is composed of a stable of stimulus–
response bonds that are best acquired and strengthened by rote drill.

Thorndike's emphasis on drill and practice has had many critics.
Notably among the earlier ones was William A. Brownell (1928), who
contended that such an emphasis would not yield an integrated comprehension of arithmetic and that what was needed was an approach
that stressed a conceptual grasp of the principles on which arithmetic
operations are based. Resnick and Ford (1981), who contrast the views
of Thorndike and Brownell, characterize the difference between them
as being in their definitions of what should be learned: "To Thorndike,
mathematical learning consisted of a collection of bonds; to Brownell, it
was an integrated set of principles and patterns" (p. 19).

Despite what appears to be the prevailing view that drill and practice methods are not likely to produce understanding of the material that
is to be learned, it is claimed that the most common form of teaching in
the U.S. schools today is based on recitation, which is a very close cousin
to drill and practice (Tharp & Gallimore, 1988). Kilpatrick, Swafford, and
Findell (2001) describe the method as follows:

> The teacher leads the class of students through the lesson material by ask-
> ing questions that can be answered with brief responses, often one word.
> The teacher acknowledges and evaluates each response, usually as right
> or wrong, and asks the next question. The cycle of question, response,
> and acknowledgement continues, often at a quick pace, until the material
> for the day has been reviewed. New material is presented by the teacher
> through telling or demonstrating. After the recitation part of the lesson,
> the students often are asked to work independently on the day's assign-
> ment, practicing skills that were demonstrated or reviewed earlier. U.S.
> readers will recognize this pattern from their own school experience
> because it has been popular in all parts of the country, for teaching all
> school subjects. (p. 48)

While this approach will undoubtedly provide students with the ability to perform many mathematical operations successfully, it is likely to fall short of giving them an understanding of why the rules they learn yield correct results. It is one thing to learn by rote that the product of two negative numbers is positive and quite another to understand why this is the case. From the teacher's perspective, the recitation method has the advantage that applying it does not require that one be able to explain the rationale for such rules.

☐ Constructivism and Discovery Learning

The mathematics curriculum in U.S. schools has changed considerably during the 20th century. Schoenfeld (1987a) describes the curriculum as being relatively stable over the first half of the century, and then experiencing a series of swings, each of which lasted about a decade, during the latter half. These swings involved, in order, the introduction of "new math" in the 1960s, largely in response to spectacular Soviet achievements in space technology; the "back to basics" movement in the 1970s; and the turn to an emphasis on mathematical problem solving in the 1980s. A similar characterization of major shifts in approaches to mathematics education during the 20th century is given in the 2001 report of the Mathematics Learning Study Committee of the National Research Council (Kilpatrick et al., 2001, p. 115). This report constitutes an extensive review of the current state of the teaching and learning of mathematics in U.S. schools through the eighth grade and concludes with numerous recommendations for improvement.

Another significant change that occurred during the 20th century, in part due to the influence of theoretical work of Swiss philosopher-psychologist Jean Piaget (1928, 1952; Inhelder & Piaget, 1958, 1964) and Russian psychologist Lev Vygotsky (1962, 1978, 1986), was a shift from a nearly exclusive dependence on rote instruction at the beginning of the century to an increasing emphasis on participatory learning, in which the student is seen as an active participant in the construction of his or her own knowledge. During the latter part of the 20th century, *constructivism*, broadly defined, became widely adopted by educational researchers (Anderson, 1981, 1982; Schoenfeld, 1987a; Wenger, 1987).

That people are more likely to remember and use what they discover than to remember and use what they are told is generally acknowledged to be a fact. From the constructivist's perspective, learning is most effective—perhaps occurs only—when learners construct their knowledge (Belmont, 1989; Steffe & Wood, 1990). Some argue that drill and practice methods, especially if introduced too early, are likely to kill the natural interest that children have in numbers and mathematics. The role of the teacher, in this view, is to facilitate this knowledge–construction process—to structure environments and situations that make it easy and natural for students to discover, invent, and construct (Cobb, Yackel, & Wood, 1992). This does not mean that the teacher is indifferent to what knowledge gets constructed. "We must make explicit the nature of the knowledge that we hope is constructed and make a case that the chosen activities will promote its construction" (P. W. Thompson, 1985, p. 192). Steffe (1994) also emphasizes the importance of guidance of discovery.

That children naturally learn some things by discovering them has been argued by Baroody and Gannon (1984), who make the case for the principle of commutativity in addition. When first learning to add, children are very likely to notice, Baroody and Gannon contend, that one gets the same sum independently of the order of the addends. Of course, an adequate understanding of the principle as applied to addition must include knowledge that it does not hold for subtraction; misapplication in the latter case can account for some of the errors that have been found in beginners' performance of subtraction tasks (e.g., always subtracting the smaller number from the larger, independently of the order of the terms). Resnick (1983) notes the possibility that children naturally assume that all arithmetic operations are commutative and have to learn that this is not so.

Many of the computer-based microworlds that have been developed are intended to make it easy for children to explore physical processes and mathematical relationships (Dugdale, 1982; Feurzeig, 1988, 2006; Resnick & Johnson, 1988). Such systems may provide environments well suited to facilitate discovery, and some children may make substantive discoveries by interacting with such systems completely on their own. However, if such systems are to be maximally effective agents of learning for most children, there is probably a need for some guidance in their use from a teacher who has specific learning goals in mind and who knows how to steer the exploration in directions that are likely to lead to the desired discoveries.

In the absence of much more powerful discovery learning tools than have yet been developed, what can be accomplished by discovery learning is bound to be limited and seems likely to fall short of producing what is generally considered a good understanding of a significant chunk of mathematics. It is at once exciting and intimidating to realize that humankind took millennia to develop many of the mathematical concepts and relationships that make up today's elementary school curriculum. That children could make, even with help and guidance, all the discoveries that define elementary mathematics—recapitulate the history of the development of the discipline, as it were—does not seem remotely possible. On the other hand, knowledge of the history of mathematics, and especially of the conceptual struggles that have attended many of the key developments, such as the many extensions of the concept of number (see Chapter 3), provides some insight into the struggles that a child making a similar conceptual journey over a few short years is likely to experience.

☐ Need for a Synthesis

While the constructivist view is attractive to many, it is seen by others to be unrealistic in its assumption that all children will benefit from this approach and that there is no need for rote, or "mechanical," learning at any phase of a mathematical education (Geary, 1994; Sweller, Mawer, & Ward, 1983). Geary argues that the acquisition of an understanding of mathematical concepts and skill in performing mathematical procedures may require different approaches to instruction, and that procedural skill, in contrast to conceptual understanding, may require considerable drill and practice. Pointing out that there was a precipitous decline in mathematical competence of public school students following the introduction of the "new math" in the 1960s, Brainerd (1979) takes issue with some of the assumptions underlying the discovery learning approach. "The assumption that answering leading questions [his characterization of the Socratic method] is in some meaningful sense self-discovery and the assumption that self-discovery is the best way to learn mathematics are both open to serious doubt" (p. 207).

Resnick and Ford (1981) give an account of the rise of interest in new methods of teaching mathematics that put more stress on conceptual learning than on the rote teaching of computational procedures, and they note that as of the time of their writing, most educators acknowledged the need for both of the types of learning experiences called for by the contrasting views—drill and meaningful instruction—but that it was not clear how the two should be integrated. Resnick and Ford review in some detail innovative approaches, promoted by Dienes (1960, 1963, 1967) and others, that make heavy use of concrete materials (e.g., attribute blocks, Cuisenaire rods) in the teaching of elementary arithmetic with an emphasis on conceptual understanding. As to whether these approaches produce better learning and deeper understanding of mathematics than do more traditional approaches that emphasize the acquisition of computational skills, they note that the available evidence is largely indirect and not decisive. Their conclusion is that instructional planning almost certainly should include opportunities for the learning of both concepts and computational skills.

> The relationship between computational skill and mathematical understanding is one of the oldest concerns in the psychology of mathematics. It is also one that has consistently eluded successful formulation as a research question. Over the years, the issue has been posed in a manner that made it unlikely that fruitful research could be carried out. Instead of focusing on the *interaction* between computation and understanding, between practice and insight, psychologists and mathematics educators have been busy trying to demonstrate the superiority of one over the other.... What is needed, and what now seems a possible research agenda, is to focus on *how* understanding influences the acquisition of computational routines and, conversely, on how, with extended practice in a computational skill, an individual's mathematical understanding may be modified. (Resnick & Ford, 1981, p. 246)

The argument that computational skills should be taught—not left to be discovered—is not, of course, a claim that they must be taught strictly by rote. Several investigators have provided evidence that learning to calculate can be facilitated by experiences designed to promote understanding of the procedures used (Bezuk & Cramer, 1989; Hiebert & Wearne, 1996; Mack, 1990, 1995). Conversely, rule-based instruction that is not accompanied by efforts to ensure that students gain a conceptual understanding of the procedures that are being taught seems unlikely to provide a solid basis for the acquisition of higher mathematical knowledge and skill. I think it is safe to say that few, if any, researchers or educators would argue against conceptual understanding as a primary goal of the teaching of mathematics at all levels. The question of precisely how best to develop that understanding is still wanting an answer.

One point on which there appears to be general agreement is that children should not be treated as *tabulae rasae* at the outset of their introduction to formal mathematical education. They come to school with a considerable body of concepts and beliefs relating to counting and arithmetic, and failure to recognize this and to build on what children already know, or believe, more or less ensures confusion and impedance of the development of mathematical competence. As Dehaene (1997) puts it, "The child's brain, far from being a sponge, is a structured organ that acquires facts only insofar as they can be integrated into its previous knowledge.... Thus, bombarding the juvenile brain with abstract axioms is probably useless. A more reasonable strategy for teaching mathematics would appear to go through a progressive enrichment of children's

intuitions, leaning heavily on their precocious understanding of quantitative manipulations and of counting" (p. 241).

☐ Order of Instruction

The idea of hierarchical structure has guided the teaching of mathematics from elementary school on. Number concepts are taught first, then
arithmetic, then algebra, then calculus, and so on. However, this progression is somewhat misleading. Some understanding of number concepts is certainly a prerequisite for learning to do arithmetic, but what
one learns about numbers before starting to learn arithmetic is but a tiny
bit of what there is to know about numbers; number theory is a very
active area of mathematical research. Some mathematicians spend their
lives studying number theory and generally raise more questions than
they answer as a consequence.

   Even within a relatively well-defined area of mathematics and
at a well-specified level of complexity—like elementary arithmetic—
the question of the order in which concepts should be introduced has
been a focus of interest to researchers. Arithmetic during the primary
grades has meant a focus mainly on addition and subtraction (Baroody &
Standifer, 1993; Coburn, 1989). Geometry, in contrast, has received relatively
little emphasis (Clements 2004; Clements & Battista, 1992; Clements,
Swaminathan, Hannibal, & Sarama, 1999; Fuys & Liebov, 1993; Porter,
1989). The potential for making connections between number concepts
and geometry—by making use of the number line, for instance—appears
not to have been much exploited (Kilpatrick et al., 2001). The results
of some attempts to teach children to locate fractions on the number
line suggest that this is a difficult task (Behr, Lesh, Post, & Silver, 1983;
Gelman, 1991; Novillis-Larson, 1980).

   The order generally followed in teaching arithmetic in the United
States is addition, subtraction, multiplication, and division, the latter two
beginning only in the third grade (National Council of Teachers of
Mathematics, 2000), but there are many views as to what the best order
is (Dienes & Golding, 1971; Gagné, 1968; Gagné, Mayor, Garstens, &
Paradise, 1962; Resnick, Wang, & Kaplan, 1973). Some researchers see
the presentation of multiplication as based on counting or repeated addition as problematic (Confrey, 1994; Steffe, 1994).

   Timing has also been an issue. Some investigators hold that the
introduction of addition and subtraction should be delayed until children
have developed a firm foundation of number concepts (Van de Walle &
Watkins, 1993). However, what constitutes a firm foundation of number

concepts in this context is open to question. The rational number concept, which many would argue is fundamental to basic mathematics, can be problematic even for secondary school students. The difficulty appears to be due, at least in part, as Behr et al. (1983) note, to rational numbers being interpretable in at least six ways: "a part-to-whole comparison, a decimal, a ratio, an indicated division (quotient), an operator, and a measure of continuous or discrete quantities" (p. 93). A complete understanding of the rational number concept, it has been claimed, requires an understanding of all six of these interpretations and their interrelations (Kieren, 1976).

When equations or algorithms should be introduced—how much work with less formal concepts should precede their introduction—is also a matter of debate (Thompson & Van de Walle, 1980; Thornton, 1990). Familiarity with algebra has generally been considered a prerequisite for the learning of calculus, but a case has been made for introducing some of the ideas that are fundamental to calculus before a student has acquired any competence in algebra (Confrey & Smith, 1994; Kaput, 1994).

Given the key role that the concept of proof plays in mathematics, the question of when it should be introduced in mathematics education is a particularly interesting one. My sense is that relatively little emphasis is given to this concept in the teaching of elementary mathematics. An exception is the work of Maher and Martino (1996) and Davis and Maher (1997). These investigators have studied the ability of beginning students to deal with the concept and to use it effectively.

Determining the most effective order of introducing mathematical concepts to students is complicated by the finding that the ability or inability of students to use specific concepts linguistically is not always a good indication of whether they understand the concepts in a more than superficial way (Brainerd, 1973c, 1973d). In some cases children may be able to use a concept appropriately in context and yet not be able to show a semantic comprehension of it by answering questions about it correctly. For many practical purposes, this perhaps does not matter—the important thing is to get a correct solution to the mathematical problem—but to the extent that understanding is a goal of education, it matters greatly.

☐ Children's Strategies in Learning Elementary Arithmetic

Strategies is likely to bring to mind sophisticated approaches to complex problems. However, the term is sometimes used to refer to approaches that children adopt in trying to cope with tasks they are given even in learning basic arithmetic. Interest in identifying strategies that children use when beginning to learn to add and subtract goes back at least to Brownell (1928). Other relatively early investigators of the subject include Ilg and Ames (1951) and Gibb (1956). The interest has increased considerably in more recent years, and there now exists a very large literature on this topic.

Children use a variety of strategies to do addition or subtraction (Baroody, 1987; Fuson, 1992a, 1992b; Hamann & Ashcraft, 1985; Siegler, 1987, 1989; Siegler & Shrager, 1984). Among those they use to add are some based on counting. To add 4 and 3, they may start with *four* and count up three additional numbers, *five, six, seven*, and take the number arrived at as the answer (Fuson, 1982; Groen & Parkman, 1972; Groen & Resnick, 1977; Steffe, Thompson, & Richards, 1982; Suppes & Groen, 1967). A counting-up procedure may also be used when the task is to identify the number that must be added to a specified number to yield a specified total, for example, $4 + ? = 7$ (Case, 1978).

A counting-up strategy for doing addition works whether one starts the count from the larger or the smaller addend, but it is more efficient to start with the larger one, and the more so the greater the difference of the addends in size. Several studies have shown that many children spontaneously discover and adopt this strategy, which has been called the *min* strategy, because it requires the minimum count (Geary, 1990; Groen & Resnick, 1977; Siegler & Jenkins, 1989; Svenson & Broquist, 1975). Counting up requires the ability to start a count with a number other than 1, which is a skill that may take some time to acquire after a child has learned to count from 1 (Fuson, Richards, & Briars, 1982).

Carpenter and Moser (1982, 1984; Carpenter, 1985) distinguish three strategies that can be seen in children's performance of addition tasks involving two addends: *direct modeling* (in which the child represents each of the two sets with physical objects or fingers and counts their union), *counting* (just described, in which the child starts with one of the numbers and counts from there the number of units represented by the second number), and *number facts* (in which the sum of the numbers is retrieved from memory). The three represent a progression; the use of number facts depends on having already committed sums of specific pairs of addends to memory. Steinberg (1985) proposes a similar, but not identical, progression of phases in dealing with addition and subtraction problems: counting, reasoning, and recall. The reasoning phase, in this model, involves the discovery of ways to apply what one knows to figure out answers, by means other than counting.

Counting strategies similar to those used for addition are also used by many children for doing subtraction (Carpenter & Moser, 1984; Siegler & Shrager, 1984; Svenson & Hedenborg, 1979; Woods, Resnick,

& Groen, 1975). Resnick (1976) found that children who used counting strategies to do subtraction learned to use either a decrementing strategy (subtracting n from m by starting with m and counting down n units) or an incrementing strategy (starting with n and counting up to m), depending on which was the more efficient for given values of m and n, although counting down appears to be the more difficult for some children. Counting strategies of various types appear to be exceedingly common among young learners of arithmetic, and their use has been observed to persist sometimes considerably beyond the primary grades (Lankford, 1972).

Counting strategies can even be used to good effect in doing basic arithmetic with fractions. If a child understands that a fraction, say 3/5, can be thought of as composed of three unit fractions—1/5, 1/5, and 1/5—this knowledge can provide the basis for a counting strategy for adding fractions with the same denominator. Thus, the problem 4/5 + 3/5 can be solved by counting up from 4/5—5/5, 6/5, 7/5 (Post, Wachsmuth, Lesh, & Behr, 1985). Thinking of a fraction, n/m, as n units of $1/m$ each reflects a comprehension of a fraction as two numbers, one of which (the bottom one) indicates the number of parts into which a whole has been divided and the other of which (the top one) indicates the number of such parts in hand (Skypek, 1984). Comparable distinctions between strategies used to solve ratio and proportion problems have been noted, especially as they are found in teaching approaches used in different countries (Karplus, Karplus, Formisano, & Paulson, 1979).

Geary (1994) describes five classes of strategies that children use in solving simple addition problems: "using manipulatives; finger counting; verbal counting without the use of manipulatives (i.e., mentally); derived facts; and fact retrieval" (p. 49). (Manipulatives include blocks, Cuisenaire rods, tiles, gears, and other physical objects that can be used to illustrate mathematical concepts and relationships.) This taxonomy appears to divide each of the first and third of Carpenter and Moser (1982) into two, but otherwise is very similar. Carpenter and Moser combine the use of manipulatives (other than fingers) and finger counting, whereas Geary treats them as separate strategies. In the normal course of development, children gradually replace strategies that require counting, the use of fingers or other manipulatives, or derived facts with reliance on fact retrieval from memory, which depends on having accumulated a cache of facts (e.g., the sums or products of specified pairs of numbers) that are available for retrieval (Delaney, Reder, Staszewski, & Ritter, 1998). The relative inability to use retrieval-based strategies to solve math problems appears to be one of the more consistent indicators of mathematical disability (Geary, 1993; Temple & Sherwood, 2002).

Manipulatives are often provided by teachers to facilitate early instruction in arithmetic, and the results of numerous studies have been interpreted as evidence of the effectiveness of their use (Carpenter & Moser, 1982; Fennema, 1972; Kieren, 1969; Steffe, 1970; Steffe & Johnson, 1971), although some researchers have cautioned that without proper guidance in their use and significance, manipulatives by themselves may simply complicate the learning, or worse (Ball, 1992; Fuson & Briars, 1990). When learning to add and subtract, children appear to do better if required to think about the addition or subtraction of real objects, even if there are no objects to be seen or handled, than when they are required to think about adding or subtracting numbers in the abstract (Jordan, Huttenlocher, & Levine, 1992). According to national studies, however, despite decades of exhortation by mathematics educators, the use of manipulatives is not common in the classroom (Kilpatrick et al., 2001; National Mathematics Advisory Panel, 2008).

The use of finger counting as a means of doing simple arithmetic appears to be spontaneous with many children (Folsom, 1975; Geary & Brown, 1991; Geary, Brown, & Samaranayake, 1991; Geary & Burlingham-Dubree, 1989; Ginsburg, 1989), although there are culture-based differences in the details regarding how numbers are represented manually (Fuson & Kwon, 1992a). Interestingly, digital agnosia (the inability to identify one's fingers) has been associated with difficulty in learning mathematics (Benson & Geschwind, 1970; Kinsbourne & Warrington, 1962, 1963; Temple, 1989). Finger agnosia and dyscalculia are two of four primary symptoms of Gerstmann's (1940) syndrome, a rare, and somewhat controversial, neurological disorder. Steffe, Spikes, and Hirstein (1976) have raised the question of whether the observed superior performance with the use of manipulatives could be due, at least in part, to children often being encouraged to use them while being discouraged from using finger counting. Children who learn to do arithmetic with an abacus may later imagine manipulating one when performing mathematical operations mentally (Hatano, Miyake, & Binks, 1977; Stigler, 1984; Stigler, Chalip, & Miller, 1986).

Most children do not learn to multiply and divide until they have acquired some proficiency with addition and subtraction. It is widely held that multiplication skills are built from existing abilities to add and subtract (Cooney, Swanson, & Ladd, 1988; Siegler, 1988), and that in learning to divide, children use not only what they know about addition and subtraction, but what they know about multiplication as well (Vergnaud, 1983). Thinking of multiplication and division as successive addition and subtraction works to a point, but it does not suffice to make clear what is required by problems like $\sqrt{6} \times \sqrt{13}$ and $\sqrt[3]{90}/\sqrt[4]{23}$.

Evidence that students and teachers have more difficulty with multiplication and division than with addition and subtraction (Mullis et al., 1991; Simon & Blume, 1992) has motivated special attention by some researchers on these operations under the rubric of the *multiplicative conceptual field* (Harel & Confrey, 1994), which, as described by Vergnaud (1983), includes not only the concepts of multiplication and division, but closely related ones such as fraction, ratio, and similarity. Vergnaud identifies numerous strategies, or *procedures*, both correct and incorrect, that students apply to the solution of these types of problems.

Many children develop informal nonquantitative approaches to problems of ratio and proportion before receiving instruction regarding how to deal with such problems quantitatively (Kaput & West, 1994; Lamon, 1994; Tourniaire, 1986; Van den Brink & Streefland, 1979). As with addition and subtraction, they rely on more than one strategy in solving multiplication problems when they first encounter them (Cooney et al., 1988; Koshmider & Ashcraft, 1991).

Based in large part on results from teaching interviews with children, Confrey (1988, 1991, 1994) has proposed a model of multiplication that takes *splitting* as a primitive concept that children acquire naturally in social contexts from experiences of sharing and dividing in half. It is not necessary to rely on counting, she points out, to verify that a division in half has been made correctly. "Equal shares of a discrete set can be justified by appealing to the use of a one-to-one correspondence and in the continuous case, appeals to congruence of parts or symmetries can be made" (p. 292). We should note, however, the difference between the concept of an equal number of shares and that of shares of equal size. There is some evidence that the emergence of the former concept precedes that of the latter in the normal course of development (Empson, 1999; Hiebert & Tonnessen, 1978; Pothier & Sawada, 1983). Both, however, can be useful in introductory treatment of fractions (Streefland, 1993).

Splitting leads naturally to consideration of exponential and logarithmic functions (Confrey, 1988, 1994), an understanding of which is seen to be essential for an educated citizenry in the modern world. Smith and Confrey (1994) give an account of the involvement of Joost Burgi, Thomas Bradwardine, Nicole Oresme, and John Napier in the development of logarithms that they see as supportive of the idea that "multiplication will be a primitive action that in many situations cannot be based on addition, that multiplicative (or splitting) worlds are based on an operational concept of number significantly different from number concepts giving rise to counting structures, and that constructing an understanding of these multiplicative worlds prior to the study of functions may facilitate students' understanding of exponential and lograthmic functions" (p. 355).

Confrey contends that the intuitions and insights that children naturally develop into splitting are ignored or inhibited by the conventional curriculum, and that this is a mistake. She argues that rather than waiting until children have made considerable progress with addition and subtraction before introducing them to multiplicative concepts, splitting should be developed as a complement to counting. "We make the basic conjecture that, by developing the construct of splitting in the curriculum, one can establish a more adequate and robust approach to such traditionally thorny topics as ratio and proportion, multiplicative rate of change, exponential functions and so on. Therefore we contend that multiple constructions of numeric quantity are possible and that uniformity in the treatment of the concept of number is detrimental to quantitative maturity of an educated citizenry" (p. 298). Confrey argues that if her conjecture about the primitive nature of splitting is borne out, early instruction that capitalizes on it could make multiplication and division as elementary as addition and subtraction rather than derivative from them.

☐ Error Analysis

Everyone makes errors when learning to do mathematical operations. Presumably some errors are due to carelessness and to guesswork, but of special interest for pedagogical purposes are errors that arise from basic misunderstandings of rules or principles of math. Such misunderstandings may be flagged by the consistency with which specific types of errors occur. Consistent errors, or error types, were the subject of research at least as early as 1926 (Buswell, 1926; Buswell & John, 1926). Another relatively early study with this focus was that of Brueckner (1930).

Much of the early work on the acquisition of elementary mathematical skills and knowledge by children focused on attempting to determine the relative difficulty of arithmetic problems. One result was the publication of lists of specific arithmetic problems (2 + 5, 7 + 9, ...) rank ordered according to difficulty for the purpose of informing instruction (Resnick & Ford, 1981). Carpenter and Moser (1983) argue that, although there were relatively high correlations among the rankings produced by several studies, there also were conflicting results, and they contend that the only clear conclusion to be drawn from the results in the aggregate is that the difficulty of addition and subtraction problems increases with the sizes of the numbers involved. The finding that the time taken by children to add two digits increases with the size of the smaller digit could be seen as evidence of adding by counting, but this interpretation is called into question by the finding that the same relationship, albeit with a smaller increase per unit of digit size, is found with college students (Groen & Parkman, 1972). Groen and Parkman consider the relationship with college students to be a consequence of addition from memory mixed with reliance on counting for a small percentage of the problems. Ashcraft (1995) proposes an alternative explanation that attributes the effect to the time required to access the memorized addition table increasing with the sizes of the addends.

Failure to understand operations and relationships at a relatively elementary level is seen in that what might appear to be incidental differences in problem representation can be consequential for children learning arithmetic. Generally, number-sentence problems of the form $? + a = b$ or $a + ? = b$ are more difficult than those of the form $a + b = ?$. Other consistent differences in difficulty among the various types of number-sentence addition and subtraction problems that can be generated have also been found (Rosenthal & Resnick, 1974; Weaver, 1971). One account of such findings is that children naturally think in terms of forms in which the result is the unknown ($a + b = ?$ and $a - b = ?$), and the greater difficulty of the other forms stems from solving them by first translating them into one of the more natural forms (Rosenthal & Resnick, 1974). The evidence suggests that the idea that the equals sign (=) represents equivalence—that the expression to the right of it is equivalent in value to the expression to the left of it, and vice versa—can be more difficult for a child to grasp than is generally assumed (Baroody & Ginsburg, 1983).

Sometimes children who have no trouble with an equation like $2 + 4 = 6$ will balk at $6 = 2 + 4$ on the grounds that the latter is backwards (Ginsburg & Baron, 1993). They are especially likely to have difficulty with equations that have mathematical operations on both sides of the equation ($4 + 2 + 1 = 3 + \_\_$) (McNeil & Alibali, 2000; Perry, Church, & Goldin-Meadow, 1988; Rittle-Johnson & Alibali, 1999). One proposed explanation of these difficulties is that, when learning basic arithmetic facts, children encounter only forms in which the arithmetic operation(s) is on the left of the equals sign and a single number ("the answer") is on the right; thus, they are unprepared to perceive departures from these forms as legitimate (McNeil & Alibali, 2002, 2004). More generally, McNeil (2007; McNeil & Alibali, 2005) sees this as an instance, of which there are others, of the learning of a new mathematical concept being impeded by existing knowledge. A more than superficial understanding of single-digits addition and how addition relates to subtraction should provide a basis for figuring out that the answer to $6 - 4$ is whatever has to be added to 4 to make 6. Apparently many children who can get the correct answer to $4 + 2$ fail to see the relevance of this relationship to the subtraction problem (Baroody, 1987; Kamii, 1985).

Numerous studies aimed at identifying common error types and determining their causes have been conducted (Anderson, 1989; Ashlock, 1976; Brown & Burton, 1978; Burton, 1981; Campbell & Graham, 1985; Cox, 1975; Davis & McKnight, 1980; Erlwanger, 1973, 1975; Ginsburg, 1977; Lankford, 1972; Matz, 1982; VanLehn, 1983, 1986). Consistent errors are assumed to occur because of conceptual misunderstandings; those who commit them believe they are using procedures correctly when they are not. Such errors are to be distinguished from errors, sometimes referred to as *slips* (Norman, 1981), that occur occasionally, but not consistently, and not because of basic misunderstandings of the procedures involved, but because of carelessness or inattentiveness. Slips are likely to be recognized immediately as erroneous when pointed out to those who have committed them, whereas consistent, or *rule-based*, errors are not. Misconceptions that underlie the latter often prove to be highly resistant to change (Rosnick & Clement, 1984).

One goal of research on consistent errors is to determine the conceptual bases for them. They are generally believed to occur as the consequences of misapplying one or another learned procedure in a systematic way. Resnick (1983) argues that many of the misapplied procedural rules were learned without an understanding of their rationale. Brown and VanLehn (1980; VanLehn, 1983) take a similar position in contending that many of the consistent errors that occur are the results of attempts by their users to "repair" poorly understood procedures when their application in specific contexts has led to an impasse. Generally, repairs of ineffective procedures get users beyond impasses, but often they yield consistent errors rather than correct solutions. Studying the types of repairs that users make provides insights into the conceptual difficulties that students have with specific rules and procedures and also into the nature of their mathematical thinking.

Systematic errors that appear to be relatively common include failing to "carry" or carrying inappropriately in addition (Fuson & Briars, 1990; Suppes & Morningstar, 1972), and giving the absolute difference between two numbers as the answer to a subtraction problem independently of whether the smaller number is to be subtracted from the larger or the larger is to be subtracted from the smaller. The latter error is sufficiently common to have been given a name—the *symmetric subtraction frame* (Davis, 1983). Not surprisingly, the concept of borrowing can give rise to misconceptions, and especially borrowing from zero or across two or more zeros (Davis & McKnight, 1980; Davis, Young, & McLoughlin, 1982). Much of the work on consistent errors has focused on multidigit subtraction.

Several investigators have catalogued the types of errors that are commonly encountered. One of these efforts was made by Bennett (1976), who analyzed nearly 1,600 subtractions performed by 33 ten-year-old school children. In a subsequent analysis of Bennett's data, Young and O'Shea (1981) classified many of the errors that were made as algorithmic. These were subclassified into several categories, such as the following: (1) borrow when the subtrahend digit is less than (rather than greater than) the minuend digit; (2) subtract the smaller number from the larger; and (3) always borrow.

The strategy used by Young and O'Shea to account for the consistent errors they identified in subtraction problems was to define a set of rules that would produce correct answers and then selectively delete rules from this set or add rules that were appropriate to other arithmetical tasks. They were able to account for about two-thirds of the procedural errors in this fashion. There was, they argue, no need for "wrong" rules to account for the types of errors studied. Of course, that one can define an algorithmic procedure that will generate the same errors that students generate does not prove that students are using that procedure, but it does identify a way in which the errors *could* have been produced. Also, correct rules incorrectly applied have the same undesirable effects as do wrong rules, but the remediation that is called for may differ in the two cases.

Fractions lend themselves to consistent errors of various types (Langford & Sarullo, 1993; Smith, Solomon, & Carey, 2005). One such is the taking of the sums of the numerators and the sums of the denominators separately when adding (Lankford, 1972; Silver, 1986). Another is the tendency to misjudge the relative sizes of two fractions with the same numerator, taking the one with the larger denominator to be the larger fraction (Behr, Wachsmuth, Post, & Lesh, 1984; Post et al., 1985). A particularly interesting error, reported by Mack (1993), was that of a child who believed that taking 1/4 from 4/4 leaves 3/3. The belief is incorrect, of course, from a conventional point of view, but from another point of view, perhaps that of the child, what one has left when one removes one of 4 equal pieces is 3 equal pieces, each of which is 1/3 of the new whole.

Students who have little or no difficulty in adding fractions with the same denominator often are unable to add correctly those that have different denominators. According to U.S. national assessment results, while most 13- and 17-year-olds can add fractions with the same denominator, only about one-third of the former and two-thirds of the latter can find the sum of 1/2 and 1/3 (Carpenter et al., 1980). Many 13-year-olds also have difficulty ordering fractions (Kouba, Carpenter, & Swafford, 1989; Wearne & Kouba, 2000). In one study a majority of those students who could add fractions correctly could not explain the rules for doing so (Peck & Jencks, 1981).

In view of the fact that children generally learn to add and subtract before they learn to multiply and divide, a finding that may appear to be paradoxical is that many children are able to multiply fractions correctly before they are able to add them correctly (Byrnes & Wasik, 1991). We have noted already that a common error in adding fractions is to add the numerators and denominators separately, getting, for example, an incorrect result such as 2/3 + 4/5 = 6/8. Following the same rule when multiplying—multiplying numerators and denominators separately—gives the correct answer, thus 2/3 × 4/5 = 8/15. Following the same rule in both cases shows a lack of conceptual understanding of the rationales for the procedure, but the procedural knowledge itself is correct in the case of multiplication. Rittle-Johnson and Siegler (1998) conclude from this type of finding that conceptual knowledge is acquired before procedural in the case of addition, but the reverse is true of multiplication.

Specific types of errors are common when children begin dealing with decimals. Often the problem appears to be that of ignoring or misunderstanding the role of the decimal point and treating the decimal as though it were a whole number. Thus, one might judge .25 to be greater than .5 because 25 is greater than 5 (Behr et al., 1984). When adding numbers with different numbers of digits following the decimal point, some children are likely to arrange them so their right-most digits are aligned instead of the decimal points and get an incorrect answer as a result (Hiebert & Wearne, 1985). Putting decimals in the correct order is problematic for many children (Carpenter, Corbitt, Kepner, Lindquist, & Reys, 1981; Moss & Case, 1999).

Zero, because of its unique status among numbers, can readily be the basis of misconceptions. One opportunity comes from its use in the solution of equations of the form $(x - a)(x - b) = c$. When $c = 0$, it must be the case that either $x - a = 0$ or $x - b = 0$, from which it follows that either $x = a$ or $x = b$. But when c is any number other than 0, a similar rule does not apply. An erroneous belief that has been found to be fairly common is that if $(x - a)(x - b) = c$, it must be the case that either $x - a = c$ or that $x - b = c$, so, for example, if $(x - 4)(x - 2) = 8$, either $x - 4 = 8$ or $x - 2 = 8$, so $x =$ either 12 or 10, but in fact it equals either 6 or 0. This error appears to be difficult to eradicate (Davis, 1983; Matz, 1980).

Many of the types of consistent errors that have been found appear to result from the induction and consistent application of an incorrect rule of operation or execution—a "buggy" algorithm (Brown & VanLehn, 1980; VanLehn, 1983, 1986, 1990). Ben-Zeev (1995, 1996, 1998) refers to such errors as *rational errors*, and points out that they can be the result of a process of generalization from examples of solved problems, the same process that often underlies the learning of correct rules and procedures. Ben-Zeev (1998) proposes a taxonomy of such errors that classifies them in terms of a few hypothesized mechanisms that could produce them.

Consistent, or rational, errors have been found not only in elementary arithmetic but also in other areas of mathematics, including geometry (Anderson, 1993), algebra (Matz, 1982; Payne & Squibb, 1990;

Wenger, 1987), and calculus (Davis & Vinner, 1986). Maurer (1987) gives several examples of incorrect generalizations of the distributive rule of multiplication, according to which, as one learns in high school algebra, $3(x + y) = 3x + 3y$. Unthinking mechanical application of the distributive rule can yield such inapt equations as $(a + b)^2 = a^2 + b^2$ and $\sin (a + b) = \sin a + \sin b$.

Other misconceptions that appear to be common among children learning arithmetic (and sometimes among teachers) include the belief that multiplication always makes bigger and division always makes smaller, which is referred to widely as MMBDMS (Bell, Fischbein, & Greer, 1984; Bell, Swan, & Taylor, 1981; Graeber & Tirosh, 1988, 1990; Greer, 1987, 1988; Vergnaud, 1983). Closely related is the belief that division is always division of a larger number by a smaller one. Such misconceptions can complicate learning how to multiply and divide fractions and decimals (Bell et al., 1981; De Corte, Verschaffel, & Van Coillie, 1988; Harel, Behr, Post, & Lesh, 1994; Resnick et al., 1989). They also can impede mathematical performance more generally, as evidenced by the fact that the facility with which many children can identify the mathematical operation(s) a word problem calls for is determined in part by the numbers that occur in the problem statement, even though they are irrelevant to the operation(s) required (De Corte et al., 1988).

Some misconceptions can be difficult to detect, or may even be reinforced by instruction in the classroom. Interpretation of the equals sign is a case in point. Students may interpret this sign when it appears in an equation as a signal that a calculation is required, and in particular, the calculation that is represented by the operational sign that most immediately precedes it (Behr, Erlwanger, & Nichols, 1980; Saenz-Ludlow & Walgamuth, 1998). Thus, solution to the problem $3 + 4 = \_\_ + 2$ would be considered to be 7. The misconception may persist because of being reinforced both by the way the equals sign often is presented to students and by experience with numerous exercises for which the interpretation works (Kilpatrick et al., 2001).

☐ Teaching Experiments

One of the effects of the emergence of the view that children construct their own knowledge has been to increase the effort that educational researchers, especially those who espouse a constructivist view, have been making to determine what children know, or believe, at various points along their educational path, and especially at the beginning of it. Much of what has been learned about how students acquire mathematical knowledge and skill during their formative years has been learned by analyzing protocols taken while they are working on problems (Davis, 1984). An investigative method that has been used increasingly is the teaching experiment (Behr et al., 1984; Cobb & Steffe, 1983; Kaput & West, 1994; Steffe, 1991, 1994; Thompson, 1982, 1994b; Vergnaud, 1983).

Steffe (1994) refers to the teaching experiment, which she sees as a replacement of the Piagetian clinical interview, as "the fundamental research methodology in mathematics education" (p. 12). The basic idea is to tailor instruction to the current knowledge (and misconceptions) of the student, which the teacher-experimenter hopes to discover as the teaching experiment proceeds. The hope is that the instruction will be effective by virtue of meeting the student where he or she is, conceptually, and that it will yield a better understanding of how mathematics is best taught and learned. Thompson cautions that the teaching experiment is characteristically opportunistic and that the teacher-experimenter must rely continually on serendipity.

Davis (1983) describes the "paradigm shift" that appears to have taken place, at least among many mathematics educators, as a shift from a focus on "teaching and learning" to one on "the processes of thinking about mathematical problems. In the new paradigm, learning is regarded as definable in terms of transitions between one form of mathematical thinking and another and is in this sense not fundamental, but is a kind of 'derived' or 'second-order' concept. The fundamental task is to describe the various forms of mathematical thought themselves, that form the 'before' and the 'after' of any change" (p. 254). Perhaps the most common source of data regarding the transitions of interest, Davis suggests, is the task-based interview of the sort pioneered by Piaget, which has become, in more contemporary terminology, the teaching experiment.

☐ Remedial Math

Most of the research on the teaching of mathematics has taken for granted an explicit or tacit model of the development of mathematical knowledge and skills over the first few years of life and the first few grades of schooling. According to this model, most children have some basic mathematical intuitions and acquire a fair amount of number knowledge before attending school. During the first few years of schooling they learn to do elementary arithmetic, begin dealing with word problems, move on to increasingly difficult arithmetic operations, begin learning a bit of algebra, and so on.

Unhappily assessment data show that many students do not acquire the knowledge and skills that this model suggests they should on the assumed time schedule. Many arrive in high school, or even college, with a very poor grasp of basic mathematics. Some college students, for example, have difficulty with such a fundamental aspect of math as dealing with negative numbers, perhaps because they learned rules by rote in elementary school without understanding the rationale for them. This poses a special challenge to teachers and mathematics educators. Henderson (1987) sees the question of how to deal with this problem as one that has yet to be answered. It is not to be assumed, she argues, that the teaching procedures that are appropriate for first grade students are appropriate for fifth grade students who do not have the knowledge and skills they should have acquired years before. And the question pertains to students at higher levels of education as well. "More and more young adults are entering college having mastered fewer and fewer of the mathematical skills they were supposed to have learned during their school years. Is more of the same instruction that led to failure in the first place going to help? Will something else help? We just don't know, and there is an urgent need to address such issues" (p. 152).

☐ Technology in the Teaching of Mathematics

The application of technology to education generally and to the teaching of mathematics in particular has been of interest for some time, but the interest has intensified considerably as computer power has become increasingly available to students at all levels. Much has been written about the use for educational purposes of calculators, computers, spreadsheets, and information technology more generally (Campbell & Stewart, 1993; Corbitt, 1985; Fey, 1989; Kelman et al., 1983; Nickerson & Zodhiates, 1988; Perkins, Schwartz, West, & Wiske, 1995; Sutherland, 1993; Sutherland & Rojano, 1993). Examples of applications of technology in the classroom, and especially for the teaching of mathematics and science abound (Campbell & Stewart, 1993; Clements, 1980, 1999a; Feurzeig, 2006; Hembree & Dessart, 1986; Lehrer, Randle, & Sancilio, 1989; Psotka, Massey, & Mutter, 1988; Sarama, 2000, 2004; Schwartz, Yerushalmy, & Wilson, 1993; Shoenfeld, 1988; Sleeman & Brown, 1982; Spiker & Kurtz, 1987; Thompson, 1994b).

Pea (1987) makes the point that because technologies can change the thinking activities in which they play a part, their effects can be complex and indirect. One of the effects may be to change some of the goals of instruction by changing the importance that is attached to the acquisition of specific knowledge and skills. There is concern, for example, that the use of calculators to perform mathematical operations interferes with the acquisition of basic computational skills that students should have (Fey, 1989), but there are also studies indicating that their use generally has not been detrimental in the feared way (Hembree & Dessart, 1992). Who learns an algorithm to find square roots these days? Is it still important for children to learn multiplication tables? It does not follow from the ease with which they can do multiplication on a calculator that knowing the multiplication tables is unimportant, but some will contend that this skill is less essential now than it was before calculators became so widely available.

Wenger (1987) notes that the availability of software for microcomputers that is capable of doing symbolic algebra raises the question of whether the amount of manipulative practice in algebra should be reduced; if such manipulations can be performed by readily accessible technology, why should people learn to make them? Wenger argues that whether such a reduction should be made is not yet clear, and suggests that we should be asking what types of skill or knowledge are needed to support the use of powerful symbolic algebra environments with understanding.

The proviso "with understanding" is important. Doing algebra requires skill in representing relationships with a symbol system—setting up algebraic equations—and performing operations according to rules on those representations. There is considerable evidence that students often have difficulty with the first of these tasks. An illustration of this that has received considerable attention is the finding that, when asked to write an equation to represent the proposition "At a certain university, there are six times as many students as professors," many college students have written the equation $6S = P$, where S and P are understood to represent students and professors, respectively (Clement, 1982a; Clement & Kaput, 1979; Clement & Rosnick, 1980; Lochhead, 1980; Soloway, Lochhead, & Clement, 1982).

Software packages that are able to do complex analyses are unquestionably powerful tools in the hands of competent users; however, they also lend themselves to misapplication by users who do not understand enough about them or of the rationales for the tests they perform to preclude using them inappropriately. It is easy to find instances of mindless uses of statistical analysis packages. As an example, I offer the following statement, from a manuscript I once received as an editor of a psychological journal: "The mean proportions of targets correctly selected were 0.56 (sd = 0.14), 0.56 (sd = 0.14), and 0.56 (sd = 0.17). A one-way within-participants ANOVA showed that there was no significant difference between the three conditions, $F(2, 70) = 0.019$, $p > 0.1$."

Although not all educators favor extensive use of calculators and computers in the classroom, many believe that their use does little or no harm and that it can be an effective aid to learning (Campbell & Clements, 1990; Schultz, Colarusso, & Strawderman, 1989). Moreover, there is evidence that students can learn to understand the results of calculations even if they are unable to perform the calculations themselves (Huntley, Rasmussen, Villarubi, Sangtong, & Fey, 2000).

The authors of a National Research Council (1989) report refer to the use of calculators as essential to mathematics education, but acknowledge that there is the potential for abuse. The integration of their use with other problem-solving experiences has been stressed as a condition of effectiveness (Brown, 1990; Campbell & Stewart, 1993). Pea (1987) makes a distinction between using technology to "get an answer" and using it to facilitate understanding, and notes that technology can be used to promote purpose or process; in the former case, it engages students to think mathematically, while in the latter it serves as an aide or tool in problem solving.

The use of calculators and computers in instruction is perhaps the most obvious way in which technology has impacted, or may in the future impact, education, but it is not the only way. The study of artificial intelligence (AI) may impact education, perhaps less directly, by providing some insights into learning processes. AI programs are developed to do "intelligent" things sometimes for the simple practical reason of getting them done, perhaps to relieve people of the necessity of doing them. And sometimes such programs are developed in order to shed light on how people manage to do them. Rissland (1985) argues that independently of what one may think about AI programs, "they do give us a place to begin, even if only in reaction and criticism" (p. 175). She notes that while AI programs that can solve the same problems that humans can solve are not demonstrations of the necessity of the approaches the AI programs use, they do demonstrate their sufficiency. And they are a useful medium for testing ideas.

Attempts to write programs that allow computers to "understand" speech and natural language revealed much about the complexity of these abilities that could easily be missed in studies of people's uses of them. Just so, as has often been pointed out, people learned much more about aerodynamics by attempting to build machines that could fly than by centuries of observing birds in flight. One suspects that the same principle may apply to an understanding of what is involved in much of the doing of mathematics. In writing a program to simulate how a student might go about trying to prove a theorem in geometry, for example, one is forced to explicate assumptions or hypotheses about the process that might not be apparent from observation of students attempting to prove the theorem (Greeno, 1985). Several programs have been developed either to simulate how people solve mathematical problems or to provide existence proofs that specified techniques suffice—whether or not people typically use those techniques—to solve them (Anderson, 1982; Bobrow, 1968; Briars & Larkin, 1984; Newell & Simon, 1972; Riley, Greeno, & Heller, 1983).

☐ **Teachers of Mathematics**

> Unfortunately, little is known from existing high-quality research about what effective teachers do to generate greater gains in student learning. (National Mathematics Advisory Panel, 2008, p. xxi)

That the teacher is a critical factor in mathematics instruction seems too apparent to warrant mention. Nevertheless, Silver (1985a) identifies the role of the teacher as one of three central themes that have received far less attention in the research literature than they deserve. (The other two underrepresented themes that Silver mentions are sensitivity to individual differences and acquisition of problem-solving expertise.) Silver contends that although the research literature on the teaching of mathematics provides many suggestions regarding teaching practices, there is little in the way of evidence, pro or con, of their effectiveness. A. G. Thompson (1985) also argues that relatively few studies have been done of the influence of teachers' conceptions of their subject matter on their instructional work. Notable recent calls for more attention to approaches and procedures for teaching mathematics (and teaching more generally) include those of Bruner (1996) and Stigler and Hiebert (1999).

Thompson (1994a) points out that if all the mathematics a teacher knows is ritualized performance, then that is what he or she can be expected to teach. An enthusiast for the use of technology as a vehicle for teaching mathematics, and a developer of computer programs for this purpose, he rejects the idea that curricula of the sort he advocates are teacher independent and argues that they are more teacher dependent than are conventional curricula (P. W. Thompson, 1985).

The results of studies of the depth of the mathematical knowledge of teachers of mathematics in primary and secondary schools in the United States present a troubling picture. In many cases, teachers fail to give evidence of a clear understanding of fundamental concepts they are attempting to teach (Ball, 1990; Graeber, Tirosh, & Glover, 1989; Leinhardt & Smith, 1985; Ma, 1999; Mayberry, 1983; Post, Harel, Behr, & Lesh, 1991). The Mathematics Learning Study Committee of the National Research Council describes the preparation of U.S. preschool to middle school teachers as often falling far short of equipping them with the knowledge they need for helping students develop mathematical proficiency (Kilpatrick et al., 2001). "Many students in grades pre-K to 8 continue to be taught by teachers who may not have appropriate certification at that grade and who have at best a shaky grasp of mathematics" (p. 4).

Some teachers have difficulty, for example, explaining the rationale for such basic operations as dividing fractions (Askey, 1999). Woodward and colleagues note the almost total dependence of teachers of early mathematics on textbook material (Tyson & Woodward, 1989; Woodward & Elliot, 1990). Perry, Van der Stoep, and Yu (1993) compared teaching styles used in the teaching of addition and subtraction in Taiwan, Japan, and the United States and found that Asian teachers more than American teachers tend to ask questions intended to develop conceptual understanding of the processes being taught. Ma (1999) also found that U.S. teachers compared unfavorably with their Chinese counterparts. Hammond (1978) sums up the situation in the United States with the claim: "Until the university, mathematics is usually taught by people who do not really understand it themselves on more than a very superficial level" (p. 26).

Some writers contend that there is little or no evidence that the effectiveness of teaching of elementary mathematics depends on the level of mathematical competence of the teacher (Battista, 1986; Begle, 1979; Bush, 1989; Monk, 1994). This is not very reassuring. It is easy to imagine that a high level of competence in mathematics does not guarantee that one will be an effective teacher, there being much more to teaching than knowing a lot about one's subject. But it is hard to imagine that a teacher who has serious misunderstandings of basic mathematical concepts, or who is anxious about mathematics, could be very effective in much more than a drill and practice context; it is not reasonable to expect such a teacher to inspire a passion for the subject.

In contrast to the view that evidence of the importance of teachers' knowledge of mathematics as a determinant of their effectiveness is lacking, the National Mathematics Advisory Panel (2008) contends that research confirms the importance of teachers' knowledge. It acknowledges, however, the need for more precise measures of the relationship among the mathematical knowledge that elementary and middle school teachers have, their instructional skill, and what students learn.

The knowledge required to teach algorithmic procedures by rote is not necessarily adequate to provide explanations of why the procedures work (Borko et al., 1992; Putnam, Heaton, Prawat, & Remillard, 1992). Some studies suggest that, while there is little evidence that students' learning of elementary mathematics benefits from completion of advanced courses by their teachers, it suffers when teachers have not attained a certain threshold of mathematical competence (Begle, 1979; Monk, 1994).

In addition to teachers' knowledge of mathematics, other likely contributors to their effectiveness are their beliefs and attitudes toward mathematics and the teaching/learning of it. A teacher who believes that relatively few people are innately endowed with the potential to do well with mathematics is likely to treat students differently than is one who works on the assumption that, if properly taught, most students can do well, if not excel, at it. A teacher who attributes students' success or failure primarily to their effort or lack thereof is likely to get results that differ from those of one who looks first for possible ways to make his or her teaching more effective (Clark & Peterson, 1986). One who believes that boys have a greater ability to learn mathematics than do girls is likely to pay more attention to boys and to get greater achievement from them as a consequence (Fennema, 1990b; Good & Findley, 1985). Clearly there is a need for a better understanding of how teachers' knowledge of, and attitudes toward, mathematics affect learning in the classroom.

Mathematical Problem Solving

To learn mathematics is to learn mathematical problem solving. (P. W. Thompson, 1985, p. 190)

Educational researchers and mathematics educators have given increasing attention in recent years to mathematical problem solving (Charles & Silver, 1988; Greeno, 1978, 1983; Hembree, 1991; Hembree & Marsh, 1993; Mayer, 1989, 1992; Schoenfeld, 1983b, 1985b, 1992, 1994a; Silver, 1985b; Stengel, LeBlanc, Jacobson, & Lester, 1977). This is not to suggest that a problem-solving emphasis in the teaching of mathematics is a totally new idea. Notable among psychologists and mathematicians who urged such an emphasis many decades ago are Duncker (1945), Wertheimer (1945/1959), and Polya (1965). But motivation for this focus appears to have increased in recent years, perhaps at least partly in response to the results of assessments of mathematical achievement that have shown many students to be deficient in their ability to deal effectively with problem-solving aspects of mathematics, even when they have good computational skills (Carpenter, Lindquist, Matthews, & Silver, 1983; Kilpatrick, Swafford, & Findell, 2001; Kouba et al., 1988; Mullis, Dossey, Owen, & Phillips, 1991; National Council of Teachers of Mathematics, 1989, 2000). The renewed emphasis on problem solving was promoted in 1980 by the National Council of Teachers of Mathematics, which made a focus on problem solving in school mathematics the number one recommendation in its *Agenda for Action* and provided supportive material in its *1980 Yearbook* (Krulik, 1980).

Problem has many connotations in the literature on mathematical research and mathematical education. In its most inclusive usage, it refers to essentially any mathematical question that is posed. What is the sum of 2 and 3? What is the tangent of an angle of 30 degrees? What is the surface area of a sphere of radius 1? Examples like these, although commonly referred to as mathematical problems—I have used *problem* in this inclusive sense in preceding chapters—are generally not what researchers and educators have in mind when they argue that the teaching of mathematics should be more focused on problem solving. What they are likely to have in mind are problems the solving of which requires some creative or critical reasoning—something beyond the tapping of memory and the mechanical application of learned algorithms.

What is a *bona fide* problem to one person may not be one for another, or even for the same individual at a different time. Addition and subtraction tasks can constitute real problems for students who are beginning to learn the relevant rules and procedures of basic arithmetic (Carpenter & Moser, 1983; Riley, Greeno, & Heller, 1983), whereas they would not be problematic in the same sense for older students who had mastered this area of mathematics. What may constitute a very challenging problem to even an advanced student of higher mathematics may be an incidental exercise to an expert in the relevant area.

Early in the 20th century Morton (1927) made a distinction between *problems* and *examples*, or what might be called *exercises*. For the latter, procedures that would lead to solution were provided beforehand, whereas for the former they were not. In a meta-analysis of studies of mathematical problem solving, Hembree and Marsh (1993) partitioned the problems studied into two major types: "*standard* (word or story) problems requiring translation of verbal statements into mathematical operations; and *process* or open-search problems for which the solvers possess no routine procedures for finding an answer" (p. 152).

Several taxonomies have been proposed for mathematical problems. Threadgill-Sowder (1985) notes that problems are sometimes grossly categorized by such terms as "'routine,' 'non-routine,' 'word problems,' '*aha* problems,' and 'applied problems'" (p. 335), as well as more formal classification schemes, some of which make quite fine distinctions, as illustrated by a taxonomy that recognizes over 250 types of "missing value" problems of ratios and proportions (Harel & Behr, 1989; Harel, Behr, Post, & Lesh, 1992). (A missing value ratios or proportions task is one in which a missing element of one ratio has to be supplied in order to make the ratio match another in which both elements are given.) Considerable work on the acquisition of skill in proportional reasoning, mainly by middle schoolers, has been done by R. Karplus and colleagues (Karplus, Adi, &

Lawson, 1980; Karplus & Karplus, 1972; Karplus, Karplus, & Wollman, 1974; Karplus & Peterson, 1970; Karplus, Pulos, & Stage, 1983).

Problems dealing with addition and subtraction have been classified as a variety of types depending on how they are structured, or what they require one to do to solve them (Baroody & Standifer, 1993; Carpenter, 1985; Carpenter & Moser, 1984; Fuson, 1992b; Riley et al., 1983). Multiplication and division problems have also been partitioned into categories in different ways (Nesher, 1988; Usiskin & Bell, 1983; Vergnaud, 1983). For present purposes, the important point is the recognition that two problems that have the same name (e.g., addition) and that may appear to an adult to be the same type of problem, may differ in significant ways to the learning child, and evidence suggests that such differences can be important determinants of problem difficulty (Bell, Greer, Grimson, & Mangan, 1989; Kouba, 1989).

It is generally recognized that effective problem solving draws on a variety of different factors, including a store of organized knowledge about the problem domain, techniques for problem representation, and metacognitive processes to manage problem-solving performance (Kilparick, 1985). Other factors that are seen to be important to mathematical problem solving include attitudes, beliefs, emotion, and motivation (Renga & Dalla, 1993).

☐ Problem Comprehension

That students have difficulties learning to deal with word problems when they first encounter them, and often later as well, is well documented (Carpenter, Corbitt, Kepner, Lindquist, & Reys, 1980; Clement, 1982a; Clement, Lochhead, & Monk, 1981; Hinsley, Hayes, & Simon, 1977; Lochhead, 1980; Lochhead & Mestre, 1988; Mayer & Hegarty, 1996; Reusser, 1988; Clement & Rosnick, 1980; Wollman, 1983). Surely the first step in problem solving must be comprehension of the problem that one is to attempt to solve. Evidence abounds that students often fail to accomplish this step. This can be so of children in primary grades dealing with word problems involving simple arithmetic (Greeno, 1980; Riley et al., 1983) as well as of high school or college students working with problems of geometry, algebra, or higher math (Greeno, 1978, 1983).

Word problems differ considerably in difficulty, and what determines their difficulty has been a focus of research (De Corte & Verschaffel, 1987; De Corte, Verschaffel, & De Win, 1985; Jerman, 1973–1974, Jerman & Mirman, 1974; Jerman & Rees, 1972). A change in wording can make what was a difficult problem considerably easier

(Riley et al., 1983). The length and grammatical complexity of problem statements are prominent among the variables that contribute to difficulty, but examples of presumably simple word problems that give students trouble abound.

Mayer and Hegarty (1996) give as their reason for studying problem comprehension processes "growing evidence that most problem solvers have more difficulty in constructing a useful problem representation than in executing a problem solution" (p. 31). They note that there are numerous examples from their work (Hegarty, Mayer, & Green, 1992; Hegarty, Mayer, & Monk, 1995; Lewis & Mayer, 1987), as well as from that of others, that students often fail to give the correct answer to simple word problems even though they demonstrably are able to do the required computations correctly when given the appropriate formulas. The difficulty appears to be in getting from the verbal representation of a problem to a prescription of the needed calculations in the form of mathematical equations (although some children have been able to solve some word problems before they have had experience with equations at all [Carpenter, Hiebert, & Moser, 1981]).

Mayer and Hegarty (1996) contend that the main difference between successful and unsuccessful word problem solvers is the tendency of the former, but not of the latter, to try to ensure they understand the problem conceptually before trying to solve it mathematically. A symptom of the failure of students to understand word problems before trying to solve them is the frequently reported tendency to perform immediately an operation (addition, division, etc.) that is cued by certain key words in the problem statement—"in all" or "more" may trigger addition; "left" or "less," subtraction; "each," division; and so on (Lester, 1985; Linville, 1976; Nesher & Teubal, 1974; Sowder, 1988). Citing Briars and Larkin (1984), Mayer and Hegarty contend that a review of mathematics textbooks reveals that most of the word problems in them can be solved by the latter approach and do not require a conceptual understanding of the situation described. In the results of their own program of research on mathematical understanding, Mayer and Hegarty see "converging evidence that students often emerge from K–12 mathematics education with adequate problem execution skills—that is, the ability to accurately carry out arithmetic and algebraic procedures—but inadequate problem representation skills—that is the ability to understand the meaning of word problems" (p. 50). Bransford, Zech, Schwartz, Barron, and Vye (1996) support this idea with the observation that in their studies of fifth and sixth graders who were having difficulties, especially in reading and mathematics, nearly everyone with whom they worked took a mechanical approach to word problems rather than attempting to understand the problems.

Comprehension failure can have many causes, including lack of attention, interest, or motivation, but among the more serious of them is lack of adequate reading ability (Muth, 1984). That many students in U.S. schools lack the ability to read comprehendingly at their appropriate grade level has been documented in numerous national assessments and widely publicized in such reports as *A Nation at Risk* (National Commission on Excellence in Education, 1983). Undoubtedly, poor reading ability can be a severe impediment to the acquisition of competence in mathematical problem solving.

A particularly thought-provoking finding is that of Carpenter and Moser (1983), who concluded from a longitudinal study that students who had just completed third grade were more inclined to try to understand problems before trying to do any calculations than were students who had just completed fifth grade. After a few years of formal instruction in mathematics, students were more likely to attempt to solve problems by attending only to their surface features (Carpenter et al., 1980). Carpenter (1985) sums up the findings this way: "After several years of mathematics, children abandon a reasonably good general problem-solving approach for mechanical application of arithmetic skills.... From the beginning they are learning that mathematics is just an exercise in symbol manipulation and is not related to real problem solving. This suggests that initial instruction in addition and subtraction may be a critical point in developing problem-solving skills and that children's later deficiencies may be traced to this point in the curriculum" (p. 37).

One plausible explanation for this unwelcome finding is the amount of emphasis that is put on computational aspects of mathematics—as distinct from problem comprehension—during early school instruction. Kouba and Franklin (1993) contend that current practices in mathematics education reward computations and answers more than exploring, analyzing, and interpreting, and ask what incentive children have to engage in critical thinking, that being the case. Feinberg (1988) advocates that word problems be part of the teaching of arithmetic from the very beginning. Whether this would prove to be motivating is an open question.

A compelling demonstration that problem statements are sometimes processed only superficially and not understood was produced by Paige and Simon (1966), who found that some students attempted to solve problems that were impossible to solve, and that should have been recognized to be so with comprehending reading of the problem statements. Susan Butler and I got a similar result in a set of tests of students' ability to select data in an efficient way. College students had to say which of several sets of data they would want to check in order to determine the truth or falsity of statements regarding the composition of the 104th U.S. Senate, which had 47 Democrats and 53 Republicans, 8 women and 92 men. They were to indicate which of the following items of information they would need to check (sex of Democrats, sex of Republicans, party affiliation of females, party affiliation of males) in order to determine, most efficiently, the truth or falsity of claims of the sort "All the females were Democrats." Among the assertions was one—"All the males were Republicans"—which could not possibly be true, given the numbers, so no checking of the data was necessary. Only about 25% of the students noticed the mathematical impossibility of the truth of the assertion; most of them requested some data.

Mayer (1982) conceptualizes the solving of word problems as a two-step process, the first step, translation, being the process of getting from the problem statement to a mathematical representation—an equation or set of equations—that can be manipulated. The second step is the application of appropriate mathematical rules to solve the equations. (Of course, not all math problems require equations for solution; here we are concerned with those that do.) This view captures the idea, promoted vigorously by many writers, that there is important work to be done in mathematical problem solving before one gets to the point of actually manipulating equations, and if that work is not done effectively, the equations one manipulates are unlikely to lead to a correct solution. Unfortunately, it appears that many would-be mathematical problem solvers give insufficient attention to the first step.

Lesh (1985) makes a distinction between the tendency of some people, when trying to solve problems, to focus immediately on procedures ("What shall I *do* next?") and that of others to focus on the question of how to think about the problem. Students in his studies who were preoccupied with the question of what to do typically did not do as well as those who concentrated first on understanding the situation conceptually. Numerous studies of expertise, not only in mathematics but other domains as well, have shown that a characteristic difference between experts and nonexperts in problem solving is that the former pay more attention than the latter to ensuring a good qualitative understanding of a problem before attempting to derive or deduce a solution for it.

Sometimes a problem can appear to have a simple straightforward solution and that solution turns out to be wrong. Consider, for example, the following problem:

> Two towns, A and B, are 200 miles apart. Tom and Harry leave Town A at the same time and drive to Town B. Tom drives at an average speed of 40 mph for the first 100 miles and at an average speed of 60 mph for the

second 100 miles. Harry drives at an average speed of 50 mph for the entire trip. Which driver arrives at Town B first?

One might be tempted to reason that because Tom went half of the distance at 40 mph and half at 60 mph, his average speed was 50 mph, the same as Harry's, so the two cars arrived at Town B at the same time. In fact, however, Harry made the trip in 4 hours, whereas it took Tom 4.167 hours (4 hours and 10 minutes). Traveling at 40 mph for 100 miles, it took Tom 100/40 = 2.5 hours to get to the halfway point, and traveling at 60 mph, it took him 100/60 = 1.667 hours to go the rest of the way. What is wrong with the reasoning that led to the conclusion that both cars took the same amount of time? The false step was in taking Tom's average speed to be 50 mph; in fact it was 48 mph (the harmonic mean of 40 and 60). The key to seeing the situation accurately is the realization that Tom traveled at the slower speed for a longer time than he traveled at the higher one. (See Falk, 1993, p. 15, for an analogous problem with the up-the-mountain and down-the-mountain round-trip of a skier.)

The situation may be made more intuitively compelling by consideration of a more extreme case. Suppose that Tom were to drive the first 100 miles at 1 mph and the second 100 miles at 99 mph. The average of the two speeds would again be 50 mph, but clearly, it would take much longer than four hours to make the trip. (In fact it would take 100 + 100/99 = 101.01 hours.)

But suppose the original problem were modified just a bit:

Two towns, A and B, are 200 miles apart. Tom and Harry leave Town A at the same time and drive to Town B. Tom drives at an average speed of 40 mph for two hours and at an average speed of 60 mph thereafter. Harry drives at an average speed of 50 mph for the entire trip. Which driver arrives at Town B first?

In this case, the answer is that Tom and Harry will arrive at the same time. Both will take four hours to make the trip. Tom will cover 80 miles in the first two hours and 120 miles in the second two. Harry, of course, will cover 50 miles in each of the four hours.

Now suppose the problem were stated as follows:

Two towns, A and B, are 200 miles apart. Tom and Harry leave Town A at the same time and drive to Town B. Tom drives at an average speed of 40 mph for the first half of the trip and at an average speed of 60 mph for the second half. Harry drives at an average speed of 50 mph for the entire trip. Which driver arrives at Town B first?

This statement of the problem is ambiguous because "half of the trip" can be interpreted to mean half of the distance (100 miles) or half of the time (2 hours). Only those who take the time to comprehend the problem before attempting to solve it are likely to notice this ambiguity and call for clarification.

Many of the "trick" problems that one finds in recreational math books are difficult primarily because they offer the reader a solution that appears to be obvious, but upon careful reflection would prove not to be correct. The following problem, which appears in a variety of versions in different books, illustrates the point. This version is a paraphrase of one in Friend (1954, p. 71). A merchant has two baskets of apples for sale. One basket contains 300 apples and the sale price is 5 cents each. The other basket has 300 lower-quality apples for sale at 3 for 10 cents. The merchant is offered $24.00 for the lot by a buyer who calculates that inasmuch as the apples in one basket are 2 for 10c and those in the other are 3 for 10c, the average is 5 for 20c, or 4c apiece, and $600 \times 4c = \$24.00$. The merchant refuses the offer on the grounds that it is not fair. Does the merchant have a case?

The would-be buyer's offer seems fair, if one does not think carefully about it. But from the merchant's point of view, if she sells 300 apples at 5c each and 300 at 3 for 10c, she will take in $300 \times 5c = \$15.00$ for the first basket and $(300/3) \times 10c = \$10.00$ for the second, or $25.00 in all. So she does well to refuse the offer. But what is wrong with the argument that the would-be buyer made to justify his offer of $24.00? I leave it to the reader to find the missing dollar.

Problems involving the computation of a mean can easily yield incorrect answers. A common misstep is to compute the wrong type of mean, usually an arithmetic mean when a different type—geometric, harmonic, weighted—is appropriate. Failure to make these distinctions can produce unjustified conclusions about relationships between such variables as the average number of children per family and the average number of siblings per child, or the average number of pupils per class and the average number of classmates per pupil (Jenkins & Tuten, 1992). Several enlightening examples of possible confusions involving means are provided by Falk (1993; see also Falk & Lann, 2008; Lann & Falk, 2006).

☐ Problem-Solving Heuristics

The term *heuristics* or *heuristic strategies* is generally used to connote ways of facilitating problem solving that are applicable in a broad range of contexts, which is to say that they are relatively domain independent.

Such strategies include decomposing the problem into a set of simpler component problems, making a diagram, considering an analogous problem that is simpler or for which one knows the solution, and considering extreme cases. Illustrations of the application of these and other heuristics in various problem contexts can be found in numerous publications (Adams, 1974; Bransford & Stein, 1984; Halpern, 1989; Hayes, 1989; Lewis, 1989; Polya, 1957; Whimbey & Lochhead, 1982; Wickelgren, 1974).

The teaching of such domain-independent heuristics has been promoted by some researchers and educators and discouraged by others (Nickerson, 1994). Evidence regarding the effectiveness of teaching them—in terms of demonstratively enhanced problem solving—has been spotty. While some investigators have reported success, others have not. Many have argued that general-purpose heuristics are unlikely to enable effective problem solving in a specific domain (physics, geometry, medicine) in the absence of substantive knowledge particular to that domain (Charles & Lester, 1982; Elstein, Shulman, & Sprafka, 1978; Glaser, 1984; McPeck, 1981).

Schoenfeld (1979) has shown that the teaching of problem-solving heuristics that are specifically intended for application in mathematics can be effective in improving performance of mathematical problem solving. Unlike novice mathematical problem solvers, experts give evidence of a strategic attitude by the relatively large amount of time they spend analyzing and representing the problem situation and considering possible strategies for attacking it before attempting computational solutions (Schoenfeld, 1987b). Lesh (1985) reports having found, in his work on mathematical problem solving, that successful problem solvers tend to use content-dependent processes rather than general (and generally weaker) content-independent processes. Examples of strategies that are specific to attempts to solve mathematical equations include simplifying mathematical expressions by collecting like terms and replacing "nasty" terms with "nicer" ones (Bundy, 1975). Illustrations of both are given by Schoenfeld (1987a, p. 15). Studies of people solving algebra word problems have shown that people who are very good at it are much more likely than those who are less good at it to use the same strategies on problems that are superficially different but similar at a relatively deep level (Novick, 1988; Novick & Holyoak, 1991).

It seems reasonably clear from the results of research in the aggregate that different strategies can be effective in mathematical problem solving at all levels of difficulty or complexity, from elementary arithmetic to advanced mathematics. However, perhaps more important than familiarity with specific strategies that can be employed is the habit of approaching problems with a strategic attitude, which means attempting to understand the nature of specific problems at a qualitative level before beginning to seek quantitative solutions.

☐ Mathematical Knowledge

Among the earlier writers to stress the importance of basic subject matter knowledge for problem solving, and to argue the inadequacy of domain-independent methods by themselves, were Bloom and Broder (1950). But having the knowledge that is sufficient to solve a mathematical problem does not guarantee that one will solve the problem. There is, as Whitehead (1929) pointed out, such a thing as inert knowledge—knowledge that one has but does not use when appropriate. Schoenfeld (1985a) makes the same distinction in arguing that there are two major components of expertise: "(1) the mastery of facts and procedures that allows the expert to dispatch routine problems quickly and accurately, and (2) the ability to use the knowledge at one's disposal to solve difficult nonstandard problems when others with at least as much specific knowledge fail to do so" (p. 185).

A distinction that has important implications for the teaching of mathematics, even at the levels of counting and elementary arithmetic, is that between *conceptual* and *procedural* knowledge (Baroody & Ginsberg, 1986; Greeno, Riley, & Gelman, 1984). In a review of work on the role of these two kinds of knowledge in the teaching or learning of mathematics, Rittle-Johnson and Siegler (1998) define conceptual knowledge as "understanding the principles that govern the domain" and procedural knowledge as "action sequences for solving problems" (p. 77). The latter type of knowledge is often referred to in the literature as *skills* or *algorithms*. Both types of knowledge are seen to be essential to a more than superficial grasp of even elementary mathematics. The distinction is buttressed by studies revealing instances of children being able to perform mathematical calculations effectively without understanding the reasons for the operations they are performing (Bryant, 1995; Byrnes & Wasik, 1991) and instances of children showing conceptual understanding unaccompanied by the ability to calculate effectively (Dowker, 1995; Jordan & Montani, 1997).

The question of how conceptual and procedural knowledge is acquired—in particular whether one type is acquired before the other, and if so, which type is acquired first—has been the focus of some research. Rittle-Johnson and Siegler (1998) note that researchers are not all agreed on the answer, even in the case of counting. They point to some work suggesting that young children understand principles of counting, such as those identified by Gelman and Gallistel (1978), when they begin to count, and to other work suggesting that children learn to apply counting procedures before they understand the principles that underlie them. Their own conclusion, based on their review, is that procedural knowledge precedes understanding of the underlying concepts. They argue that most children acquire some conceptual understanding of multidigit numbers before acquiring correct procedures for adding and subtracting them, but note that, as shown by results obtained by Hiebert and Wearne (1996), a significant minority do not. The possibility of learning to use a procedure effectively before acquiring an understanding of why it works is not limited to children learning elementary mathematics; there are many examples in history of mathematicians using a procedure effectively before anyone understood why the procedure worked, the infinitesimal calculus being a notable case in point.

Sometimes children appear to apply specific knowledge to one task to which it is relevant while failing to apply it to another to which it is also relevant (Trabasso et al., 1978). Sometimes they fail to apply in school contexts mathematical knowledge they have acquired outside of school (Ginsburg, 1989; Ginsburg, Posner, & Russell, 1981c). Similarly, students frequently fail to use outside of the classroom the mathematical procedures and methods they have learned in school that would be relevant to the problems they are trying to solve (Hiebert, 1989). One explanation of the latter failure is that the learning did not result in an understanding of the real-world usefulness of the procedures and methods involved.

Knowledge can contribute to problem solving in the form of memory for past encounters with the same problem or similar ones. People for whom mathematical problem solving is an important part of their work (engineers, economists, teachers of mathematics) are likely to see the same or similar problems many times, and therefore to be able to retrieve from memory the approaches they took in solving them, and perhaps even the solutions. That people often draw on knowledge of past encounters with similar problems has been demonstrated with problem-solving tasks in the laboratory (Ross, 1987) and in simulations of practical real-world problems (Klein, 1998; Klein, Orasanu, Calderwood, & Zsambok, 1993). The strategy of applying solutions that one remembers having used advantageously in the past is often effective, but it also can lead one down a garden path if the similarity between the current and past problems is a "surface" similarity (as distinct from a "structural" similarity) only (Dixon, 2005; Novick, 1988; Ross, 1989).

Some writers have argued that the word problems that appear in mathematics textbooks can be classified into a modest number of different types—distance-rate-time, river current, triangle—that often the type of a given problem can be identified by the first few words that describe it, and that students sometimes learn to deal with problems on the basis of the types they identify (Hayes, Waterman, & Robinson, 1977; Hinsley et al., 1977; Mayer, 1982; Robinson & Hayes, 1978). This can be a useful strategy—learning an approach that is effective for a whole class of problems is a generally efficient way to proceed—but the effectiveness depends on the accuracy of type recognition and can produce errors when problems are misclassified or procedural schemata are inappropriately applied (Davis, 1982; Dixon, 2005; Matz, 1982).

A common source of errors in mathematical problem solving is the application of principles in contexts in which they are not appropriate. Some of the types of errors noted in the preceding chapter illustrate the point. According to the commutative rule of addition and multiplication, $4 + 2 = 2 + 4$ and $4 \times 2 = 2 \times 4$. Application of the rule to subtraction and division, $4 - 2 = 2 - 4$ and $4 \div 2 = 2 \div 4$, obviously does not work. Principles that are appropriate to finite mathematics are easily applied inappropriately to mathematics involving infinity; similarly, principles that pertain to deterministic relationships are easily misapplied to relationships that are probabilistic.

☐ Metacognition

Most of the research that has been done on mathematical problem solving has focused on discrete skills and procedures and comparatively little attention has been given to managerial aspects and to the total environment in which problem solving takes place. Lester (1985), who makes this point, sees this neglect of metacognitive aspects as responsible for the limited success of efforts to teach problem solving. "The failure of most efforts to improve students' problem-solving performance may be due in large part to the fact that instruction has overemphasized the development of heuristic skills and has virtually ignored the managerial skills necessary to regulate one's activity" (p. 62).

Schoenfeld (1983a, 1983b, 1985c, 1987b) stresses the importance of metacognition to competent mathematical problem solving and has proposed approaches to the teaching of such metacognitive skills as control and management of one's problem-solving thought processes. He (1987b) distinguishes three foci of research on metacognition: (1) knowledge about one's own thought processes, (2) control or self-regulation, and (3) one's beliefs and intuitions about mathematics. Greeno (1987) speaks of the importance of "strategic knowledge" to problem solving—"the knowledge that organizes problem-solving activity, including the processes that set goals and choose plans" (p. 81).

Other calls for more attention to metacognitive issues in the teaching of mathematics have come from Lester and Garofalo (1982) and Silver (1985a, 1987). Silver notes that, although processes such as planning, monitoring, and evaluation, which are generally considered aspects of metacognition, are acknowledged to be important components of mathematical problem solving, they are seldom made the explicit focus of mathematical instruction. This, he contends, is a mistake.

The importance of metacognitive skills even in doing simple arithmetic is illustrated by an example, provided by Davis and McKnight (1980), of a third-grade student who, when subtracting 28 from 7,002, got 5,084 as the answer. Apparently this student applied an algorithm for borrowing inappropriately, and failed to notice that when a small number is subtracted from a much larger one, the answer should be close in magnitude to the larger number. The point is that it is easy to get nonsensical answers even to relatively easy problems when one applies a learned procedure mechanically and fails to consider whether the result makes sense. Multiplication and division of decimal numbers often produce erroneous answers that could be detected readily by a consideration of the plausibility of the answer; it should be apparent, for example, that $6.23 \times 3.84 = 2,392.32$ cannot be correct.

The occurrence of errors of the sorts illustrated suggests the importance of calling students' attention to a variety of principles that might be expected to be discovered in the process of doing mathematics: that the sum of two positive numbers is always larger than either of the addends; that the product of two integral numbers has as many digits as the sum of the digits in the numbers being multiplied or that number minus 1; that the product of two fractions, each less than 1, is always smaller than either of the fractions being multiplied; that the integral part of the square root of an integer has approximately half the integer's number of digits; and so on. These are all principles that should be apparent to anyone who has done very much basic arithmetic, but it is not clear that students pick them up spontaneously, or that they use them to check the plausibility of the results of computations. They should be encouraged explicitly to do so. A very useful habit that should be cultivated is that of estimating the solutions to math problems when that is feasible, and using those estimates to determine whether the results of calculations are in the ballpark.

☐ Beliefs, Expectations, and Affect

> Limiting one's research perspective to the purely cognitive seems accept-
> able for those interested mainly in the performance of machines; however,
> researchers who are interested in human performance need to go beyond
> the purely cognitive if their theories and investigations are to be important
> for problem solving in mathematics classrooms. (McLeod, 1985, p. 269)

Beliefs can have important effects—both beneficial and detrimental—on cognitive performance generally (Baron, 1985b, 1991; Bloom
& Broder, 1950; D'Andrade, 1981; Schoenfeld, 1987b; Shaughnessy,
1985). Beliefs about intelligence—whether it is flexible and enhance-
able through learning or genetically determined and unchanging—can
influence the way students approach, or avoid, challenging cognitive
work (Dweck, 1975; Dweck & Bempechat, 1983; Dweck & Elliot, 1983;
Heyman & Dweck, 1998). A fatalistic belief that one lacks the ability to
do mathematics can be debilitating, especially if reinforced by early failures on mathematical tasks (Haladyna, Shaughnesssy, & Shaughnessy,
1983).

Beliefs about the usefulness, or uselessness, of what is being studied can affect performance (Lampert, 1986; Schoenfeld, 1983a), as can
beliefs about the causes of success and failure on cognitively demanding tasks—especially whether self or external factors, such as chance
(Andrews & Debus, 1978; Deci & Ryan, 1985; Reid, 1987; Rowe, 1983;
Weiner, 1979). Beliefs about the nature of mathematics seem likely to
be important determinants of one's attitude toward, or interest in, the
subject. One who sees it as nothing but number crunching is less likely
to get excited about it than is one who sees it as a means of exploring an
unlimited universe of intrinsically fascinating relationships. Students'
beliefs about the nature of mathematics can be shaped in part by the
types of exercises they are given in mathematics classes (Schoenfeld,
1989). The belief that every solvable mathematical problem can be solved
in one and only one way can impede the development of mathematical problem-solving expertise; happily students can be dissuaded from
this belief and their problem solving improved if they are required to
solve problems in different ways (Sweller, Mawer, & Ward, 1983). Other
beliefs that can impede the learning of mathematics include the beliefs
that learning mathematics is largely a matter of rote memorization and
that if a problem can be solved, it can be solved in a relatively few minutes (Schoenfeld, 1985c). The belief that the mathematics one learns in
school has little relevance to life outside the classroom can be demotivat-
ing (Saxe, 1991).

Some of the attitudes and beliefs that students hold about mathematics presumably are acquired from experiences in school. Failure to identify with mathematics, or to see its relevance to their personal lives, could stem from an exclusive focus on problems invented by, and perhaps of interest only to, others. As Pea (1987) puts it, "The student rarely sees significance in the learning; someone else has made all the decisions about scope and sequence, about the lesson for the day. The learning is meant to deal not with the student's problem or a problematic situation the teacher has helped highlight, but with someone else's. And the knowledge used to solve the problem is someone else's as well, something that someone else might have found useful at some other time" (p. 101). Pea contends that students seldom are informed of the past utility of problems with which they are presented.

Teachers' beliefs are at least as important as students' beliefs as determinants of what goes on in the classroom and of the effectiveness of the learning experience. A teacher who sees mathematics as a static body of knowledge is likely to have a different type of class than one who sees it as a dynamic discipline in which discoveries are constantly being made. One who sees students as vessels into which knowledge is to be poured will treat them differently than will one who sees them as thinking beings trying to make sense of their experiences in and out of the mathematics classroom.

Separating the effects of motivation from those of differences in ability can be difficult, but when differences in ability are accounted for, a large proportion of the remaining individual differences in performance may be motivation based (Dweck & Elliot, 1983; Elliot & Dweck, 1988; McLeod, 1985; Sternberg, 1983). How much effort one is willing to make to perform well is determined both by motivation and by expectations: If one is highly motivated to do well and expects to be able to do so, one is much more likely to try hard than if one has little interest in doing well or expects not to be able to do so even if one tries (Bandura, 1977, 1986, 1991, 1994, 1997). Moreover, an expectation of doing poorly may lower motivation (Pyszczynski & Greenberg, 1983), thereby more or less guaranteeing poor performance.

That expectations can affect performance in school has been known for a long time. Among the more famous demonstrations of this is the *Pygmalion effect* described by Rosenthal and Jacobsen (1968/1992), who showed that when teachers were led to expect superior performance from some students, that is what they obtained from those students. A phenomenon that has attracted considerable attention from educational researchers recently is that of *stereotype threat*, which is closely related to the Pygmalion effect, but with the opposite polarity. The phenomenon in this case is underperformance of a member of a stereotyped group when reminded that members of that group are expected to perform poorly.

Negative effects of stereotype threat have been observed with members of various stigmatized groups (Aronson et al., 1999; Steel & Aronson, 1995; Stone, Lynch, Sjomeling, & Darley, 1999). Its role in the performance of mathematical tasks has been reviewed by Ben-Zeev, Duncan, and Forbes (2005). Although attention has been focused primarily on negative effects of stereotype threat, a complementary (positive) effect can be obtained when a member of a group that, according to its stereotype, is expected to do well is reminded of that affiliation—as, for example, when the mathematical performance of males is enhanced by being told that they are expected to do better than females on the task (Spencer, Steele, & Quinn, 1999). In a study by Shih, Pittinsky, and Ambady (1999), Asian women performed better on a math test when reminded of their positive stereotype (Asians are good at math) and poorer when reminded of their negative one (women are poor at math). A closely related finding is that people who are not members of a stigmatized group may get a *stereotype lift*—improvement in performance—by being reminded of the stereotype to which they do not conform (Walton & Cohen, 2003).

McLeod (1985), who notes that affective issues arise whenever research on the teaching of problem solving focuses on students in mathematics classes, includes as matters of interest the emotional reactions of students to problem solving—anxiety or enjoyment—and their willingness to engage in this activity. Lester (1980) includes students' confidence and their willingness to take risks. Negative affect regarding mathematics is sometimes identified as *mathematics anxiety,* which manifests itself as apprehensive avoidance of the need to engage in mathematical activities or that interferes with the doing of mathematics (Ramirez & Dockweiler, 1987; Richardson & Suinn, 1972), including avoidance of math as a college major or of pursuit of careers in the physical sciences or engineering (Chipman, Krantz, & Silver, 1992, 1995). Math anxiety can vary widely in degree, from mild apprehension or dislike to genuine fear or dread.

Although some measure of anxiety is a common reaction to the need to cope with any cognitively demanding task, anxiety with respect to mathematics is seen by some researchers to be distinct (Ashcraft & Faust, 1994; Ashcraft, Kirk, & Hopko, 1998), and tests designed explicitly to measure it have been developed, such as the Mathematics Anxiety Rating Scale (Richardson & Suinn, 1972) and an abbreviated version of the same (Alexander & Martray, 1989). "Learned helplessness" (Seligman, 1975; Seligman & Maier, 1967) has been noted as a source of poor performance in mathematics as in other types of cognitive tasks (Gentile & Monaco, 1986; Parsons, Meece, Adler, & Kaczala, 1982).

Identifying the origins of math anxiety is obviously important for educational purposes. Especially is this so if, as has been claimed, a major source of it is early unhappy experiences in math classes (Stodolsky, 1985). Ashcraft, Kirk, and Hopko (1998) point out that very little is known about the origins of math anxiety inasmuch as very few studies have focused on its occurrence with children below about the sixth grade. They note too that the relationship between math anxiety and aptitude for math is not well understood. They argue the possibility that the aptitude of people with math anxiety may be underestimated because of the adverse effect of the anxiety on test-taking performance. They hypothesize that the effect occurs because the anxious person is essentially performing a dual task, working on the test while trying to control intrusive thoughts prompted by the anxiety, thus overloading working memory.

A distinction that is commonly found in the literature on cognitive styles is that between impulsiveness and reflectiveness as characteristic ways of approaching problems (Kagan, 1966; Kagan, Rosman, Day, Albert, & Philips, 1964; Sternberg & Zhang, 2001). Lesh (1985) sees these characteristics as controllable functions of other variables rather than as invariant properties of individuals. This view gets support from reports of successful attempts to improve children's performance of cognitive tasks by training them to approach the tasks more attentively and reflectively (Parish & Ericksen, 1981).

☐ Individual Differences

As noted in Chapter 15, Silver (1985a) identifies sensitivity to individual differences as one of three central themes that have received far less attention in the research literature than they deserve. I take his point to be that too little attention has been paid to the question of how the teaching of mathematics can best take individual differences in the abilities (perhaps also interests, motivation, etc.) of students into account. One might wonder too if enough attention has been given to the question of how best to account for the individual differences in the acquisition of mathematical knowledge and problem-solving competence that clearly exist. What is the nature of the causes of these differences? How much is genetic, how much due to differences in exposure and opportunity to learn? How much can be attributed to differences in interest, motivation, teaching techniques, family and cultural values and expectations?

Geary (1994) suggests, tentatively, that individual differences in mathematical reasoning ability are probably related to differences in four sets of cognitive skills: the ability to set up representations of math word problems, the ease with which one develops problem-solving schemas, working memory capacity, and the speed with which basic mathematical operations can be executed. All of these abilities are presumably highly correlated with general intelligence, and whether anything predicts mathematical performance better than scores on general intelligence tests is questionable (Aiken, 1973). As to the extent to which differences in numerical facility and mathematical reasoning are genetically determined or acquired, Geary, citing Vandenberg's (1966) review of twin studies, among other sources, estimates that about half of the variability in numerical facility (basic arithmetic) is due to genetic differences. He estimates genetically based variability in mathematical reasoning ability—which is less well understood—to be probably less than that.

Many researchers consider limited working memory capacity to be a major source of individual differences in learning mathematics (Fletcher, 1985; Geary, 1993; Siegel & Linder, 1984; Wilson & Swanson, 2001), perhaps especially in doing mental arithmetic (Adams & Hitch, 1997, 1998; Ashcraft, 1995; Hitch, 1978a, 1978b). LeFevre, DeStefano, Coleman, and Shanahan (2005) contend that, despite this, relatively little research on the role of working memory in mathematical cognition has been done. Some studies have shown a correlation between working memory capacity and mathematical ability in children (Adams & Hitch, 1997; Gathercole & Pickering, 2000), but others have found little evidence of general working memory deficit in children classified as dyscalculic (Landerl, Bevan, & Butterworth, 2004; McLean & Hitch, 1999; Temple & Sherwood, 2002). Following a review of work on the topic, LeFevre et al. (2005) conclude that the evidence supports the importance of short-term and working memory in constraining mathematical performance. They note too evidentiary reasons to believe that the need for working memory capacity may decrease as children acquire, and retain in long-term memory, factual knowledge upon which they can draw in order to perform mathematical operations that, in the absence of that knowledge, are likely to be performed with counting strategies that require working memory capacity.

An aspect of individual differences that has received considerable attention is that of differences in how males and females relate to mathematics (Chipman, Brush, & Wilson, 1985; Fox, Fennema, & Sherman, 1977; Fox, Brody, & Tobin, 1980; Gallagher & Kaufman, 2005; Halpern, 2000; Henrion, 1997; Oakes, 1990; Shoemaker, 1977). The vast majority of notable mathematicians of the past—those who have been responsible for major advances in the field—have been males. There have been notable exceptions—Theano (student and wife of Pythagoras), Hypatia of Alexandria, Maria Agnesi, Sophie Germain, Sofia (aka Sonya) Kovalevskaya, Emmy

Noether, Gertrude Cox, and Florence Nightingale David—but they represent a small fraction of the mathematical pantheon. And they are not prominent in histories of mathematics; only two of those just mentioned—Kovalevskaya and Noether—appear in the index of mathematicians in *The VNR Concise Encyclopedia of Mathematics* (Gellert, Küstner, Hellwich, & Kästner (1977). Indexes of biographies (http://www-groups.dcs.st-and. ac.uk/~history/BiogIndex.html) list approximately 2,500 mathematicians, ancient and modern, only about 100 of whom are females. This is roughly the same female-to-male ratio that is found in *Remarkable Mathematicians* (James, 2002), where 3 of the 60 biographies are of women.

In the past, women with the ability to do creative mathematics typically could exercise this ability only by overcoming very considerable obstacles—chauvinistic biases and social taboos. Some who did manage to do this took male pseudonyms and kept their identities secret. This meant being deprived of the type of intellectual stimulation that most mathematicians got from interactions with colleagues at conferences and other gatherings. Sophie Germain, who learned mathematics on her own against strong opposition from her family, won a prize from the French Academy for work done under the pseudonym Antoine LeBlanc, which she used until her true identity came to light as a consequence of a correspondence with Gauss. Upon learning that Antoine was really Sophie, Gauss was gracious and extravagant in his praise of her accomplishments despite the biases of the times that worked against her.

If females are born with no less potential for mathematical competence than are males—and I am unaware of compelling evidence that they are not—the advance of mathematical knowledge has been severely hampered by the failure of society to facilitate the development of the potential of half the members of the species. This does not distinguish mathematics from philosophy, most sciences, the fine arts, and numerous other fields; consider the relative sparseness of women among the notable composers, painters, and sculptors of the past. Even in literature, where women are more in evidence, female authors have often found it advantageous to take male pseudonyms as a means of avoiding gender bias. The challenge to research is to determine the factors that have collectively produced gender disparity in fields in which it is prominent.

For many years the lack of women in mathematics was widely attributed to what were presumed to be innate neurological or psychological differences between men and women: Women had different temperaments, less capacity for logical thinking, less tolerance for focused effort. Most such "explanations" have been thoroughly discredited, but the question of whether there are any innate differences that would help explain the relative lack of participation by women in mathematics remains incompletely resolved. An extensive review of research on factors that have been hypothesized to be determinants of females' participation in mathematics has been provided by Chipman and Wilson (1985).

One hypothesis that has had some advocates in relatively recent times is that males and females differ in their spatial, or visualization, skills (Connor & Serbin, 1985; Fruchter, 1954; Nuttall, Casey, & Pezaris, 2005; Robinson, Abbott, Beringer, & Busse, 1996; Royer & Garofoli, 2005; Sherman, 1967; Smith, 1964), the difference possibly being mediated to some extent by sex-related hormonal differences (Casey, Pezaris, & Nuttall, 1992; Hines, 1982; Nyborg, 1983) or differences in brain organization (Annett, 1985, 1995; Nuttall et al., 2005). If visual-spatial ability is either essential or even especially helpful to the learning of mathematics, individual differences in this ability should translate to individual differences in mathematical achievement. And if, as hypothesized by some, males have better spatial or visualization (right-brain) skills than females, while females exceed males in verbal (left-brain) abilities, this could perhaps be a basis of some advantage for males in at least those areas of mathematics (geometry, trigonometry, topology) that involve visualization to a significant degree.

There are two questions to consider. Is visual-spatial ability especially important to the learning and doing of mathematics? Do males and females differ with respect to this type of ability? Whether visual-spatial ability is an asset in acquiring competence in mathematics is a contested issue. Chipman (2005) notes that several reviewers of early literature on the relationship between such ability and performance in mathematics (Fruchter, 1954; Very, 1967; Werdelin, 1961) concluded that such a relationship had not been shown (see also Tartre, 1990). Nevertheless, several researchers have promoted the ability to visualize and manipulate (e.g., rotate) objects mentally as an important factor in the development of mathematical competence (Burnett, Lane, & Dratt, 1979; Casey, Nuttal, & Pezaris, 1997, 2001; Casey, Nuttal, Pezaris, & Benbow, 1995; Smith, 1964) and a contributor to performance on standardized math tests like the SAT-Q and GRE-Q (Gallagher & De Lisi, 1994; Gallagher et al., 2000).

Do males and females differ with respect to visual-spatial abilities? Again, the picture is less than crystal clear. On the basis of findings of high-performance boys doing better than high-performance girls on spatial tasks and also being faster at retrieving math facts (Geary, Saults, Liu, & Hoard, 2000; Royer, Rath, Tronsky, Marchant, & Jackson, 2002; Royer, Tronsky, Chan, Jackson, & Marchant, 1999), Royer and Garofoli (2005) concluded that these two abilities account for much of the sex-based difference in math achievement. Spatial ability, they argue, affects primarily the process of identifying and representing a problem, while the speed of fact retrieval affects the subsequent process of finding a computational solution to the problem. The greater facility of males in both spatial cognition and speed of fact retrieval, Royer and Garofoli contend, gives them an edge over females in both problem representation and solution stages of problem solving. They argue too that this edge will be apparent in performance on high-level tests where time pressure is in play, but not in performance in math courses.

On the other hand, in one large sample of students, Armstrong (1985) found that among 13-year-olds, females were better than males at both computation and visualization, and that by the end of high school, the differences between females and males in these respects had disappeared. There is evidence that visual-spatial skill can be improved by training, even in relatively brief sessions (Connor & Serbin, 1985), and some attempts to improve such skills through training have had better success with girls than with boys (Connor, Schackman, & Serbin, 1978; Connor, Serbin, & Schackman, 1977; Goldstein & Chance, 1965).

Whatever the role of visual-spatial ability in mathematics, and whether males and females differ in this type of ability, there is a very large literature documenting that for a long time males have been more likely than females to study mathematics and to do well with it. The disparity is widely recognized and generally referred to as the *gender gap* in mathematics (American Association of University Women, 1998; Catsambis, 2005; Friedman, 1989; Walters & Brown, 2005). Despite the relative freedom in modern Western society for females to enroll in courses and to pursue careers of their choice, girls have been considerably less likely than boys, at least until very recently, to engage in mathematical activities, to take advanced math courses (trigonometry and calculus) in high school, to major in math in college, or to seek careers in fields requiring advanced knowledge of mathematics (Armstrong, 1981; Aronson, Quinn, & Spencer, 1998; Benbow, Lubinski, Shea, & Eftekhari-Sanjani, 2000; Catsambis, 1994; Centra, 1974; Chipman & Thomas, 1985; Dick & Rallis, 1991; Eccles & Jacobs, 1987; Fennema, 1977; Geary, 1996a; LeFevre, Kulak, & Hyemans, 1992; McCarthy & Wolfe, 1975; Meece, Parsons, Kaczala, Goff, & Futterman, 1982; Sells, 1973, 1980; Travers & Westbury, 1989). Boys generally have obtained higher scores on standardized mathematics reasoning tests, on average, than have girls, from about the time of entering high school (ages 14–15) (Brown & Josephs, 1999; Geary, 1996b; Grandy, 1994; Harnisch, Steinkamp, Tsai, & Walberg, 1986; Maccoby & Jacklin, 1974), and this appears to be an international phenomenon (National Center for Education Statistics, 2002b). (Such differences have been less likely to be found in presecondary school children [Fennema & Carpenter, 1981; Meyer, 1989].) Boys have outnumbered girls by a wide margin among students who demonstrate exceptional mathematical ability (by, for example, scoring above 700 on the mathematical portion of SAT) (Benbow & Stanley, 1980, 1983; Fox & Cohn, 1980; Mills, Ablard, & Stumpf, 1993). They have also been more likely than girls to be at the bottom of the performance distribution, which is to say their test scores have generally been more variable and distributed over a wider range (Hedges & Nowell, 1995; Willingham, Cole, Lewis, & Leung, 1997). What is the explanation of such differences?

There can be no doubt that cultural variables—social mores, gender role stereotypes, differences in beliefs regarding career opportunities (with careers requiring extensive training in mathematics being less readily open to women), differences in interests, aspirations, and expectations developed in boys and girls because of early upbringing, gender biases in the presentation of mathematics in textbooks, parental and teacher attitudes—have made it much more difficult for women than for men to enter certain fields. Whether the disparity between male and female participation and achievement in mathematics is fully accounted for by such variables remains a debated question (Boswell, 1985; Eccles, 1987; Eccles & Jacobs, 1986; Eccles, Jacobs, & Harold, 1990; Fox, 1977; Graf & Ruddell, 1972; Halpern, Wai, & Saw, 2005; Nosek, Banaji, & Greenwald, 2002).

We have already noted that beliefs, expectations, and affect can have significant effects on the performance of cognitively demanding tasks. Some research on gender differences in mathematics has focused on whether males and females differ with respect to the beliefs, expectations, or attitudes they have regarding mathematics and, if so, whether such differences help account for gender-based differences in participation or achievement in mathematics (Leder, 1990a; Meyer & Koehler, 1990). In one questionnaire study, Chipman et al. (1992) found a strong association between the attitudes of female college students regarding math and their openness to careers in science, when QSAT scores were held relatively constant, but no association between QSAT scores and openness to scientific careers, when attitude was held relatively constant. The implication is that attitude toward math appears to be a better predictor of interest in the possibility of pursuing a career in science than is mathematical ability.

There is at least suggestive evidence of a positive relationship between the perceived usefulness of mathematics, which tends to be higher among males, and both mathematical participation and achievement (Armstrong & Price, 1982; Fennema & Sherman, 1977, 1978; Hilton & Berglund, 1974; Pedro, Wolleat, Fennema, & Becker, 1981; Sherman & Fennema, 1977). A positive relationship has also been found between expressed confidence in one's mathematical ability or potential and mathematical participation and achievement (Fennema & Sherman, 1977, 1978; Lantz & Smith, 1981). Not surprisingly, students who expect to do well at math are more likely than those who do not to express the intention to take optional math courses (Eccles et al., 1985). Confidence has been found to be a stronger predictor of achievement for girls than for boys (Wise, 1985); males tend to express greater confidence in their mathematical ability than do females, sometimes relatively independently of the extent to which it is justified by higher achievement (Catsambis, 1994; Crandall, 1969; Eccles et al., 1990; Ewers & Wood, 1993; Leder, 1990b; Lussier, 1996; Pajares, 2005; Pajares & Miller, 1994; Robitaille, 1977; Seegers & Boekaerts, 1996; Wigfield et al., 1997; Williams, 1994).

Females are likely to score higher than males on math anxiety tests or otherwise admit to greater anxiety about math (Brush, 1985; Dew, Galassi, & Galassi, 1983; Hyde, Fennema, Ryan, Frost, & Hopp, 1990; Levitt & Hutton, 1983; Lussier, 1996; Pajares & Kranzler, 1995). The results of some studies suggest that females tend to be more anxious—worrisome—about test taking than are males, independently of the subject of the tests (Hong, O'Neil, & Feldon, 2005). General test anxiety, coupled with anxiety about mathematics, could help explain why females do more poorly than males on math tests despite getting better grades in math courses. Efforts to reduce math anxiety have been prominent among attempts to obtain greater parity in male-female math achievement (Tobias, 1978; Tobias & Donady, 1977; Tobias & Weissbrod, 1980). The results of some studies suggest that boys and girls tend to account differently for their successes and failures; boys are likely to credit themselves for successes and to attribute failures to external causes, while girls tend to do the opposite (Bar-Tal, 1978; Dweck & Bush, 1976; Kloosterman, 1990; Nicholls, 1975).

As noted above, *stereotype threat* has been identified as a possible contributor to poor performance of members of groups who are stereotyped as poor performers of the task of interest. Considerable attention has been focused on the role that stereotype threat may play in lowering the performance of females on mathematics tests (Davies & Spencer, 2005; Spencer et al., 1999). The possibility gets support from evidence of the pervasiveness of the stereotype of math as a male strength and female weakness (Brown & Josephs, 1999; Crocker, Major, & Steele, 1998).

In one study, female college students who were told that women typically perform more poorly than men on a math test they were about to take actually did perform more poorly, whereas female students for whom the stereotype was not made salient (who were not given the same story) did equally as well as the male students (Spencer et al., 1999). In a follow-up study, simply telling women that a math test they were about to take was not diagnostic of their mathematical ability was enough to eliminate the typical male-female performance difference (Davies, Spencer, Quinn, & Gerhardstein, 2002). A similar result was obtained by Brown and Josephs (1999), who found that women who focused on the possibility that a math test would reveal weakness in math did more poorly than those who focused on the possibility that it would reveal whether they were especially good at math. In still another study, women with high gender identification performed less well on math tests than did women with lower gender identification (Schmader, 2002). That an explicit verbal reminder of the stereotype is not essential to get a stereotype threat effect is suggested by the results of an experiment in which people took a difficult math test in three-person groups. Women did more poorly on the tests when the groups were composed of one woman and two men than when they were composed of two women and one man or three women (Inzlicht & Ben-Zeev, 2000).

An obvious question that consideration of stereotype threat raises is whether girls would do better at math in all-female than in mixed-sex classes. There is some evidence that they do (Lee & Bryk, 1986; Mael, 1998; Riordan, 1990), but generally the sexual composition of the class is only one of several factors that distinguish single-sex from coeducational situations, and, in any case, despite the gender gap in standardized test results, females typically outperform males in mathematics classes, even in coed contexts.

Regarding expectations, Henrion (1997) argues that two sets of expectations about women should be recognized: those held by society at large and those held by the mathematical community. She argues too that there are tensions between what is expected of women and what is expected of mathematicians (presumably by both the community at large and the community of mathematicians). Of special relevance to this discussion is the possibility, for which several investigators claim evidence, that teachers generally have different expectations for males and females in mathematics classes and sometimes reflect that difference in their interactions with students, by paying more attention to boys than to girls, asking boys more questions that require more than a yes or no answer, and so on (Brophy & Good, 1974; Dweck & Reppucci, 1973; Eccles & Blumenthal, 1985; Koehler, 1990; Leder, 1990b, 1992; Leinhardt, Seewald, & Engel, 1979; Reyes & Stanic, 1988; Stallings, 1979; Wilkinson & Marrett, 1985).

There is evidence too that many parents share the view that more is to be expected of sons than of daughters in their school mathematics and are more supportive of sons' pursuit of math (Casserly, 1980; Eccles, 1994; Eccles et al., 1985; Hill, 1967; Jacobs, Davis-Kean, Bleeker, Eccles, & Malanchuk, 2005; Poffenberger & Norton, 1959; Ruble & Martin, 1998; Yee & Eccles, 1988), but encouragement from both teachers and parents can be especially effective in influencing females to take advanced math courses (Casserly & Rock, 1985; Maple & Stage, 1991; Muller, 1998). Jacobs (1991) found that students' confidence in their mathematical abilities was influenced more by parents' expectations than by their own mathematical performance. Fox, Brody, and Tobin (1985) have shown that early intervention programs designed to encourage girls to take advanced courses and to do well with them can also be effective.

People, perhaps especially adolescents, are painfully sensitive to what they think their peers think of them. It is not surprising to learn that choices of what elective courses to take in high school can be influenced by the assumed effect that one's selections could have on the image one hopes to project. To the extent that interest in, or competence with, mathematics is perceived as a predominantly male trait, girls might be reluctant, for that reason, to elect optional math courses (Coleman, 1961; Ernest, 1976; Lavach & Lanier, 1975; Nash, 1979) or to pursue careers that require advanced mathematical knowledge (Hawley, 1971, 1972; Levine, 1976). Though students may deny that peers have much influence on their decision making, the likelihood that they will enroll in elective math courses has been found to correlate highly with their expectation that valued peers would approve of their doing so (Lantz & Smith, 1981; Zeldin & Pajares, 2000). Girls appear to be especially influenced by their understanding of the attitudes of male peers (Ellis & Bentler, 1973; Hawley, 1971).

There has been a recent trend toward something closer to gender parity with respect to interest in math courses in both high school and college (Chipman, 2005; Eisenberg, Martin, & Fabes, 1996; Gallagher & Kaufman, 2005; Lantz, 1985; Xie & Shauman, 2003). Also, the magnitude of the difference between male and female test scores has been decreasing (American Association of University Women, 1998; Catsambis, 1994; Goldstein & Stocking, 1994; Langenfeld, 1997); several studies have found gender differences that are small to insignificant (Fan, Chen, & Matsumoto, 1997; Friedman, 1989; Hyde, 2005; Hyde, Fennema, & Lamon, 1990; Mau & Lynn, 2000; Sadker, Sadker, & Klein, 1991; Willingham, Cole, Lewis, & Leung, 1997). Differences that are found can often be accounted for by the performance of a small fraction of the tested population (Favreau, 1997). And whatever the situation with respect to performance on standardized tests, girls consistently tend to get better grades than boys in math courses, especially in the first several years of school, but also sometimes in college and with advanced courses such as calculus (Astin, 1993; Benbow & Stanley, 1982; Bridgeman & Lewis, 1996; Dwyer & Johnson, 1997; Gallagher & Kaufman, 2005; Kessel & Linn, 1996; Kimball, 1989; Xie & Shauman, 2003). Most recently, the claims of substantive differences in math ability of male and female students have been disputed on the basis of analysis of data from 7 million

U.S. students from 10 states (Hyde, Lindberg, Linn, Ellis, & Williams, 2008). Park (2008) titled her report of the findings of this study in *Time Magazine* "The Myth of the Math Gender Gap." Whether the "gap" has been completely closed, or reduced to the status of a myth, may remain a matter of debate for some time, but if there is a gap, it appears to be much smaller now than it once was, or was believed to be.

Based on an extensive review of demographic data (Chipman & Thomas, 1985; Chipman & Wilson, 1985), Chipman (2005) concluded that the representation of women among college math majors has been as close to their representation among all recipients of BA degrees as is the case in any field. As she points out, this casts doubt on the widely held view that women have an aversion to mathematics and discredits one generally accepted explanation of why they are underrepresented in mathematically oriented professions. The parity in mathematics at the baccalaureate level does not carry over to graduate training; a 1989 survey showed women receiving 46% of the baccalaureates in mathematics, but only 35% and 17% of the master's and PhD degrees, respectively (National Research Council, 1989). A more recent report of the National Center for Education Statistics (2002a) gives 25 as the percentage of doctorates in math going to women as of 2001. And although the difference between the numbers of males and females pursuing careers that require advanced mathematical training has been decreasing (Mervis, 2003), that gap has not been closed.

Byrnes (2005) has proposed a model that attributes mathematical achievement to the combined influence of three factors: "The 3C [three conditions] model suggests that ... children are most likely to acquire high levels of proficiency in a given subject area (e.g., math) if *all three* of the following conditions hold: (1) they *regularly* find themselves in contexts that provide them with *genuine* opportunities to enhance their skills (the exposure condition), (2) they are *willing* to take advantage of these opportunities (the motivation condition), and (3) they are *able* to take advantage of these opportunities (the aptitude condition)" (p. 87). Gender differences in math achievement, Byrnes contends, can be largely explained by appeal to these three factors. I believe that the prevailing opinion among researchers is that the question of the third of these factors has been pretty much settled and that the evidence that males and females differ significantly in their potential to acquire advanced mathematical knowledge and skill, if given the same incentives and opportunities to do so, is far from compelling (Chipman, 2005; Spelke, 2005). Reports of the absence of gender-based differences in mathematical ability or achievement of students in some countries (Byrnes, Li, & Shaoying, 1997; Cherian & Siweya, 1996; Taal 1994) undermine the assumption that any such differences that have been observed are biologically based.

Spelke reviews data that challenge three claimed sex-based differences in intrinsic aptitude relating to the potential to develop mathematical competence: (1) that male infants are predisposed to learn about objects and mechanical relationships while females are more likely to learn about people, emotions, and personal relationships; (2) that boys have better command than girls over cognitive systems important to mathematical reasoning; and (3) that males show greater variability than females in mathematical talent and are therefore likely to have greater representation among the more mathematically gifted. Spelke argues that there is little evidence for any of these claims, and that while research on children and adults has revealed sex-based differences in specific cognitive tasks (Geary, 1998; Halpern, 2000), it does not provide evidence of differences between males and females in overall aptitude for mathematics at any point in development. (Spelke [2005] notes that inasmuch as mathematical reasoning tests typically assess a complex mix of capacities and strategies, and different items show different performance disparities by sex, tests can be made to favor either males or females through item selection.)

Caplan and Caplan (2005) contend that some of the claims of sex-based differences in mathematical ability have been the consequences of biased interpretation of relatively noisy data so as to favor preexisting beliefs. They also note exceptions to some of the generally reported findings, to wit the finding by Pajares (1996) of mathematically gifted girls surpassing mathematically gifted boys in mathematical problem solving, which is contrary to the more usual finding of a difference in the opposite direction. They conclude from their review of the literature that "it has simply never been established that there is any meaningful and substantial sex differences in mathematics ability that is not massively confounded with factors related to individual experience" (p. 42). Byrnes (2005) makes a distinction between natural aptitude and acquired aptitude. He suggests that there is little evidence of a sex-based difference in the first type of aptitude but does not rule out such a difference in the second type.

In view of the full complex of factors that have served, and in many cases still serve, as impediments to females pursuing excellence in mathematics, that there exists a gap between male and female performance in this area is unsurprising. Koehler (1990) puts it this way:

> When one considers that females endure remarks from teachers or texts indicating that mathematics is not a female domain, are involved in far

fewer interactions with their teachers involving mathematics, are rarely asked high-cognitive-level questions in mathematics, are encouraged to be dependent rather than independent thinkers, spend more time helping their peers and not getting helped in return, and are often not placed in groups that are appropriate to their level, it is amazing that the gap is not considerably larger. (p. 145)

Clearly, more research is needed to understand better the reasons for the documented differences in male and female participation in mathematics beyond the baccalaureate and in mathematically demanding occupations. Decisions that open or close doors to educational and professional opportunities are made on the basis of evidence of mathematical ability or potential, and it would be good to get it right. The belief that girls are simply not as good at mathematics as boys can become a self-fulfilling prophecy by decreasing the likelihood that girls will try. Extensive discussions of the subject may be found in Chipman et al. (1985), Geary (1994), and Gallagher and Kaufman (2005). Chipman (2005) notes some of the factors that can complicate an understanding of sex-based differences in performance on mathematical achievement tests, perhaps the most disturbing among them being a tendency of some researchers to distort, presumably unwittingly, the interpretation of their results in the direction of stereotyped expectations. Another limitation is the fact that most of the studies of variables that relate to sex-based differences in math have been correlational in nature, and when a noteworthy relationship has been found, it generally has been impossible to distinguish cause from effect.

The stories of successful contemporary female mathematicians demonstrate that the field is open to women, if not yet as welcoming as it is to males of comparable ability (Chipman & Thomas, 1985; Henrion, 1997; Osen, 1974). There are many existence proofs that women can succeed and excel as professional mathematicians in modern society, although probably not without extraordinary commitment and effort. Henrion's (1997) interviews of successful contemporary female mathematicians make it clear that even unusual productivity and acclaim do not ensure that a female mathematician will feel fully accepted among male peers. On the other hand, the opportunities for women in mathematics have increased considerably in the recent past, and it seems likely that they will continue to do so. In reviewing the literature on the learning and teaching of mathematics for this book, it has been a pleasant surprise to note what a substantial fraction of the recent relevant research has been done by women. At most, any differences that may exist between males and females with respect to mathematical ability are very small relative to the within-gender differences that exist, which means that sex alone is, at best, a poor predictor of mathematical potential.

Another dimension on which differences in mathematical ability might be expected to differ is age. It is widely held that mathematics is primarily a young person's domain. The prevailing view is that most of the great mathematicians—those who had the largest impact on the development of mathematics—did their most creative work while relatively young. It is also widely held that ordinary mortals tend to hit their peak of mathematical competence, whatever that peak is, well before middle age, say by age 40 or before (Lehman, 1953). While this belief may be correct as a general rule, the relationship between age and mathematical ability is complex and still not very well understood (Duverne & Lemaire, 2005). Even whether the belief is correct as a general rule has been challenged (Hersh, 2001; Stern, 1978). Some of the observed decline in mathematical ability is presumably a corollary, or consequence, of a general decline in cognitive ability that appears likely to accompany increasing age. Whether mathematical ability is especially susceptible to the insults of the aging process is less clear, although there is suggestive evidence that it is (Duverne & Lemaire, 2005).

Older people may find it more difficult than younger people to learn new mathematics, and they may have less motivation (driven, for example, by job requirements) to do so. On the other hand, if they have acquired a store of mathematical knowledge and know-how as a consequence of years of experience, they may be able to apply that knowledge to good effect for a very long time. One of the age-related changes that appears to affect mathematical performance is a general decrease in the speed with which cognitive processes proceed (Cerella, 1985; Salthouse, 1996). Duverne and Lemaire (2005) conclude, following a review of work on aging and mental arithmetic, that in the aggregate the studies suggest that age-related differences in arithmetic result from a slowing of cognitive processes: "All effects of age on calculation processes, on central as well as on peripheral processes, are completely eliminated after controlling for basic processing speed" (p. 405).

There are, of course, factors other than gender and age that have been studied as possible contributors to individual differences in participation and achievement in mathematics. These include ethnicity, socioeconomic status, parental education (especially in mathematics), and teacher education, among others. It suffices for present purposes to note that while evidence can generally be found to support the conclusion that differences exist along most of the dimensions studied, how they should be interpreted is much in dispute and, in any case, the practical significance of differences in average performance between any two contrasted groups is typically dwarfed by the degree of overlap of the distributions represented, as is true also in the cases of gender and age.

Finally, presumably people differ greatly with respect to the extent to which they enjoy the challenge of attempting to solve mathematical problems. Presumably, too, the amount of pleasure one gets from this type of challenge is not independent of how successful one's attempts to find solutions tend to be. Whether people who enjoy mathematical problem solving do so because they are good at it, or they are good at it because they enjoy it and therefore practice doing it, is an open question. It is easy to believe that it works both ways, but it seems likely that people differ with respect to the degree to which they are naturally drawn to this type of challenge.

☐ Cooperation and Group Problem Solving

There are many proponents of using groups to facilitate the acquisition of mathematical knowledge and problem-solving skills, or learning more generally (Burns, 1981; Cobb, Boufi, McClain, & Whitenack, 1997; Davidson, 1989; Natasi & Clements, 1991; Schoenfeld, 1985b; Slavin, 1980; Snyder, 1982; Thornton & Wilson, 1993). The idea is that in working collaboratively on problems, members of a group contribute to each other's thinking and learning. Kilpatrick (1987) sees group work as especially conducive to the development of problem-formulating skills because students working together are likely to identify problems that they would miss if working alone. Some researchers see in computer technology a means of facilitating collaborative learning through the interaction of students around shared computer displays and interactive software (Hawkins, Sheingold, Gearhart, & Berger, 1982).

While there is much enthusiasm for cooperative learning, which is assumed to be facilitated by the use of small group organizations in the classroom, not everyone is convinced of its merits. Noddings (1985), for example, expresses some reservations about the claimed effectiveness of small group instruction in mathematics relative to the more conventional *whole class* instruction. The question of how the groups should be composed—whether, for example, they should be made up of individuals of similar or disparate abilities—is an unresolved issue (Dawson, 1987; Linchevski & Kutscher, 1998; Slavin, 1987, 1993).

☐ Emulating Expertise

One approach to the teaching of mathematical problem solving, and problem solving more generally, that has been the focus of considerable attention is that of studying the performance of expert problem solvers, especially noting how it differs from the performance of novices, and then trying to get the novices to emulate the experts. Comparisons of the performance of experts and that of novices in various domains have revealed numerous ways in which the one characteristically differs from the other.

Not surprisingly, the results of some of this work have shown that experts typically understand and think about problems at a deeper conceptual level than do novices: When asked to sort problems in terms of similarities, novices are likely to sort on the basis of surface characteristics (objects referred to explicitly in the problem statement, such as inclined planes, pulleys, or springs; the physical configuration described), whereas experts are more likely to group problems on the basis of the fundamental physical or mathematical principles to which they relate (the law of conservation of energy; the law of conservation of momentum) (Chi, Feltovich, & Glaser, 1981; Chi, Glaser, & Rees, 1982). The categories used by experts tend to involve higher levels of abstraction than those used by novices.

When asked to specify the features of problems that led them to adopt the approaches they did for attempting to arrive at solutions, novices mention literal objects and terms contained in problem statements, whereas experts are more likely to mention states and conditions of the physical situations described by the problem statements and derived or "second order" features not explicitly mentioned (Chi et al., 1981). In solving geometry problems, experts may make fewer inferences than nonexperts, but are likely to be more selective in the inferences they make (Koedinger & Anderson, 1995).

Expert problem solvers tend to be more reflective and more aware of their own problem-solving processes than are novices (Champagne, Klopfer, & Anderson, 1980). In general, experts tend to make more use than do nonexperts of certain heuristic strategies in problem solving, such as considering problems that are analogous to the ones they are trying to solve, breaking complex problems down into manageable components, and making qualitative representations of the problems (Larkin, 1979; Lesgold, 1984; Smith & Good, 1984; VanLehn, 1989). And they are likely to spend time in planning a strategy for approaching a problem before launching an attack on it (Koedinger & Anderson, 1995).

Many researchers emphasize the important role that representations play in expert problem solving. Experts are more likely than novices to construct one or more qualitative representations of a problem before attempting to apply formulas and equations or other quantitative techniques (Larkin, McDermott, Simon, & Simon, 1980). Voss, Greene, Post, and Penner (1983) claim that one can see in the problem-solving behavior of experts within a given domain a clear distinction between finding a way to represent a problem and working on the production and justification of a solution, whereas this distinction is much less clear in the behavior of novices.

Given the findings from studies of the differences between the performance of experts and that of novices, the objective would seem to be to get the novices to emulate experts—to represent problems in useful ways, to break them up into appropriate subproblems, to think deeply about them, and so on. But how does one do that? If novices *could* do these things, they would not be novices.

Consider an analogy in athletics. Suppose one wants to transform a high jumper who cannot jump very high into a record-breaking performer. What one should do, according to the logic of the emulate experts approach, is to determine how the performance of the champion high jumper differs from that of the aspirant. One does the study and discovers that the champion exercises greater control in approaching the bar, has greater acceleration just before take-off, gets more force from his leap, coordinates leg and arm movements better, and rotates more gracefully around the bar. The problem with telling the novice to do all these things is that he simply does not know how to follow that advice, assuming he has the potential to do so. If he did know how, he would not need the instructions. While a detailed understanding of how his performance differs from that of the champion may be of some help, by itself it will not suffice to produce championship performance.

There is a cart–horse issue here. The strategy of teaching people how experts approach problems assumes that experts are experts because of the approaches they use. An alternative assumption is that experts use the approaches they do because they are experts. Given the latter assumption, the question of how they became experts is another matter entirely. And although many researchers have studied expertise, the fact is, as Holyoak (1995) points out, research has revealed a lot about how the performance of experts differs from that of novices in a variety of situations, but it has revealed less about how novices can be turned into experts.

Schoenfeld (1987a) makes a similar point in noting that although mathematicians generally agree that Polya's characterization of expert problem solving is descriptive of what they do, the approach of trying to teach students to apply the strategies identified in his books has not been very successful. Schoenfeld attributes this to the fact that Polya's characterizations are accurate descriptions of what expert problem solvers do, but they do not constitute prescriptions of how to teach novices to do the same. "There is a huge difference between *description*, which merely characterizes a procedure in sufficient detail for it to be recognized, and *prescription*, which characterizes a procedure in precise enough detail so that the characterization serves as a guide for implementing the strategy" (p. 18). Schoenfeld suggests that general heuristics like those described by Polya can sometimes be reformulated as several more specific—and more prescriptive—strategies. He illustrates this possibility by reformulating the heuristic "To understand an unfamiliar problem better, you may wish to exemplify the problem by considering various special cases" into five specific strategies that can be applied in specific mathematical contexts, the first of which is: "If there is an integer parameter n in a problem statement, consider the values $n = 1, 2, 3, 4,\ldots$. You may see a pattern that suggests an answer, and the calculations themselves may suggest the mechanism for an inductive proof that the answer is correct" (p. 19). General heuristics of the sort described by Polya are treated as generalizations of more specific strategies, and Schoenfeld's suggestion of a way to make them prescriptive is to explicate and contextualize the specific strategies they encompass.

Research has also failed to shed a lot of light on how experts became experts. Silver (1985a) challenges the assumption that there is a continuous path between novice and expert, which he considers more likely than not to be false. Further, he argues that a distinction should be made between *experts* and *EXPERTS*, the latter category being reserved for the relatively few people—the Fields Medal winners and the Nobel laureates—who are recognized to be truly extraordinary in their areas. Perhaps, Silver suggests, it would be good to recognize the possibility that there are fundamental differences in the ways that problems are approached by *EXPERTS*, *experts*, the *highly competent*, the *competent*, and so on.

Heller and Hungate (1985) express some reservations about the goal of teaching novices to be experts, and advocate the alternative goal of teaching novices to perform as "expert novices," which they consider to be more realistic. An expert novice, in this view, would be able to solve problems competently, but not necessarily in the same way as true experts. Heller and Hungate argue that what is needed to make this approach work is an identification of the knowledge that novices need to achieve good performance, which presumably need not include all the knowledge that a *bona fide* expert is likely to bring to the task.

Anderson (1990) makes the observation that much of the laboratory research on problem solving has focused on puzzles, games, and a variety of academic activities that may have little adaptive value, and he surmises that perhaps our fascination with such tasks—the reason we find them challenging—is because we are not adapted to succeed with them. An advantage of using such tasks for experimentation is that they, presumably, tend to minimize the influence that domain knowledge can have on performance, and therefore facilitate the investigation of reasoning per se. A disadvantage is that what is learned from performance on these tasks may have limited generalizability to problem solving in everyday life.

The conclusion I draw with respect to the question of how to foster the development of expertise in mathematics, or in other domains, is that, despite considerable research focused on exploring the differences between the performance of experts and novices on various problem-solving tasks, and substantial progress in this regard, no one has yet found an effective way to turn novices into experts that does not require a significant commitment of time and effort. Generally speaking, experts get to be experts—at least on complex tasks—gradually over time and as a consequence of intensive study and practice. This is not to argue that efforts to facilitate the acquisition of expertise are bound to be futile, but only to claim that dramatic effective shortcuts to the goal have yet to be found. It is an endorsement of Lester's (1985) suggestion that, independently of other factors, an essential to the development of problem-solving expertise is experience in solving many problems over a long time.

In the foregoing discussion of expertise, *expert* has generally meant "expert in some specific domain," although the domain has not always been identified explicitly. Almost all the experimental work on the development of expertise has dealt with domain-specific expertise. Expertise in mathematics is generally limited to expertise in a specific area of mathematics. An important question for research is that of the extent to which it is possible to facilitate the development of expertise that extends across many domains—the kind of expertise that will facilitate problem solving in domains in which one has only limited domain-specific knowledge and know-how. I believe domain-independent approaches to problem solving can be taught to good effect, but that only modest progress toward this goal has yet been made (Nickerson, 1994, 2004b).

☐ Effectiveness of Teaching of Mathematical Problem Solving

How effective has the emphasis on problem solving in the teaching of mathematics been? Opinions differ. Lester's (1985) assessment is harsh: "It is my view that most problem-solving instruction not only does not enable students to use their heads, but in fact it does more harm than good" (p. 41). He sees a danger that the type of mathematical instruction that students are receiving is fostering rigidity rather than flexibility and adaptability of thinking, that students are being taught how to perform procedures, but not when and why they should be performed. He argues that none of several approaches to the teaching of problem solving has proved to be clearly superior to the others, but that extensive experience in solving a wide range of problems is essential.

Hembree (1991; also summarized in Hembree & Marsh, 1993) did a meta-analysis of a large number of studies of mathematical problem solving by children in grades K through 4. With respect to various techniques for problem solving on which instruction was focused, he found that the most pronounced effect on performance was obtained from the development of skill with diagrams. The failure of performance to be improved as a result of practice without direct instruction is strong evidence of the essential role that instruction plays in the acquisition of problem-solving skills.

Most of the mathematical problems that students encounter in their school experiences with mathematics are problems that are presented to them, already formulated in someone else's terms. Students can generally assume that the problems they encounter in school can be solved—textbooks seldom present problems that are not solvable. Life outside the classroom does not have the same helpful restriction. An important aspect of mathematical competence that is neglected in mathematical education is the ability to find and formulate problems on one's own. Kilpatrick (1987), who makes this point, contends that problem formulating should be not only a goal of instruction but a means of instruction as well. Skill in problem formulation should become an increasingly important focus of the mathematics curriculum, Kilpatrick argues, as more and more of the mechanical aspects of mathematics—numerical and algebraic manipulations—are performed by computers, and consequently are less critically important skills for people to possess. Essential to the development of that skill, he contends, is a classroom environment in which the exploration of ideas is encouraged.

☐ Street Mathematics

Street mathematics is used here to refer to mathematical techniques that have been observed in use by people—often children—who have not learned them in school or another formal educational context. The users of such homespun mathematics include unschooled tailors and cloth merchants in the Ivory Coast (Petitto, 1979; Petitto & Ginsburg, 1982), apprentice tailors in Liberia (Lave, 1977; Reed & Lave, 1981), teenage and preteen street vendors in Brazil (Carraher, Carraher, & Schliemann, 1985; Nunes, Schliemann, & Carraher, 1993; Saxe, 1988), unschooled Nigerian rice traders (Gay & Cole, 1967), Micronesian navigators (Gladwin, 1970), and grocery shoppers (Lave, Murtaugh, & de la Rocha, 1999). A finding common to many of these studies is that people typically show considerable skill in doing the mathematics required of their chosen activities even if they do poorly on academic tests of mathematical abilities.

Forsyth (1928/1963) describes a process an illiterate Russian peasant may use to make out an invoice for the goods—wool, furs, timber, produce—he may be selling:

> He can multiply by the number two, but no other number; he can divide by the number two, but by no other number; and he can perform simple addition. By these three operations he effects all that he needs for business transactions, and he effects it accurately. But he could not give you one word of explanation of a process which to him has become a habit; and if you explained the mathematical argument that justifies his action, he could not understand one word of the explanation. (p. 32)

In effect Forsyth's peasant is doing his calculations in base 2 arithmetic, without an understanding of the rationale for his procedure. Of course, making effective use of a calculating algorithm without understanding its rationale is not unique to Russian peasants; most of us do it all the time when we use calculating procedures that we learned only by rote.

The idea that the ability to perform complicated mathematical tasks does not necessarily rest on school mathematics, or even on unusually high intelligence, as measured by IQ tests, gets support from a study by Ceci and Liker (1986b) of the performance of harness-racing handicappers. These investigators concluded, on the basis of a variety of analyses of handicapping performance, that expert handicapping, which is a cognitively demanding process, is not related to IQ.

The moral of this story is that people are quite good at learning, or in some cases even inventing, the mathematics they need in order to do things that they very much want to do. Without any formal schooling, many people, in a variety of cultural contexts, become adept at mental arithmetic of the kind that is useful in their daily pursuits. More generally, there is considerable evidence that people, other things being equal, do better at mathematical problem solving when problems are framed in terms of meaningful content or contexts than in abstract terms (Carraher & Schliemann, 1985: Lave, 1988; Pettito & Ginsburg, 1982).

While unschooled people often acquire surprisingly well the mathematics they need in order to conduct their daily business, whatever that may be, the evidence also suggests that the mathematics they acquire in this way is limited, for the most part, to relatively elementary levels and to what is directly applicable to practical ends. One is not likely to acquire much in the way of advanced mathematics in the absence of proper instruction or access to the knowledge to be found in expository books.

☐ **Research-Practice Disconnect**

> The search for a solid foundation for mathematics is still on; the search for the best approach to the teaching of mathematics is still in progress. (Hellman, 2006, p. 217)

Educational researchers and teachers live in different worlds. Their day-to-day lives present different challenges. Some individuals live in both worlds, but they are relatively few in number. Despite laudable efforts to bridge the gap—Crosswhite (1987) calls it a chasm—between the worlds, it remains unbridged. Even teachers who have a strong interest in research and who wish to apply the results of research to the improvement of their teaching may see some of the urgings of researchers to be unrealistic or insufficiently sensitive to classroom practicalities (Henderson, 1987).

Teachers want practical advice—scientifically derived information that will increase the effectiveness of classroom instruction. Of the several ways that have been proposed to explain to students why the product of two negative numbers is positive, for example, teachers would like to know which, if any, of them work (Pollak, 1987).

Noting that the problems that are of interest to scientists doing cognitive research are not necessarily those that are likely to be of interest to classroom teachers, Crosswhite (1987) recommends, as a way of ensuring that some attention is given to problems of practical significance for the classroom, having practitioners become partners of researchers in research projects. "The direct involvement of practitioners in research may help scientists to know when they are close to a problem of practical significance and help them design the translation studies that will bridge the gap between research and practice" (p. 272).

****

This chapter and the two preceding ones have covered only a small fraction of the very large literature on the teaching and learning of mathematics. Most of the studies cited have dealt with counting and elementary arithmetic or, to a much lesser extent, with algebra. Almost nothing has been said about geometry, trigonometry, calculus, differential equations, statistics, probability theory, or any of the numerous other areas of mathematics that are taught in high school and college. What has been covered is perhaps enough, however, to illustrate the enormity of the challenge that designing and delivering an effective program of mathematical education represents.

Final Thoughts

The idea that mathematical thinking differs fundamentally from that in other fields is a pernicious one. (Paulos, 1992, p. 233)

Although, thanks to the work of George Polya, Imre Lakatos, and others who have thought and written about the doing of mathematics, we know more about the nature of mathematical reasoning than can be gleaned from mathematics textbooks, we are still in the dark about many aspects of this subject. Much about mathematical reasoning, and about mathematics per se, remains obscure, even—perhaps especially—to those who have thought most deeply about these subjects. And it is not just about the more arcane forms of "higher" mathematics that this is true. The basic concepts and operations have been, and continue to be, subjects of much thought and discussion. What is a number? What does it mean to be able to count? Do nonhuman species have a sense of number? Or, in what sense do they have a sense of number? What comes first in human development, ordinality or cardinality? To what extent is the ability to manipulate numbers—to calculate—innate and to what extent is it acquired by learning? It may come as a surprise to the nonmathematician that arithmetic, to which we were all introduced at the beginning of our formal schooling, if not before, is not thoroughly understood from a theoretical point of view, but that is the case. As Newman (1956c) puts it, "The fundamental rules and operations of arithmetic are extraordinarily hard to define" (p. 497).

Much of the reasoning involved in the doing of mathematics remains hidden from view. We can lay out on paper a mathematical proof or an extensive informal argument supporting some conclusion or conjecture. We can, as Polya and Lakatos have done, give an account of the many guesses—correct and incorrect—made in the course of developing a proof. But any such account is bound to be incomplete. It will not explain how one gets from one step to another in the process. It will not always reveal what evoked the guesses. Or why the reasoning goes off in one direction rather than another. It will fail to tell us what goes on between the lines. It will probably leave unexplained how one knows whether one is making progress.

We are inclined to think of mathematics as relatively unchanging over time, but this is a gross misconception. Creative work in mathematics is being done all the time, and the borders of the discipline are continually being extended in various directions. Although one might assume that new developments would become increasingly rare as more and more of mathematics becomes understood, just the opposite appears to be the case. The more that is understood, the greater becomes the realization of how very much there remains to be discovered, or invented. According to King (1992), more mathematics was created during the last half of the 20th century than in all previous time, and this after Bell's (1937) claim that the 19th century contributed about five times as much to mathematical knowledge as did all of preceding history. There is little to suggest that things will slow down in the foreseeable future (Arnold, Atiyah, Lax, & Mazur, 2000; Engquist & Schmid, 2001). Peterson (1988) describes mathematics as a wilderness with a few scattered well-mapped settlements linked by a skimpy network of highways and trails. This metaphor invites the thought that a major effect of increasing the settlement—pushing the frontier—is to increase awareness of the size of the wilderness that remains unexplored.

☐ Progress

The rigor of mathematics is not absolute; it is in a process of continual development; the principles of mathematics have not congealed once and for all but have a life of their own and may even be the subject of scientific quarrels. (Aleksandrov, 1956/1963, p. 3)

Mathematics has certainly changed greatly over the centuries and continues to change. Has it progressed? And if so, toward what? These questions lead to others. To what extent are the ways in which mathematics has changed been determined—might it have gone in quite different directions if some of the major contributors to its development had not been born, or would others have made more or less the same contributions? This question can probably not be answered with any conclusiveness; opinions on what the answer is are likely to differ depending on one's views about what mathematics *really is* and whether the business of mathematicians is primarily discovery or invention. Those who consider it to be discovery are likely to believe that the development would be pretty much the same, independently of who the players were; those who consider it to be invention are likely to believe that a different set of contributors would have produced a different outcome.

People who write about the history of mathematics appear to accept the idea that mathematics has progressed without necessarily making an effort to justify the belief. Perhaps progress is tacitly equated with increasing complexity or increasing usefulness. Few would question that both its complexity and its usefulness have increased manyfold. In any case, the prevailing opinion seems to be that progress is the correct way to consider what has taken, and is taking, place, even if the goal toward which progress is being made is vague.

A question that historians of mathematics have addressed explicitly is that of whether the progress that mathematics has made has been continuous, as opposed to being characterized by new developments that bear little relationship to what preceded them. Boyer and Merzbach (1991) claim that continuity is so much the rule that when a discontinuity appears, the possibility should be considered that documents that would have filled the gap have been lost. However, to suggest that the development of mathematics has been relatively continuous is not to suggest that progress has been steady in all places and at all times. Dantzig (1930/2005) insists that history shows that "the progress of mathematics has been mostly erratic, and that intuition has played a predominant role in it" (p. 188). It has not been a record of uninterrupted progress from the more primitive to the more refined, from unfounded conjectures to rigorous proofs, from occluded glimpses of possibilities to clear apprehensions of mathematical realities. It has been along a tortuously winding path, not without a general direction, but full of excursions down alleys that led nowhere, backtrackings, course corrections, and retracing of already traversed paths.

Bell (1945/1992) describes the trip somewhat uncharitably: "Its history is not the record of one brilliant victory after another. Rather it is a somewhat sobering chronicle of intelligence fighting desperately against tremendous odds to overcome the all but ineluctable stupidity of the human mind. That such progress as has been made should have been possible at all is the miracle of the ages" (p. 66). If this sounds like a serious indictment of human rationality, it should not escape our notice that the intelligence that is credited with doing the fighting is equally as human as the stupidity that it allegedly must fight against. Bell sees the origins of modern mathematics in major advances that occurred in the 17th century. Prominent among those responsible for these advances were Galileo, Descartes, Fermat, Pascal, Newton, and Leibniz. This period, in Bell's view, represents the second great age of mathematics, the first being that of Archimedes, Euclid, and Apollonius; the distinguishing characteristic of the earlier age was a spirit of synthesis, that of the later, a spirit of analysis.

An obvious fact that needs an explanation is that many cultures—Egyptian, Mesopotamian, Chinese, Greek, Indian, Arabian, Western European—have had periods of exceptional mathematical creativity but none has been able to sustain these periods indefinitely. What causes the rise and fall of such periods is a question of considerable interest. In particular, why did the ancient Greeks follow a thousand years of extraordinary mathematical creativity with another thousand that were almost devoid of original work? Why did the Romans contribute almost nothing to theoretical science or mathematics during the heyday of the Roman empire? What caused the blossoming of both science and the arts in Western Europe during the period we refer to as the Renaissance?

The idea of continuity gets some support from the fact that mathematicians, like scientists, build upon what their predecessors have done, as already noted, and typically a new discovery (invention) can be seen, in retrospect, to have been foreshadowed by, or latent in, the preceding work. Occasionally there have been developments in mathematics that have caused a sufficiently abrupt rethinking as to be the exceptions that prove the rule. The invention of non-Euclidean geometries, which destroyed the millennia-old belief in self-evident truth of Euclid's axioms, is a case in point. Given the revolutionary impact of this event on the mathematicians' attitudes about the foundations of their discipline, references to Lobachevsky as the Copernicus of geometry seem entirely appropriate. However, even this revolutionary discovery or invention was made at about the same time independently by several people, suggesting the importance of the preceding progress that brought mathematics to the point where this event was possible, if not inevitable.

The work of Lobachevsky (and the other developers of non-Euclidean geometries) and its affect on mathematics are relatively well known to people with a modest interest in science or mathematics. Another development, mentioned in Chapter 6, probably far less well known except to specialists, but considered by some to have been a discovery perhaps on a par with the conception of non-Euclidean geometry, was Hamilton's discovery—or invention—of an algebra for which the commutative law of multiplication does not apply, which he called an algebra of *quaternions*. Although the revolutionary impact of this development is not so apparent to nonmathematicians, it encouraged algebraists to explore other possibilities and stimulated the development of abstract algebras of many forms.

Boyer and Merzbach (1991) contrast how mathematics makes progress with how science does so in the following way: "Mathematics grows by accretions, with very little need to slough off irrelevancies, whereas science grows largely through substitutions when better replacements are found" (p. 336). I doubt that this contrast will bear scrutiny. Both mathematical knowledge and scientific knowledge are cumulative in the sense that mathematicians and scientists alike stand on the shoulders of their predecessors. Neither mathematics nor science is "complete" in the sense that it has no need of assumptions or presuppositions. Both are subject to major reorientations and changes in perspective—paradigm shifts—from time to time. Each is driven by both esthetic and pragmatic concerns. In both cases, progress is seen not only in the solutions of existing problems and the answering of existing questions, but also in the identification of new challenges that could not have been conceived before. As Maor (1987) puts it, "Like every science, mathematics has a refreshing air of incompleteness about it; no sooner has one mystery been solved, than a new one is already being introduced" (p. viii). Both must appeal, in the final analysis, to human intuition regarding what should compel one to accept some claim—theorem or theory—as adequately demonstrated. Both are fallible human endeavors aimed at understanding reality, and each is subject to all the limitations of the human cognitive apparatus.

Despite the lack of agreement regarding its foundations, new areas of mathematics continue to be developed and old areas extended. Both the new and the old continue to find countless applications, and the types of problems that can be solved by mathematical means keep increasing in number. Even the murder of certitude, effected by Gödel's undecidability demonstration, at worst disquieted mathematicians for a time, but did not curtail the practice of mathematics; the theorems just kept coming. Dantzig (1930/2005) notes, with some bemusement, that even those mathematicians who are most vocal about the insecure foundations of mathematics often participate in its continuing development: "We see these gloomy deans forsaking from time to time their own counsels of alarum to join in the feverish activity of extending the empire, of pushing further and further the far-flung battle-line" (p. 237).

To what should we attribute such productivity despite the often expressed concerns about the discipline's shaky foundations? The answer, in the view of some, is that mathematics is rooted in human intuition. From intuition come the creative advances that lead to the establishment of new areas, from intuition comes the recognition of inadequacies in specific mathematical arguments, and from intuition come the insights that allow replacement of or improvements of those arguments. "What then is mathematics if it is not a unique, rigorous, logical structure? It is a series of great intuitions carefully sifted, refined, and organized by the logic men are willing and able to apply at any time. The more they attempt to refine the concepts and systematize the deductive structure of mathematics, the more sophisticated are its intuitions. But mathematics rests upon certain intuitions that may be the product of what our sense organs, brains, and the external world are like" (Kline, 1980, p. 312).

It is important to note, however, that intuitions change. What appears to be counterintuitive at one point in time may be widely accepted as intuitively palatable, if not obvious, at another. Many of the mathematical concepts that are easily accepted today were considered counterintuitive in the extreme when first introduced. The history of probability theory, for example, has many cases of conflicting intuitions that had to be resolved during the theory's formative years. Some of the relationships that we accept easily today were debated vigorously by the early developers of this discipline, which is not to suggest that now all probability problems are easily solved. I have said very little about probabilistic reasoning in this book, but have written about aspects of it elsewhere (Nickerson, 1996, 2004a, 2007).

Of course, not all the new concepts that have been introduced into mathematics have eventually been accepted. Wallace (2003) comments on the fact that some originally counterintuitive ideas become accepted whereas others do not. "The truth is that all manner of strange, non-directly-observable entities such as 0, negative integers, irrational numbers, etc. originally entered math under the same sort of insanity/incoherence cloud but are now totally accepted, even essential. At the same time, there have been plenty of other innovations that really were insane or unworkable and got laughed out of town, mathematically speaking, and we laymen never heard of them" (p. 41).

Wallace goes on to make the important point that just because a line is a fine one—for example, the fine line between only apparently insane and truly insane ideas—does not mean that it does not exist, and notes that the reality of the line in the case of mathematical ideas tends to be decided on pragmatic grounds. "The difference between a brilliant, revolutionary mathematical theory, and a wacko one lies, therefore, in what-all can be done with it, in whether or not it yields significant results" (p. 41).

From a pragmatic point of view, progress in mathematics can be seen in both the steady increase in the range of practical problems to which mathematical techniques can be applied to advantage and the increasing ease with which mathematical problems can be solved. Thanks to the accumulation of mathematical knowledge and the development of mathematical tools and techniques, unexceptional mathematics students today can solve problems with ease that would have been beyond the abilities of the vast majority of the brightest of the bright a few centuries ago. That is not to suggest, of course, that today's average students are smarter than leading mathematicians of the past, but rather to note that they have a much greater body of accumulated knowledge and much more powerful tools with which to work.

What is the future of mathematical reasoning? Are there limits to how mathematical knowledge or know-how can be extended? There is a curious paradox lurking in this question, which suggests that, if there is a limit, we cannot know what it is. Moore (2001) puts it this way: "If our understanding really cannot reach beyond certain narrow confines (if only for the time being), then, in particular, it cannot reach to the fact that it cannot reach beyond those confines. We simply cannot make sense of our being subject to limitations in this way" (p. 165). This is, I believe, a happy state of affairs; we should not wish it otherwise.

☐ The Persistent Allure of Mathematics

What is it that attracts people to mathematics? Why do some people commit their lives to attaining a better understanding of mathematics, or some aspect of mathematics, and perhaps to the possibility of contributing to its progress? And why do some others, who have little expectation of making any original contribution to the field, still find it an enormously appealing subject for thought? These are deep psychological questions, and I do not think the answers to them are obvious.

Knowledge of mathematics unquestionably has great practical value. Without it we could not have the technology on which modern society rests. Unquestionably, too, some people study mathematics because an understanding of it, or some aspects of it, is essential to certain types of work. Its practicality is surely one of the allures of mathematics, and it is one that is much emphasized as a reason—sometimes *the* reason—for including mathematics in the curriculum from the beginning of formal schooling.

Intellectual curiosity arguably accounts for some of the attention mathematics gets beyond, or instead of, that which it receives because of practical demands. People are curious; given a question—any question—we generally would like to know the answer. One might say that mathematicians are driven by an insatiable curiosity, an intense desire to know the answers to certain types of questions. Singh (1997) suggests this to be the case. "The desire for a solution to any mathematical problem is largely fired by curiosity, and the reward is the simple but enormous satisfaction derived from solving any riddle" (p. 147).

While practicality and curiosity can account for some of the allure of mathematics, I strongly doubt that even in combination they can account for all of it. When Paul Erdös wished to compliment the producer of an especially elegant proof of a theorem, he is said to have described it as worthy of inclusion in "the Book," the Book being an imaginary repository of the best and most elegant proofs of all mathematical theorems, including those not yet discovered. Mere mortals cannot, of course, look at the Book, but imagine that by some magic Erdös's Book could be made real and readily accessible to doers of mathematics. Or indulge an even more extravagant fantasy that the Book, or a supercomputer analog of it, could not only produce an elegant proof of any theorem, but could answer any question about mathematics one could ask. There would be no more practical need for mathematical knowledge—just as one with a hand calculator does not need to know how to find square roots—one with the Book could simply look up answers to mathematical questions that arise in practical contexts. Possession of the Book would meet the needs of curiosity that is not motivated by practical concerns as well.

Availability of the Book would, of course, put professional mathematicians out of work. Andrew Wiles would not have had to spend all those years developing a proof to Fermat's last theorem; he could have simply looked it up. More generally, had the Book been available a long time ago, all the countless hours that an army of mathematicians has spent over the years developing proofs for myriad theorems, solving practical and theoretical problems, satisfying their curiosity, winning Fields Medals, getting university chairs, earning accolades and awards of various sorts—not to mention salaries—would all have been unnecessary, and presumably would have been put to some different (better?) use.

I very much doubt that mathematicians would welcome the availability of anything close to the Book or its supercomputer analog that could provide answers to mathematical questions for the asking. I strongly suspect that the second part of Singh's suggestion—"the enormous satisfaction derived from solving any riddle"—has much more to do with the attraction of mathematics than does simple curiosity regarding the answers to mathematical questions. And the amount of satisfaction they get—I am guessing—increases directly with the difficulty of the problem and the number of others who have tried, and failed, to solve it.

In any attempt to consider the nature of mathematical reasoning, it is very easy to glamorize mathematicians and the work they do. The natural tendency is to focus on the work and lives of the famous—those who have made noteworthy contributions to their field. And undoubt-edly the people who have shaped their fields are deserving of the atten-tion they receive. But it is easy to forget that this focus reveals only the tip of the iceberg. There are many mathematicians at work in the world at the present time, only a very small percentage of whom will be remem-bered 100 years from now.

It has been said that one asks of a scientist "What did he or she discover?" and of a mathematician "What did he or she prove?" But it is not to be assumed that every scientist spends his or her time making momentous discoveries or that every mathematician devotes himself or herself to finding new proofs. Any history of science or mathematics is necessarily a condensation of the real story—a complete story could not possibly be told—and it describes even the major developments only in greatly abbreviated form.

Nagel (1995) notes that most philosophers and their works are quickly forgotten, and he repeats a question attributed to Bernard Williams: "What is the point of doing philosophy if you're not extraordinarily good at it? ... If you're not extraordinary, what you do in philosophy will be either unoriginal (and therefore unnecessary) or inadequately supported (and therefore useless). More likely, it will be both unoriginal and wrong. That is why most of philosophy of the past is not worth studying" (p. 10). Nagel's tentative answer to Williams's question points out the possibil-ity that what emerges from the collective enterprise that philosophical inquiry represents may be of greater value than what would have been produced if the giants in the field—those few whose works are read cen-turies after they were produced—had been the only ones thinking and writing about the issues.

Williams's question can be asked also of mathematics: What is the point of learning math unless one is extraordinarily good at it? And I think Nagel's tentative answer to Williams's question may pertain to math as well as to philosophy. Although the contributions, direct and indirect, of the large majority of mathematicians to their fields are likely to be invisible to posterity, they are nonetheless real. And without them it is doubtful that the results that do get lasting attention would have been obtained. The lineages of ideas are impossible to trace completely. We shall never learn of the critical roles that many unknown predeces-sors played in shaping the thinking of Euclid, Newton, Euler, or Gauss. Moreover, many people study, and do, mathematics not in the expecta-tion of having an impact on the future development of the discipline, but for the simple intellectual satisfaction from the activity.

Paulos (1995) contends that there are three broad classes of reasons to study mathematics. The first is practical, pertaining to job skills; the second relates to what one should understand in order to be an informed and effective citizen; and the third involves "considerations of curiosity, beauty, playfulness, perhaps even transcendence and wisdom" (p. 165).

Not everyone can be, or would want to be, a professional mathematician, or even to hold a job that requires proficiency in higher mathematics. Arguably, in order to be an effective citizen and everyday decision maker, it is helpful to have some competence in mathematics—familiarity with mathematical concepts—beyond elementary arithmetic. But for many people, the supreme reason for acquiring some competence in mathematics is the door it opens to an immensely attractive and rewarding workspace—or playground—for the mind.

REFERENCES

Aczel, A. D. (1996). *Fermat's last theorem: Unlocking the secret of an ancient mathematical problem*. New York: Bantam Doubleday.

Aczel, A. D. (2000). *The mystery of the aleph: Mathematics, the Kabbalah, and the search for infinity*. New York: Washington Square Press.

Adams, J. L. (1974). *Conceptual blockbusting: A guide to better* ideas. San Francisco: Freeman.

Adams, J. W., & Hitch, G. J. (1997). Working memory and children's mental addition. *Journal of Experimental Child Psychology, 67*, 21–38.

Adams, J. W., & Hitch, G. J. (1998). Children's mental arithmetic and working memory. In C. Donlan (Ed.), *The development of mathematical skills* (pp. 153–173). Hove, UK: Psychology Press.

Adey, P., & Shayer, M. (1994). *Really raising standards: Cognitive intervention and academic achievement*. London: Routledge.

Adler, M. J. (1981). *Six great ideas*. New York: Collier/Macmillan.

Aiken, L. R., Jr. (1972). Language factors in learning mathematics. *Review of Educational Research, 42*, 359–385.

Aiken, L. R., Jr. (1973). Ability and creativity in math. *Review of Educational Research, 43*, 405–432.

Alarcon, M., Defries, J., Gillis Light, J., & Pennington, B. (1997). A twin study of mathematics disability. *Journal of Learning Disabilities, 30*, 617–623.

Aleksandrov, A. D. (1963). A general view of mathematics. In A. D. Aleksandrov, A. N. Kolmogorov, & M. A. Lavrent'ev (Eds.), *Mathematics: Its content methods and meaning* (S. H. Gould & T. Bartha, Trans.). Cambridge, MA: MIT Press. (Original work published 1956)

Alexander, L., & Martray, C. (1989). The development of an abbreviated version of the Mathematical Anxiety Rating Scale. *Measurement and Evaluation in Counseling and Development, 22*, 143–150.

Allardice, B. S., & Ginsberg, H. P. (1983). Children's psychological difficulties in mathematics. In H. P. Ginsburg (Ed.), *The development of mathematical thinking* (pp. 319–350). Orlando, FL: Academic Press.

American Association for the Advancement of Science. (1989). *Project 2061: Science for all Americans*. Washington, DC: Author.

American Association of University Women. (1998). *Gender gaps: Where schools still fail our children*. Washington, DC: American Association of University Women Educational Foundation.

Anderson, J. R. (Ed.). (1981). *Cognitive skills and their acquisition*. Hillsdale, NJ: Erlbaum.

Anderson, J. R. (1982). Acquisition of cognitive skill. *Psychological Review, 89,* 396–406.

Anderson, J. R. (1989). The origin of errors in problem solving. In D. Klar & K. Kotovsky (Eds.), *Complex information processing: The impact of Herbert A. Simon* (pp. 343–371). Mahwah, NJ: Erlbaum.

Anderson, J. R. (1990). *The adaptive character of thought.* Hillsdale, NJ: Erlbaum.

Anderson, J. R. (1993). *Rules of the mind.* Hillsdale, NJ: Erlbaum.

Anderson, S. W., Damasio, A. R., & Damasio, H. (1990). Troubled letters but not numbers: Domain specific cognitive impairments following focal damage in frontal cortex. *Brain, 113,* 749–766.

Andrade, E. N. Da C. (1956). Isaac Newton. In J. R. Newman (Ed.), *The world of mathematics* (Vol. 1, pp. 255–276). New York: Simon and Schuster. (Original work published 1947)

Andrews, G. R., & Debus, R. I. (1978). Persistence and the causal perception of failure: Modifying cognitive attributions. *Journal of Educational Psychology, 70,* 154–166.

Annett, M. (1985). *Left, right, hand, and brain: The right shift theory.* London: Erlbaum.

Annett, M. (1995). The right shift theory of a genetic balanced polymorphism for cerebral dominance and cognitive processing. *Cahiers de Psycholigie Cognitive, 14,* 623–650.

Antell, S. R., & Keating, D. (1983). Perception of numerical invariance by neonates. *Child Development, 54,* 695–701.

Appel, K., & Haken, W. (1977a). Every planar map is four colorable. Part I. Discharging. *Illinois Journal of Mathematics, 21,* 429–490.

Appel, K., & Haken, W. (1977b). The solution of the four-color-map problem. *Scientific American, 237,* 108–121.

Appel, K., & Haken, W. (1978). The four-color problem. In L. A. Steen (Ed.), *Mathematics today: Twelve informal essays* (pp. 153–180). New York: Springer-Verlag.

Appel, K., Haken, W., & Koch, J. (1977). Every planar map is four colorable. Part II. Reducibility. *Illinois Journal of Mathematics, 21,* 491–567.

Arkes, H. R., & Hammond, K. R. (Eds.) (1986). *Judgment and decision making: An interdisciplinary reader.* New York: Cambridge University Press.

Armstrong, J. M. (1981). Achievement and participation of women in mathematics: Results from two national surveys. *Journal for Research in Mathematics Education, 12,* 356–372.

Armstrong, J. M. (1985). A national assessment of participation and achievement of women in mathematics. In S. F. Chipman, L. R. Brush, & D. M. Wilson (Eds.), *Women and mathematics: Balancing the equation* (pp. 59–94). Hillsdale, NJ: Erlbaum.

Armstrong, J. M., & Price, R. A. (1982). Correlates and predictors of women's mathematics participation. *Journal for Research in Mathematics Education, 13,* 99–109.

Arnold, V., Atiyah, M., Lax, P., & Mazur, B. (Eds.). (2000). *Mathematical frontiers and perspectives.* Providence, RI: American Mathematical Society.

Aronson, J., Lustina, M. J., Good, C., Keough, K., Steele, C. M., & Brown, J. (1999). When white men can't do math: Necessary and sufficient factors in stereotype threat. *Journal of Experimental Social Psychology, 35,* 29–46.

Aronson, J., Quinn, D., & Spencer, S. (1998). Stereotype threat and the academic underperformance of minorities and women. In J. K. Swim & C. Stangor (Eds.), *Prejudice: The target's perspective* (pp. 83–103). San Diego: Academic Press.

Ashcraft, M. H. 1992). Cognitive arithmetic: A review of data and theory. *Cognition, 44*, 75–106.

Ashcraft, M. H. (1995). Cognitive psychology and simple arithmetic: A review and summary of new directions. *Mathematical Cognition, 1*, 3–34.

Ashcraft, M. H., & Battaglia, J. (1978). Cognitive arithmetic: Evidence for retrieval and decision processes in mental addition. *Journal of Experimental Psychology: Human Learning and Memory, 4*, 527–538.

Ashcraft, M. H., & Faust, M. W. (1994). Mathematics anxiety and mental arighmetic performance: An exploratory investigation. *Cognition and Emotion, 8*, 97–125.

Ashcraft, M. H., Kirk, E. P., & Hopko, D. (1998). On the cognitive consequences of mathematics anxiety. In C. Donlan (Ed.), *The development of mathematical skills* (pp. 175–196). Hove, UK: Psychology Press.

Ashcraft, M. H., & Ridley, K. S. (2005). Math anxiety and its cognitive consequences: A tutorial review. In J. I. D. Campbell (Ed.), *Handbook of mathematical cognition* (pp. 315–327). New York: Psychology Press.

Ashcraft, M. H., & Stazyk, E. H. (1981). Mental addition: A test of three verification models. *Memory and Cognition, 9*, 185–196.

Asher, M., & Asher, R. (1981). *Code of the Quipu*. Ann Arbor: University of Michigan Press.

Ashlock, R. B. (1976). *Error patterns in computation*. Columbus, OH: Bell and Howell.

Asimov, I. (1972). *Asimov's biographical encyclopedia of science and technology: The lives and achievements of 1195 great scientists from ancient times to the present*. Garden City, NY: Doubleday.

Asimov, I. (1989). The relatively of wrong. *The Skeptical Inquirer, 14*, 35044.

Askey, R. (1999, Fall). Knowing and teaching elementary mathematics. *American Educator, 23*, 6–9, 12–13, 49.

Astin, A. W. (1993). *What matters in college? Four critical years revisited*. San Francisco: Jossey-Bass.

Atkins, P. W. (1994). *Creation revisited: The origin of space, time and the universe*. London: Penguin. (Original work published 1992)

Backman, K. (2007). *Conversations with Fermat*. Gilroy, CA: Bookstand Publishing.

Badian, N. A. (1983). Dyscalculia and nonverbal disorders of learning. In H. R. Myklebust (Ed.), *Progress in learning disabilities* (Vol. 5, pp. 235–264). New York: Stratton.

Balakrishnan, J. D., & Ashby, F. G. (1992). Subitizing: Magical numbers or mere superstition. *Psychological Research, 54*, 80–90.

Ball, D. L. (1990). Prospective elementary and secondary teachers' understanding of division. *Journal for Research in Mathematics Education, 21*, 132–144.

Ball, D. L. (1992). Magical hopes: Manipulatives and the reform of mathematics education. *American Educator, 16*, 14–18, 46–47.

Bamber, D. (1969). Reaction times and error rates for "same"-"different" judgments of multidimensional stimuli. *Perception and Psychophysics, 6*, 169–174.

Bandura, A. (1977). Self-efficacy: Toward a unifying theory of behavioral change. *Psychological Review, 84*, 191–215.

Bandura, A. (1986). *Social foundations of thought and action: A social cognitive theory.* Englewood Cliffs, NJ: Prentice Hall.

Bandura, A. (1991). Self-regulation of motivation through anticipatory and self-regulatory mechanisms. In R. A. Dienstbier (Ed.), *Perspectives on motivation: Nebraska symposium on motivation* (Vol. 38, pp. 69–164). Lincoln: University of Nebraska Press.

Bandura, A. (1994). Self-efficacy. In V. S. Ramachaudran (Ed.), *Encyclopedia of human behavior* (Vol. 4, pp. 71–81). New York: Academic Press.

Bandura, A. (1997). *Self-efficacy: The exercise of control.* New York: Freeman.

Banks, W. P., Fujii, M., & Kayra-Stuart, F. (1976). Semantic congruity effect in comparative judgments. *Journal of Experimental Psychology: Human Perception and Performance, 2,* 435–447.

Barnes, M. A., Smith-Chant, B., & Landry, S. H. (2005). Number processing in neurodevelopmental disorders: Spina bifida myelomeningocele. In J. I. D. Campbell (Ed.), *Handbook of mathematical cognition* (pp. 299–313). New York: Psychology Press.

Baron, J. (1985a). *Rationality and intelligence.* New York: Cambridge University Press.

Baron, J. (1985b). What kinds of intelligence components are fundamental? In S. F. Chipman, J. W. Segal, & R. Glaser (Eds.), *Thinking and learning skills: Research and open questions* (Vol. 2, pp. 365–390). Hillsdale, NJ: Lawrence Erlbaum Associates.

Baron, J. (1991). Beliefs about thinking. In J. F. Voss, D. N. Perkins, & J. W. Segal (Eds.), *Informal reasoning and education* (pp. 169–186). Hillsdale, NJ: Erlbaum.

Baroody, A. J. (1987). *Children's mathematical thinking: A developmental framework for pre-school, primary, and special education teachers.* New York: Teachers College Press.

Baroody, A. J. (1990). How and when should place-value concepts and skill be taught? *Journal for Research in Mathematical Education, 4,* 281–286.

Baroody, A. J. (1992). The development of preschoolers' counting skills and principles. In J. Bideaud, C. Meljac, & J.-P. Fischer (Eds.), *Pathways to number: Children's developing numerical abilities* (pp. 99–126). Hillsdale, NJ: Erlbaum.

Baroody, A. J. (2000). Does mathematics instruction for three- to five-year-olds really make sense? *Young Children, 55,* 61–67.

Baroody, A. J., & Gannon, K. E. (1984). The development of the communtativity principle and economical addition strategies. *Cognition and Instruction, 1,* 321–339.

Baroody, A. J., Gannon, K. E., Berent, R., & Ginsburg, H. P. (1983). The development of basic formal math abilities. *Acta Paedologica, 1,* 133–151.

Baroody, A. J., & Gatzke, M. R. (1991). The estimate of set size by potentially gifted kindergarten-age children. *Journal for Research in Mathematics Education, 22,* 59–68.

Baroody, J. D., & Ginsburg, H. P. (1983). The effects of instruction on children's understanding of the "equals" sign. *The Elementary School Journal, 84,* 199–212.

Baroody, J. D., & Ginsburg, H. P. (1986). The relationship between initial meaningful and mechanical knowledge of arithmetic. In J. Hiebert (Ed.), *Conceptual and procedural knowledge: The case of arithmetic* (pp. 75–112). Hillsdale, NJ: Erlbaum.

Baroody, A. J., & Standifer, D. J. (1993). Addition and subtraction in the primary grades. In R. J. Jensen (Ed.), *Research ideas for the classroom: Early childhood mathematics* (pp. 72–102). New York: Macmillan.

Barrett, W. (1958). *Irrational man: A study in existential philosophy.* New York: Doubleday.

Barrow, J. D. (1988). *The world within the world.* New York: Oxford University Press.

Barrow, J. D. (1990). *World within the world.* New York: Oxford University Press.

Barrow, J. D. (1991). *Theories of everything: The quest for ultimate explanation.* New York: Oxford University Press.

Barrow, J. D. (1992). *Pi in the sky: Counting, thinking, and being.* New York: Oxford University Press.

Barrow, J. D. (1995a). *The artful universe.* New York: Oxford University Press.

Barrow, J. D. (1995b). Theories of everything. In J. Cornwell (Ed.), *Nature's imagination: The frontiers of scientific vision* (pp. 45–63). New York: Oxford University Press.

Barrow, J. D. (1998). *Impossibility: The limits of science and the science of limits.* New York: Oxford University Press.

Barrow, J. D. (2005). *The infinite book: A short guide to the boundless, timeless and endless.* New York: Vintage Books.

Bar-Tal, D. (1978). Attributional analysis of achievement-related behavior. *Review of Educational Research, 48,* 259–271.

Batchelder, W. H. (2000). Mathematical psychology. In A. E. Kazdin (Ed.), *Encyclopedia of psychology* (Vol 5, pp. 120–123). Washington, DC: American Psychological Association/Oxford University Press.

Battista, M. T. (1986). The relationship of mathematics anxiety and mathematical knowledge to the learning of mathematical pedagogy by preservice elementary teachers. *School Science and Mathematics, 86,* 10–19.

Beard, R. M. (1963). The order of concept development: Studies in two fields. Part I. Number concepts in the infant school. *Educational Review, 15,* 105–117.

Beckman, P. (1971). *A history of pi.* New York: St. Martin's Press.

Beckwith, M., & Restle, F. (1966). Process of enumeration. *Psychological Review, 73,* 437–444.

Begle, E. G. (1979). *Critical variables in mathematics education: Findings from a survey of the empirical literature.* Washington, DC: Mathematics Association of America.

Behr, M. J., Erlwanger, S., & Nichols, E. (1980). How children view the equal sign. *Mathematics Teaching, 92,* 13–15.

Behr, M. J., Lesh, R., Post, T. R., & Silver, E. A. (1983). Rational-number concepts. In R. Lesh & M. Landau (Eds.), *Acquisition of mathematical concepts and processes* (pp. 91–126). New York: Academic Press.

Behr, M. J., Wachsmuth, I., Post, T. R., & Lesh, R. (1984). Order and equivalence of rational numbers: A clinical teaching experiment. *Journal for Research in Mathematics Education, 15,* 323–341.

Behr, M. J., Harel, G., Post, T., & Lesh, R. (1994). Units of quantity: A conceptual basis common to additive and multiplicative structures. In G. Harel & J. Confrey (Eds.), *The development of multiplicative reasoning in learning mathematics* (pp. 121–176). Albany, NY: State University of New York Press.

Beilen, H., & Gillman, I. S. (1967). Number language and number reversal learning. *Journal of Experimental Child Psychology, 5,* 263–277.

Beiler, A. H. (1966). *Recreations in the theory of numbers* (2nd ed.). New York: Dover.

Bell, A., Fischbein, E., & Greer, B. (1984). Choice of operation in verbal arithmetic problems: The effect of number size, problem structure and context. *Educational Studies in Mathematics, 15,* 129–147.

Bell, A., Greer, B., Grimson, L., & Mangan, C. (1989). Children's performance on multiplicative word problems: Elements of a descriptive theory. *Journal for Research in Mathematics Education, 20,* 434–449.

Bell, A., Swan, M., & Taylor, G. (1981). Choice of operations in verbal problems with decimal numbers. *Educational Studies in Mathematics, 12,* 399–420.

Bell, E. T. (1933). *Numerology.* Baltimore: Williams & Wilkins.

Bell, E. T. (1937). *Men of mathematics: The lives and achievements of the great mathematicians from Zeno to Poincare.* New York: Dover.

Bell, E. T. (1956). The prince of mathematicians. In J. R. Newman (Ed.), *The world of mathematics* (Vol. 1, pp. 295–339). New York: Simon & Schuster. (Original work published 1937)

Bell, E. T. (1991). *The magic of numbers.* New York: Dover. (Original work published 1946)

Bell, E. T. (1992). *The development of mathematics* (2nd ed.). New York: Dover. (Original work published 1945)

Belmont, J. M. (1989). Cognitive strategies and strategic learning: The socio-instructional approach. *American Psychologist, 44,* 142–148.

Benacerraf, P. (2001). Tasks, super-tasks, and the modern Eleatics. In W. C. Salmon (Ed.), *Zeno's paradoxes* (pp. 103–129). Indianapolis: Hackett Publishing. (Original work published 1962)

Benbow, C. P., Lubinski, D., Shea, D. L., & Eftekhari-Sanjani, H. (2000). Sex differences in mathematical-reasoning ability at age 13: 20 years later. *Psychological Science, 11,* 474–479.

Benbow, C. P., & Stanley, J. C. (1980). Sex differences in mathematical ability: Fact or fiction? *Science, 210,* 1262–1264.

Benbow, C. P., & Stanley, J. C. (1982). Consequences in high school and college of sex differences in mathematical reasoning ability: A longitudinal study. *American Educational Research Journal, 19,* 598–622.

Benbow, C. P., & Stanley, J. C. (1983). Sex differences in mathematical reasoning ability in intellectually talented preadolescents: Their nature, effects, and possible causes. *Behavioral Brain Sciences, 11,* 169–232.

Bender, B. G., Linden, M. G., & Robinson, A. (1993). Neuropsychological impairments in 42 adolescents with sex chromosome abnormalities. *American Journal of Medical Genetics (Neuropsychiatric Genetics), 48,* 169–173.

Bennett, M. (1976). *SUBSTITUTOR: A teaching program.* Unpublished project report, Department of Artificial Intelligence, University of Edinburgh.

Benson, D. F., & Denckla, M. B. (1969). Verbal paraphasia as a source of calculation disturbance. *Archives of Neurology, 21,* 96–102.

Benson, D. F., & Geschwind, N. (1970). Developmental Gerstmann syndrome. *Neurology, 20,* 293–298.

Bentham, J. (1879). *An introduction to the principles of morals and legislation.* Oxford: Clarendon Press. (Original work published 1789)

Ben-Zeev, T. (1995). The nature and origin of rational errors in arithmetic thinking: Induction from examples and prior knowledge. *Cognitive Science, 19,* 341–376.

Ben-Zeev, T. (1996). When erroneous mathematical thinking is just as "correct": The oxymoron of rational errors. In R. J. Sternberg & T. Ben-Zeev (Eds.), *The nature of mathematical thinking* (pp. 55–79). Mahwah, NJ: Erlbaum.

Ben-Zeev, T. (1998). Rational errors and the mathematical mind. *Review of General Psychology, 2,* 366–383.

Ben-Zeev, T., Duncan, S., & Forbes, C. (2005). Stereotypes and math performance. In J. I. D. Campbell (Ed.), *Handbook of mathematical cognition* (pp. 235–249). New York: Psychology Press.

Beran, M. J. (2007). Rhesus monkeys (*Macaca mulatta*) enumerate large and small sequentially presented sets of items using analog numerical representations. *Journal of Experimental Psychology: Animal Behavior Processes, 33,* 42–54.

Bergamini, D. (1963). *Mathematics.* New York: Time Incorporated.

Bergson, H. (2001). The cinematographic view of becoming. In W. C. Salmon (Ed.), *Zeno's paradoxes* (pp. 59–66). Indianapolis: Hackett Publishing. (Original work published 1911)

Berkeley, B. (1956). The analyst. In J. R. Newman (Ed.), *The world of mathematics* (Vol. 1, pp. 288–293). New York: Simon & Schuster. (Original work published 1734)

Berlinski, D. (1997). *A tour of the calculus.* New York: Vintage.

Berlinski, D. (2000). *Newton's gift: How Sir Isaac Newton unlocked the system of the world.* New York: Simon & Schuster.

Berman, M. (1984). *The reenchantment of the world.* New York: Bantam Books.

Bernstein, J. (1993). *Cranks, quarks, and the cosmos.* New York: Basic Books.

Bernstein, P. L. (1996). *Against the gods: The remarkable story of risk.* New York: Wiley.

Bezuk, N., & Cramer, K. (1989). Teaching about fractions: What, when and how? In P. Trafton (Ed.), *New directions for elementary school mathematics: 1989 yearbook* (pp. 156–157). Reston, VA: National Council of Teachers of Mathematics.

Bialystok, E., & Codd, J. (1996). Developing representations of quantity. *Canadian Journal of Behavioural Science, 28,* 281–291.

Bideaud, J., Meljac, C., & Fischer, J.-P. (Eds.). (1992). *Pathways to number: Children's developing numerical abilities.* Hillsdale, NJ: Erlbaum.

Binet, A. (1981). *Psychologie des grands calculateurs et joueurs d'échecs.* Paris: Slatkine. (Original work published 1894)

Birkhoff, G. D. (1933). *Aesthetic measure.* Cambridge, MA: Harvard University Press.

Bisanz, J., Sherman, J. L., Rasmussen, C., & Ho, E. (2005). Development of arithmetic skills and knowledge in preschool children. In J. I. D. Campbell (Ed.), *Handbook of mathematical cognition* (pp. 143–162). New York: Psychology Press.

Black, M. (2001). Achilles and the tortoise. In W. C. Salmon (Ed.), *Zeno's paradoxes* (pp. 67–81). Indianapolis: Hackett Publishing. (Original work published 1950)

Blank, R. K., Manise, J., & Braithwaite, B. C. (2000). *State education indicators with a focus on Title I: 1999.* Washington, DC: Council of Chief State School Officers.

Blay, M. (1998). *Reasoning with the infinite: From the closed world to the mathematical universe* (M. B. DeBevoise, Trans.). Chicago: University of Chicago Press.

Bloom, B. S., & Broder, L. J. (1950). *Problem solving processes of college students.* Chicago: University of Chicago Press.

Blumenthal, L. M., & Menger, K. (1970). *Studies in geometry.* San Francisco: Freeman.

Bobrow, D. (1968). Natural language input for a computer problem solving system. In M. Minsky (Ed.), *Semantic information processing*. Cambridge, MA: MIT Press.

Bolanzo, B. (1921). *Paradoxien des Unendlichen*. Leipzig: Meiner. (Original work published 1851)

Booth, L. R. (1981). Child methods in secondary school mathematics. *Educational Studies in Mathematics, 12*, 29–41.

Boring, E. G. (1957). *A history of experimental psychology* (2nd ed.). New York: Appleton-Century-Croft.

Borko, H., Eisenhart, M., Brown, C. A., Underhill, R. G., Jones, D., & Agard, P. C. (1992). Learning to teach hard mathematics: Do novice teachers and their instructors give up too easily? *Journal for Research in Mathematics Education, 23*, 194–222.

Boswell, S. L. (1985). The influence of sex-role stereotyping on women's attitudes and achievements in mathematics. In S. F. Chipman, L. R. Brush, & D. M. Wilson (Eds.), *Women and mathematics: Balancing the equation* (pp. 175–197). Hillsdale, NJ: Erlbaum.

Bouleau, C. (1963). *The painter's secret geometry*. New York: Harcourt, Brace and World.

Boulger, W. (1989). Pythagoras meets Fibonacci. *Mathematics Teacher, 82*, 277–282.

Bourbon, B. (1908). Sur le temps nécessaire pour nommer les nombres. *Revue Philosophique de la France et de l'etranger, 65*, 426–431.

Boyer, C. B. (1959). *History of the calculus and its conceptual development*. New York: Dover.

Boyer, C. B., & Merzbach, U. C. (1991). *A history of mathematics* (2nd ed.). New York: Wiley.

Boysen, S. T. (1992). Counting as the chimpanzee views it. In G. Fetterman & W. K. Honig (Eds.), *Cognitive aspects of stimulus control* (pp. 367–383). Hillsdale, NJ: Erlbaum.

Boysen, S. T. (1993). Counting in chimpanzees: Nonhuman principles and emergent properties of number. In S. T. Boysen & E. J. Capaldi (Eds.), *The development of numerical competence: Animal and human models* (pp. 39–59). Hillsdale, NJ: Erlbaum.

Boysen, S. T., & Berntson, G. G. (1989). Numerical competence in a chimpanzee (*Pan troglodytes*). *Journal of Comparative Psychology, 103*, 23–31.

Boysen, S. T., & Berntson, G. G. (1996). Quality-based interference and symbolic representations in chimpanzees (*Pan troglodytes*). *Journal of Experimental Psychology: Animal Behavior Processes, 22*, 76–86.

Boysen, S. T., & Capaldi, E. J. (Eds.). (1993). *The development of numerical competence*. Mahwah, NJ: Erlbaum.

Bradis, V. M., Minkovskii, V. L., & Kharcheva, A. K. (1999). *Lapses in mathematical reasoning* (J. J. Schorr-Kon, Trans.; E. A. Maxwell Mineola, Ed.). New York: Dover. (Original work published 1938)

Brainerd, C. J. (1973a). Mathematical and behavioral foundations of number. *Journal of General Psychology, 88*, 221–281.

Brainerd, C. J. (1973b). The origins of number concepts. *Scientific American, 228*, 101–109.

Brainerd, C. J. (1973c). Order of acquisition of transitivity, conservation, and class inclusion of length and weight. *Developmental Psychology, 8,* 105–116.

Brainerd, C. J. (1973d). Judgments and explanations as criteria for the presence of cognitive structures. *Psychological Bulletin, 79,* 172–179.

Brainerd, C. J. (1979). *The origins of the number concept.* New York: Wiley.

Brannon, E. M. (2002). The development of ordinal numerical knowledge in infancy. *Cognition, 83,* 223–240.

Brannon, E. M. (2005). What animals know about numbers. In J. I. D. Campbell (Ed.), *Handbook of mathematical cognition* (pp. 85–107). New York: Psychology Press.

Brannon, E. M., Cantlon, J. F., & Terrace, H.S. (2006). The role of reference points in ordinal numerical comparisons by rhesus macaques (*Macaca mulatta*). *Journal of Experimental Psychology: Animal Behavior & Processes, 32,* 120–134.

Brannon, E. M., & Roitman, J. D. (2003). Nonverbal representations of time and number in animals and human infants. In W. H. Meck (Ed.), *Functional and neural mechanisms of interval timing* (pp. 143–182). Boca Raton, FL: CRC Press.

Brannon, E. M., & Terrace, H. S. (1998). Ordering of the numerosities 1–9 by monkeys. *Science, 282,* 746–749.

Brannon, E. M., & Terrace, H. S. (2000). Representation of the numerosities 1–9 by rhesus monkeys. *Journal of Experimental Psychology: Animal Behavioral Processes, 25,* 31–49.

Brannon, E. M., & Terrace, H. S. (2002). The evolution and ontogeny of ordinal numerical ability. In M. Bekoff, C. Allen, & G. M. Burghardt (Eds.), *The cognitive animal: Empirical and theoretical perspectives on animal cognition* (pp. 197–204). Cambridge, MA: MIT Press.

Brannon, E. M., Wustoff, C. J., Gallistel, C. R., & Gibbon, J. (2001). Numerical subtraction in the pigeon: Evidence for a linear subjective number scale. *Psychological Science, 112,* 238–243.

Bransford, J. D., & Stein, B. S. (1984). *The ideal problem solver: A guide for improving thinking, learning, and creativity.* New York: Freeman.

Bransford, J. D., Zech, L., Schwartz, D., Barron, B., Vye, N., & the Cognition and Technology Group at Vanderbilt. (1996). Fostering mathematical thinking in middle school students: Lessons from research. In R. J. Sternberg & T. Ben-Zeev (Eds.), *The nature of mathematical thinking* (pp. 203–250). Mahwah, NJ: Erlbaum.

Briars, D., & Siegler, R. S. (1984). A featural analysis of preschooler' counting knowledge. *Developmental Psychology, 20,* 607–618.

Briars, D. J., & Larkin, J. H. (1984). An integrated model of skill in solving elementary word problems. *Cognition and Instruction, 1,* 245–296.

Bridgeman, B., & Lewis, C. (1996). Gender differences in college mathematics grades and SAT-M scores: A reanalysis of Wainer and Steinberg. *Journal of Educational Measurement, 33,* 257–270.

Broadbent, H. A., Church, R. M., Meck, W. H., & Rakitin, B. C. (1993). Qualitative relationships between timing and counting. In S. T. Boysen & E. J. Capaldi (Eds.), *The development of numerical competence: Animal and human models* (pp. 171–187). Hillsdale, NJ: Erlbaum.

Brophy, J. E., & Good, T. (1974). *Teacher-student relationships: Causes and consequences.* New York: Holt, Reinhart and Winston.

Browder, F. E. (1976). *Mathematical developments arising from Hilbert problems.* Providence, RI: American Mathematical Society.

Browder, F. E., & Lane, S. M. (1978). The relevance of mathematics. In L. A. Steen (Ed.), *Mathematics today: Twelve informal essays* (pp. 324–350). New York: Springer-Verlag.

Brown, J. S., & Burton, R. R. (1978). Diagnostic models for procedural bugs in basic mathematical skills. *Cognitive Science, 2,* 155–192.

Brown, J. S., & VanLehn, K. (1980). Repair theory: A generative theory of bugs in basic mathematical skills. *Cognitive Science, 4,* 379–426.

Brown, M. (1996). FIMS and SIMS: The first two IEA international mathematics surveys. *Assessment in Education, 3,* 181–200.

Brown, R. P., & Josephs, R. A. (1999). A burden of proof: Stereotype relevance and gender differences in math performance. *Journal of Personality and Social Psychology, 76,* 246–257.

Brown, S. (1990). Integrating manipulatives and computers in problem-solving experiences. *Arithmetic Teacher, 38,* 8–10.

Brownell, W. A. (1928). *The development of children's number ideas in the primary grades.* Chicago: University of Chicago Press.

Brownell, W. A. (1941). *Arithmetic in grades I and II: A critical summary of new and previously reported research* (Duke University Research Studies in Education, No. 5). Durham, NC: Duke University Press.

Brueckner, L. J. (1930). *Diagnostic and remedial teaching in arithmetic.* Philadelphia: John C. Winston.

Bruner, J. (1996). *The culture of education.* Cambridge, MA: Harvard University Press.

Brush, L. R. (1985). Cognitive and affective determinants of course preferences and plans. In S. F. Chipman, L. R. Brush, & D. M. Wilson (Eds.), *Women and mathematics: Balancing the equation* (pp. 123–150). Hillsdale, NJ: Erlbaum.

Bryant, P. E. (1995). The distinction between knowing when to do a sum and knowing how to do it. *Educational Psychology, 5,* 207–215.

Brysbaert, M. (1995). Arabic number reading: On the nature of the numerical scale and the origin of phonological recoding. *Journal of Experimental Psychology: General, 124,* 434–452.

Bullock, M., & Gelman, R. (1977). Numerical reasoning in young children: The ordering principle. *Child Development, 48,* 427–434.

Bunch, B. (1982). *Mathematical fallacies and paradoxes.* Mineola, NY: Dover Publications.

Bundy, A. (1975). *Analysing mathematical proofs (or reading between the lines).* Research Report 2. Edinburgh: University of Edinburgh, Department of Artificial Intelligence.

Burau, W. (1975). Schweikart, Ferdinand Karl. In *Dictionary of scientific biography* (Vol. 13, p. 225). New York: Scribners.

Burger, E. B., & Starbird, M. (2005). *Coincidences, chaos, and all that math jazz: Making light of weighty ideas.* New York: Norton.

Burnett, S. A., Lane, D. M., & Dratt, L. M. (1979). Spatial visualization and sex differences in quantitative ability. *Intelligence, 3,* 345–354.

Burns, M. (1981, September). Groups of four: Solving the management problem. *Learning*, 48–51.

Burton, R. B. (1981). Debuggy: Diagnosis of errors in basic mathematical skills. In D. H. Sleeman & J. S. Brown (Eds.), *Intelligent tutoring systems* (pp. 157–183). London: Academic Press.

Bush, W. S. (1989). Mathematics anxiety in upper elementary school teachers. *School Science and Mathematics, 89*, 499–509.

Bush, W. S., & Kincer, L. A. (1993). The teacher's influence on the classroom learning environment. In R. J. Jensen (Ed.), *Research ideas for the classroom: Early childhood mathematics* (pp. 311–328). New York: Macmillan.

Buswell, G. T. (1926). *Diagnostic studies in arithmetic.* Chicago: Chicago University Press.

Buswell, G. T., & John, L. (1926). *Diagnostic studies in arithmetic* (Supplementary Educational Monograph 30). Chicago: University of Chicago Press.

Butler, C. (1970). *Number symbolism.* London: Routledge and Kegan Paul.

Butterworth, B. (1999). *The mathematical brain.* London: Macmillan.

Butterworth, B. (2005). Developmental dyscalculia. In J. I. D. Campbell (Ed.), *Handbook of mathematical cognition* (pp. 455–467). New York: Psychology Press.

Butterworth, B., Cappelletti, M., & Kopelman, M. (2001). Category specificity in reading and writing: The case of number words. *Nature Neuroscience, 4*, 784–786.

Butterworth, B., Granà, A., Piazza, M., Girelli, L., Price, C., & Skuse, D. (1999). Language and the origins of number skills: Karyotypic differences in Turner's syndrome. *Brain and Language, 69*, 486–488.

Byers, W. (2007). *How mathematicians think: Using ambiguity, contradiction, and paradox to create mathematics.* Princeton, NJ: Princeton University Press.

Bynner, J., & Parsons, S. (1997). *Does numeracy matter?* London: The Basic Skills Agency.

Byrne, G. (1989). U.S. students flunk math, science. *Science, 243*, 729.

Byrnes, J. P. (2005). Gender differences in math. In A. M. Gallagher & J. C. Kaufman (Eds.), *Gender differences in mathematics* (pp. 73–98). New York: Cambridge University Press.

Byrnes, J. P., Li, H., & Shaoying, X. (1997). Gender differences on the math subtest of the scholastic aptitude test may be culture-specific. *Educational Studies in Mathematics, 34*, 49–66.

Byrnes, J. P., & Wasik, B. A. (1991). Role of conceptual knowledge in mathematical procedural learning. *Developmental Psychology, 27*, 777–786.

Cajori, F. (1985). *The history of mathematics* (4th ed.). New York: Chelsea. (Original work published 1893)

Cajori, F. (1994). *A history of the logarithmic slide rule and allied instruments.* Mendham, NJ: Astragal Press. (Original published in 1910)

Campbell, J. I. D. (1987). Network interference and mental multiplication. *Journal of Experimental Psychology: Learning, Memory, and Cognition, 13*, 109–123.

Campbell, J. I. D. (1994). Architectures for numerical cognition. *Cognition, 53*, 1–44.

Campbell, J. I. D. (1999). The surface form by problem size interaction in cognitive arithmetic: Evidence against an encoding locus. *Cognition, 70*, 25–33.

Campbell, J. I. D., & Clark, J. M. (1992). Numerical cognition: An encoding-complex perspective. In J. I. D. Campbell (Ed.), *The nature and origins of mathematical skills* (pp. 457–491). Amsterdam: Elsevier.

Campbell, J. I. D., & Epp, L. J. (2005). Architectures for arithmetic. In J. I. D. Campbell (Ed.), *Handbook of mathematical cognition* (pp. 347–360). New York: Psychology Press.

Campbell, J. I. D., & Graham, D. J. (1985). Mental multiplication skill: Structure, process, and acquisition. *Canadian Journal of Psychology, 39,* 338–366.

Campbell, P. F., & Clements, D. H. (1990). Using microcomputers for mathematics learning. In J. Payne (Ed.), *Teaching and learning mathematics for the young child* (pp. 265–283). Reston, VA: National Council of Teachers of Mathematics.

Campbell, P. F., & Stewart, E. L. (1993). Calculators and computers. In R. J. Jensen (Ed.), *Research ideas for the classroom: Early childhood mathematics* (pp. 251–268). New York: Macmillan.

Cantlon, J. F., & Brannon, E.M. (2005). Semantic congruity facilitates number judgments in monkeys. *Proceedings of the National Academy of Sciences, 102,* 16507–16511.

Cantlon, J. F., & Brannon, E. M. (2006). Shared system for ordering small and large numbers in monkeys and humans. *Psychological Science, 17,* 401–406.

Cantlon, J. F., & Brannon, E. M. (2007). How much does number matter to a monkey (*Macaca mulatta*). *Journal of Experimental Psychology: Animal Behavior Processes, 33,* 32–41.

Cantor, G. (1955). *Contributions to the founding of the theory of transfinite numbers* (P. E. Jordain, Trans.). New York: Courier Dover. (Original work published 1911)

Capaldi, E. J. (1964). Effects of N-length, number of N-lengths, and number of reinforcements on resistance to extinction. *Journal of Experimental Psychology, 68,* 230–239.

Capaldi, E. J. (1966). Partial reinforcement: A hypothesis of sequential effects. *Psychological Review, 73,* 459–477.

Capaldi, E. J. (1993). Animal number abilities: Implications for a hierarchical approach to instrumental learning. In S. T. Boysen & E. J. Capaldi (Eds.), *The development of numerical competence: Animal and human models* (pp. 191–209). Hillsdale, NJ: Erlbaum.

Capaldi, E. J., & Miller, D. J. (1988a). Counting in rats: Its functional significance and the independent cognitive processes which comprise it. *Journal of Experimental Psychology: Animal Behavior Processes, 14,* 3–17.

Capaldi, E. J., & Miller, D. J. (1988b). Number tags applied by rats to reinforcers are general and exert powerful control over responding. *Quarterly Journal of Experimental Psychology, 40B,* 279–297.

Caplan, J. B., & Caplan, P. J. (2005). The perseverative search for sex differences in mathematics abilities. In A. M. Gallagher & J. C. Kaufman (Eds.), *Gender Differences in Mathematics. An Integrative Approach* (pp. 25–47). New York: Cambridge University Press.

Caplan, N., Choy, M. H., & Whitmore, J. K. (1992). Indochinese refugee families and academic achievement. *Scientific American, 266,* 36–42.

Cappelletti, M., Butterworth, B., & Kopelman, M. (2001). Spared numerical abilities in a case of semantic dementia. *Neuropsychologica, 39,* 1224–1239.

Carey, S. (2001). Cognitive foundations of arithmetic: Evolution and ontogenesis. *Mind and Language, 16*, 37–55.

Carey, S. (2004). Bootstrapping and the origins of concepts. *Daedalus, 133*, 59–68.

Carpenter, T. P. (1985). Learning to add and subtract: An exercise in problem solving. In E. A. Silver (Ed.), *Teaching and learning mathematical problem solving: Multiple research perspectives* (pp. 17–40). Hillsdale, NJ: Erlbaum.

Carpenter, T. P., Corbitt, M. K., Kepner, H. S., Jr., Lindquist, M. M., & Reys, R. E. (1980). National assessment: A perspective of students' mastery of basic mathematics skills. In M. M. Lindquist (Ed.), *Selected issues in mathematics education* (pp. 215–257). Berkeley, CA: McCuchan.

Carpenter, T. P., Corbitt, M. K., Kepner, H. J., Lindquist, M. M., & Reys, R. E. (1981). Decimals: Results and implications from the Second NAEP Mathematics Assessment. *Arithmetic Teacher, 28*, 34–37.

Carpenter, T. P., Franke, M. L., Jacobs, V. R., Fennema, E., & Empson, S. B. (1998). A longitudinal study of invention and understanding in children's multidigit addition and subtraction. *Journal for Research in Mathematics Education, 29*, 3–20.

Carpenter, T. P., Hiebert, J., & Moser, J. (1981). The effect of problem structure on first-grader's initial solution processes for simple addition and subtraction problems. *Journal for Research in Mathematics Education, 12*, 27–39.

Carpenter, T. P., Lindquist, M. M., Matthews, W., & Silver, E. A. (1983). Results of the Third NAEP Mathematics Assessment: Secondary school. *Mathematics Teacher, 76*, 652–659.

Carpenter, T. P., & Moser, J. M. (1982). The development of addition and substraction problem solving skills. In T. P. Captenter, J. M. Moser, & T. A. Romberg (Eds.), *Addition and subtraction: A cognitive perspective* (pp. 9–24). Hillsdale, NJ: Erlbaum.

Carpenter, T. P., & Moser, J. M. (1983). The acquisition of addition and subtraction concepts. In R. Lesh & M. Landau (Eds.), *Acquisition of mathematical concepts and processes* (pp. 7–44). New York: Academic Press.

Carpenter, T. P., & Moser, J. M. (1984). The acquisition of addition and subtraction concepts in grades one through three. *Journal for Research in Mathematics Education, 15*, 179–202.

Carraher, T. N., Carraher, D., & Schliemann, A. D. (1985) Mathematics in the streets and the schools. *British Journal of Developmental Psychology, 3*, 21–29.

Carraher, T. N., & Schliemann, A. S. (1985). Computation routines prescribed by schools: Help or hindrance? *Journal for Research in Mathematics Education, 16*, 37–44.

Carraher, T. N., Schliemann, A. S., & Carraher, D. W. (1988). Mathematical concepts in everyday life. In G. B. Saxe & M. Gearhart (Eds.), *Children's mathematics: New directions for child development*, (pp. 71–87). San Francisco: Jossey Bass.

Carroll, J. B. (1967). Instructional methods and individual differences. In R. M. Gagné (Ed.), *Learning and individual differences*. Columbus, OH: Charles E. Merrill.

Carroll, J. B. (1996). Mathematical abilities: Some results from factor analysis. In R. J. Sternberg & T. Ben-Zeev (Eds.), *The nature of mathematical thinking* (pp. 3–25). Mahwah, NJ: Erlbaum.

Case, R. (1978). Piaget and beyond: Toward a developmentally based theory and technology of instruction. In R. Glaser (Ed.), *Advances in instructional psychology* (Vol. 1, pp. 167–228). Hillsdale, NJ: Erlbaum.

Casey, M. B., Nuttall, R. L., & Pezaris, E. (1997). Mediators of gender differences in mathematics college entrance test scores: A comparison of spatial skills with internalized beliefs and anxieties. *Developmental Psychology, 33,* 669–680.

Casey, M. B., Nuttall, R. L., & Pezaris, E. (2001). Spatial-mechanical reasoning skills versus mathematics self-confidence as mediators of gender differences on mathematics subtests using cross-national gender-based items. *Journal for Research in Mathematics Education, 32,* 28–57.

Casey, M. B., Nuttall, R. L., Pezaris, E., & Benbow, C. P. (1995). The influence of spatial ability on gender differences in mathematics college entrance test scores across diverse samples. *Developmental Psychology, 31,* 697–705.

Casey, M. B., Pezaris, E., & Nuttall, R. L. (1992). Spatial ability as a predictor of math achievement: The importance of sex and handedness patterns. *Neuropsychologia, 30,* 345.

Casserly, P. L. (1980). Factors affecting female participation in advanced placement programs in mathematics, chemistry and physics. In L. Fox, L. Grody, & D. Tobin (Eds.), *Women and the mathematical mystique* (pp. 138–163). Baltimore: Johns Hopkins University Press.

Casserly, P. L., & Rock, D. (1985). Factors related to young women's persistence and achievement in advanced placement mathematics. In S. F. Chipman, L. R. Brush, & D. M. Wilson (Eds.), *Women and mathematics: Balancing the equation* (pp. 225–247). Hillsdale, NJ: Erlbaum.

Cassirer, E. (1923). *Substance and function.* Chicago: Open Court. (Original work published 1910)

Casti, J. L. (1996). Confronting science's logical limits. *Scientific American, 275,* 102–105.

Casti, J. L. (2001). *Mathematical mountaintops: The five most famous problems of all time.* New York: Oxford University Press.

Catsambis, S. (1994). The path to math: Gender and racial-ethnic differences in mathematics participation from middle school to high school. *Sociology of Education, 67,* 199–215.

Catsambis, S. (2005). The gender gap in mathematics: Merely a step function? In A. M. Gallagher & J. C. Kaufman (Eds.), *Gender differences in mathematics* (pp. 220–245). New York: Cambridge University Press.

Cauley, M. D. (1988) Construction of logical knowledge: Study of borrowing in subtraction. *Journal of Educational Psychology, 80,* 202–205.

Ceci, S. J., & Liker, J. K. (1986a). Academic and nonacademic intelligence: An experimental separation. In R. J. Sternberg & R. K. Wagner (Eds.), *Practical intelligence* (pp. 119–142). Cambridge, UK: Cambridge University Press.

Ceci, S. J., & Liker, J. K. (1986b). A day at the races: A study of IQ, expertise, and cognitive complexity. *Journal of Experimental Psychology: General, 115,* 255–266.

Centra, J. A. (1974). *Women, men, and the doctorate.* Princeton, NJ: Educational Testing Service.

Cerella, J. (1985). Information processing rates in the elderly. *Psychological Bulletin, 98,* 67–83.

Chaitin, G. J. (1995). Randomness in arithmetic and the decline and fall of reductionism in mathematics. In J. Cornwell (Ed.), *Nature's imagination: The frontiers of scientific vision* (pp. 27–44). New York: Oxford University Press.

Champagne, A., Klopfer, L., & Anderson, J. (1980). Factors influencing the learning of classical mechanics. *American Journal of Physics, 48,* 1074–1079.

Charles, R., & Silver, E. A. (Eds.). (1988). *The teaching and assessing of mathematical problem solving.* Hillsdale, NJ: Erlbaum.

Charles, R. I., & Lester, F. K. (1982). *Teaching problem solving: What, why and how.* Palo Alto, CA: Dale Seymour.

Chechile, R. A. (2005). Mathematical psychology. In B. S. Everitt & D. C. Howell (Eds.), *Encyclopedia of statistics in behavioral science* (pp. 1158–1164). Chichester, UK: Wiley,

Chechile, R. A. (2006). Mathematical psychology. In L. Nadel (Ed.), *Encyclopedia of cognitive science* (Vol. 2). New York: Wiley.

Cheetam, B. (1978). Counting and number in Huli. *Papua New Guinea Journal of Education, 14,* 16–27.

Cherian, V. I., & Siweya, J. (1996). Gender and achievement in mathematics by indigenous African students majoring in mathematics. *Psychological Reports, 78,* 27–34.

Chesterton, G. K. (1959). *Orthodoxy.* Garden City, NY: Doubleday. (Original work published 1908)

Chi, M. T. H., Feltovich, P. J., & Glaser, R. (1981). Categorization and representation of physics problems by experts and novices. *Cognitive Science, 5,* 121–152.

Chi, M. T. H., Glaser, R., & Rees, E. (1982). Expertise in problem solving. In R. Sternberg (Ed.), *Advances in the psychology of human intelligence* (Vol. 1, pp. 7–75). Hillsdale, NJ: Erlbaum.

Chi, M. T. H., & Klahr, D. (1975). Span and rate of apprehension in children and adults. *Journal of Experimental Child Psychology, 19,* 434–439.

Chipman, S. F. (2005). Research on the women and mathematics issue: A personal case history. In A. M. Gallagher & J. C. Kaufman (Eds.), *Gender differences in mathematics* (pp. 1–24). New York: Cambridge University Press.

Chipman, S. F., Brush, L. R., & Wilson, D. M. (Eds.). (1985). *Women and mathematics: Balancing the equation.* Hillsdale, NJ: Erlbaum.

Chipman, S. F., Krantz, D. H., & Silver, R. (1992). Mathematics anxiety and science careers among able college women. *Psychological Science, 3,* 292–295.

Chipman, S. F., Krantz, D. H., & Silver, R. (1995). Mathematics anxiety/confidence and other determinants of college major selection. In B. Grevholm & G. Hanna (Eds.), *Gender and mathematics education: An ICMI Study* (pp. 113–120). Lund, Sweden: Lund University Press.

Chipman, S. F., & Thomas, V. G. (1985). Women's participation in mathematics: Outlining the problem. In S. F. Chipman, L. R. Brush, & D. M. Wilson (Eds.), *Women and mathematics: Balancing the equation* (pp. 1–24). Hillsdale, NJ: Erlbaum.

Chipman, S. F., & Wilson, D. (1985). Understanding mathematics course enrollment and mathematics achievement: A synthesis of the research. In S. F. Chipman, L. Brush, & D. Wilson (Eds.), *Women and mathematics: Balancing the equation* (pp. 275–328). Hillsdale, NJ: Erlbaum.

Chittka, L., & Geiger, K. (1995). Can honeybees count landmarks? *Animal Behavior, 49,* 159–164.

Church, R. M., & Gibbon, J. (1982). Temporal generalization. *Journal of Experimental Psychology: Animal Behavior Processes, 8,* 165–186.

Church, R. M., & Meck, W. H. (1984). The numerical attribute of stimuli. In H. L. Roitblat, T. G. Bever, & H. S. Terrace (Eds.), *Animal cognition* (pp. 445–464). Hillsdale, NJ: Erlbaum.

Cipolotti, L. (1995). Multiple route for reading words, why not numbers? Evidence from a case of Arabic dyslexia. *Cognitive Neuropsychology, 12,* 313–342.

Cipolotti, L., & Delacycostello, A. (1995). Selective impairment for simple division. *Cortex, 31,* 433–449.

Cipra, B. (1993). Fermat's last theorem finally yields. *Science, 261,* 32–33.

Cipra, B. (1994a). Fermat proof hits a stumbling block. *Science, 262,* 1967–1968.

Cipra, B. (1994b). Is the fix in on Fermat's last theorem? *Science, 266,* 725.

Cipra, B. (1995). At math meetings, enormous theorem eclipses Fermat. *Science, 267,* 794–795.

Clark, C. M., & Peterson, P. L. (1986). Teachers' thought processes. In M. C. Wittrock, *Second handbook of research on teaching* (pp. 255–296). New York: Macmillan.

Clegg, B. (2003). *Infinity: The quest to think the unthinkable.* New York: Carroll and Graf.

Clement, J. (1982a). Algebra word problems: Thought processes underlying a common misconception. *Journal for Research in Mathematics Education, 13,* 16–30.

Clement, J. (1982b). Students' preconceptions in introductory mechanics. *American Journal of Physics, 50,* 66–71.

Clement, J., & Kaput, J. J. (1979). Letter to the editor. *Journal of Children's Mathematical Behavior, 2,* 208.

Clement, J., Lochhead, J., & Monk, G. S. (1981). Translation difficulties in learning mathematics. *American Mathematical Monthly, 88,* 286–290.

Clement, J., & Rosnick, P. (1980). Learning without understanding: The effect of tutoring strategies on algebra misconceptions. *Journal of Mathematical Behavior, 3,* 3–27.

Clements, D. H. (1980). *Computers in elementary mathematics instruction.* Engelwood Cliffs, NJ: Prentice Hall.

Clements, D. H. (1999a). Subitizing: What is it? Why teach it? *Teaching Children Mathematics, 5,* 400–405.

Clements, D. H. (1999b). Young children and technology. In G. D. Nelson (Ed.), *Dialogue on early childhood science, mathematics, and technology education* (pp. 92–105). Washington, DC: American Association for the Advancement of Science.

Clements, D. H. (2001). Mathematics in the preschool. *Teaching Children Mathematics, 7,* 270–275.

Clements, D. H. (2004). Geometric and spatial thinking in early childhood education. In D. H. Clements, J. Sarama, & A.-M. DiBiase (Eds.), *Engaging young children in mathematics: Standards for early childhood education* (pp. 267–298). Mahwah, NJ: Erlbaum.

Clements, D. H., & Battista, M. T. (1992). Geometry and spatial reasoning. In D. A. Grouws (Ed.), *Handbook of research on mathematics teaching and learning* (pp. 420–464). New York: Macmillan.

Clements, D. H., Swaminathan, S., Hannibal, M.-A., & Sarama, J. (1999). Young children's concepts of shape. *Journal for Research in Mathematics Education, 30,* 192–212.

Clifford, W. K. (1956). The postulates of the science of space. In J. R. Newman (Ed.), *The world of mathematics* (Vol. 1, pp. 552–567). New York: Simon & Schuster. (Original lecture was given in 1873 before the Royal Institution)

Cobb, P., Boufi, A., McClain, K., & Whitenack, J. (1997). Reflective discourse and collective reflection. *Journal for Research in Mathematics Education, 28,* 258–277.

Cobb, P., & Steffe, L. (1983). The constructivist researcher as teacher and model builder. *Journal for Research in Mathematics Education, 14,* 83–94.

Cobb, P., Yackel, E., & Wood, T. (1992). A constructivist alternative to the representational view of mind in mathematics education. *Journal for Research in Mathematics Education, 23,* 2–33.

Coburn, T. G. (1989). The role of computation in the changing mathematics curriculum. In P. R. Trafton (Ed.), *New directions for elementary school mathematics* (pp. 43–56). Reston, VA: National Council of Teachers of Mathematics.

Cohen, L., & Dehaene, S. (1994). Amnesia for arithmetic facts: A single case study. *Brain and Language, 47,* 214–232.

Cohen, L., & Dehaene, S. (1995). Number processing in pure alexia: The effect of hemisphere asymmetries and task demands. *Neurocase, 1,* 121–137.

Cohen, L., & Dehaene, S. (2000). Calculating without reading: Unsuspected residual abilites in pure alexia. *Cognitive Neuropsychology, 17,* 563–583.

Cohen, P. J. (1963). The independence of the continuum hypothesis. *Proceedings of the National Academy of Sciences, 50,* 1143–1148.

Cohen, P. J. (1964). The independence of the continuum hypothesis. Part II. *Proceedings of the National Academy of Sciences, 51,* 105–110.

Coleman, J. (1961). *The adolescent society: The social life of the teenager and its impact on education.* New York: The Free Press.

Collins, G. P. (2004). The shape of space. *Scientific American, 291,* 94–103.

Columbia Associates in Philosophy. (1923). *An introduction to reflective thinking.* New York: Houghton Mifflin.

Conant, L. L. (1956). Counting. In J. R. Newman (Ed.), *The world of mathematics* (Vol. 1, pp. 432–441). New York: Simon & Schuster. (Original work published c. 1906)

Confrey, J. (1988). Multiplication and splitting: Their role in understanding exponential functions. In M. Behr (Ed.), *Proceedings of the Tenth Annual Meeting of the North American Chapter of the International Group for the Psychology of Mathematics Education* (pp. 250–259). DeKalb: Northern Illinois University Press.

Confrey, J. (1991). The concept of exponential functions: A student's perspective. In L. Steffe (Ed.), *Epistemological foundations of mathematical experience* (pp. 124–159). New York: Springer-Verlag.

Confrey, J. (1994). Slitting, similarity, and rate of change: A new approach to multiplication and exponential functions. In G. Harel & J. Confrey (Eds.), *The development of multiplicative reasoning in the learning of mathematics* (pp. 291–330). Albany: State University of New York Press.

Confrey, J., & Smith, E. (1994). Exponential functions, rates of change, and the multiplicative unit. *Educational Studies in Mathematics, 26,* 135–164.

Connor, J. M., Schackman, M., & Serbin, L. A. (1978). Sex-related differences in response to practice on a visual-spatial test and generalization to a related test. *Child Development, 49,* 24–29.

Connor, J. M., & Serbin, L. A. (1985). Visual-spatial skill: Is it important for mathematics? Can it be taught? In S. F. Chipman, L. R. Brush, & D. M. Wilson (Eds.), *Women and mathematics: Balancing the equation* (pp. 151–174). Hillsdale, NJ: Erlbaum.

Connor, J. M., Serbin, L. A., & Schackman, M. (1977). Sex differences in response to training on a visual-spatial test. *Developmental Psychology, 13,* 293–295.

Conway, J. H. (1976). *On numbers and games.* New York: Academic Press.

Conway, J. H. (1980). Monsters and moonshine. *The Mathematical Intelligencer, 2,* 165–171.

Cook, S. A. (1971). The complexity of theorem proving procedures. *Proceedings, Third Annual ACM Symposium on the Theory of Computing* (pp. 151–158). New York: ACM.

Coolidge, J. L. (1950). The number *e. American Mathematical Monthly, 57,* 591–602.

Cooney, J. B., Swanson, H. L., & Ladd, S. F. (1988). Acquisition of mental multiplication skill: Evidence for the transition between counting and retrieval strategies. *Cognition and Instruction, 5,* 323–345.

Cooper, R. G., Jr. (1984). Early number development: Discovering number space with addition and subtraction. In C. Sophian (Ed.), *Origins of cognitive skills: The Eighteenth Annual Carnegie Symposium on Cognition* (pp. 157–192). Hillsdale, NJ: Erlbaum.

Corbitt, M. K. (1985). The impact of computing technology on school mathematics: Report of an NCTM conference. *Mathematics Teacher, 78,* 243–250.

Cordes, S., & Gelman, R. (2005). The young numerical mind. In J. I. D. Campbell (Ed.), *Handbook of mathematical cognition* (pp. 127–142). New York: Psychology Press.

Court, N. A. (1961). *Mathematics in fun and in earnest.* New York: Signet. (Original work published 1935)

Couturat, L. (1975). *De l'infini mathématique* [*On mathematical infinity*]. New York: Georg Olms. (Original work published 1896)

Cox, L. S. (1975). Diagnosing and remediating systematic errors in addition and subtraction computations. *The Arithmetic Teacher, 22,* 151–157.

Crandall, V. C. (1969). Sex differences in expectancy of intellectual and academic reinforcement. In C. P. Smith (Ed.), *Achievement related behavior in children* (pp. 11–45). New York: Russell Sage Foundation.

Crocker, J., Major, B., & Steele, C. M. (1998). Social stigma. In D. Gilbert, S. T. Fiske, & G. Lindzey (Eds.), *Handbook of social psychology* (4th ed., pp. 504–553). Boston: McGraw-Hill.

Crosswhite, F. J. (1987). Cognitive science and mathematics education: A mathematics educator's perspective. In A. H. Schoenfeld (Ed.), *Cognitive science and mathematics education* (pp. 265–277). Hillsdale, NJ: Erlbaum.

Crosswhite, F. J., Dossey, J. A., Swafford, J. O., KcKnight, C. C., & Cooney, T. J. (1985). *Second International Mathematics Study summary report for the United States.* Champaign, IL: Stipes.

Crump, T. (1990). *The anthropology of number*. New York: Cambridge University Press.

Dagenbach, D., & McCloskey, M. (1992). The organization of arithmetic facts in memory: Evidence from a brain-damaged patient. *Brain and Cognition, 20*, 345–366.

D'Andrade, R. G. (1981). The cultural part of cognition. *Cognitive Science, 5*, 179–195.

Dantzig, T. (2005). *Number, the language of science* (4th ed.). New York: Pi Press. (Original work published 1930)

Daston, L. J. (1988). *Classical probability in the enlightment*. Princeton, NJ: Princeton University Press.

David, F. N. (1962). *Games, gods, and gambling: The origins and history of probability*. New York: Hafner.

Davidson, N. (Ed.). (1989). *Cooperative learning in mathematics: A handbook for teachers*. Menlo Park, CA: Addison-Wesley.

Davies, P. C. W. (1992). *The mind of God: The scientific basis for a rational world*. New York: Simon & Schuster.

Davies, P. G., & Spencer, S. J. (2005). The gender-gap artifact: Women's underperformance in quantitative domains through the lens of stereotype threat. In A. M. Gallagher & J. C. Kaufman (Eds.), *Gender differences in mathematics* (pp. 172–188). New York: Cambridge University Press.

Davies, P. G., Spencer, S. J., Quinn, D. M., & Gerhardstein, R. (2002). Consumer images: How television commercials that elicit stereotype threat can restrain women academically and professionally. *Personality and Social Psycholoy Bulletin, 28*, 1615–1628.

Davis, H. (1984). Discrimination of the number three by a raccoon (*Procyon lotor*). *Animal Learning and Behavior, 14*, 57–59.

Davis, H. (1993). Numerical competence in animals: Life beyond Clever Hans. In S. T. Boysen & E. J. Capaldi (Eds.), *The development of numerical competence: Animal and human models* (pp. 109–125). Hillsdale, NJ: Erlbaum.

Davis, H., & Albert, M. (1986). Numerical discrimination by rats using sequential auditory stimuli. *Animal Learning and Behavior, 14*, 57–59.

Davis, H., & Braford, S. A. (1986). Counting behavior by rats in a simulated natural environment. *Ethology, 73*, 265–280.

Davis, H., & Memmott, J. (1982). Counting behavior in animals: A critical evaluation. *Psychological Bulletin, 92*, 547–571.

Davis, H., & Pérusse, R. (1988a). Numerical competence in animals: Definition issues, current evidence, and a new research agenda. *Behavior and Brain Sciences, 11*, 561–579.

Davis, H., & Pérusse, R. (1988b). Numerical competence: From backwater to mainstrean of comparative psychology. *Behavior and Brain Sciences, 11*, 602–615.

Davis, M. (1978). What is a computation? In L. A. Steen (Ed.), *Mathematics today: Twelve informal essays* (pp. 241–267). New York: Springer-Verlag.

Davis, P. J., & Hersh, R. (1972). Nonstandard analysis. *Scientific American, 226*, 78–86.

Davis, P. J., & Hersh, R. (1981). *The mathematical experience*. Boston: Houghton Mifflin.

Davis, R. B. (1982). The postulation of certain specific, explicit, commonly-shared frames. *Journal of Mathematical Behavior, 3*, 167–201.

Davis, R. B. (1983). Complex mathematical cognition. In H. P. Ginsburg (Ed.), *The development of mathematical thinking* (pp. 253–290). New York: Academic Press.

Davis, R. B. (1984). *Learning mathematics: The cognitive science approach to mathematics education*. Norwood, NJ: Ablex.

Davis, R. B., & Maher, C. (1997). How students think. In L. D. English (Ed.), *Mathematical reasoning: Analogies, metaphors, and images* (pp. 93–115). Mahwah, NJ: Erlbaum.

Davis, R. B., & McKnight, C. (1980). The infuence of semantic content on algorithmic behavior. *Journal of Mathematical Behavior, 3*, 39–87.

Davis, R. B., & Vinner, S. (1986). The notion of limit: Some seemingly unavoidable misconception stages. *Journal of Mathematical Behavior, 10*, 117–160.

Davis, R. B., Young, S., & McLoughlin, P. (1982). *The role of understanding in the learning of mathematics*. Urbana: University of Illinois, Curriculum Laboratory.

Dawson, M. M. (1987). Beyond ability grouping: A review of the effectiveness of ability grouping and its alternatives. *School Psychology Review, 16*, 348–369.

Dawson, M. M. (1989). Beyond ability grouping: A review of the effectiveness of ability grouping and its alternatives. *School Psychology Review, 16*, 348–369.

Deci, E. L., & Ryan, R. M. (1985). *Intrinsic motivation and self-determination in human behavior*. New York: Plenum Press.

De Corte, E., & Verschaffel, L. (1987). The effect of semantic structure on first graders' strategies for solving addition and substraction word problems. *Journal for Research in Mathematics Education, 18*, 363–381.

De Corte, E., Verschaffel, L., & De Win, L. (1985). Influence of rewording verbal problems on children's problem representations and solutions. *Journal of Educational Psychology, 77*, 460–470.

De Corte, E., Verschaffel, L., & Van Coillie, V. (1988). Influence of number size, problem structure and response mode on children's solutions of multiplication word problems. *Journal of Mathematical Behavior, 7*, 197–216.

Dedekind, R. (1901). *Essays on the theory of numbers*. La Salle, IL: Open Court. (Original work published 1872)

Dehaene, S. (1992). Varieties of numerical abilities. *Cognition, 44*, 1–42.

Dehaene, S. (1996). The organization of brain activations in number comparison: Event-related potentials and the additive-factor methods. *Journal of Cognitive Neuroscience, 8*, 47–68.

Dehaene, S. (1997). *The number sense: How the mind creates mathematics*. New York: Oxford University Press.

Dehaene, S., & Akhavein, R. (1995). Attention, automaticity and levels of representation in number processing. *Journal of Experimental Psychology: Learning, Memory and Cognition, 21*, 314–326.

Dehaene, S., Bossini, S., & Giraux, P. (1993). The mental representation of parity and number magnitude. *Journal of Experimental Psychology: General, 122*, 371–396.

Dehaene, S., & Changeux, J. (1993). Development of elementary numerical abilities: A neuronal model. *Journal of Cognitive Neuroscience, 5*, 390–407.

Dehaene, S., & Cohen, L. (1991). Two mental calculation systems: A case study of severe acalculia with preserved approximation. *Neuropsychologia, 29*, 1045–1074.

Dehaene, S., & Cohen, L. (1995). Towards an anatomical and functional model of number processing. *Mathematical Cognition, 1*, 83–120.

Dehaene, S., & Cohen, L. (1997). Cerebral pathways for calculation: Double dissociation between rote verbal and quantitative knowledge of arithmetic. *Cortex, 33*, 219–250.

Dehaene, S., Dupoux, E., & Mehler, J. (1990). Is numerical comparison digital: Analogical and symbolic effects in two-digit number comparison. *Journal of Experimental Psychology: Human Perception and Performance, 16*, 626–641.

Dehaene, S., Piazza, M., Pinel, P., & Cohen, L. (2005). Three parietal circuits for number processing. In J. I. D. Campbell (Ed.), *Handbook of mathematical cognition* (pp. 433–453). New York: Psychology Press.

Dehaene, S., Spelke, E., Pinel, P., Stanescu, R., & Tviskin, S. (1999). Sources of mathematical thinking: Behavioral and brain-imaging evidence. *Science, 284*, 970–974.

Dehaene, S., Tzourio, N., Frak, V., Raynaud, L., Cohen, L., Mehler, J., et al. (1996). Cerebral activations during number multiplication and comparison: A PET study. *Neuropsychologia, 34*, 1097–1106.

Delahaye, J.-P. (2006). The science behind sudoku. *Science, 294*, 80–87.

Delaney, P. F., Reder, L. M., Staszewski, J. J., & Ritter, F. E. (1998). The strategy-specific nature of improvement: The power law applies by strategy within task. *Psychological Science, 9*, 1–7.

Delazer, M., & Benke, T. (1997). Arithmetic facts without meaning. *Cortex, 33*, 697–710.

Deloche, G., & Seron, X. (Eds.). (1987). *Mathematical disabilities: A cognitive neuropsychological perspective*. Hillsdale, NJ: Erlbaum.

Den Heyer, K., & Briand, K. (1986). Priming single-digit numbers: Automatic spreading activation dissipates as a function of semantic distance. *American Journal of Psychology, 99*, 315–340.

Dennis, M., & Barnes, M. A. (2002). Math and numeracy in young adults with spina bifida and hydrocephalus. *Developmental Neuropsychology, 21*, 141–156.

Descartes, R. (1644). *Principia philosophiae*. In *Oeuvres* (Vol. 9).

Descartes, R. (1981). Doubt as the starting point of knowledge. In R. D. Tweney, M. E. Doherty, & C. R. Mynatt (Eds.), *On scientific thinking* (pp. 87–88) New York: Columbia University Press. (Original work published 1637)

Devi, S. (1977). *Figuring: The joy of numbers*. New York: Harper & Row.

Devlin, K. (2000a). *The language of mathematics: Making the invisible visible*. New York: Freeman.

Devlin, K. (2000b). Snake eyes in the Garden of Eden. *The Sciences, 40*, 14–17.

Devlin, K. (2002). *The millennium problems: The seven greatest unsolved mathematical puzzles of our time*. New York: Basic Books.

Dew, K. M., Galassi, J. P., & Galassi, M. D. (1983). Mathematics anxiety: Some basic issues. *Journal of Counseling Psychology, 30*, 443–446.

Dick, T. P., & Rallis, S. F. (1991). Factors and influences on high school students' career choices. *Journal for Research in Mathematics Education, 22*, 281–292.

Dienes, Z. P. (1960). *Building up mathematics*. New York: Hutchinson Educational Ltd.

Dienes, Z. P. (1963). *An experimental study of mathematics learning*. New York: Hutchinson Educational Ltd.

Dienes, Z. P. (1967). *Fractions: An operational approach*. New York: Herder & Herder.

Dienes, Z. P., & Golding, E. W. (1971). *An approach to modern mathematics*. New York: Herder & Herder.

Dixon, J. A. (2005). Mathematical problem solving: The roles of exemplar, schema, and relational representations. In J. I. D. Campbell (Ed.), *Handbook of mathematical cognition* (pp. 379–395). New York: Psychology Press.

Dodwell, P. C. (1960). Children's understanding of number and related concepts. *Canadian Journal of Psychology, 14,* 191–205.

Dodwell, P. C. (1961). Children's understanding of number concepts: Characteristics of an individual and of a group test. *Canadian Journal of Psychology, 15,* 29–36.

Dodwell, P. C. (1962). Relations between the understanding of the logic of classes and of cardinal number in children. *Canadian Journal of Psychology, 16,* 152–160.

Donlan, C. (1998). Number without language? Studies of children with specific language impairment. In C. Donlan (Ed.), *The development of mathematical skills* (pp. 255–274). Hove, UK: Psychology Press.

Dossey, J. A., Mullis, I. V. S., Lindquist, M. M., & Chambers, D. L. (1988). *The mathematical report card: Are we measuring up?* Princeton, NJ: Educational Testing Service.

Dowker, A. (1995). Children with specific calculation difficulties. *Links, 2,* 7–12.

Dowker, A. (1998). Individual differences in normal arithmetic development. In C. Donlan (Ed.), *The development of mathematical skills* (pp. 275–302). Hove, UK: Psychology Press.

Drake, S. (1957). *Discoveries and opinions of Galileo.* New York: Doubleday.

Dreyfus, T., & Eisenberg, T. (1996). Different facets of mathematical thinking. In R. J. Sternberg & T. Ben-Zeev (Eds.), *The nature of mathematical thinking* (pp. 253–284). Mahwah, NJ: Erlbaum.

Dubinsky, E. (1994a). A theory and practice of learning college mathematics. In A. H. Schoenfeld (Ed.), *Mathematical thinking and problem solving* (pp. 221–243). Hillsdale, NJ: Erlbaum.

Dubinsky, E. (1994b). Comment in Olkin, I., & Schoenfeld, A. H. A discussion of Bruce Reznick's chapter. In A. H. Schoenfeld (Ed.), *Mathematical thinking and problem solving* (p. 47). Hillsdale, NJ: Erlbaum.

Dudeney, H. E. (1958). *Amusements in mathematics.* New York: Dover.

Dugdale, S. (1982). Green Globs: A microcomputer application for graphing of equations. *Mathematics Teachers, 75,* 208–214.

Duncan, E. M., & McFarland, C. E. (1980). Isolating the effects of symbolic distance and semantic congruity in comparative judgments: An additive-factors analysis. *Memory and Cognition, 8,* 612–622.

Duncker, K. (1945). On problem solving. *Psychological Monographs, 58,* 1–112.

Dunham, W. (1991). *Journey through genius: The great theorems of mathematics.* New York: Penguin.

Dunlap, R. A. (1997). *The golden ratio and Fibonacci numbers.* Singapore: World Scientific.

Durkin, K., Shire, B., Riem, R., Crowther, R. D., & Rutter, D. R. (1986). The social and linguistic context of early number word use. *British Journal of Developmental Psychology, 4,* 269–288.

Du Sautoy, M. (2004). *The music of the primes: Why an unsolved problem in mathematics matters.* New York: HarperCollins.

Duverne, S., & Lemaire, P. (2005). Aging and mental arithmetic. In J. I. D. Campbell (Ed.), *Handbook of mathematical cognition* (pp. 397–411). New York: Psychology Press.

Dweck, C. S. (1975). The role of expectations and attributions in the alleviation of learned helplessness. *Journal of Personality and Social Psychology, 45*, 165–171.

Dweck, C. S., & Bempechat, J. (1983). Children's theories of intelligence. In S. Paris, G. Olsen, & H. Stevenson (Eds.), *Learning and motivation in the classroom* (pp. 239–256). Hillsdale, NJ: Erlbaum.

Dweck, C. S., & Bush, E. (1976). Sex differences in learned helplessness. Part I. Differential debilitation with peer and adult evaluations. *Developmental Psychology, 12*, 147–156.

Dweck, C. S., & Eliott, E. S. (1983). Achievement motivation. In P. H. Mussen (Ed.), *Handbook of child psychology* (Vol. 4). New York: Wiley.

Dweck, C. S., & Reppucci, N. D. (1973). Learned helplessness and reinforcement responsibility in children. *Journal of Personality and Social Psychology, 25*, 109–116.

Dwyer, C. A., & Johnson, L. M. (1997). Grades, accomplishments and correlates. In W. A. Willingham & N. S. Cole (Eds.), *Gender and fair assessment* (pp. 127–156). Mahwah, NJ: Erlbaum.

Dye, N. W., & Very, P. S. (1968). Growth changes in factorial structure by age and sex. *Genetic Psychology Monographs, 78*, 55–88.

Dyson, F. (1979). *Disturbing the universe*. New York: Harper & Row.

Dyson, F. J. (1995). The scientist as a rebel. In J. Cornwell (Ed.), *Nature's imagination: The frontiers of scientific vision* (pp. 1–11). New York: Oxford University Press.

Eccles, J. (1987). Gender roles and women's achievement-related decisions. *Psychology of Women's Quarterly, 11*, 135–172.

Eccles, J. (1994). Understanding women's educational and occupational choices. *Psychology of Women's Quarterly, 18*, 585–609.

Eccles, J., Adler, T. F., Futterman, R., Goff, S. B., Kaczala, C. M., Meece, J. L., et al. (1985). In S. F. Chipman, L. R. Brush, & D. M. Wilson (Eds.), *Women and mathematics: Balancing the equation* (pp. 95–122). Hillsdale, NJ: Erlbaum.

Eccles, J. S., & Blumenthal, P. (1985). Classroom experiences and student gender: Are there differences and do they matter? In L. C. Wilkinson & C. B. Marrett (Eds.), *Gender influences in classroom interaction* (pp. 79–114). Hillsdale, NJ: Erlbaum.

Eccles, J. S., & Jacobs, J. E. (1986). Social forces shape math attitudes and performance. *Signs: Journal of Women in Culture and Society, 11*, 367–380.

Eccles, J. S., & Jacobs, J. E. (1987). Social forces shape math attitudes and performance. In M. R. Walsh (Ed.), *The psychology of women: Ongoing debates* (pp. 333–354). New Haven, CT: Yale University Press.

Eccles, J. S., Jacobs, J. E., & Harold, R. D. (1990). Gender role stereotypes, expectancy effects, and parents' socialization of gender differences. *Journal of Social Issues, 46*, 183–201.

Edwards, W., Lindman, H., & Savage, L. J. (1963). Bayesian statistical inference for psychological research. *Psychological Review, 70*, 193–242.

Egeth, H. E. (1966). Parallel versus serial processes in multidimensional stimulus discrimination. *Perception and Psychophysics, 1*, 245–252.

Ehrenberg, R. G., Brewer, D. J., Gamoran, A., & Willms, J. D. (2001). Class size and student achievement. *Psychological Science in the Public Interest, 2*, 1–30.

Eisenberg, N., Martin, C. L., & Fabes, R. A. (1996). Gender development and gender effects. In D. C. Berliner & R. C. Calfee (Eds.), *Handbook of educational psychology* (pp. 358–306). New York: Simon & Schuster Macmillan.

Ekeland, I. (1993). *The broken dice*. Chicago: University of Chicago Press. (Original work published 1991)

Ekstrom, R. B., French, J. W., & Harman, H. H. (1979). Cognitive factors: Their identification and replication. *Multivariate Behavioral Research Monographs, 79*(2).

Elliott, E. S., & Dweck, C. S. (1988). Goals: An approach to motivation and achievement. *Journal of Personality and Social Psychology, 54*, 5–12.

Ellis, K. (1978). *Number power: In nature, art, and everyday life*. New York: St. Martin's Press.

Ellis, L., & Bentler, P. M. (1973). Traditional sex-determined role standards and sex stereotypes. *Journal of Personality and Social Psychology, 25*, 28–34.

Elstein, A. S., Shulman, L. S., & Sprafka, S. A. (1978). *Medical problem solving: An analysis of clinical reasoning*. Cambridge, MA: Harvard University Press.

Empson, S. B. (1999). Equal sharing and shared meaning: The development of fraction concepts in a first-grade classroom. *Cognition and Instruction, 17*, 283–342.

Engquist, B., & Schmid, W. (Eds.). (2001). *Mathematics unlimited: 2001 and beyond*. Berlin: Springer.

Ericsson, K. A., Delaney, P. F., Weaver, G. A., & Mahadevan, S. (2004). Uncovering the structure of a memorist's superior "basic" memory capacity. *Cognitive Psychology, 49*, 191–237.

Erlwanger, S. H. (1973). Benny's concept of rules and answers in IPI mathematics. *Journal of Children's Mathematical Behavior, 1*, 7–26.

Erlwanger, S. H. (1975). Case studies of children's conception of mathematics, Part I. *Journal of Children's Mathematical Behavior, 1*, 157–183.

Ernest, J. (1976). Mathematics and sex. *American Mathematics Monthly, 83*, 595–615.

Ernest, J. (1980). Is mathematics a sexist discipline? In L. H. Fox, L. Brody, & D. Tobin (Eds.), *Women and the mathematical mystique* (pp. 57–65). Baltimore: Johns Hopkins University Press.

Eves, H. (1983). *An introduction to the history of mathematics*. Philadelphia: Saunders College Publishing. (Original work published 1964)

Ewers, C. A., & Wood, N. L. (1993). Sex and ability differences in children's math self-efficacy and prediction accuracy. *Learning and Individual Differences, 5*, 259–267.

Fairservis, W. A., Jr. (1985). The script of the Indus Valley civilization. *Scientific American, 248*, 58–66.

Falk, R. (1993). *Understanding probability and statistics: A book of problems*. Wellsley, MA: A. K. Peters.

Falk, R. (1994). Infinity: A cognitive challenge. *Theory and Psychology, 4*, 35–60.

Falk, R. (2009). Why don't we live forever? *Teaching Statistics, 31* 78–80.

Falk, R. (in press). The infinite challenge: Levels of conceiving the endlessness of numbers. *Cognition and Instruction*.

Falk, R., & Ben-Lavy, S. (1989). How big is an infinite set? Exploration of children's ideas. In G. Vergnaud, J. Rogalski, & M. Artigue (Eds.), *Actes de la 13ème conference internationale. Psychology of mathematics education* (Vol. 1, pp. 252–259). Paris: G. R. Didactique.

Falk, R., Gassner, D., Ben-Zoor, F., & Ben-Simon, K. (1986). How do children cope with the infinity of numbers? In *Proceedings of the Tenth International Conference for the Psychology of Mathematics Education* (pp. 13–18). London: University of London Institute of Education.

Falk, R., & Lann, A. (2008). The allure of equality: Uniformity in probabilistic and statistical judgment. *Cognitive Psychology, 57,* 293–334.

Falk, R., & Samuel-Cahn, E. (2001). Lewis Carroll's obtuse problem. *Teaching Statistics, 23,* 72–75.

Fan, X., Chen, M., & Matsumoto, A. R. (1997). Gender differences in mathematics achievement: Findings from the National Education Longitudinal Study of 1988. *Journal of Experimental Education, 65,* 229–242.

Farmelo, G. (2003b). Foreword. In G. Farmelo (Ed.), *It must be beautiful: Great equations of modern science* (pp. xi–xviii). New York: Granta Books.

Farnham-Diggory, S. (1980). Learning disabilities: A view from cognitive science. *Journal of the American Academy of Child Psychiatry, 19,* 570–578.

Favreau, O. E. (1997). Sex and gender comparisons: Does null hypothesis testing create a false dichotomy? *Feminism and Psychology, 7,* 63–81.

Fayol, M., & Seron, X. (2005). About numerical representations: Insights from neuropsychological, experimental, and developmental studies. In J. I. D. Campbell (Ed.), *Handbook of mathematical cognition* (pp. 3–22). New York: Psychology Press.

Feigenson, L., & Carey, S. (2003). Tracking individuals via object-files: Evidence from infants' manual search. *Developmental Science, 6,* 568–584.

Feigenson, L., Carey, S., & Hauser, M. (2002). The representations underlying infants' choice of more: Object files versus analog magnitudes. *Psychological Science, 13,* 150–156.

Feigenson, L., Carey, S., & Spelke, E. (2002). Infants' discrimination of number vs. continuous extent. *Cognitive Psychology, 44,* 33–66.

Feinberg, M. M. (1988). *Solving word problems in the primary grades: Addition and subtraction.* Reston, VA: National Council of Teachers of Mathematics.

Fennema, E. (1972). Models and mathematics. *Arithmetic Teacher, 19,* 635–640.

Fennema, E. (1977). Influences of selected cognitive, affective, and educational variables on sex-related differences in mathematics learning and studying. In L. H. Fox, E. Fennema, & J. Sherman (Eds.), *Women and mathematics: Research perspectives for change* (pp. 79–135). Washington, DC: National Institute of Education.

Fennema, E. (1990a). Justice, equity, and mathematics education. In E. Fennema & G. H. Leder (Eds.), *Mathematics and gender* (pp. 1–9). New York: Teachers College Press.

Fennema, E. (1990b). Teachers' beliefs and gender differences in mathematics. In E. Fennema & G. H. Leder (Eds.), *Mathematics and gender* (pp. 169–187). New York: Teachers College Press.

Fennema, E., & Carpenter, T. (1981). The second national assessment and sex-related differences in mathematics. *Mathematics Teacher, 74,* 554–559.

Fennema, E., & Sherman, J. (1977). Sex-related differences in mathematics achievement, spatial visualization and affective factors. *American Educational Research Journal, 14,* 51–71.

Fennema, E., & Sherman, J. (1978). Sex-related differences in mathematics achievement and related factors: A further study. *Journal for Research in Mathematics Education, 9,* 189–203.

Fernandes, D. M., & Church, R. M. (1982). Discrimination of the number of sequential events by rats. *Animal Learning and Behavior, 10,* 171–176.

Ferro, J. M., & Botelho, M. A. S. (1980). Alexia for arithmetic signs: A cause of disturbed calculation. *Cortex, 16,* 175–180.

Ferster, C. B. (1964). Arithmetic behavior in chimpanzees. *Scientific American, 210,* 98–106.

Feurzeig, W. (1988). Apprentice tools: Students as practitioners. In R. S. Nickerson & P. P. Zodhiates (Eds.), *Technology in education: Looking toward 2020* (pp. 97–120). Hillsdale, NJ: Erlbaum.

Feurzeig, W. (2006). Educational technology at BBN. *IEEE Annals of the History of Computing, 28,* 18–31.

Fey, J. T. (1989). Technology and mathematics education: A survey of recent developments and important problems. *Educational Studies in Mathematics, 20,* 237–272.

Fey, J. T. (1990). Quantity. In L. A. Steen (Ed.), *On the shoulders of giants: New approaches to numeracy* (pp. 61–94). Washington, DC: National Academy Press.

Feynman, R. P. (1989). *The character of physical law.* Cambridge, MA: MIT Press. (Original work published 1965)

Fias, W. (2001). Two routes for the processing of verbal numbers: Evidence from the SNARC effect. *Psychological Research—Psychologische Forschung, 65,* 250–259.

Fias, W., & Fischer, M. H. (2005). Spatial representation of numbers. In J. I. D. Campbell (Ed.), *Handbook of mathematical cognition* (pp. 43–54). New York: Psychology Press.

Fias, W., Lammertyn, J., Reynvoet, B., Dupont, P., & Orban, G. A. (2003). Parietal representation of symbolic and nonsymbolic magnitude. *Journal of Cognitive Neuroscience, 15,* 47–56.

Fischer, F. E. (1990). A part-part-whole curriculum for teaching number in the kindergarten. *Journal for Research in Mathematics Education, 21,* 207–215.

Fishbein, E., Tirosh, D., & Hess, P. (1979). The intuition of infinity. *Educational Studies in Mathematics, 10,* 3–40.

Fisher, A. (1988). *The logic of real arguments.* New York: Cambridge University Press.

Fitts, P. M., & Seeger, C. M. (1953). S-R compatibility: Spatial characteristics of stimulus and response codes. *Journal of Experimental Psychology, 46,* 199–210.

Fitzgerald, M., & James, I. (2007). *The mind of the mathematician.* Baltimore: Johns Hopkins University Press.

Flansburg, S. (1993). *Math magic.* New York: William Morrow and Company.

Flegg, G. (1983). *Numbers: Their history and meaning.* Mineola, NY: Dover.

Fleishner, J. E. (1994). Diagnosis and assessment of mathematics disabilities. In G. R. Lyon (Ed.), *Frames of reference for the assessments of learning disabilities: New views on measurement issues* (pp. 441–458). Baltimore: Brooks.

Fletcher, J. M. (1985). Memory for verbal and nonverbal stimuli in learning disability subgroups: Analysis by selective reminding. *Journal of Experimental Child Psychology, 40,* 244–259.

Folk, C. L., Egeth, H., & Kwak, H. W. (1988). Subitizing: Direct apprehension or serial processing? *Perception and Psychophysics, 44,* 313–320.

Folsom, M. (1975). Operations on whole number. In J. N. Payne (Ed.), *Mathematics learning in early childhood* (pp. 162–190). Reston, VA: National Council of Teachers of Mathematics.

Forsyth, A. R. (1963). Mathematics, in life and thought. In R. M. Hutchins, M. J. Adler, & C. Fadiman (Eds.), *Gateway to the great books: Mathematics* (Vol. 9, pp. 26–46). Chicago: William Benton. (Original lecture given 1928)

Fox, L. H. (1977). The effects of sex-role socialization on mathematics participation and achievement. In J. Shoemaker (Ed.), *Women and mathematics: Research perspectives for change* (pp. 1–78). Washington, DC: The National Institute of Education.

Fox, L. H., Brody, L., & Tobin, D. (Eds.). (1980). *Women and the mathematical mystique*. Baltimore: Johns Hopkins University Press.

Fox, L. H., Brody, L., & Tobin, D. (1985). The impact of early intervention programs upon course-taking and attitudes in high school. In S. F. Chipman, L. R. Brush, & D. M. Wilson (Eds.), *Women and mathematics: Balancing the equation* (pp. 249–274). Hillsdale, NJ: Erlbaum.

Fox, L. H., & Cohn, S. J. (1980). Sex differences in the development of precocious mathematical talent. In L. H. Fox, L. Brody, & D. Tobin (Eds.), *Women and the mathematical mystique* (pp. 94–111). Baltimore: Johns Hopkins University Press.

Fox, L. H., Fennema, E., & Sherman, J. (Eds.). (1977). *Women and mathematics: Research perspectives for change* (pp. 79–135). Washington, DC: National Institute of Education.

Freudenthal, H. (1983). *Didactical phenomenology of mathematical structures*. Boston: D. Reidel.

Friberg, J. (1984). Numbers and measure in the earliest written records. *Scientific American, 250,* 110–118.

Friedlander, E. (1965). *Psychology in scientific thinking*. New York: Philosophical Library.

Friedman, L. (1989). Mathematics and the gender gap: A meta-analysis of recent studies on sex differences in mathematical tasks. *Review of Educational Research, 59,* 185–213.

Friedrich, W. N., Lovejoy, M. C., Shaffer, J., Shurtleff, D., & Beilke, R. L. (1991). Cognitive abilities and achievement status of children with myelomeningocele: A contemporary sample. *Journal of Pediatric Psychology, 16,* 423–428.

Friend, J. N. (1954). *Numbers fun and facts*. New York: Charles Scribner's Sons.

Fruchter, B. (1954). Measurement of spatial abilities: History and background. *Educational and Psychological Measurement, 14,* 387–395.

Frydman, O., & Bryant, P. (1988). Sharing and the understanding of number equivalence by young children. *Cognitive Development, 3,* 323–339.

Frye, D., Braisby, N., Lowe, J., Maroudas, C., & Nicholls, J. (1989). Young children's understanding of cardinality and counting. *Child Development, 60,* 1158–1171.

Furley, D. (1967). *Two studies in the Greek atomists.* Princeton, NJ: Princeton University Press.

Fuson, K. C. (1982). An analysis of the counting-on solution procedure in addition. In T. P. Carpenter, J. M. Moser, & T. A. Romberg (Eds.), *Addition and subtraction: A cognitive perspective* (pp. 67–78). Hillsdale, NJ: Erlbaum.

Fuson, K. C. (1988). *Children's counting and concepts of number.* New York: Springer-Verlag.

Fuson, K. C. (1991). Children's early counting: Saying the number-word sequence, counting objects, and understanding cardinality. In K. Durkin & B. Shire (Eds.), *Language in mathematical education: Research and practice* (pp. 27–39). Milton Keynes, PA: Open University Press.

Fuson, K. C. (1992a). Relationships between counting and cardinality from age 2 to age 8. In J. Bideaud, C. Meljac, & J.-P. Fischer (Eds.), *Pathways to number: Children's developing numerical abilities* (pp. 127–149). Hillsdale, NJ: Erlbaum.

Fuson, K. C. (1992b). Research on whole number addition and subtraction. In D. Grouws (Ed.), *Handbook of research on mathematics teaching and learning* (pp. 243–275). New York: Macmillan.

Fuson, K., & Briars, D. (1990). Using a base-ten blocks learning/teaching approach for first- and second-grade place value and multidigit addition and subtraction. *Journal for Research in Mathematics Education, 21,* 180–206.

Fuson, K. C., De La Cruz, Y., Smith, S. B., Lo Cicero, A., Hudson, K., Ron, P., et al. (2000). Blending the best of the twentieth century to achieve a mathematics equity pedagogy in the twenty-first century. In M. J. Burke (Ed.), *Learning mathematics for a new century* (pp. 197–212). Reston, VA: National Council of Teachers of Mathematics.

Fuson, K. C., & Hall, J. W. (1983). The acquisition of early number word meanings: A conceptual analysis and review. In H. P. Ginsburg (Ed.), *The development of mathematical thinking* (pp. 49–107). New York: Academic Press.

Fuson, K. C., & Kwon, Y. (1991). Chinese-based regular and European irregular systems of number words: The disadvantages for English-speaking children. In K. Durkin & B. Shire (Eds.), *Language in mathematical education: Research and practice* (pp. 211–226). Milton Keynes, PA: Open University Press.

Fuson, K. C., & Kwon, Y. (1992a). Korean children's single-digit addition and subtraction: Numbers structured by ten. *Journal for Research in Mathematics Education, 23,* 148–165.

Fuson, K. C., & Kwon, Y. (1992b). Korean children's understanding of multidigit addition and subtracton. *Child Development, 63,* 491–506.

Fuson, K. C., Richards, J., & Briars, D. J. (1982). The acquisition and elaboration of the number word sequence. In C. Brainerd (Ed.), *Progress in cognitive development: Children's logical and mathematical cognition* (Vol. 1, pp. 33–92). New York: Springer-Verlag.

Fuson, K. C., Stigler, J. W., & Bartsch, K. (1988). Grade placement of addition and subtraction topics in Japan, Mainland China, the Soviet Union, Taiwan, and the United States. *Journal for Research in Mathematics Education, 19,* 449–456.

Fuys, D. J., & Liebov, A. K. (1993). Geometry and spatial sense. In R. J. Jensen (Ed.), *Research ideas for the classroom: Early childhood mathematics* (pp. 195–222). New York: Macmillan.

Gagné, R. M. (1968). Learning hierarchies. *Educational Psychologist, 6,* 1–9.

Gagné, R. M., Mayor, J. R., Garstens, H. L., & Paradise, N. E. (1962). Factors in acquiring knowledge of a mathematical task. *Psychological Monographs: General and Applied, 76,* Whole No. 526.

Galelei, G. (1914). *Discorsi e dimostrazioni matematiche, intorno a due nuove scienze* [*Discourses and mathematical demonstrations relating to two new sciences*] (H. Crew & A. deSalvio, Trans.). New York: Macmillan. (Original work published 1638)

Gallagher, A. M., & De Lisi, R. (1994). Gender differences in Scholastic Aptitude Test: Mathematics problem solving among high-ability students. *Journal of Educational Psychology, 86,* 204–211.

Gallagher, A. M., De Lisi, R., Holst, P. C., McGillicuddy-DeLisi, A. V., Morley, M., & Cahalan, C. (2000). Gender differences in advanced mathematical problem solving. *Journal of Experimental Child Psychology, 75,* 165–190.

Gallagher, A. M., & Kaufman, J. C. (Eds.). (2005). *Gender differences in mathematics: An integrative psychological approach.* New York: Cambridge University Press.

Gallistel, C. R. (1988). Counting versus subitizing versus the sense of number. *Behavioral and Brain Sciences, 11,* 565–586.

Gallistel, C. R. (1990). *The organization of learning.* Cambridge, MA: MIT Press.

Gallistel, C. R. (1993). A conceptual framework for the study of numerical estimation and arithmetic reasoning in animals. In S. T. Boysen & E. J. Capaldi (Eds.), *The development of numerical competence: Animal and human models* (pp. 211–223). Hillsdale, NJ: Erlbaum.

Gallistel, C. R., & Gelman, R. (1990). The what and how of counting. *Cognition, 34,* 197–199.

Gallistel, C. R., & Gelman, R. (1991). Subitizing: The preverbal counting process. In W. Kessen, A. Ortony, & F. Craik (Eds.), *Memories, thoughts, and emotions: Essays in honor of George Mandler* (pp. 65–81). Hillsdale, NJ: Erlbaum.

Gallistel, C. R., & Gelman, R. (1992). Preverbal and verbal counting and computation. *Cognition, 44,* 43–74.

Gallistel, C. R., & Gelman, R. (2000). Non-verbal numerical cognition: From reals to integers. *Trends in Cognitive Sciences, 4,* 59–65.

Gardner, M. (1956). *Mathematics, magic and mystery.* New York: Dover.

Gardner, M. (1959). *The Scientific American book of mathematical puzzles and diversions.* New York: Simon & Schuster.

Gardner, M. (1961). *The second Scientific American book of mathematical puzzles and diversions.* New York: Simon & Schuster.

Gardner, M. (1977). Mathematical games. *Scientific American, 237,* 18–28.

Garland, T. H. (2000). *Fibonacci numbers in nature.* White Plains, NY: Dale Seymour Publications.

Gathercole, S. E., & Pickering, S. J. (2000). Assessment of working memory in six- and seven-year old children. *Journal of Educational Psychology, 92,* 377–390.

Gay, J., & Cole, M. (1967). *The new mathematics and an old culture: A study of learning among the Kpelle of Nigeria.* New York: Holt, Rinehart & Winston.

Geary, D. C. (1990). A componential analysis of an early learning deficit in mathematics. *Journal of Experimental Child Psychology, 49,* 363–383.

Geary, D. C. (1993). Mathematical disabilities: Cognitive, neuropsychological, and genetic components. *Psychological Bulletin, 114,* 345–362.

Geary, D. C. (1994). *Children's mathematical development: Research and practical applications.* Washington, DC: American Psychological Association.

Geary, D. C. (1996a). Biology, culture, and cross-national differences in mathematical ability. In R. J. Sternberg & T. Ben-Zeev (Eds.), *The nature of mathematical thinking* (pp. 145–171). Mahwah, NJ: Erlbaum.

Geary, D. C. (1996b). Sexual selection and sex differences in mathematical studies. *Behavioral and Brain Science, 19,* 229–284.

Geary, D. C. (1998). *Male, female: The evolution of human sex differences.* Washington, DC: American Psychological Association.

Geary, D. C., Bow-Thomas, C. C., Fan, L., & Siegler, R. S. (1993). Even before formal instruction, Chinese children outperform American children in mental addition. *Cognitive Development, 8,* 517–529.

Geary, D. C., Bow-Thomas, C. C., Liu, F., & Siegler, R. S. (1996). Development of arithmetical competencies in Chinese and American children: Influence of age, language, and schooling. *Child Development, 67,* 2022–2044.

Geary, D. C., Bow-Thomas, C. C., & Yao, Y. (1992). Counting knowledge and skill in cognitive addition: A comparison of normal and mathematically disabled children. *Journal of Experimental Child Psychology, 554,* 372–391.

Geary, D. C., & Brown, S. C. (1991). Cognitive addition: Strategy choice and speed-of-processing differences in gifted, normal and mathematically disabled children. *Developmental Psychology, 27,* 398–406.

Geary, D. C., Brown, S. C., & Samaranayake, V. A. (1991). Cognitive addition: A short longitudinal study of strategy choice and speed-of-processing differences in normal and mathematically disabled children. *Developmental Psychology, 27,* 787–797.

Geary, D. C., & Burlingham-Dubree, M. (1989). External validation of the strategy choice model for addition. *Journal of Experimental Child Psychology, 47,* 175–192.

Geary, D. C., Fan, L., & Bow-Thomas, C. C. (1992). Numerical cognition: Loci of ability differences comparing children from China and the United States. *Psychological Science, 3,* 180–185.

Geary, D. C., & Hoard, M. K. (2005). Learning disabilities in arithmetic and mathematics: Theoretical and empirical perspectives. In J. I. D. Campbell (Ed.), *Handbook of mathematical cognition* (pp. 253–267). New York: Psychology Press.

Geary, D. C., Saults, S. J., Liu, F., & Hoard, M. K. (2000). Sex differences in spatial cognition, computational fluency and arithmetical reasoning. *Journal of Experimental Child Psychology, 77,* 337–353.

Gellert, W., Küstner, H., Hellwich, M., & Kästner, H. (Eds.). (1977). *The VNR concise encyclopedia of mathematics.* New York: Von Nostrand Reinhold.

Gelman, R. (1972). Logical capacity of very young chidren: Number invariance rules. *Child Development, 43,* 75–90.

Gelman, R. (1991). Epigenetic foundations of knowledge structures: Initial and transcendent constructions. In S. Carey & R. Gelman (Eds.), *The epigenesis of mind: Essays on biology and cognition* (pp. 293–322). Hillsdale, NJ: Erlbaum.

Gelman, R., & Gallistel, C. R. (1978). *The child's understanding of number*. Cambridge, MA: Harvard University Press.

Gentile, J. R., & Monaco, N. M. (1986). Learned helplessness in mathematics: What educators should know. *Journal of Mathematical Behavior, 5,* 159–178.

Gerstmann, J. (1940). Syndrome of finger agnosia, disorientation for right and left, agraphia, and acalculia. *Archives of Neurology: Psychiatry, 44,* 398–408.

Geschneider, G. (1997). *Psychophysics: The fundamentals* (3rd ed.). Mahwah, NJ: Erlbaum.

Ghyka, M. (1946). *Esthétique des proportions dans la nature et dans les arts* [*Aesthetics of proportions in nature and in the arts*]. New York: Sheed and Ward. (Original work published 1927)

Gibb, E. (1956). Children's thinking in the process of subtraction. *Journal of Experimental Education, 25,* 71–80.

Gibbon, J., & Church, R. M. (1990). Representation of time. *Cognition, 37,* 23–54.

Gibbs, W. W. (2003). A digital slice of pi. *Scientific American, 288,* 23–24.

Ginsburg, H. P. (1977). *Children's arithmetic: The learning process*. New York: Van Nostrand.

Ginsburg, H. P. (1982). The development of addition in the contexts of culture, social class, and race. In T. P. Carpenter, J. M. Moser, & T. A. Romberg (Eds.), *Addition and subtraction: A cognitive perspective* (pp. 191–210). Hillsdale, NJ: Erlbaum.

Ginsburg, H. P. (1989). *Children's arithmetic: The learning process* (2nd ed.). Austin, TX: Pro-Ed.

Ginsburg, H. P., & Allardice, B. S. (1999). Children's difficulties with school mathematics. In B. Rogoff & J. Lave (Eds.), *Everyday cognition: Its development in social context* (pp. 194–219). Cambridge, MA: Harvard University Press.

Ginsburg, H. P., & Baron, J. (1993). Cognition: Young children's construction of mathematics. In R. J. Jensen (Ed.), *Research ideas for the classroom: Early childhood mathematics* (pp. 3–21). New York: Macmillan.

Ginsburg, H. P., Klein, A., & Starkey, P. (1998). The development of children's mathematical thinking: From research to practice. In I. E. Siegel & K. A. Renninger (Eds.), *Child's psychology in practice* (Vol. 4 of W. Damon [Ed.], *Handbook of child psychology*, 5th ed., pp. 401–476). New York: Wiley.

Ginsburg, H. P., Posner, J. K., & Russell, R. L. (1981a). The development of knowledge concerning written arithmetic: A cross-cultural study. *International Journal of Psychology, 16,* 13–34.

Ginsburg, H. P., Posner, J. K., & Russell, R. L. (1981b). The development of mental addition as a function of schooling and culture. *Journal of Cross-Cultural Psychology, 12,* 163–178.

Ginsburg, H. P., Posner, J. K., & Russell, R. L. (1981c). Mathematics learning difficulties in African children. *The Quarterly Newsletter of the Laboratory of Comparative Human Development, 3,* 8–11.

Ginsburg, H. P., & Russell, R. L. (1981). Social class and racial influences on early mathematical thinking. *Monographs of the Society for Research in Child Development, 46,* No. 193.

Gladwin, T. (1970). *East is a big bird*. Cambridge, MA: Harvard University Press.

Glaser, R. E. (1984). Education and thinking: The role of knowledge. *American Psychologist, 39,* 93–104.

Gleick, J. (2004). *Isaac Newton*. New York: Vintage Books.

Gödel, K. (1930). Die vollständigkeit der axiome des logischen funktionkalküls. *Monatshefte für Mathematik und Physik, 37*, 349–360.

Gödel, K. (1931). Über formal unentscheidbare sätze der *Principia mathematica* und verwandter systeme I. *Monatshefte für Mathematik und Physik, 38*, 173–198.

Gödel, K. (1964). What is Cantor's continuum problem? In P. Benacerraf & H. Putnam (Eds.), *Philosophy of mathematics* (pp. 258–273). Englewoods Cliffs, NJ: Prentice Hall. (Original work published 1947)

Godkewitsch, M. (1974). The golden section: An artifact of stimulus range and measure of preference. *American Journal of Psychology, 87*, 269–277.

Goffman, E. (1963). *Behavior in public places: Notes on the social organization of gatherings*. New York: The Free Press.

Goldstein, A. G., & Chance, J. E. (1965). Effects of practice on sex-related differences in performance on embedded figures. *Psychonomic Science, 3*, 361–362.

Goldstein, D., & Stocking, V. B. (1994). TIP studies of gender differences in talented adolescents. In K. A. Heller & E. A. Hany (Eds.), *Competence and responsibility* (Vol. 2, pp. 190–203). Ashland, OH: Hofgreve.

Gonchar, A. J. (1975). *A study in the nature and development of the natural number concept: Initial and supplementary analyses*. Tech. Rep. 340, Research and Development Center for Cognitive Learning, University of Wisconsin.

Good, I. J. (1983). *Good thinking: The foundations of probability and its applications*. Minneapolis: University of Minnesota Press.

Good, T. L., & Findley, M. J. (1985). Sex role expectations and achievement. In J. B. Dusek (Ed.), *Teacher expectancies* (pp. 271–302). Hillsdale, NJ: Erlbaum.

Goodman, A. W. (1965). *The pleasures of math*. New York: Macmillan.

Goodman, C. H. (1943). A factorial analysis of Thurstone's sixteen Primary Mental Abilities tests. *Psychometrika, 8*, 141–151.

Gordon, P. (2004). Numerical cognition without words: Evidence from Amazonia. *Science, 306*, 496–499.

Gottfredson, L. S. (2005). Suppressing intelligence research: Hurting those we intend to help. In R. H. Wright & N. A. Cummings (Eds.), *Destructive trends in mental health: The well-intentioned path to harm* (pp. 155–186). New York: Taylor & Francis.

Graeber, A. O., & Tirosh, D. (1988). Multiplication and division involving decimals: Preservice elementary teachers' performance and beliefs. *Journal of Mathematical Behavior, 7*, 263–280.

Graeber, A. O., & Tirosh, D. (1990). Insights fourth and fifth graders bring to multiplication and division with decimals. *Educational Studies in Mathematics, 21*, 565–588.

Graeber, A. O., Tirosh, D., & Glover, R. (1989). Preservice teachers' misconceptions in solving verbal problems in multiplication and division. *Journal for Research in Mathematics Education, 20*, 95–102.

Graf, G., & Ruddell, J. (1972). Sex differences in problem solving as a function of problem content. *Journal of Educational Research, 65*, 451–452.

Grafman, J., Kampen, D., Rosenberg, J., Salazar, A. M., & Boller, F. (1989). The progressive breakdown of number processing and calculation ability: A case study. *Cortex, 25*, 121–133.

Grandy, J. (1994). *GRE—Trends and profiles: Statistics about general test examinees by sex and ethnicity*. Princeton, NJ: Educational Testing Service.

Gray, J. (2000). *The Hilbert challenge*. Oxford: Oxford University Press.

Green, D. M., & Swets, J. A. (1966). *Signal detecton theory and psychophysics*. New York: Wiley.

Greene, B. (2004). *The fabric of the cosmos: Space, time, and the texture of reality*. New York: Knopf.

Greeno, J. G. (1978). A study of problem solving. In R. Glaser (Ed.), *Advances in instructional psychology* (pp. 13–75). Hillsdale, NJ: Erlbaum.

Greeno, J. G. (1980). Some examples of cognitive task analysis with instructional implications. In R. E. Snow, P. Federico, & W. E. Montague (Eds.), *Aptitude, learning, and instruction* (Vol. 2, pp. 1–24). Hillsdale, NJ: Erlbaum.

Greeno, J. G. (1983). Forms of understanding in mathematical problem solving. In S. G. Paris, G. M. Olson, & H. W. Stevenson (Eds.), *Learning and motivation in the classroom* (pp. 83–111). Hillsdale, NJ: Erlbaum.

Greeno, J. G. (1985). Cognitive theory and curriculum design: A discussion of Thompson's paper. In E. A. Silver (Ed.), *Teaching and learning mathematical problem solving: Multiple research perspectives* (pp. 237–243). Hillsdale, NJ: Erlbaum.

Greeno, J. G. (1987). Instructional representations based on research about understanding. In A. H. Schoenfeld (Ed.), *Cognitive science and mathematics education* (pp. 61–88). Hillsdale, NJ: Erlbaum.

Greeno, J. G. (1994). Comments on Susanna Epp's chapter. In A. H. Schoenfeld (Ed.), *Mathematical thinking and problem solving* (pp. 270–278). Hillsdale, NJ: Erlbaum.

Greeno, J. G., Riley, M. S., & Gelman, R. (1984). Conceptual competence and children's counting. *Cognitive Psychology, 16*, 94–134.

Greer, B. (1987). Nonconservation of multiplication and division involving decimals. *Journal for Research in Mathematics Education, 18*, 37–45.

Greer, B. (1988). Nonconservation of multiplication and division: Analysis of a symptom. *Journal of Mathematical Behavior, 7*, 281–298.

Grewel, F. (1969). The acalculia. In P. J. Vinken & G. W. Bruyn (Eds.), *Handbook of clinical neurology* (Vol. 4, pp. 181–196). New York: Wiley.

Gribbin, J., & Rees, M. (1989). *Cosmic coincidences: Dark matter, mankind, and anthropic cosmology*. New York: Bantam Books.

Groen, G. J., & Parkman, J. M. (1972). A chronometric analysis of simple addition. *Psychological Review, 79*, 329–343.

Groen, G. J., & Resnick, L. B. (1977). Can preschool children invent addition algorithms? *Journal of Educational Psychology, 69*, 645–652.

Grouws, D. A. (1972). Open sentences: Some instructional considerations from research. *Arithmetic Teacher, 19*, 595–599.

Gruber, O., Indefrey, P., & Kleinschmidt, A. (2001). Dissociating neural correlates of cognitive components in mental calculation. *Cerebral Cortex, 11*, 350–359.

Grünbaum, A. (1967). *Modern science and Zeno's paradoxes*. Middletown, CN: Wesleyan University Press.

Grünbaum, A. (2001a). Modern science and refutation of the paradoxes of Zeno. In W. C. Salmon (Ed.), *Zeno's paradoxes* (pp. 164–175). Indianapolis: Hackett Publishing. (Original work published 1955)

Grünbaum, A. (2001b). The Zeonian runners. In W. C. Salmon (Ed.), *Zeno's paradoxes* (pp. 204–218). Indianapolis: Hackett Publishing. (Original work published 1969)

Hadamard, J. (1954). *The psychology of invention in the mathematical field*. New York: Dover. (Original work published 1945)

Hahn, H. (1956). Infinity. In J. R. Newman (Ed.), *The world of mathematics* (Vol. 3, pp. 1593–1611). New York: Simon & Schuster. (Original lecture given unknown, but before 1934, when Hahn died)

Haith, M. M., & Benson, J. B. (1998). Infant cognition. In D. Kuhn & R. Siegler (Eds.), *Handbook of child psychology: Cognition, perception, and language* (Vol. 2, 5th ed., pp. 199–254). New York: Wiley.

Haladyna, T., Shaughnessy, J. M., & Shaughnessy, J. M. (1983). A causal analysis of attitude toward mathematics. *Journal for Research in Mathematics Education, 14*, 19–29.

Hall, R. (1980). *Philosophers at war*. New York: Cambridge University Press.

Halmos, P. R. (1980). The heart of mathematics. *The American Mathematical Monthly, 87*, 519–524.

Halmos, P. R. (1985). *I want to be a mathematician: An automathography in three parts*. New York: Springer-Verlag.

Halpern, D. F. (1989). *Thought and knowledge: An introduction to critical thinking* (2nd ed.). Hillsdale, NJ: Erlbaum.

Halpern, D. F. (2000). *Sex differences in cognitive abilities* (3rd ed.). Mahwah, NJ: Erlbaum.

Halpern, D. F., Wai, J., & Saw, A. (2005). A psychobiosocial model: Why females are sometimes greater than and sometimes less than males in math achievement. In A. M. Gallagher & J. C. Kaufman (Eds.), *Gender differences in mathematics* (pp. 48–72). New York: Cambridge University Press.

Hamann, M. S., & Ashcraft, M. H. (1985). Simple and complex addition across development. *Journal of Experimental Child Psychology, 40*, 49–72.

Hamming, R. W. (1980). The unreasonable effectiveness of mathematics. *American Mathematics Monthly, 87*, 81–90.

Hammond, A. L. (1978). Mathematics—Our invisible culture. In L. A. Steen (Ed.), *Mathematics today: Twelve informal essays* (pp. 15–34). New York: Springer-Verlag.

Hardy, G. H. (1940). *Ramanajun*. Cambridge, UK: Cambridge University Press.

Hardy, G. H. (1989). A mathematician's apology. Cambridge, UK: Cambridge University Press. (Original work published 1940)

Harel, G., & Behr, M. (1989). Structure and hierarchy of missing value proportion problems and their representation. *Journal of Mathematical Behavior, 8*, 77–119.

Harel, G., Behr, M., Post, T., & Lesh, R. (1992). The blocks task: Comparative analyses with other proportion tasks, and qualitative reasoning skills among seventh grade children in solving the task. *Cognition and Instruction, 9*, 45–96.

Harel, G., Behr, M., Post, T., & Lesh, R. (1994). The impact of the number type on the solution of multiplication and division problems: Further considerations. In G. Harel & J. Confrey (Eds.), *The development of multiplicative reasoning in the learning of mathematics* (pp. 363–384). Albany: State University of New York Press.

Harel, G., & Confrey, J. (Eds.). (1994). *The development of multiplicative reasoning in the learning of mathematics*. Albany: State University of New York Press.

Harnisch, D. L., Steinkamp, M. W., Tsai, S. L., & Walberg, H. J. (1986). Cross-national differences in mathematics attitude and achievement among seventeen-year-olds. *International Journal of Educational Development, 6,* 233–244.

Hatano, G. (1982). Learning to add and subtract: A Japanese perspective. In T. P. Carpenter, J. M. Moser, & T. A. Romberg (Eds.), *Addition and subtraction: A cognitive perspective* (pp. 211–223). Hillsdale, NJ: Erlbaum.

Hatano, G., Miyake, Y., & Binks, M. G. (1977). Performance of expert abacus operators. *Cognition, 5,* 95–110.

Hawkins, H. L. (1969). Parallel processing in complex visual discrimination. *Perception and Psychophysics, 5,* 56–64.

Hawkins, J., Sheingold, K., Gearhart, M., & Berger, C. (1982). Microcomputers in schools: Impact on the social life of elementary classrooms. *Journal of Applied Developmental Psychology, 3,* 361–373.

Hawley, P. (1971). What women think men think: Does is affect their career choice? *Journal of Counseling Psychology, 19,* 193–199.

Hawley, P. (1972). Perceptions of male models of femininity related to career choice. *Journal of Counseling Psychology, 19,* 308–313.

Hayes, J. R. (1989). *The complete problem solver* (2nd ed.). Hillsdale, NJ: Erlbaum.

Hayes, J. R., & Simon, H. A. (1974). Understanding problem instructions. In L. W. Gregg (Ed.), *Knowledge and cognition* (pp. 161–200). Hillsdale, NJ: Erlbaum.

Hayes, J. R., Waterman, D. A., & Robinson, C. S. (1977). Identifying relevant aspects of a problem text. *Cognitive Science, 1,* 297–313.

Hécaen, H. (1962). Clinical sympotmatology in right and left hemispheric lesions. In V. B. Mountecastle (Ed.), *Interhemispheric relations and cerebral dominance* (pp. 215–243). Baltimore: Johns Hopkins University Press.

Hedges, L. V., & Nowell, A. (1995). Sex differences in mental test scores, variability, and numbers of high-scoring individuals. *Science, 269,* 41–45.

Hegarty, M., Mayer, R. E., & Green, C. (1992). Comprehension of arithmetic word problems: Evidence from students' eye fixations. *Journal of Educational Psychology, 84,* 76–84.

Hegarty, M., Mayer, R. E., & Monk, C. A. (1995). Comprehension of arithmetic word problems. *Journal of Educational Psychology, 85,* 18–32.

Heinrich, B., & Bugnyar, T. (2007). Just how smart are ravens? *Scientific American, 296,* 64–71.

Heller, J. I., & Hungate, H. N. (1985). Implications for mathematics instruction of research on scientific problem solving. In E. A. Silver (Ed.), *Teaching and learning mathematical problem solving: Multiple research perspectives* (pp. 83–112). Hillsdale, NJ: Erlbaum.

Hellman, H. (2006). *Great feuds in mathematics: Ten of the liveliest disputes ever.* New York: Wiley.

Hembree, R. (1991). Experimental and relational studies in problem solving: A meta-analysis. *Journal for Research in Mathematics Education, 23,* 242–273.

Hembree, R., & Dessart, D. J. (1986). Effects of hand-held calculators in precollege mathematics education: A meta-analysis. *Journal for Research in Mathematics Education, 17,* 83–99.

Hembree, R., & Dessart, D. J. (1992). Research on calculators in mathematics education. In J. T. Fey (Ed.), *Calculators in mathematics education, 1992 yearbook of the National Council of Teachers of Mathematics* (pp. 23–32). Reston, VA: National Council of Teachers of Mathematics.

Hembree, R., & Marsh, H. (1993). Problem solving in early childhood: Building foundations. In R. J. Jensen (Ed.), *Research ideas for the classroom: Early childhood mathematics* (pp. 151–170). New York: Macmillan.

Hempel, C. G. (1945). Studies in the logic of confirmation (I). *Mind, 54,* 1–26.

Hempel, C. G. (1956a). Geometry and empirical science. In J. R. Newman (Ed.), *The world of mathematics* (Vol. 3, pp. 1635–1646). New York: Simon & Schuster. (Original work published 1935)

Hempel, C. G. (1956b). On the nature of mathematical truth. In J. R. Newman (Ed.), *The world of mathematics* (Vol. 3, pp. 1619–1634). New York: Simon & Schuster. (Original work published 1945)

Henderson, A. (1987). From the teacher's side of the desk. In A. H. Schoenfeld (Ed.), *Cognitive science and mathematics education* (pp. 149–164). Hillsdale, NJ: Erlbaum.

Henik, A., & Tzelgov, J. (1982). Is 3 greater than 5? The relation between physical and semantic size in comparison tasks. *Memory and Cognition, 10,* 389–395.

Henrion, C. (1997). *Women in mathematics.* Bloomington: Indiana University Press.

Hermelin, B., & O'Connor, N. (1986). Idiot savant calendrical calculators: Rules and regularities. *Psychological Medicine, 16,* 885–893.

Hermelin, B., & O'Connor, N. (1990). Factors and primes: A specific numerical ability. *Psychological Medicine, 20,* 163–169.

Hersh, R. (1997). *What is mathematics really?* New York: Oxford University Press.

Hersh, R. (2001). Mathematical menopause, or, a young man's game? What is it like to be still a mathematician and no longer young? *Mathematical Intelligencer, 23*(3), 52–60.

Herz-Fischler, R. A. (1998). *A mathematical history of the golden number.* Mineola, NY: Dover.

Hess, R. D., Chang, C., & McDevitt, T. M. (1987). Cultural variations in family beliefs about children's performance in mathematics: Comparisons among People's Republic of China, Chinese-American, and Caucasian-American families. *Journal of Educational Psychology, 79,* 179–188.

Heyman, G. D., & Dweck, C. S. (1998). Children's thinking about traits: Implications for judgments of the self and others. *Child Development, 64,* 391–403.

Hicks, L. H. (1956). An analysis of number-concept formation in the rhesus monkey. *Journal of Comparative and Physiological Psychology, 49,* 212–218.

Hiebert, J. (1989). The struggle to link written symbols with understandings: An update. *Arithmetic Teacher, 36,* 38–44.

Hiebert, J., & Tonnessen, L. H. (1978). Development of the fraction concept in two physical contexts: An exploratory investigation. *Journal for Research in Mathematics Education, 9,* 374–378.

Hiebert, J., & Wearne, D. (1985). A model of students' decimal computation procedures. *Cognition and Instruction, 2,* 175–205.

Hiebert, J., & Wearne, D. (1996). Instruction, understanding, and skill in multidigit addition and subtraction. *Cognition and Instruction, 14,* 251–283.

Hill, J. P. (1967). Similarity and accordance between parents and sons in attitudes toward mathematics. *Child Development, 38,* 777–791.

Hilton, T. L., & Berglund, G. W. (1974). Sex differences in mathematical achievement: A longitudinal study. *Journal of Educational Research, 67,* 231–237.

Hines, M. (1982). Prenatal gonadal hormones and sex differences in human behavior. *Psychological Bulletin, 92,* 56–80.

Hinsley, D. A., Hayes, J. R., & Simon, H. A. (1977). From words to equations: Meaning and representation in algebra word problems. In M. A. Just & P. A. Carpenter (Eds.), *Cognitive processes in comprehension* (pp. 89–106). New York: Wiley.

Hitch, G. J. (1978a). The role of short-term working memory in mental arithmetic. *Cognitive Psychology, 10,* 302–323.

Hitch, G. J. (1978b). Mental arithmetic: Short-term storage and information processing in a cognitive skill. In A. M. Lesgold, J. W. Pellegrino, S. D. Fokkema, & R. Glaser (Eds.), *Cognitive psychology and instruction* (pp. 331–338). New York: Plenum.

Hoffman, P. (1998). *The man who loved only numbers: The story of Paul Erdös and the search for mathematical truth.* New York: Hyperion.

Hofstadter, D. R. (1979). *Gödel, Escher, Bach: An eternal golden braid.* New York: Basic Books.

Holden, A. (1971). *Shapes, space, and symmetry.* New York: Dover.

Holton, G. (1973). *Thematic origins of scientific thought.* Cambridge, MA: Harvard University Press.

Holyoak, K. J. (1995). Problem solving. In E. E. Smith & D. N. Osherson (Eds.), *Thinking: An invitation to cognitive science* (2nd ed., Vol. 3, pp. 267–296). Cambridge, MA: MIT Press.

Hong, E., O'Neil, H. F., & Feldon, D. (2005). Gender effects on mathematics achievement. In A. M. Gallagher & J. C. Kaufman (Eds.), *Gender differences in mathematics* (pp. 264–293). New York: Cambridge University Press.

Honig, W. K. (1993). Numerosity as a dimension of stimulus control. In S. T. Boysen & E. J. Capaldi (Eds.), *The development of numerical competence: Animal and human models* (pp. 61–86). Hillsdale, NJ: Erlbaum.

Honigmann, H. (1942). The number conception in animal psychology. *Biological Review, 17,* 315–337.

Hood, H. B. (1962). An experimental study of Piaget's theory of the development of number in children. *British Journal of Psychology, 53,* 273–286.

Hope, J. A. (1987). A case study of a highly skilled mental calculator. *Journal for research in mathematics education, 18,* 331–342.

Hopko, D. R., McNeil, D. W., Gleason, P. J., & Rabalais, A. E. (2002). The emotional Stroop paradigm: Performance as a function of stimulus properties and self-reported mathematics anxiety. *Cognitive Therapy and Research, 26,* 157–166.

Hopper, V. F. (2000). *Medieval number symbolism: Its sources, meaning, and influence on thought and expression.* Mineola, NY: Dover. (Original work published 1938)

Huff, D. (1973). *How to lie with statistics.* New York: Viking Penguin. (Original work published 1954)

Hughes, M. (1986). *Children and number.* Oxford: Basil Blackwell.

Hughes, M. (1991). What is difficult about learning arithmetic? In P. Light, S. Sheldon, & M. Woodhead (Eds.), *Learning to think: Child development in social context* (pp. 184–204). London: Routledge.

Huinker, D. A. (1998) Letting fraction algorithms emerge through problem solving. In L. J. Morow & M. J. Kenny (Eds.), *The teaching and learning of algorithms in school mathematics: 1998 yearbook* (pp. 170–182). Reston, VA: National Council of Teachers of Mathematics.

Hunter, I. M. L. (1962). An exceptional talent for calculative thinking. *British Journal of Psychology, 53,* 243–258.

Hunter, I. M. L. (1977). An exceptional memory. *British Journal of Psychology, 68,* 155–164.

Huntley, H. E. (1970). *The divine proportion: A study in mathematical beauty.* New York: Dover.

Huntley, M. A., Rasmussen, C. L., Villarubi, R. S., Sangtong, J., & Fey, J. T. (2000). Effects of standards-based mathematics education: A study of the Core-Plus Mathematics Project algebra and functions strand. *Journal for Research in Mathematics Education, 31,* 328–361.

Husén, T. (1967). *International study of achievement in mathematics: A comparison of twelve countries* (Vols. 1 & 2). New York: Wiley.

Huttenlocher, J., Jordan, N. C., & Levine, S. C. (1994). A mental model for early arithmetic. *Journal of Experimental Psychology: General, 123,* 284–296.

Huttenlocher, J., & Strauss, S. (1968). Comprehension and a statement's relation to the situation it describes. *Journal of Verbal Learning and Verbal Behavior, 7,* 300–304.

Hyde, J. S. (2005). The gender similarities hypothesis. *American Psychologist, 60,* 581–592.

Hyde, J. S., Fennema, E., & Lamon, S. J. (1990). Gender differences in mathematics performance: A meta-analysis. *Psychological Bulletin, 107,* 139–155.

Hyde, J. S., Fennema, E., Ryan, M., Frost, L. A., & Hopp, C. (1990). Gender difference in mathematics attitudes and affect: A meta-analysis. *Psychology of Women Quarterly, 14,* 299–324.

Hyde, J. S., Lindberg, S. M., Linn, M. C., Ellis, A. B., and Williams, C. C. (2008). Gender similarities characterize math performance. *Science, 321,* 494–495.

Ifrah, G. (1987). *From one to zero: A universal history of numbers* (L. Bair, Trans.). New York: Viking Penguin.

Ifrah, G. (2000). *The universal history of numbers: From prehistory to the invention of the computer.* New York: Wiley.

Ilg, F., & Ames, L. B. (1951). Developmental trends in arithmetic. *Journal of Genetic Psychology, 79,* 3–28.

Inhelder, B., & Piaget, J. (1958). *The growth of logical thinking from childhood to adolescence: An essay on the constuction of formal operational structures.* New York: Basic Books. (Original work published 1955)

Inhelder, B., & Piaget, J. (1964). *The early growth of logic in the child: Classification and seriation.* New York: Harper & Row. (Original work published 1959)

Inzlicht, M., & Ben-Zeev, T. (2000). A threatening intellectual environment: Why females are susceptible to experiencing problem-solving deficits in the presence of males. *Psychological Science, 11,* 365–371.

Ito, Y., & Hatta, T. (2003). Semantic processing of Arabic, Kanji, and Kana numbers: Evidence from interference in physical and numerical size judgments. *Memory and Cognition, 31,* 360–368.

Jacobs, J. E. (1991). Influence of gender stereotypes on parent and child mathematics attitudes. *Journal of Educational Psychology, 83*, 518–527.

Jacobs, J. E., Davis-Kean, P., Bleeker, M., Eccles, J. S., & Malanchuk, O. (2005). "I can, but I don't want to": The impact of parents, interests, and activities on gender differences in math. In A. M. Gallagher & J. C. Kaufman (Eds.), *Gender differences in mathematics* (pp. 246–263). New York: Cambridge University Press.

James, I. (2002). *Remarkable mathematicians.* Cambridge, UK: Cambridge University Press.

Jenkins, J. J., & Tuten, J. T. (1992). Why isn't the average child from the average family?—and similar puzzles. *American Journal of Psychology, 105*, 517–526.

Jensen, A. R. (1990). Speed of information processing in a calculating prodigy. *Intelligence, 14*, 259–274.

Jensen, E. M., Reese, E. P., & Reese, T. W. (1950). The subitizing and counting of visually presented fields of dots. *Journal of Psychology, 30*, 363–392.

Jerman, M. E. (1973–1974). Problem length as a structural variable in verbal arithmetic problems. *Educational Studies in Mathematics, 5*, 109–123.

Jerman, M. E., & Mirman, S. (1974). Linguistic and computational variables in problem solving in elementary mathematics. *Educational Studies in Mathematics, 5*, 317–362.

Jerman, M. E., & Rees, R. (1972). Predicting the relative difficulty of verbal arithmetic problems. *Educational Studies in Mathematics, 4*, 306–323.

Jones, A., Morris, S. A., & Pearson, K. R. (1991). *Abstract algebra and famous impossibilities.* New York: Springer-Verlag.

Jordan, N. C., Huttenlocher, J., & Levine, S. C. (1992). Differential calculation abilities in young children from middle and low-income families. *Developmental Psychology, 28*, 644–653.

Jordan, N. C., & Montani, T. O. (1997). Cognitive arithmetic and problem solving: A comparison of children with specific and general mathematical difficulties. *Journal of Learning Disabilities, 28*, 624–634.

Jourdain, P. E. B. (1956). The nature of mathematics. In J. R. Newman (Ed.), *The world of mathematics* (Vol. 1, pp. 4–72). New York: Simon & Schuster. (Original work published 1913)

Kac, M. (1973). Probability and related topics in physical sciences. In J. Mehra (Ed.), *The physicist's conception of nature* (pp. 380–403). Dordrecht: Reidel.

Kac, M. (1985). *Enigmas of chance: An autobiography.* New York: Harper & Row.

Kafatos, M., & Nadeau, R. (1990). *The conscious universe: Part and whole in modern physical theory.* New York: Springer-Verlag.

Kagan, J. (1966). Reflection-impulsivity: The generality and dynamics of conceptual tempo. *Journal of Abnormal Psychology, 71*, 17–24.

Kagan, J., Rosman, B., Day, D., Albert, J., & Philips, W. (1964). Information processing and the child: Significance of analytic and reflective attitudes. *Psychological Monographs, 78*, Whole No.

Kail, R., & Hall, L. K. (1999). Sources of developmental change in children's word-problem performance. *Journal of Educational Psychology, 91*, 660–668.

Kamii, C. K. (1985). *Young children reinvent arithmetic.* New York: Teachers College Press.

Kamii, C. K. (1986). Place value: An explanation of its difficulty and educational implications for the primary grades. *Journal of Research on Childhood Education, 1,* 75–85.

Kamlah, A. (1987). The decline of the Laplacian theory of probability: A study of Stumpf, von Kries, and Meinong. In L. Krüger, L. J. Daston, & M. Heidelberger (Eds.), *The probabilistic revolution: Ideas in history* (Vol. 1, pp. 91–116). Cambridge, MA: MIT Press.

Kaplan, A. (1956). Sociology learns the language of mathematics. In J. R. Newman (Ed.). *The world of mathematics* (Vol. 2, pp. 1294–1313). New York: Simon & Schuster.

Kaplan, R. (1999). *The nothing that is: A natural history of zero.* New York: Oxford University Press.

Kaplan, R., & Kaplan, E. (2003). *The art of the infinite: The pleasure of mathematics.* New York: Oxford University Press.

Kaplan, R. G., Yamamoto, T., & Ginsburg, H. P. (1989). Teaching mathematical concepts. In L. B. Resnick & L. E. Klopfer (Eds.), *Toward the thinking curriculum: Current cognitive research* (pp. 59–82). Alexandria, VA: Association for Supervision and Curriculum Development.

Kaput, J. J. (1985). Representation and problem solving: Methodological issues related to modeling. In E. A. Silver (Ed.), *Teaching and learning mathematical problem solving: Multiple research perspectives* (pp. 381–398). Hillsdale, NJ: Erlbaum.

Kaput, J. J. (1994). Democratizing access to calculus: New routes to old roots. In A. H. Schoenfeld (Ed.), *Mathematical thinking and problem solving* (pp. 77–156). Hillsdale, NJ: Erlbaum.

Kaput, J. J., & Dubinsky, E. (Eds.). (1994), *Research issues in undergraduate mathematics learning* (MAA Notes Series, Vol. 33, pp. 19–26). Washington, DC: Mathematical Association of America.

Kaput, J. J., & West, M. M. (1994). Missing-value proportional reasoning problems: Factors affecting informal reasoning patterns. In G. Harel & J. Confrey (Eds.), *The development of multiplicative reasoning in the learning of mathematics* (pp. 235–287). Albany: State University of New York Press.

Karplus, R., Adi, H., & Lawson, A. E. (1980). Intellectual development beyond elementary school: Proportional, probabilistic, and correlational reasoning. *School Science and Mathematics, 890,* 673–683.

Karplus, R., & Karplus, E. F. (1972). Intellectual development beyond elementary school: Ratio, a longitudinal study. *School Science and Mathematics, 72,* 735–742.

Karplus, R., Karplus, E. F., Formisano, M., & Paulson, A. C. (1979). Proportional reasoning and control of variables in seven countries. In J. Lochhead & J. Clement (Eds.), *Cognitive process instruction* (pp. 47–104). Philadelphia: The Franklin Institute Press.

Karplus, R., Karplus, E. F., & Wollman, W. (1974). Intellectual development beyond elementary school: Ratio, the influence of cognitive style. *School Science and Mathematics, 74,* 476–482.

Karplus, R., & Peterson, R. W. (1970). Intellectual development beyond elementary school II: Ratio, a survey. *School Science and Mathematics, 70,* 813–820.

Karplus, R., Pulos, S., & Stage, E. K. (1983). Proportional reasoning of early adolescents. In R. Lesh & M. Landau (Eds.), *Acquisition of mathematics concepts and processes* (pp. 45–90). New York: Academic Press.

Kasner, E., & Newman, J. R. (1940). *Mathematics and imagination.* New York: Simon & Schuster.

Kasner, E., & Newman, J. R. (1956). Pastimes of past and present times. In J. R. Newman (Ed.), *The world of mathematics* (Vol. 4, pp. 2416–2438). New York: Simon & Schuster. (Original work published 1940)

Kaufman, E. L., Lord, M. W., Reese, T. W., & Volkmann, J. (1949). The discrimination of visual number. *American Journal of Psychology, 62,* 498–525.

Kaufmann, L., Montanes, P., Jacquier, M., Matallana, D., Eibl, G. K., & Delazer, M. (2002). About the relationship between numerical processing and arithmetic in early Alzheimer disease: A follow up study. *Brain and Cognition, 48,* 398–405.

Kaye, D. B. (1986). The development of mathematical cognition. *Cognitive Development, 1,* 157–170.

Kelman, P., Bardige, A., Choate, J., Hanify, G., Richards, J., Roberts, N., et al. (1983). *Computers in teaching mathematics.* Reading, MA: Addison-Wesley.

Kepler, J. (1975). *The harmonies of the world* (C. G. Wallis, Trans.). Chicago: Chicago University Press. (Original work published 1619)

Kessel, C., & Lynn, M. C. (1996). Grades or scores: Predicting future college mathematics performance. *Educational Measurement: Issues and Practice, 15,* 10–14.

Keynes, J. M. (1956). Newton, the man. In J. R. Newman (Ed.), *The world of mathematics* (Vol. 1, pp. 277–285). New York: Simon & Schuster. (Original lecture given 1946)

Kieran, C. (1989). The early learning of algebra: A structural perspective. In S. Wagner & C. Kieran (Eds.), *Research issues in the learning and teaching of algebra* (pp. 33–56). Hillsdale, NJ: Erlbaum.

Kieren, T. E. (1969). Activity learning. *Review of Educational Research, 39,* 509–522.

Kieren, T. E. (1976). On the mathematical, cognitive, and instructional foundations of rational numbers. In R. Lesh (Ed.), *Number and measurement: Papers from a research workshop* (pp. 101–144). Columbus, OH: ERIC/SMEAC.

Kilian, A., Yaman, S., von Fersen, L., & Güntürkün, O. (2003). A bottlenose dolphin discriminates visual stimuli differing in numerosity. *Learning and Behavior, 31,* 133–142.

Kilpatrick, J. (1985). A retrospective account of the past 25 years of research on teaching mathematical problem solving. In E. A. Silver (Ed.), *Teaching and learning mathematical problem solving: Multiple research perspectives* (pp. 1–15). Hillsdale, NJ: Erlbaum.

Kilpatrick, J. (1987). Problem formulating: Where do good problems come from? In A. H. Schoenfeld (Ed.), *Cognitive science and mathematics education* (pp. 123–147). Hillsdale, NJ: Erlbaum.

Kilpatrick, J., Swafford, J., & Findell, B. (Eds.). (2001). *Adding it up: Helping children learn mathematics.* Washington, DC: National Academy Press.

Kim, A., & Zaidel, E. (2003). Plasticity in the SNARC effect during manipulation of order in response conditions. *Journal of Cognitive Neuroscience* (Suppl.), *134.*

Kimball, M. M. (1989). A new perspective on women's math achievement. *Psychological Bulletin, 105,* 198–214.

King, J. P. (1992). *The art of mathematics.* New York: Fawcett Columbine.

Kinsbourne, M., & Warrington, E. (1962). A study of finger agnosia. *Brain, 85,* 47–66.

Kinsbourne, M., & Warrington, E. (1963). The development of Gerstmann syndrome. *Archives of Neurology, 8,* 490–501.

Kirova, A., & Bhargava, A. (2002). Learning to guide preschool children's mathematical understanding: A teacher's professional growth. *Early Childhood Research and Practice, 4,* 1–20.

Klahr, D., & Wallace, J. G. (1973). The role of quantification operators in the development of conservation of quantity. *Cognitive Psychology, 4,* 301–327.

Klahr, D., & Wallace, J. G. (1976). *Cognitive development: An information processing view.* Hillsdale, NJ: Erlbaum.

Klein, A., & Starkey, P. (1988). Universals in the development of early arithmetic cognition. In G. B. Saxe & M. Gearhart (Eds.), *Children's mathematics: New directions for child development.* (pp. 5–26). San Francisco: Jossey Bass.

Klein, G. A. (1998). *Sources of power: How people make decisions.* Cambridge, MA: MIT Press.

Klein, G. A., Orasanu, J., Calderwood, R., & Zsambok, C. E. (Eds.). (1993). *Decision making in action: Models and methods.* Norwood, NJ: Ablex.

Kleiner, I., & Movshovitz-Hadar, N. (1997). Proof: A many splendored thing. *The Mathematical Intelligencer, 19,* 16–26.

Kline, M. (1953). *Mathematics in Western culture.* New York: Oxford University Press.

Kline, M. (1956). Projective geometry. In J. R. Newman (Ed.), *The world of mathematics* (Vol. 1, pp. 622–641). New York: Simon & Schuster. (Original work published 1953)

Kline, M. (1980). *Mathematics: The loss of certainty.* New York: Oxford University Press.

Kloosterman, P. (1990). Attributions, performance following failure, and motivation in mathematics. In E. Fennema & G. H. Leder (Eds.), *Mathematics and gender* (pp. 96–127). New York: Teachers College Press.

Knorr, W. R. (1982). Infinity and continuity: The interaction of mathematics and philosophy in antiquity. In N. Kretzmann (Ed.), *Infinity and continuity in ancient and medieval thought* (pp. 112–145). Ithaca, NY: Cornell University Press.

Knuth, D. E. (1985). Algorithmic thinking and mathematical thinking. *American Mathematical Monthly, 92,* 170–181.

Koechlin, E., Dehaene, S., & Mehler, J. (1997). Numerical transformations in five-month-old human infants. *Mathematical Cognition, 3,* 89–104.

Koedinger, K. R., & Anderson, J. R. (1995). Abstract planning and perceptual chunks. In J. Glasgow, N. H. Narayanan, & B. Chandrasekaran (Eds.), *Diagrammatic reasoning: Cognitive and computational perspectives* (pp. 577–625). Cambridge, MA: MIT Press.

Koehler, M. S. (1990). Classrooms, teachers, and gender differences in mathematics. In E. Fennema & G. H. Leder (Eds.), *Mathematics and gender* (pp. 128–148). New York: Teachers College Press.

Koehler, O. (1937). Können Tauben "zahlen"? [Can pigeons "count"?] *Zeitschrift für Tierpsychologie, 1*, 39–48.

Koehler, O. (1943). "Zahl" Versuche an einem Kolkraben und Vergleichsversuche an Menschen ["Counting" study with a raven and comparative research with people]. *Zeitschrift für Tierpsychologie, 5*, 575–712.

Koehler, O. (1950). The ability of birds to "count." *Bulletin of Animal Behavior, 9*, 41–45.

Kornblum, S., Hasbroucq, T., & Osman, A. (1990). Dimensional overlap: Cognitive basis for stimulus-response compatibility—A model and taxonomy. *Psychological Review, 97*, 253–270.

Kosc, L. (1974). Developmental dyscalculia. *Journal of Learning Disabilities, 7*, 164–177.

Koshmider, J. W., III, & Ashcraft, M. H. (1991). The development of children's mental multiplication skills. *Journal of Experimental Child Psychology, 51*, 53–89.

Kouba, V. L. (1989). Children's solution strategies for equivalent set multiplication and division word problems. *Journal for Research in Mathematics Education, 20*, 147–158.

Kouba, V. L., Brown, C. A., Carpenter, T. P., Lindquist,M. M., Silver, E. A., & Swafford, J. O. (1988). Results of the fourth NAEP assessment of mathematics: Number, operations, and word problems. *Arithmetic Teacher, 35*, 10–16.

Kouba, V. L., Carpenter, T. P., & Swafford, J. O. (1989). Number and operations. In M. M. Lindquist (Ed.), *Results from the Fourth Mathematics Assessment of the National Assessment of Educational Progress* (pp. 64–93). Reston, VA: National Council of Teachers of Mathematics.

Kouba, V. L., & Franklin, K. (1993). Multiplication and division: Sense making and meaning. In R. J. Jensen (Ed.), *Research ideas for the classroom: Early childhood mathematics* (pp. 103–126). New York: Macmillan.

Kouba, V. L., & Wearne, D. (2000). Whole number properties and operations. In E. A. Silver & P. A. Kennedy (Eds.), *Results from the Seventh Mathematics Assessment of the National Assessment of Educational Progress* (pp. 141–161). Reston, VA: National Council of Teachers of Mathematics.

Kraitchik, M. (1942). *Mathematical recreations*. New York: Norton.

Kretzmann, N. (1982a). Continuity, contrariety, contradiction, and change. In N. Kretzmann (Ed.), *Infinity and continuity in ancient and medieval thought* (pp. 270–296). Ithaca, NY: Cornell University Press.

Kretzmann, N. (Ed.). (1982b). *Infinity and continuity in ancient and medieval thought.* Ithaca, NY: Cornell University Press.

Krueger, A. (1999). Experimental estimates of education production functions. *Quarterly Journal of Economics, 114*, 497–532.

Krueger, L. E. (1982). Single judgments of numerosity. *Perception and Psychophysics, 31*, 175–182.

Krulik, S. (Ed.). (1980). *Problem solving in school mathematics*. Reston, VA: National Council of Teachers of Mathematics.

Lakatos, I. (1976). *Proofs and refutations: The logic of mathematical discovery* (J. Worrall & E. Zahar, Eds.). New York: Cambridge University Press.

Lakoff, G. (1987). *Women, fire, and dangerous things: What categories reveal about the mind*. Chicago: University of Chicago Press.

Lakoff, G., & Núñez, R. E. (2000). *Where mathematics comes from: How the embodied mind brings mathematics into being.* New York: Basic Books.

Laming, D. (1973). *Mathematical psychology.* New York: Academic Press.

Lamon, S. J. (1994). Ratio and proportion: Cognitive foundations in unitizing and norming. In G. Harel & J. Confrey (Eds.), *The development of multiplicative reasoning in the learning of mathematics* (pp. 89–120). Albany: State University of New York Press.

Lampert, M. (1986). Teaching multiplication. *Journal of Mathematical Behavior, 5,* 241–280.

Lampl, Y., Eshel, Y., Gilad, R., & Sarova-Pinhas, I. (1994). Selective acalculia with sparing of the subtraction process in a patient with left parietotemporal hemorrhage. *Neurology, 44,* 1759–1781.

Lancy, D. F. (Ed.). (1978). The indigenous mathematics project. *Papua New Guinea Journal of Education, 14* (Special issue).

Landerl, K., Bevan, A., & Butterworth, B. (2004). Developmental dyscalculia and basic numerical capacities: A study of 8–9-year-old students. *Cognition, 93,* 99–125.

Langenfeld, T. E. (1997). Test fairness: Internal and external investigations of gender bias in mathematics testing. *Educational Measurement: Issues and Practice, 16,* 20–26.

Langford, K., & Sarullo, A. (1993). Introductory common and decimal fraction concepts. In R. J. Jensen (Ed.), *Research ideas for the classroom: Early childhood mathematics* (pp. 223–247). New York: Macmillan.

Lankford, F. G. (1972). Some computational strategies for seventh-grade pupils. ERIC Reports, School of Education, Virginia University.

Lann, A., & Falk, R. (2006). Tell me the method, I'll give you the mean. *The American Statistician, 60,* 322–327.

Lantz, A. (1985). Strategies to increase mathematics enrollments. In S. F. Chipman, L. R. Brush, & D. M. Wilson (Eds.), *Women and mathematics: Balancing the equation* (pp. 329–354). Hillsdale, NJ: Erlbaum.

Lantz, A. E., & Smith, G. P. (1981). Factors influencing the choice of nonrequired mathematics courses. *Journal of Educational Psychology, 73,* 825–837.

Laplace, P. S. (1951). *A philosophical essay on probabilities.* F. W. Truscott & F. L. Emory (Trans.). New York: Dover. (Original work published 1814)

Lapointe, A. E., Mead, N. A., & Askew, J. M. (1992). *Learning mathematics.* Princeton, NJ: Educational Testing Service.

Lappan, G., Fey, J. T., Fitzgerald, W. M., Friel, S. N., & Phillips, E. D. (1996). *Connected mathematics.* Needham, MA: Prentice Hall

Larkin, J., McDermott, J., Simon, D. P., & Simon, H. A. (1980). Expert and novice performance in solving physics problems. *Science, 208,* 1335–1342.

Larkin, J., & Simon, H. A. (1987). Why a diagram is (sometimes) worth ten thousand words. *Cognitive Science, 11,* 65–100.

Larkin, J. H. (1979). Information processing models and science instruction. In J. Lochhead & J. Clement (Eds.), *Cognitive process instruction: Research on teaching thinking skills.* Philadelphia: Franklin Institute Press.

Larkin, J. H. (1980). Skilled problem solving in physics: A hierarchical planning approach. *Journal of Structural Learning, 6,* 121–130.

Larkin, J. H. (1983). The role of problem representation in physics. In D. Gentner & A. Stevens (Eds.), *Mental models* (pp. 75–99). Hillsdale, NJ: Erlbaum.

Larson, L. C. (1994). Comments on Bruce Reznick's chapter. In A. H. Schoenfeld (Ed.), *Mathematical thinking and problem solving* (pp. 30–38). Hillsdale, NJ: Erlbaum.

Lavach, J. F., & Lanier, H. B. (1975). The motive to avoid success in 7th, 8th, 9th, and 10th grade in high achieving girls. *Journal of Educational Research, 68,* 216–218.

Lave, J. (1977). Cognitive consequences of traditional apprenticeship training in West Africa. *Anthropology and Education Quarterly, 8,* 177–180.

Lave, J. (1988). *Cognition in practice: Mind, mathematics and culture in everyday life,* Cambridge, UK: Cambridge University Press.

Lave, J., Murtaugh, M., & de la Rocha, O. (1999). The dialectic of arithmetic in grocery shopping. In B. Rogoff & J. Lave (Eds.), *Everyday cognition: Its development in social context* (pp. 67–94). Cambridge, MA: Harvard University Press.

Layzer, D. (1990). *Cosmogenesis: The growth of order in the universe.* New York: Oxford University Press.

Le Corbeiller, P. (1956). Crystals and the future of physics. In J. R. Newman (Ed.), *The world of mathematics* (Vol. 2, pp. 871–881). New York: Simon & Schuster. (Original work published 1943)

Leder, G. C. (1990a). Gender differences in mathematics: An overview. In E. Fennema & G. H. Leder (Eds.), *Mathematics and gender* (pp. 10–26), New York: Teachers College Press.

Leder, G. C. (1990b). Teacher/student interactions in the mathematics classroom: A different perspective. In E. Fennema & G. H. Leder (Eds.), *Mathematics and gender* (pp. 149–168). New York: Teachers College Press.

Leder, G. C. (1992). Mathematics and gender: Changing perspectives. In D. A. Grouws (Ed.), *Handbook of research on mathematical teaching and learning* (pp. 597–624). New York: Macmillan.

Lee, K. M. (2000). Cortical areas differentially involved in multiplication and subtraction: A functional magnetic resonance imaging study and correlation with a case of selective acalculia. *Annals of Neurology, 48,* 657–661.

Lee, V. E., & Bryk, A. S. (1986). Effects of single-sex secondary schools on student achievement and attitudes. *Journal of Educational Psychology, 78,* 381–395.

LeFevre, J., Bisanz, J., Daley, K. E., Buffone, L., Greenham, S. L., & Sadesky, G. S. (1996). Multiple routes to solution of single-digit multiplication problems. *Journal of Experimental Psychology: General, 125,* 284–306.

LeFevre, J., Kulak, A., & Heymans, S. (1992). Factors influencing the selection of university majors varying in mathematical content. *Canadian Journal of Behavioral Sciences, 24,* 276–289.

LeFevre, J.-A., DeStefano, D., Coleman, B., & Shanahan, T. (2005). Mathematical cognition and working memory. In J. I. D. Campbell (Ed.), *Handbook of mathematical cognition* (pp. 361–377). New York: Psychology Press.

Lehman, H. C. (1953). *Age and achievement.* Princeton, NJ: Princeton University Press.

Lehrer, R., Randle, L., & Sancilio, L. (1989). Learning pre-proof geometry with Logo. *Cognition and Instruction, 6,* 159–184.

Leinhardt, G., Seewald, A. M., & Engel, M. (1979). Learning what's taught: Sex differences in instruction. *Journal of Educational Psychology, 71,* 432–439.

Leinhardt, G., & Smith, D. A. (1985). Expertise in mathematics instruction: Subject matter knowledge. *Journal of Educational Psychology, 77,* 247–271.

Lesgold, A. M. (1984). Acquiring expertise. In J. R. Anderson & S. M. Kosslyn (Eds.), *Tutorials in learning and memory.* San Francisco: Freeman.

Lesh, R. (1985). Conceptual analyses of problem-solving performance. In E. A. Silver (Ed.), *Teaching and learning mathematical problem solving: Multiple research perspectives* (pp. 309–329). Hillsdale, NJ: Erlbaum.

Lester, E. A. (1985). Methodological considerations in research on mathematical problem-solving instruction. In E. A. Silver (Ed.), *Teaching and learning mathematical problem solving: Multiple research perspectives* (pp. 41–69). Hillsdale, NJ: Erlbaum.

Lester, F. K. (1980). Research on mathematical problem solving. In R. J. Shumway (Ed.), *Research in mathematics education* (pp. 286–323). Reston, VA: National Council of Teachers of Mathematics.

Lester, F. K., & Garofalo, J. (1982). *Mathematical problem solving: Issues in research.* Philadelphia: Franklin Press.

Levine, M. (1976). *Identification of reasons why qualified women do not pursue mathematical careers.* Washington, DC: National Science Foundation.

Levine, S. C., Jordan, N. C., & Huttenlocher, J. (1992). Development of calculation abilities in young children. *Journal of Experimental Child Psychology, 53,* 72–103.

Levitt, E. E., & Hutton, L. A. (1983). Correlates and possible causes of mathematics anxiety. In C. D. Spielberger & J. N. Butcher (Eds.), *Advances in personality assessment* (Vol. 3, pp. 129–140). Hillsdale, NJ: Erlbaum.

Lewandowsky, S., & Thomas, J. L. (In press). Expertise: Acquisition, limitations, and control. In F. T. Durso (Ed.), *Reviews of human factors and ergonomics* (Vol. 5).

Lewis, A. B. (1989). Training students to represent arithmetic word problems. *Journal of Educational Psychology, 81,* 521–531.

Lewis, A. B., & Mayer, R. E. (1987). Students' miscomprehension of relational statements in arithmetic word problems. *Journal of Educational Psychology, 79,* 363–371.

Lewis, C. I., & Langford, C. H. (1956). History of symbolic logic. In J. R. Newman (Ed.), *The world of mathematics* (Vol. 3, pp. 1859–1877). New York: Simon & Schuster. (Original work published 1932)

Linchevski, L., & Kutscher, B. (1998). Tell me with whom you're learning and I'll tell you how much you've learned: Mixed-ability versus same-ability grouping in mathematics. *Journal for Research in Mathematics Education, 29,* 533–554.

Lindley, D. V. (1993). *The end of physics: The myth of a unified theory.* New York: Basic Books.

Lindquist, M. M. (Ed.). (1980). *Selected issues in mathematics education* (pp. 215–257). Berkeley, CA: McCuchan.

Linville, W. J. (1976). Syntax, vocabulary, and the verbal arithmetic problem. *School Science and Mathematics, 76,* 152–158.

Lipton, J. S., & Spelke, E. S. (2003). Origins of number sense: Large number discrimination in human infants. *Psychological Science, 14,* 396–401.

Livio, M. (2002). *The golden ratio: The story of phi, the world's most astonishing number.* New York: Broadway Books.

Lochhead, J. (1980). Faculty interpretations of simple algebraic statements: The professor's side of the equation. *Journal of Mathematical Behavior, 3*, 29–37.

Lochhead, J., & Mestre, J. (1988). From words to algebra: Mending misconceptions. In A. Coxford (Ed.), *The ideas of algebra: K-12* (pp. 127–135). Reston, VA: National Council of Teachers of Mathematics.

Logan, G. D., & Zbrodoff, N. J. (2003). Subitizing and similarity: Toward a pattern-matching theory of enumeration. *Psychonomic Bulletin and Review, 10*, 676–682.

Loomis, E. S. (1968). *The Pythagorean proposition*. Reston, VA: National Council of Teachers of Mathematics.

Love, W. P. (1989). Infinity: The twilight zone of mathematics. *Mathematics Teacher, 82*, 284–292.

Lubbock, J. (1885). On the intelligence of the dog. *Nature, 33*, 45–46.

Lubinski, D., & Benbow, C. P. (1994). The study of mathematically precocious youth: The first three decades of a planned 50-year study of intellectual talent. In R. F. Subotnik & K. D. Arnold (Eds.), *Beyond Terman: Contemporary longitudinal studies of giftedness and talent* (pp. 255–281). Norwood, NJ: Ablex.

Lubinski, D., & Humphreys, L. G. (1990). A broadly based analysis of mathematical giftedness. *Intelligence, 14*, 327–355.

Luce, R. D. (Ed.). (1960). *Developments in mathematical psychology*. Glencoe, IL: The Free Press.

Luce, R. D., Bush, R. R., & Galanter, E. (Eds.). (1963a). *Handbook of mathematical psychology* (3 vols). New York: Wiley.

Luce, R. D., Bush, R. R., & Galanter, E. (Eds.). (1963b). *Readings in mathematical psychology* (2 vols). New York: Wiley.

Luce, R. D., & Narens, L. (1992). Instrinsic Archimedeanness and the continuum. In C. W. Savage, & P. Ehrlich (Eds.), *Philosophical and foundations issues in measurement theory* (pp. 15–38). Hillsdale, NJ: Erlbaum.

Luce, R. D., & Raiffa, H. (1957). *Games and decisions: Introduction and critical survey*. New York: Wiley.

Lussier, G. (1996). Sex and mathematical background as predictors of anxiety and self-efficacy in mathematics. *Psychological Reports, 79*, 827–833.

Ma, L. (1999). *Knowing and teaching elementary mathematics: Teachers' understanding of fundamental mathematics in China and the United States*. Mahwah, NJ: Erlbaum.

Macaruso, P., McCloskey, M., & Aliminosa, D. (1993). The functional architecture of the cognitive numerical-processing system: Evidence from a patient with multiple impairments. *Cognitive Neuropsychology, 10*, 341–376.

Macaruso, P., & Sokol, S. M. (1998). Cognitive neuropsychology and developmental dyscalculia. In C. Donlan (Ed.), *The development of mathematical skills* (pp. 201–225). Hove, UK: Psychology Press.

Maccoby, E., & Jacklin, C. (1974). *The psychology of sex differences*. Palo Alto, CA: Stanford University Press.

Mach, E. (1956). The economy of science. In J. R. Newman (Ed.), *The world of mathematics* (Vol. 3, pp. 1787–1795). New York: Simon & Schuster. (Original work published 1883)

Mach, E. (1960). *The science of mechanics*. LaSalle, IL: Open Court. (Original work published in 1893)

Mach, E. (1974). *The science of mechanics: A critical and historical account of its development* (6th ed., J. McCormack, Trans.). Lasalle, IL: Open Court. (Original work published 1906)

Mack, N. K. (1990). Learning fractions with understanding: Building on informal knowledge. *Journal for Research in Mathematics Education, 21,* 16–32.

Mack, N. K. (1993). Learning rational numbers with understanding: The case of informal knowledge. In T. P. Carpenter, T. A. Romberg, & E. Fennema (Eds.), *Rational numbers: An integration of research* (pp. 85–105). Hillsdale, NJ: Erlbaum.

Mack, N. K. (1995). Confounding whole-number and fraction concepts when building on informal knowledge. *Journal for Research in Mathematics Education, 26,* 422–441.

Mael, F. A. (1998). Single-sex and coeducational schooling: Relationships to socioemotional and academic development. *Review of Educational Research, 68,* 101–129.

Maertens, N. W., Jones, R. C., & Waite, A. (1977). Elemental groupings help children perceive cardinality: A two-phase research study. *Journal for Research in Mathematics Education, 8,* 181–193.

Maeterlinck, M. (1975). *The unknown guest.* New Hyde Park, NY: University Press. (Original work published 1914)

Maher, C. A., & Martino, A. M. (1996). The development of the idea of mathematical proof: A 5-year case study. *Journal for Research in Mathematics Education, 27,* 194–214.

Mandelbrot, B. B. (1977). *Fractals: Form, chance, and dimension.* San Francisco: Freeman.

Mandler, G., & Shebo, B. J. (1982). Subitizing: An analysis of its component processes. *Journal of Experimental Psychology: General, 111,* 1–21.

Maor, E. (1987). *To infinity and beyond: A cultural history of the infinite.* Princeton, NJ: Princeton University Press.

Maor, E. (1994). *e: The story of a number.* Princeton, NJ: Princeton University Press.

Maor, E. (1998). *Trigonometric delights.* Princeton, NJ: Princeton University Press.

Maple, S. A., & Stage, F. K. (1991). Influences on the choice of math/science major by gender and ethnicity. *American Educational Research Journal, 28,* 37–60.

Margolis, H. (1987). *Patterns, thinking, and cognition.* Chicago: University of Chicago Press.

Mariotte, E. (1992). *Essai de logique. Suivi de: Les principes du devoir et des connaissances humaines.* Paris: Fayard. (Original work published 1678 under the title *Essai de logique, contenant les principes des sciences et la manière de s'en servir pour faire de bons raisonnemens*)

Mason, J., Burton, L., & Stacey, K. (1985). *Thinking mathematically* (rev. ed.). Reading, MA: Addison-Wesley.

Matsuzawa, T. (1985). Use of numbers by a chimpanzee. *Nature, 315,* 57–59.

Matsuzawa, T., Asano, T., Kubota, K., & Murofushi, K. (1986). Acquisition and generalization of numerical labeling by a chimpanzee. In D. M. Taub & F. A. King (Eds.), *Current perspectives in primate social dynamics* (pp. 416–430). New York: Van Nostrand Reinhold.

Matz, M. (1980). Towards a computational model of algebraic competence. *Journal of Mathematical Behavior, 3,* 93–166.

Matz, M. (1982). Towards a process model for high school algebra errors. In D. Sleeman & J. S. Brown (Eds.), *Intelligent tutoring systems* (pp. 25–49). New York: Academic Press.

Mau, W. C., & Lynn, R. (2000). Gender differences in homework and test scores in mathematics, reading and science at tenth and twelfth grade. *Psychology, Evolution and Gender, 2,* 119–125.

Maurer, S. B., (1987), New knowledge about errors and new views about learners: What they mean to educators and more educators would like to know. In A. H. Schoenfeld (Ed.), *Cognitive science and mathematics education* (pp. 165–188). Hillsdale, NJ: Erlbaum.

Maxim, G. W. (1989). Developing preschool mathematical concepts. *Arithmetic Teacher, 37,* 36–41.

May, K. O. (1965). The origin of the four-color conjecture. *Isis, 56,* 346–348.

Mayberry, J. (1983). The van Hiele levels of geometric thought in undergraduate preservice teachers. *Journal for Research in Mathematics Education, 15,* 84–95.

Mayer, R. E. (1982). Memory for algebra story problems. *Journal of Educational Psychology, 74,* 199–216.

Mayer, R. E. (1983). *Thinking, problem solving, and cognition.* San Francisco: Freeman.

Mayer, R. E. (1989). Cognition and instruction in mathematics. *Journal of Educational Psychology, 81,* 452–456.

Mayer, R. E. (1992). *Thinking, problem solving, cognition* (2nd ed.). New York: Freeman.

Mayer, R. E., & Hegarty, M. (1996). The process of understanding mathematical problems. In R. J. Sternberg & T. Ben-Zeev (Eds.), *The nature of mathematical thinking* (pp. 29–53). Mahwah, NJ: Erlbaum.

Mayer, R. E., Sims, V., & Tajika, H. (1995). A comparison of how textbooks teach mathematical problem solving in Japan and the United States. *American Educational Research Journal, 32,* 443–460.

Mazur, B. (2003). *Imagining numbers (particularly the square root of minus fifteen).* New York: Picador.

Mazzocco, M. M. M. (1998). A process approach to describing mathematics difficulties in girls with Turner syndrome. *Pediatrics* (Suppl. 3), 492–496.

Mazzocco, M. M. M., & McCloskey, M. (2005). Math performance in girls with Turner or Fragile X syndrome. In J. I. D. Campbell (Ed.), *Handbook of mathematical cognition* (pp. 269–297). New York: Psychology Press.

McCarthy, J. L., & Wolfe, D. (1975). Doctorates granted to women and minority group members. *Science, 189,* 856–859.

McCloskey, M. (1992). Cognitive mechanisms in numerical processing: Evidence from acquired dyscalculia. *Cognition, 44,* 107–157.

McCloskey, M., Caramazza, A., & Basili, A. (1985). Cognitive mechanisms in number processing and calculation: Evidence from dyscalculia. *Brain and Cognition, 4,* 171–196.

McCloskey, M., & Macaruso, P. (1995). Representing and using numerical information. *American Psychology, 59,* 351–363.

McCloskey, M., Sokol, S. M., & Goodman, R. A. (1986). Cognitive processes in verbal number production: Inferences from the performance of brain-damaged subjects. *Journal of Experimental Psychology: General, 115,* 307–330.

McCulloch, W. S. (1965). What is a number that a man may know it, and a man, that he may know a number? In W. S. McCulloch (Ed.), *Embodiments of mind* (pp. 1–18). Cambridge, MA: The MIT Press. (Original work published 1961)

McKnight, C. C., Crosswhite, F. J., Dossey, J. A., Kifer, E., Swafford, J. O., Travers, K. T., et al. (1987). *The underachieving curriculum: Assessing U.S. school mathematics from an international perspective.* Champaign, IL: Stipes Publishing.

McLaughlin, W. I. (1994). Resolving Zeno's paradoxes. *Scientific American, 271,* 84–89.

McLean, J. F., & Hitch, G. J. (1999). Working memory impairment in children with specific arithmetic learning disabilities. *Journal of Experimental Child Psychology, 74,* 240–260.

McLeish, J. (1994). *The story of numbers: How mathematics has shaped civilization.* New York: Fawcett Columbine. (Original work published 1991 under the title *Number*)

McLeod, D. B. (1985). Affective issues in research on teaching mathematical problem solving. In E. A. Silver (Ed.), *Teaching and learning mathematical problem solving: Multiple research perspectives* (pp. 267–279). Hillsdale, NJ: Erlbaum.

McNeil, J. E., & Warrington, E. K. (1993). A modality-specific case of dyscalculia. *Journal of Clinical and Experimental Psychology, 15,* 415.

McNeil, J. E., & Warrington, E. K. (1994). A dissociation between addition and subtraction with written calculation. *Neuropsychologia, 32,* 717–728.

McNeil, N. M. (2007). U-shaped development in math: 7-year-olds outperform 9-year-olds on equivalence problems. *Developmental Psychology, 43,* 687–695.

McNeil, N. M., & Alibali, M. W. (2000). Learning mathematics from procedural instruction: Externally imposed goals influence what is learned. *Journal of Educational Psychology, 92,* 734–744.

McNeil, N. M., & Alibali, M. W. (2002). A strong schema can interfere with learning: The case of children's typical addition schema. In C. D. Schunn & W. Gray (Eds.), *Proceedings of the Twenty-Fourth Annual Conference of the Cognitive Science Society.* Mahwah, NJ: Lawrence Erlbaum Associates.

McNeil, N. M., & Alibali, M. W. (2004). You'll see what you mean: Students encode equations based on their knowledge of arithmetic. *Cognitive Science 28,* 451–466.

McNeil, N. M., & Alibali, M. W. (2005). Why won't you change your mind? Knowledge of operational patterns hinders learning and performance on equations. *Child Development, 76,* 1–17.

McPeck, J. (1981). *Critical thinking and education.* Oxford: Martin Robinson.

Mechner, F. (1958). Probability relations within response sequences under ratio reinforcement. *Journal of Experimental Analysis of Behavior, 1,* 109–121.

Mechner, F., & Guevrekian, L. (1962). Effects of deprivation upon counting and timing in rats. *Journal of the Experimental Analysis of Behavior, 5,* 463–466.

Meck, W. H., & Church, R. M. (1983). A mode control model of counting and timing processes. *Journal of Experimental Psychology: Animal Behavior Processes, 9,* 320–334.

Meck, W. H., Church, R. M., & Gibbon, J. (1985). Temporal integration in duration and number discrimination. *Journal of Experimental Psychology: Animal Behavior Processes, 11,* 591–597.

Meece, J. L., Parsons, J. E., Kaczala, C., Goff, S. B., & Futterman, R. (1982). Sex differences in math achievement: Toward a model of academic choice. *Psychological Bulletin, 91*, 324–348.

Mehler, J., & Bever, T. G. (1967). Cognitive capacity of very young children. *Science, 158*, 141–142.

Menger, K. (1937). What is calculus of variations and what are its applications? *The Scientific Monthly, 45*, 250–253.

Menninger, K. (1992). *Number words and number symbols: A cultural history of numbers.* New York: Dover. (Original work published 1958)

Merton, R. K. (1968). *Social theory and social structure.* New York: Free Press.

Mervis, J. (2003). Down for the count? *Science, 300*, 1070–1074.

Meyer, M. R. (1989). Gender differences in mathematics. In M. M. Lindquist (Ed.), *Results from the Fourth Mathematics Assessment of Mathematical Progress* (pp. 149–159). Reston, VA: National Council of Teachers of Mathematics.

Meyer, M. R., & Koehler, M. S. (1990). Internal influences on gender differences in mathematics. In E. Fennema & G. H. Leder (Eds.), *Mathematics and gender* (pp. 60–95). New York: Teachers College Press.

Meyer, R. A. (1980). Mathematical problem-solving performance and intellectual abilities of fourth-grade children. In J. G. Harvey & T. A. Romberg (Eds.), *Problem solving studies in mathematics* (pp. 179–198). Madison: Wisconsin Research and Development Center Monograph Series.

Meyerson, M. (2002). *Political numeracy.* New York: W.W. Norton.

Miller, D. J. (1993). Do animals subitize? In S. T. Boysen & E. J. Capaldi (Eds.), *The development of numerical competence: Animal and human models* (pp. 149–169). Hillsdale, NJ: Erlbaum.

Miller, F. D., Jr. (1982). Aristotle against the atomists. In N. Kretzmann (Ed.), *Infinity and continuity in ancient and medieval thought* (pp. 87–111). Ithaca, NY: Cornell University Press.

Miller, K. F., Kelly, M., & Zhou, X. (2005). Learning mathematics in China and the United States: Cross-cultural insights into the nature and course of preschool mathematical development. In J. I. D. Campbell (Ed.), *Handbook of mathematical cognition* (pp. 163–178). New York: Psychology Press.

Miller, K. F., & Paredes, D. R. (1996). On the shoulders of giants: Cultural tools and mathematical development. In R. J. Sternberg & T. Ben-Zeev (Eds.), *The nature of mathematical thinking* (pp. 83–117). Mahwah, NJ: Erlbaum.

Miller, K. F., Perlmutter, M., & Keating, D. (1984). Cognitive arithmetic: Comparison of operations. *Journal of Experimental Psychology: Learning, Memory, and Cognition, 10*, 46–60.

Miller, K. F., Smith, C. M., Zhu, J., & Zhang, H. (1995). Pre-school origins of cross-national differences in mathematical competence. *Psychological Science, 6*, 56–60.

Miller, K. F., & Stigler, J. W. (1987). Counting in Chinese: Cultural variation in a basic cognitive skill. *Cognitive Development, 2*, 279–305.

Mills, C. J., Ablard, K. E., & Stumpf, H. (1993). Gender differences in academically talented young students' mathematical reasoning: Patterns across age and subskills. *Journal of Educational Psychology, 85*, 340–346.

Miura, I. T., Okamoto, Y., Kim, C. C., Chang, C. M., Steere, M., & Fayol, M. (1994). Comparisons of cognitive representation of number: China, France, Japan, Korea, Sweden, and the United States. *International Journal of Behavioral Development, 17*, 401–411.

Miura, I. T., Okamoto, Y., Kim, C. C., Steere, M., & Fayol, M. (1993). First graders' cognitive representation of number and understanding of place value: Cross-national comparisons—France, Japan, Korea, Sweden, and the United States. *Journal of Educational Psychology, 85*, 24–30.

Mix, K. S., Huttenlocher, J., & Levine, S. C. (2002a). Multiple cues for quantification in infancy: Is number one of them? *Psychological Bulletin, 128*, 278–294.

Mix, K. S., Huttenlocher, J., & Levine, S. C. (2002b). *Quantitative development in infancy and early childhood.* New York: Oxford University Press.

Mix, K., Levine, S., & Huttenlocher, J. (1999). Early fraction calculation ability. *Developmental Psychology, 35*, 164–174.

Monk, D. H. (1994). Subject area preparation of secondary mathematics and science teachers and student achievement. *Economics of Education Review, 13*, 125–145.

Montague-Smith, A. (1997). *Mathematics in nursery education.* London: David Fulton Publishers.

Moomaw, S., & Hieronymus, B. (1995). *More than counting. Whole math activities for preschool and kindergarten.* St. Paul, MN: Redleaf Press.

Moore, A. W. (1995). A brief history of infinity. *Scientific American, 272*, 112–116.

Moore, A. W. (2001). *The infinite* (2nd ed.). New York: Routledge.

Moreno, L. E. A., & Waldegg, G. (1991). The conceptual evolution of actual mathematical infinity. *Educational Studies in Mathematics, 22*, 211–231.

Morris, R. (1987). *The nature of reality.* New York: The Noonday Press.

Morton, R. L. (1927). *Teaching arithmetic in the primary grades.* New York: Silver, Burdett and Co.

Moss, J., & Case, R. (1999). Developing children's understanding of rational numbers: A new model and an experimental curriculum. *Journal for Research in Mathematics and Education, 48*, 239–257.

Moyer, R. S., & Landauer, T. K. (1967). Time required for judgments of numerical inequality. *Nature, 215*, 1519–1520.

Moyer, R. S., & Landauer, T. K. (1973). Determinants of reaction time for digit inequality judgments. *Bulletin of the Psychonomic Society, 1*, 167–168.

Muller, C. (1998). Gender differences in parental involvement and adolescents' mathematics achievement. *Sociology of Education, 71*, 336–356.

Mullis, I. V., & Jenkins, L. B. (1988). *The science report card: Elements of risk and recovery.* Princeton, NJ: Educational Testing Service.

Mullis, I. V. S., Dossey, J. A., Owen, E. H., & Phillips, G. W. (1991). *The state of mathematics achievement: NAEP's 1990 assessment of the nation and the trial assessment of the states.* Washington, DC: U.S. Department of Education.

Munn, P. (1998). Symbolic function in pre-schoolers. In C. Donlan (Ed.), *The development of mathematical skills* (pp. 47–71). Hove, UK: Psychology Press.

Murdoch, J. E. (1982). William of Ockham and the logic of infinity and continuity. In N. Kretzmann (Ed.), *Infinity and continuity in ancient and medieval thought* (pp. 165–206). Ithaca, NY: Cornell University Press.

Murofushi, K. (1997). Numerical matching behavior by a chimpanzee (*Pan troglo-dytes*): Subitizing and analogue magnitude estimation. *Japanese Psychological Research, 39*, 140–153.

Muth, K. D. (1984). Solving arithmetic word problems: Roles of reading and computational skills. *Journal of Educational Psychology, 76*, 205–210.

Nagel, E. R. (1956). The meaning of probability. In J. R. Newman (Ed.), *The world of mathematics* (Vol. 2, pp. 1398–1414). New York: Simon & Schuster. (Original work published 1936)

Nagel, E. R., & Newman, J. R. (2001). *Gödle's proof* (rev. ed., D. Hofstadter, Ed.). New York: New York University Press. (Original work published 1958)

Nagel, T. (1995). Chomsky: Linguistics and epistemology. In T. Nagel, *Other minds: Critical essays 1969–1994* (pp. 26–44). New York: Oxford University Press. (Original essays published 1969–1994)

Nahin, P. (1998). *An imaginary tale: The story of $\sqrt{-1}$*. Princeton, NJ: Princeton University Press.

Nasar, S. (1998). *A beautiful mind*. New York: Simon & Schuster.

Nasar, S., & Gruber, D. (2006, August 28). Manifold destiny: A legendary problem and the battle over who solved it. *New Yorker*.

Nash, S. C. (1979). Sex role as a mediator of intellectual functioning. In M. A. Wittig & A. C. Petersen (Eds.), *Sex-related differences in cognitive functioning: Developmental issues* (pp. 263–302). New York: Academic Press.

Nastasi, B. K., & Clements, D. H. (1991). Research on cooperative learning: Implications for practice. *School Psychology Review, 20*, 110–131.

National Center for Education Statistics. (2002a). *Digest of education statistics 2001*. Washington, DC: Department of Education, Office of Educational Research and Development.

National Center for Education Statistics. (2002b). *Outcomes of learning: Results from the 2000 Program for International Student Assessment for 15-year olds in reading, mathematics, and science literacy* (NCES 2002–115). Washington, DC: U.S. Government Printing Office.

National Commission on Excellence in Education (NCEE). (1983). *A nation at risk: The imperative for educational reform*. Washington, DC: U.S. Government Printing Office.

National Council of Teachers of Mathematics (NCTM). (1989). *Curriculum and evaluation standards for school mathematics*. Reston, VA: Author.

National Council of Teachers of Mathematics (NCTM). (2000). *Principles and standards for school mathematics*. Reston, VA: Author.

National Mathematics Advisory Panel. (2008). *Foundations for success: The final report of the National Mathematics Advisory Panel*. Washington, DC: U.S. Department of Education.

National Research Council. (1989). *Everybody counts: A report to the nation on the future of mathematics education*. Washington, DC: National Academy Press.

Nelson, E. (1977). A new approach to nonstandard analysis. *Bulletin of the American Mathematical Society, 83*, 1165–1198.

Nelson, T. M., & Bartley, S. H. (1961). Numerosity, number, arithmetization, measurement and psychology. *Philosophy of Science, 28*, 178–203.

Nersessian, N. J. (1992). Capturing the dynamics of conceptual change in science. In R. N. Giere (Ed.), *Cognitive models of science* (pp. 3–45). Minneapolis: University of Minnesota Press.

Nesher, P. (1988). Multiplicative school word problems: Theoretical approaches and empirical findings. In J. Hiebert & M. Behr (Eds.), *Number concepts and operations in the middle grades* (pp. 19–40). Reston, VA: National Council of Teachers of Mathematics.

Nesher, P., & Teubal, E. (1974). Verbal cues as an interfering factor in verbal problem solving. *Educational Studies in Mathematics, 6,* 41–51.

Nesselmann, G. H. F. (1842). *Versuch einer kritischen Geschichte der Algebra: Die Algebra der Griechen* (Vol. 1). Berlin: Reimer.

Neves, D. M., & Anderson, J. R. (1981). Knowledge compilation: Mechanisms for the automatization of cognitive skills. In J. R. Anderson (Ed.), *Cognitive skills and their acquisition* (pp. 57–84). Hillsdale, NJ: Erlbaum.

Newell, A., & Simon, H. A. (1972). *Human problem solving.* Englewood Cliffs, NJ: Prentice Hall.

Newman, J. R. (1956a). Commentary on certain important abstractions. In J. R. Newman (Ed.), *The world of mathematics* (Vol. 3, pp. 1534–1537). New York: Simon & Schuster.

Newman, J. R. (1956b). Commentary on Hermann von Helmholtz. In J. R. Newman (Ed.), *The world of mathematics* (Vol. 1, pp. 642–646). New York: Simon & Schuster.

Newman, J. R. (1956c). Commentary on the mysteries of arithmetic. In J. R. Newman (Ed.), *The world of mathematics* (Vol. 1, p. 497). New York: Simon & Schuster.

Nicholls, J. G. (1975). Causal attributions and other achievement-related cognitions: Effects of task outcome, attainment value, and sex. *Journal of Personality and Social Psychology, 31,* 379–389.

Nickerson, R. S. (1967). "Same"-"different" response times with multi-attribute stimulus differences. *Perceptual & Motor Skills, 24,* 543–554.

Nickerson, R. S. (1972). Binary-classification reaction time: A review of some studies of human information-processing capabilities. *Psychonomic Monographs Supplement, 4,* Whole No. 65.

Nickerson, R. S. (1988). *Practical Intelligence,* edited by R. J. Sternberg & R. K. Wagner [Book review]. *American Journal of Psychology, 101,* 293–302.

Nickerson, R. S. (1994). The teaching of thinking and problem solving. In R. J. Sternberg (Ed.), *Thinking and problem solving* (Vol. 12 of E. C. Carterette & M. Friedman [Eds.], *Handbook of perception and cognition,* pp. 409–449). San Diego: Academic Press.

Nickerson, R. S. (1996). Ambiguities and unstated assumptions in probabilistic reasoning. *Psychological Bulletin, 120,* 410–433.

Nickerson, R. S. (1997). Cognitive technology: Reflections on a long history and promising future. *Cognitive Technology, 2,* 6–20.

Nickerson, R. S. (1999). Why are there twelve inches in a foot? *Cognitive Technology, 4,* 18–25.

Nickerson, R. S. (2004a). *Cognition and chance: The psychology of probabilistic reasoning.* Mahwah, NJ: Erlbaum.

Nickerson, R. S. (2004b). Teaching reasoning. In J. P. Leighton & R. J. Sternberg (Eds.), *The nature of reasoning* (pp. 410–442). New York: Cambridge University Press.

Nickerson, R. S. (2005). Technology and cognitive amplification. In R. J. Sternberg & D. D. Preiss (Eds.), *Intelligence and technology: The impact of tools on the nature and development of human abilities* (pp. 3–28). Mahwah, NJ: Erlbaum.

Nickerson, R. S. (2007). Penney ante: Counterintuitive probabilities in coin tossing. *Undergraduate Mathematics and its Applications (UMAP)*, *28*, 503–532.

Nickerson, R. S., & Zodhiates, P. P. (1988). *Technology in education: Looking toward 2020*. Hillsdale, NJ: Erlbaum.

Noddings, N. (1985). Small groups as a setting for research on mathematical problem solving. In E. A. Silver (Ed.), *Teaching and learning mathematical problem solving: Multiple research perspectives* (pp. 345–359). Hillsdale, NJ: Erlbaum.

Noel, M.-P., & Seron, X. (1993). Arabic number reading deficit: A single case study of when 236 is read (2306) and judged superior to 1258. *Cognitive Neuropsychology*, *10*, 317–339.

Norman, D. A. (1981). Categorization of action slips. *Psychological Review*, *88*, 1–15.

Normore, C. G. (1982). Walter Burley on continuity. In N. Kretzmann (Ed.), *Infinity and continuity in ancient and medieval thought* (pp. 258–269). Ithaca, NY: Cornell University Press.

Nosek, B. A., Banaji, M. R., & Greenwald, A. G. (2002). Math = male, me = female, therefore math ≠ me. *Journal of Personality and Social Psychology*, *83*, 44–59.

Novick, L. R. (1988). Analogical transfer, problem similarity and expertise. *Journal of Experimental Psychology: Learning, Memory and Cognition*, *14*, 510–520.

Novick, L. R., & Holyoak, K. J. (1991). Mathematical problem solving by analogy. *Journal of Experimental Psychology: Learning, Memory and Cognition*, *17*, 398–415.

Novillis-Larson, C. (1980). Locating proper fractions. *School Science and Mathematics*, *53*, 423–428.

Nozick, R. (1981). *Philosophical explanations*. Cambridge, MA: Harvard University Press.

Nuerk, H. C., Iversen, W., & Willmes, K. (2004). Notational modulation of the SNARC and the MARC (linguistic markedness of response codes) effect. *Quarterly Journal of Experimental Psychology*, *57A*, 835–863.

Nunes, T., Schliemann, A. D., & Carraher, D. W. (1993). *Street mathematics and school mathematics*. Cambridge, UK: Cambridge University Press.

Núñez, R., & Lakoff, G. (1997). The metaphorical structure of mathematics: Sketching out cognitive foundations for a mind-based mathematics. In L. English (Ed.), *Mathematical reasoning: Analogies, metaphors, and images* (pp. 21–89). Mahwah, NJ: Erlbaum.

Núñez, R., & Lakoff, G. (2005). The cognitive foundations of mathematics: The role of conceptual metaphor. In J. I. D. Campbell (Ed.), *Handbook of mathematical cognition* (pp. 109–124). New York: Psychology Press.

Nuttall, R. L, Casey, B., & Pezaris, E. (2005). Spatial ability as a mediator of gender differences on mathematics tests. In A. M. Gallagher & J. C. Kaufman (Eds.), *Gender differences in mathematics* (pp. 121–142). New York: Cambridge University Press.

Nyborg, H. (1983). Spatial ability in men and women: Review and new theory. *Advances in Behavior Research and Therapy*, *5*, 89–140.

Oakes, J. (1990). Opportunities, achievement and choice: Women and minority students in science and mathematics. *Review of Research in Education, 16,* 153–222.

O'Connor, N., & Hermelin, B. (1984). Idiot savant calendrical calculators: Maths or memory? *Psychological Medicine, 14,* 801–806.

Ogburn, W. F. (1923). *Social change.* New York: Dell Publishing.

Ogilvy, C. S. (1984). *Excursions in mathematics.* Mineola, NY: Dover. (Original work published 1956)

Olthof, A., Iden, C. M., & Roberts, W. A. (1997). Judgments of ordinality and summation of number symbols by squirrel monkeys (*Saimiri sciureus*). *Journal of Experimental Psychology: Animal Behavior Processes, 23,* 325–339.

Oreskes, N., Shrader-Frechette, K., & Belitz, K. (1994). Verification, validation, and confirmation of numerical models in the earth sciences. *Science, 263,* 641–646.

Osborne, R. T., & Lindsey, J. M. (1967). A longitudinal investigation of change in the factorial composition of intelligence with age in young school children. *Journal of Learning Disabilities, 22,* 339–348.

Osen, L. (1974). *Women in mathematics.* Cambridge, MA: MIT Press.

Osgood, C. E. (1953). *Method and theory in experimental psychology.* New York: Oxford University Press.

Owen, G. E. L. (2001). Zeno and the mathematicians. In W. C. Salmon (Ed.), *Zeno's paradoxes* (pp. 139–163). Indianapolis, IN: Hackett Publishing. (Original work published in 1957.)

Pacioli, L. (1988). *Divine proportion.* Paris: Librairie do Compagnonnage. (Original work published 1509)

Pagels, H. R. (1991). *Perfect symmetry: The search for the beginning of time.* New York: Bantam Books.

Paige, J. M., & Simon, H. A. (1966). Cognitive processes in solving algebra word problems. In B. Kleinmuntz (Ed.), *Problem solving: Research, method, and theory* (pp. 51–119). New York: Wiley.

Paik, M. K. (1983). Empty or infinitely full? *Mathematics Magazine, 56,* 221–223.

Pajares, F. (1996). Self-efficacy beliefs and mathematical problem-solving of gifted students. *Contemporary Educational Psychology, 21,* 325–344.

Pajares, F. (2005). Gender differences in mathematics: Self-efficacy beliefs. In A. M. Gallagher & J. C. Kaufman (Eds.) *Gender differences in mathematics: An integrative psychological approach.* New York: Cambridge University Press.

Pajares, F., & Kranzler, J. (1995). Self-efficacy beliefs and general mental ability in mathematical problem solving. *Contemporary Educational Psychology, 20,* 426–443.

Pajares, F., & Miller, M. D. (1994). Role of self-efficacy and self-concept beliefs in mathematical problem solving: A path analysis. *Journal of Educational Psychology, 86,* 193–203.

Pansky, A., & Algom, D. (1999). Stroop and Garner effects in comparative judgment of numerals: the role of attention. *Journal of Experimental Psychology: Human Perception and Performance, 25,* 39–58.

Papert, S. (1972). Teaching children to be mathematicians versus teaching about mathematics. *International Journal for Mathematical Education, Science, and Techology, 3,* 249–262.

Pappas, T. (1989). *The joy of mathematics: Discovering mathematics all around you* (rev. ed.). San Carlos, CA: World Wide Publishing/Tetra.

Pappas, T. (1993). *More joy of mathematics: Exploring mathematical insights and concepts.* San Carlos, CA: World Wide Publishing/Tetra.

Pappas, T. (1997). *Mathematical scandals.* San Carlos, CA: Wide World Publishing/Tetra.

Parish, J. M., & Ericksen, M. T. A. (1981). A comparison of cognitive strategies in modifying the cognitive style of impulsive third-grade children. *Cognitive Therapy and Research, 5,* 71–78.

Park, A. (2008, July 24). The myth of the math gender gap. *Time Magazine.*

Parkman, J. M. (1971). Temporal aspects of digit and letter inequality judgments. *Journal of Experimental Psychology, 91,* 191–205.

Parsons, J. E., Meece, J. L., Adler, T. F., & Kaczala, C. M. (1982). Sex differences in attributions and learned helplessness. *Sex Roles, 8,* 421–432.

Pascal, B. (1947). *Pensées.* In S. Commins & R. N. Linscott (Eds.), *Man and spirit: The speculative philosophers.* New York: Random House. (Original work published 1670)

Paulos, J. A. (1992). *Beyond numeracy.* New York: Vintage Books.

Paulos, J. A. (1995). *A mathematician reads the newspaper.* New York: Anchor Books.

Pavese, A., & Umilta, C. (1998). Symbolic distance between numerosity and identity modulates stroop interference. *Journal of Experimental Psychology: Human Perception and Performance, 24,*1535–1545.

Payne, J. N. (1990). *Mathematics for the young child.* Reston, VA: National Council of Teachers of Mathematics.

Payne, J. N., & Huinker, D. M. (1993). Early number and numeration. In R. J. Jensen (Ed.), *Research ideas for the classroom: Early childhood mathematics* (pp. 43–71). New York: Macmillan.

Payne, S., & Squibb, H. (1990). Algebra mal-rules and cognitive accounts of error. *Cognitive Science, 14,* 445–481.

Pea, R. D. (1987). Cognitive technologies for mathematics education. In A. H. Schoenfeld (Ed.), *Cognitive science and mathematics education* (pp. 89–122). Hillsdale, NJ: Erlbaum.

Peacocke, A. (1993). *Theology for a scientific age: Being and becoming—natural, divine, and human.* Minneapolis: Fortress Press.

Peak, L. (1996). *Pursuing excellence: A study of U.S. eighth-grade mathematics and science achievement in international context.* Washington, DC: U.S. Government Printing Office.

Peak, L. (1997). *Pursuing excellence: A study of U.S. fourth-grade mathematics and science achievement in international context.* Washington, DC: U.S. Government Printing Office.

Peck, D. M., & Jencks, S. M. (1981). Conceptual issues in the teaching and learning of fractions. *Journal for Research in Mathematics Education, 12,* 339–348.

Pedro, J. D., Wolleat, P., Fennema, E., & Becker, A. D. (1981). Election of high school mathematics by females and males: Attributions and attitudes. *American Educational Research Journal, 18,* 207–218.

Peirce, C. S. (1956a). The essence of mathematics. In J. R. Newman (Ed.), *The world of mathematics* (Vol. 3, pp. 1773–1783). New York: Simon & Schuster. (Original work published 1902)

Peirce, C. S. (1956b). The probability of induction. In J. R. Newman (Ed.), *The world of mathematics* (Vol. 2, pp. 1341–1354). New York: Simon & Schuster.

Peirce, C. S. (1956c). The red and the black. In J. R. Newman (Ed.), *The world of mathematics* (Vol. 2, pp. 1334–1340). New York: Simon & Schuster. (Original work published unknown)

Penrose, R. (1978). The geometry of the universe. In L. A. Steen (Ed.), *Mathematics today: Twelve informal essays* (pp. 83–125). New York: Springer-Verlag.

Penrose, R. (1989). *The emperor's new mind: Concerning computers, minds, and the laws of physics.* New York: Oxford University Press.

Pepperberg, I. M. (1987). Evidence for conceptual quantitative abilities in the African grey parrot: Labeling of cardinal sets. *Ethology, 75,* 37–61.

Pepperberg, I. M. (1988). Comprehension of "absence" by an African grey parrot: Learning with respect to questions of same/different. *Journal of Experimental Analysis of Behavior, 50,* 553–564.

Pepperberg, I. M. (1994). Numerical competence in an African grey parrot (*Psittacus erithacus*). *Journal of Comparative Psychology, 108,* 36–44.

Pepperberg, I. M. (1999). *The Alex studies: Cognitive and communicative abilities of grey parrots.* Cambridge, MA: Harvard University Press.

Pepperberg, I. M. (2008). *Alex and me.* New York: HarperCollins.

Perkins, D. (2000). *Archimedes bathtub: The art and logic of breakthrough thinking.* New York: W.W. Norton.

Perkins, D. N., Schwartz, J. L., West, M. M., & Wiske, M. S. (Eds.). (1995). *Software goes to school: Teaching for understanding with new technologies.* New York: Oxford University Press.

Perry, M., Church, R. B., & Goldin-Meadow, S. (1988). Transitional knowledge in the acquisition of concepts. *Cognitive Development, 3,* 359–400.

Perry, M., Van der Stoep, S. W., & Yu, S. L. (1993). Asking questions in first grade mathematics classes: Potential influences on mathematical thought. *Journal of Educational Psychology, 85,* 31–40.

Pérusse, R., & Rumbaugh, D. M. (1990). Summation in chimpanzees (*Pan troglodytes*): Effects of amounts, number of wells and finer ratios. *International Journal of Primatology, 11,* 425–437.

Pesenti, M. (2005). Calculation abilities in expert calculators. In J. I. D. Campbell (Ed.), *Handbook of mathematical cognition* (pp. 413–430). New York: Psychology Press.

Pesenti, M., Seron, X., Samson, D., & Duoux, B. (1999). Basic and exceptional calculation abilities in a calculating prodigy: A case study. *Mathematical Cognition, 5,* 97–148.

Péter, R. (1976). *Playing with infinity: Mathematical explorations and excursions* (Z. P. Dienes, Trans.). New York: Dover. (Original work published 1961)

Peterson, I. (1988). *The mathematical tourist: Snapshots of modern mathematics.* New York: Freeman.

Petitto, A. L. (1979). *Knowledge of arithmetic among schooled and unschooled African tailors and cloth merchants.* Doctoral dissertation, Cornell University.

Petitto, A. L., & Ginsburg, H. P. (1982). Mental arithmetic in Africa and America: Strategies, principles and explanations. *International Journal of Psychology, 17,* 81–102.

Pfungst, O. (1965). *Clever Hans: The horse of Mr. Von Osten*. New York: Holt, Rinehart and Winston. (Original work published 1911)

Piaget, J. (1928). *Judgment and reasoning in the child*. London: Routledge and Kegan Paul.

Piaget, J. (1942). *Classes, relations et nombre*. Paris: Vrin.

Piaget, J. (1952). *The child's conception of number*. New York: Norton. (Original work published 1941)

Piaget, J., & Inhelder, B. (1969). *The psychology of the child*. New York: Basic Books.

Piaget, J., Inhelder, B., & Szeminska, A. (1960). *The child's conception of geometry*. New York: Basic Books. (Original work published 1948)

Pickover, C. A. (2000). *Wonder of numbers: Adventures in math, mind and meaning*. New York: Oxford University Press.

Platt, J. R., & Johnson, D. M. (1971). Localization of position within a homogeneous behavior chain: Effects of error contingencies. *Learning and Motivation, 2*, 386–414.

Poffenberger, T. M., & Norton, D. A. (1959). Factors in the formation of attitudes toward mathematics. *Journal of Educational Research, 52*, 171–176.

Poincaré, H. (1913). Mathematical creation. In *The foundations of science* (G. B. Halstead, Trans.). New York: The Science Press.

Polkinghorne, J. (1991). *Reason and reality: The relationship between science and theology*. Valley Forge, PA: Trinity Press.

Polkinghorne, J. (1998). *Belief in God in an age of science*. New Haven, CT: Yale University Press.

Polkinghorne, J. C. (2006). *Science and creation: The search for understanding*. West Conshohocken, PA: Templeton Foundation Press.

Pollak, H. O. (1987). Cognitive science and mathematics education: A mathematician's perspective. In A. H. Schoenfeld (Ed.), *Cognitive science and mathematics education* (pp. 253–264). Hillsdale, NJ: Erlbaum.

Polster, B. (2004). *Q.E.D.: Beauty in mathematical proof*. New York: Walker and Company.

Polya, G. (1954a). *Mathematics and plausible reasoning: Induction and analogy in mathematics* (Vol. 1). Princeton, NJ: Princeton University Press.

Polya, G. (1954b). *Mathematics and plausible reasoning: Patterns of plausible inference* (Vol. 2). Princeton, NJ: Princeton University Press.

Polya, G. (1957). *How to solve it: A new aspect of mathematical method*. Garden City, NY: Doubleday. (Original work published 1945)

Polya, G. (1965). *Mathematical discovery: On understanding, learning, and teaching problem solving*. New York: Wiley.

Porter, A. (1989). A curriculum out of balance: A case study of elementary mathematics. *Educational Researcher, 18*, 9–15.

Posner, J. (1982). The development of mathematical knowledge in two West African societies. *Child Development, 53*, 200–208.

Post, T. R., Harel, G., Behr, M., & Lesh, R. (1991). Intermediate teachers' knowledge of rational number concepts. In E. Fennema, T. P. Carpenter, & S. L. Lamon (Eds.), *Integrating research on teaching and learning mathematics* (pp. 177–198). Ithaca, NY: SUNY Press.

Post, T. R., Wachsmuth, I., Lesh, R., & Behr, M. J. (1985). Order and equivalence of rational numbers: A cognitive analysis. *Journal for Research in Mathematics Education, 16*, 18–36.

Pothier, Y., & Sawada, D. (1983). Partitioning: The emergence of rational number ideas in young children. *Journal for Research in Mathematics Education 14*, 307–317.

Program for International Student Assessment. (2006). PISA 2006 summary. Retrieved August 3, 2009 from http://nces.ed.gov/surveys/pisa/index.asp

Psotka, J., Massey, L. D., & Mutter, S. A. (Eds.). (1988). *Intelligent tutoring systems: Lessons learned*. Hillsdale, NJ: Erlbaum.

Putnam, R. T., Heaton, R. M., Prawat, R. S., & Remillard, J. (1992). Teaching mathematics for understanding: Discussing case studies of four fifth-grade teachers. *Elementary School Journal, 93*, 213–228.

Pylyshyn, Z. (2001). Visual indexes, preconceptual objects, and situated vision. *Cognition, 80*, 127–158.

Pyszczynski, T., & Greenberg, J. (1983). Determinants of reduction in effort as a strategy for coping with anticipated failure. *Journal of Research in Personality, 17*, 412–422.

Quinn, D. M., & Spencer, S. J. (2001). The interence of stereotype threat with women's generation of mathematical problem-solving strategies. *Journal of Social Issues, 57*, 55–71.

Raiffa, H., & Schlaifer, R. (1961). *Applied statistical decision theory*. Cambridge, MA: MIT Press.

Raine, C. W. (1948). Pythagorean triples from the Fibonacci series. *Scripta Mathematica, 14*, 164.

Ramirez, O. M., & Dockweiler, C. J. (1987). Mathematics anxiety: A systematic review. In R. Schwarzer, H. M. van der Ploeg, & C. D. Spielberger (Eds.), *Advances in test anxiety research* (Vol. 5, pp. 157–175). Berwyn, PA: Swets North America.

Rapoport, A. (1960). *Fights, games and debates*. Ann Arbor: University of Michigan Press.

Reed, H. J., & Lave, J. (1979). Arithmetic as a tool for investigating relations between culture and cognition. *American Ethologist, 6*, 568–582.

Reed, H. J., & Lave, J. (1981). Arithmetic as a tool for investigating relations between culture and cognition. In R. W. Casson (Ed.), *Language, culture, and cognition: Anthropological perspectives* (pp. 437–455). New York: Macmillan.

Reid, W. A. (1987). Institutions and practices: Professional education reports and the language of reform. *Educational Researcher, 16*, 10–15.

Renga, S., & Dalla, L. (1993). Affect: A critical component of mathematical learning in early childhood. In R. J. Jensen (Ed.), *Research ideas for the classroom: Early childhood mathematics* (pp. 22–39). New York: Macmillan.

Resnick, L. B. (1976). Task analysis in instructional design: Some cases from mathematics. In D. Klahr (Ed.), *Cognition and instruction* (pp. 51–80). Hillsdale, NJ: Erlbaum.

Resnick, L. B. (1983). A developmental theory of number understanding. In H. P. Ginsburg (Ed.), *The development of mathematical thinking* (pp. 109–151). New York: Academic Press.

Resnick, L. B., & Ford, W. W. (1981). *The psychology of mathematics for instruction*. Hillsdale, NJ: Erlbaum.

Resnick, L. B., & Johnson, A. (1988). Intelligent machines for intelligent people: Cognitive theory and the future of computer-assisted learning. In R. S. Nickerson & P. P. Zodhiates (Eds.), *Technology in education: Looking toward 2020* (pp. 139–168). Hillsdale, NJ: Erlbaum.

Resnick, L. B., Nesher, P., Leonard, F., Magone, M., Omanson, S., & Peled, I. (1989). Conceptual bases of arithmetic errors: The case of decimal fractions. *Journal for Research in Mathematics Education, 20,* 8–27.

Resnick, L. B., & Omanson, S. F. (1987). Learning to understand arithmetic. In R. Glaser (Ed.), *Advances in instructional psychology* (Vol. 3, pp. 41–95). Hillsdale, NJ: Erlbaum.

Resnick, L. B., Wang, M. C., & Kaplan, J. (1973). Task analysis in curriculum design: A hierarchically sequenced introductory mathematics curriculum. *Journal of Applied Behavior Analysis, 6,* 679–710.

Resnik, M. D. (1997). *Mathematics as a science of patterns.* Oxford: Clarendon Press.

Reusser, K. (1988). Problem solving beyond the logic of things: Contextual effects on understanding and solving word problems. *Instructional Science, 17,* 309–338.

Reyes, L. H., & Stanic, G. M. A. (1988). Race, sex, socioeconomic status, and mathematics. *Journal of Research in Mathematics Education, 19,* 26–43.

Reynolds, D., & Farrell, S. (1996). *Worlds apart? A review of international surveys of educational achievement involving England.* London: Her Majesty's Stationery Office.

Reynvoet, B., Brysbaert, M., & Fias, W. (2002). Semantic priming in number naming. *Quarterly Journal of Experimental Psychology, 55A,* 1127–1139.

Reys, R. E., Rybolt, J. F., Bestgen, B. J., & Wyatt, J. W. (1982). Processes used by good computational estimators. *Journal for Research in Mathematics Education, 13,* 183–201.

Richards, I. (1978). Number theory. In L. A. Steen (Ed.), *Mathematics today: Twelve informal essays* (pp. 37–64). New York: Springer-Verlag.

Richardson, F. C., & Suinn, R. M. (1972). The Mathematics Anxiety Rating Scale. *Journal of Counseling Psychology, 19,* 551–554.

Rickard, T. C., & Bourne, L. E., Jr. (1996). Some tests of an identical elements model of basic ability in arithmetic skills. *Journal of Experimental Psychology: Learning, Memory, and Cognition, 22,* 1281–1295.

Riley, M. S., Greeno, J. G., & Heller, J. I. (1983). Development of children's problem-solving ability in arithmetic. In H. P. Ginsburg (Ed.), *The development of mathematical thinking* (pp. 153–196). New York: Academic Press.

Rilling, M. E. (1967). Number of responses as a stimulus in fixed interval and fixed ratio schedules. *Journal of Comparative and Physiological Psychology, 63,* 60–65.

Rilling, M. E. (1993). Invisible counting animals: A history of contributions from comparative psychology, ethology, and learning theory. In S. T. Boysen & E. J. Capaldi (Eds.), *The development of numerical competence: Animal and human models* (pp. 3–37). Hillsdale, NJ: Erlbaum.

Rilling, M. E., & McDiarmid, C. (1965). Signal detection in fixed-ratio schedule. *Science, 148,* 526–527.

Riordan, C. (1990). *Girls and boys in school: Together or separate?* New York: Teachers College Press.

Rissland, E. L. (1985). Artificial intelligence and the learning of mathematics: A tutorial sampling. In E. A. Silver (Ed.), *Teaching and learning mathematical problem solving: Multiple research perspectives* (pp. 147–176). Hillsdale, NJ: Erlbaum.

Rittle-Johnson, B., & Alibali, M. W. (1999). Conceptual and procedural knowledge of mathematics: Does one lead to the other? *Journal of Educational Psychology, 91,* 175–189.

Rittle-Johnson, B., & Siegler, R. S. (1998). The relation between conceptual and procedural knowledge in learning mathematics: A review. In C. Donlan (Ed.), *The development of mathematical skills* (pp. 75–110). Hove, UK: Psychology Press.

Roberts, W. A., & Mitchell, S. (1994). Can a pigeon simultaneously process temporal and numerical information? *Journal of Experimental Psychology: Animal Behavior Processes, 20,* 137–150.

Robinson, A. (1969). The metaphysics of the calculus. In J. Hintikka (Ed.), *The philosophy of mathematics* (pp. 153–163). London: Oxford University Press.

Robinson, C. S., & Hayes, J. R. (1978). Making inferences about relevance in understanding problems. In R. Revlin & R. E. Mayer (Eds.), *Human reasoning* (pp. 195–206). Washington, DC: Winston.

Robinson, N. M., Abbott, R. D., Beringer, V. W., & Busse, J. (1996). The structure of abilities in math-precocious young children: Gender similarities and differences. *Journal of Educational Psychology, 88,* 341–352.

Robitaille, D. A. (1977). A comparison of boys' and girls' feelings of self-confidence in arithmetic computation. *Canadian Journal of Education, 2,* 15–22.

Romberg, T. A. (1994). Classroom instruction that fosters mathematical thinking and problem solving: Connections between theory and practice. In A. H. Schoenfeld (Ed.), *Mathematical thinking and problem solving* (pp. 287–304). Hillsdale, NJ: Erlbaum.

Rosenthal, D. J. A., & Resnick, L. B. (1974). Children's solution processes in arithmetic word problems. *Journal of Educational Psychology, 66,* 817–825.

Rosenthal, R. (1965). *Clever Hans* (C. L. Rahn, Trans.). New York: Hold, Rinehart & Winston. (Original work published 1911)

Rosenthal, R., & Jacobson, L. (1992). *Pygmalion in the classroom* [Expanded edition]. New York: Irvington. (Original work published 1968)

Rosnick, P., & Clement, J. (1984). Learning without understanding: The effect of tutoring strategies on algebra misconceptions. *Journal of Mathematical Behavior, 3,* 3–27.

Ross, B. H. (1987). This is like that: The use of earlier problems and the separation of similarity effects. *Journal of Experimental Psychology: Learning, Memory, and Cognition, 13,* 629–639.

Ross, B. H. (1989). Distinguishing types of superficial similarities: Different effects on access and use of earlier problems. *Journal of Experimental Psychology: Learning, Memory, and Cognition, 15,* 456–468.

Ross, S. (1976). *A first course in probability.* New York: Macmillan.

Rota, G.-C. (1981). Introduction to P.J. Davis and R. Hirsch. In *The mathematical experience.* Boston: Houghton-Mifflin.

Rouse Ball, W. W. (1892). *Mathematical recreations and problems of past and present times.* London: Macmillan.

Rouse Ball, W. W., & Coxeter, H. S. M. (1987). *Mathematical recreations and essays* (13th ed.). New York: Courier Dover. (Original work published 1982)

Rovet, J. F. (1993). The psychoeducational characteristics of children with Turner syndrome. *Journal of Learning Disabilities, 26,* 333–341.

Rovet, J. F., Szekely, C., & Hockenberry, M. N. (1994). Specific arithmetic calculation deficits in children with Turner syndrome. *Journal of Clinical and Experimental Neuropsychology, 16,* 820–839.

Rowe, M. B. (1983). Science education: A framework for decision makers. *Daedalus, 112,* 123–142.

Royer, J. M., & Garofoli, L. M. (2005). Cognitive contributions to sex differences in math performance. In A. M. Gallagher & J. C. Kaufman (Eds.), *Gender differences in mathematics* (pp. 99–120). New York: Cambridge University Press.

Royer, J. M., Rath, K., Tronsky, L. N., Marchant, H., & Jackson, S. (2002). *Spatial cognition and math fact retrieval as the causes of sex differences in math test performance.* Paper presented at the Annual Meeting of the American Educational Research Association, New Orleans.

Royer, J. M., Tronsky, L. N., Chan, Y., Jackson, S., & Marchant, H. (1999). Math-fact retrieval as the cognitive mechanism underlying gender differences in math test performance. *Contemporary Educational Psychology, 24,* 181–266.

Rozeboom, W. W. (1997). Good science is abductive, not hypothetico-deductive. In L. L. Harlow, S. A. Mulaik, & J. H. Steiger (Eds.), *What if there were no significance tests?* (pp. 335–391). Mahwah, NJ: Erlbaum.

Ruble, D., & Martin, C. (1998). Gender development. In N. Eisenberg (Ed.), *Handbook of child psychology* (Vol. 3, 5th ed., pp. 933–1016). New York: Wiley.

Rucker, R. (1980). *White light; or, what is Cantor's continuum problem?* New York: Ace Books.

Rucker, R. (1982). *Infinity and the mind: The science and philosophy of the infinite.* Boston: Birkhäuser.

Ruelle, D. (1991). *Chance and chaos.* Princeton, NJ: Princeton University Press.

Rumbaugh, D. M. (1990). Comparative psychology and the great apes: Their competence in learning, language, and numbers. *The Psychological Record, 40,* 15–39.

Rumbaugh, D. M., Hopkins, W. D., Washburn, D. A., & Savage-Rumbaugh, E. S. (1989). Lana chimpanzee learns to count by 'Numath': A summary of a videotaped experimental report. *Psychological Record, 39,* 459–470.

Rumbaugh, D. M., Savage-Rumbaugh, E. S., & Hegel, M. (1987). Summation in the chimpanzee (*Pan troglodytes*). *Journal of Experimental Psychology: Animal Behavior Processes, 13,* 107–115.

Rumbaugh, D. M., Savage-Rumbaugh, E. S., Hopkins, W. D., Washburn, D. A., & Runfeldt, S. A. (1989). *Lana chimpanzee (Pan troglodytes) counts by Numath* [Videotape]. University Park: Pennsylvania State University.

Rumbaugh, D. M., & Washburn, D. A. (1993). Counting by chimpanzees and ordinality judgments by macaques in video-formatted tasks. In S. T. Boysen & E. J. Capaldi (Eds.), *The development of numerical competence: Animal and human models* (pp. 87–106). Hillsdale, NJ: Erlbaum.

Runion, G. E. (1990). *The golden section.* Palo Alto, CA: Dale Seymour.

Russell, B. (1903). *The principles of mathematics.* Cambridge, UK: Cambridge University Press.

Russell, B. (1910). *The study of mathematics: Philosophical essays.* London: Unwin Books.

Russell, B. (1919). *Introduction to mathematical philosophy.* London: Allen and Unwin.

Russell, B. (1926). *Our knowledge of the external world.* London: Allen and Unwin.

Russell, B. (1956a). Mathematics and the metaphysicians. In J. R. Newman (Ed.), *The world of mathematics* (Vol. 3, pp. 1576–1590). New York: Simon & Schuster. (Original work published 1901)

Russell, B. (1956b). My mental development. In J. R. Newman (Ed.), *The world of mathematics* (Vol. 1, pp. 381–394). New York: Simon & Schuster. (Original work published 1944)

Russell, B. (1994). The greatness of Albert Einstein. In M. Gardner (Ed.), *Great essays in science* (pp. 408–412). New York: Prometheus Books. (Original work published 1955)

Russell, B. (2001). The problem of infinity considered historically. In W. C. Salmon (Ed.), *Zeno's paradoxes* (pp. 45–58). Indianapolis: Hackett Publishing. (Original work published in 1929)

Sadker, M., Sadker, D., & Klein, S. (1991). The issue of gender in elementary and secondary education. *Review of Research in Education, 17,* 269–334.

Saenz-Ludlow, A., & Walgamuth, C. (1998). Third graders interpretations of equality and the equal symbol. *Educational Studies in Mathematics, 35,* 153–187.

Salman, D. H. (1943). Note on the number conception in animal psychology. *British Journal of Psychology, 33,* 209–219.

Salmon, W. C. (2001). Introduction. In W. C. Salmon (Ed.), *Zeno's paradoxes* (pp. 5–44). Indianapolis: Hackett Publishing.

Salomon, F. (2004). *The cord keepers: Khipus and cultural life in a Peruvian village.* Durham, NC: Duke University Press.

Salomon, G. (Ed.). (1993). *Distributed cognitions: Psychological and educational considerations.* New York: Cambridge University Press.

Salsburg, D. (2001). *The lady tasting tea: How statistics revolutionized science in the twentieth century.* New York: W. H. Freeman.

Salthouse, T. A. (1996). The processing-speed theory of adult age differences in cognition. *Psychological Review, 103,* 403–428.

Sarama, J. (2000). Toward more powerful computer environments: Developing mathematics software on research-based principles. *Focus on Learning Problems in Mathematics, 22,* 125–147.

Sarama, J. (2004). Technology in early childhood mathematics: Building Blocks™ as an innovative technology-based curriculum. In D. H. Clements, J. Sarama, & A.-M. DiBiase (Eds.), *Engaging young children in mathematics: Standards for early childhood education* (pp. 361–376). Mahwah, NJ: Erlbaum.

Sarnecka, B. W., & Gelman, S. A. (2004). Six does not just mean a lot: Preschoolers see number words as specific. *Cognition, 92,* 329–352.

Saxe, G. B. (1981). Body parts as numerals: A developmental analysis of numeration among a village population in Papua New Guinea. *Child Development, 52,* 306–316.

Saxe, G. B. (1982). Developing forms of arithmetic operations among the Oksapmin of Papua New Guinea. *Developmental Psychology, 18*, 583–594.

Saxe, G. B. (1985). The effects of schooling on arithmetical understandings: Studies with Oksapmin children in Papua New Guinea. *Journal of Educational Psychology, 77*, 503–513.

Saxe, G. B. (1988). The mathematics of child street vendors. *Child Development, 59*, 1415–1425.

Saxe, G. B. (1991). *Culture and cognitive development: Studies in mathematical understanding.* Hillsdale, NJ: Erlbaum.

Saxe, G. B., Dawson, V., Fall, R., & Howard, S. (1996). Culture and children's mathematical thinking. In R. J. Sternberg & T. Ben-Zeev (Eds.), *The nature of mathematical thinking* (pp. 119–144). Mahwah, NJ: Erlbaum.

Saxe, G. B., & Posner, J. (1983). The development of numerical cognition: Cross-cultural perspectives. In H. P. Ginsburg (Ed.), *The development of mathematical thinking* (pp. 291–317). New York: Academic Press.

Saxe, G. B., Guberman, S. R., & Gearhart, M. (1987). Social processes in early number develoment. *Monographs of the Society for Research in Child Development, 52*, No. 2, Serial No. 216.

Schaeffer, B., Eggleston, V. H., & Scott, J. L. (1974). Number development in young children. *Cognitive Psychology, 6*, 357–379.

Schattschneider, D. (1994). Escher's metaphors. *Scientific American, 271*, 66–71.

Schechter, B. (1998). *My brain is open: The mathematical journeys of Paul Erdös.* New York: Simon & Schuster.

Schimmel, A. M. (1993). *The mystery of numbers.* New York: Oxford University Press.

Schmader, T. (2002). Gender identification moderates stereotype threat effects on women's math performance. *Journal of Experimental Social Psychology, 38*, 194–201.

Schmandt-Besserat, D. (1978). The earliest precursors of writing. *Scientific American, 238*, 50–59.

Schmandt-Besserat, D. (1981). From tokens to tablets: A re-evaluation of the so-called "numerical tablets." *Visible Language, 15*, 321–344.

Schmandt-Besserat, D. (1982). The emergence of recording. *The American Anthropologist, 84*, 871–878.

Schmandt-Besserat, D. (1984). Before numerals. *Visible Language, 18*, 48–60.

Schmandt-Besserat, D. (1992). *From counting to cuneiform* (Vol. 1 of *Before writing*). Austin, TX: University of Texas Press.

Schmidt, W. H., McKnight, C. C., Cogan, L. S., Jakwerth, P. M., & Houang, R. T. (1999). *Facing the consequences: Using TIMSS for a closer look at mathematics and science education.* Dordrecht: Kluwer.

Schmidt, W. H., McKnight, C. C., & Raizen, S. A. (1997). *A splintered vision: An investigation of U.S. science and mathematics education.* Dordrecht: Kluwer.

Schoenfeld, A. H. (1979). Explicit heuristic training as a variable in problem-solving performance. *Journal of Research in Mathematical Education, 10*, 173–187.

Schoenfeld, A. H. (1983a). Beyond the purely cognitive: Belief systems, social cognitions, and metacognitions as driving forces in intellectual performance. *Cognitive Science, 7*, 329–363.

Schoenfeld, A. H. (1983b). *Problem solving in the mathematics curriculum: A report, recommendations and an annotated bibliography.* Washington, DC: Mathematical Association of America.

Schoenfeld, A. H. (1985a). Artificial intelligence and mathematics education: A discussion of Rissland's paper. In E. A. Silver (Ed.), *Teaching and learning mathematical problem solving: Multiple research perspectives* (pp. 177–187). Hillsdale, NJ: Erlbaum.

Schoenfeld, A. H. (1985b). *Mathematical problem solving.* San Diego: Academic Press.

Schoenfeld, A. H. (1985c). Metacognitive and epistemological issues in mathematical understanding. In E. A. Silver (Ed.), *Teaching and learning mathematical problem solving: Multiple research perspectives* (pp. 361–379). Hillsdale, NJ: Erlbaum.

Schoenfeld, A. H. (1987a). Cognitive science and mathematics education: An overview. In A. H. Schoenfeld (Ed.), *Cognitive science and mathematics education* (pp. 1–31). Hillsdale, NJ: Erlbaum.

Schoenfeld, A. H. (1987b). What's all the fuss about metacognition? In A. H. Schoenfeld (Ed.), *Cognitive science and mathematics education* (pp. 189–215). Hillsdale, NJ: Erlbaum.

Schoenfeld, A. H. (1988). Mathematics, technology, and higher order thinking. In R. S. Nickerson & P. P. Zodhiates (Eds.), *Technology in education: Looking toward 2020* (pp. 67–96). Hillsdale, NJ: Erlbaum.

Schoenfeld, A. H. (1989). Explorations of students' mathematical beliefs and behavior. *Journal for Research in Mathematics Education, 20,* 338–355.

Schoenfeld, A. H. (1992). Learning to think mathematically: Problem solving, metacognition, and sense-making in mathematics. In D. Grouws (Ed.), *Handbook for research on mathematic teaching and learning* (pp. 334–370). New York: Macmillan.

Schoenfeld, A. H. (Ed.). (1994a). *Mathematical thinking and problem solving.* Hillsdale, NJ: Erlbaum.

Schoenfeld, A. H. (1994b). Reflections on doing and teaching mathematics. In A. H. Schoenfeld (Ed.), *Mathematical thinking and problem solving* (pp. 53–70). Hillsdale, NJ: Erlbaum.

Schultz, K. A., Colarusso, R. P., & Strawderman, V. W. (1989). *Mathematics for every young child.* Columbus, OH: Merrill.

Schwartz, J. L. (1994). The role of research in reforming mathematics education: A different approach. In S. H. Schoenfeld (Ed.), *Mathematical thinking and problem solving* (pp. 1–7). Hillsdale, NJ: Erlbaum.

Schwartz, J. L., Yerushalmy, M., & Wilson, B. (1993). *The geometric supposer: What is it a case of?* Hillsdale, NJ: Erlbaum.

Schwartz, J. T. (1978). Mathematics as a tool for economic understanding. In L. A. Steen (Ed.), *Mathematics today: Twelve informal essays* (pp. 269–295). New York: Springer-Verlag.

Scott, J. F. (1958). *A history of mathematics from antiquity to the beginning of the nineteenth century.* London: Taylor & Francis.

Seegers, G., & Boekaerts, M. (1996). Gender-related differences in self-referenced cognitions in relation to mathematics. *Journal for Research in Mathematics Education, 27,* 215–240.

Seife, C. (2000). *Zero: The biography of a dangerous idea*. New York: Penguin.

Sekuler, R., & Mierkiewicz, D. (1977). Children's judgment of numerical inequality. *Child Development, 48*, 630–633.

Selden, A., Selden, J., & Mason, A. (1994). Even good calculus students can't solve nonroutine problems. In J. Kaput & E. Dubinsky (Eds.), *Research issues in undergraduate mathematics learning* (MAA Notes Series, Vol. 33, pp. 19–26). Washington, DC: Mathematical Association of America.

Seligman, M. E. P. (1975). *Helplessness: On depression, development, and death*. San Francisco: Freeman.

Seligman, M. E. P., & Maier, S. F. (1967). Failure to escape traumatic shock. *Journal of Experimental Psychology, 74*, 1–9.

Sells, L. (1973). High school mathematics as the critical filter in the job market. In R. T. Thomas (Ed.), *Developing opportunities for minorities in graduate education* (pp. 37–39). Berkeley: University of California Press.

Sells, L. (1980). The mathematics filter and the education of women and minorities. In L. H. Fox, L. Brody, & D. Tobin (Eds.), *Women and the mathematical mystique* (pp. 66–75). Baltimore: Johns Hopkins Press.

Seo, K., & Ginsburg, H. P. (2003). "You've got to carefully read the math sentence ...": Classroom context and childrens' interpretations of the equals sign. In A. J. Baroody & A. Dowker (Eds.), *The development of arithmetic concepts and skills: Constructing adaptive expertise* (pp. 161–187). Mahwah, NJ: Erlbaum.

Shalev, R. S., & Gross-Tsur, V. (1993). Developmental dyscalculia and medical assessment. *Journal of Learning Disabilities, 26*, 134–137.

Shalev, R. S., & Gross-Tsur, V. (2001). Developmental dyscalculia. *Pediatric Neurology, 24*, 337–342.

Shalev, R. S., Auerbach, J., Manor, O., & Gross-Tsur, V. (2000). Developmental dyscalculia: Prevalence and prognosis. *European Child and Adolescent Psychiatry, 9* (Suppl. 2), 1158–1164.

Shalev, R. S., Manor, O., Kerem, B., Ayali, M., Badichi, N., & Friedlander, Y. (2001). Developmental dyscalculia is a family learning disability. *Journal of Learning Disabilities, 34*, 59–65.

Shalin, V., & Bee, N. V. (1985a). *Analysis of the semantic structure of a domain of word problems*. Pittsburgh, PA: University of Pittsburgh, Learning Research and Development Center.

Shalin, V., & Bee, N. V. (1985b). *Structural differences between two-step word problems*. Pittsburgh, PA: University of Pittsburgh, Learning Research and Development Center.

Shanks, D. R. (1999). Outstanding performers: Created, not born? *Science Spectra, 18*, 28–34.

Shaughnessy, J. M. (1985). Problem-solving derailers: The influence of misconceptions on problem-solving performance. In E. A. Silver (Ed.), *Teaching and learning mathematical problem solving: Multiple research perspectives* (pp. 399–415). Hillsdale, NJ: Erlbaum.

Sheffer, H. (1913). A set of five independent postulates for Boolean algebras, with application to logical constants. *Transactions of the American Mathematical Society, 14*, 481–488.

Shepard, R. N. (1995). Mental universals: Toward a twenty-first century science of mind. In R. L. Solso & D. W. Massaro (Eds.), *The science of the mind: 2001 and beyond* (pp. 50–62). New York: Oxford University Press.

Sherman, J. A. (1967). Problems of sex differences in space perception and aspects of intellectual functioning. *Psychological Review, 74,* 290–299.

Sherman, J. A., & Fennema, E. (1977). The study of mathematics by high school girls and boys, grades 8–11. *American Educational Research Journal, 14,* 159–168.

Shih, M., Pittinsky, T. L., & Ambady, N. (1999). Stereotype susceptibility: Identity salience and shifts in quantitative performance. *Psychological Science, 10,* 80–83.

Shmukler, A. (1980). *Introduction to set theory* [in Hebrew]. Jerusalem: Academon.

Shoemaker, J. (Ed.). (1977). *Women and mathematics: Research perspectives for change.* Washington, DC: The National Institute of Education.

Siegel, L. S. (1971a). The sequence of development of certain number concepts in preschool children. *Developmental Psychology, 5,* 357–361.

Siegel, L. S. (1971b). The development of the understanding of certain number concepts. *Developmental Psychology, 5,* 362–363.

Siegel, L. S. (1974). Development of number concepts: Ordering and correspondence operations and the role of length cues. *Developmental Psychology, 10,* 907–912.

Siegel, L. S., & Linder, B. A. (1984). Short-term memory processes in children with reading and arithmetic learning disabilities. *Developmental Psychology, 20,* 200–207.

Siegler, R. S. (1987). The perils of averaging data over strategies: An example from children's addition. *Journal of Experimental Psychology: General, 116,* 250–264.

Siegler, R. S. (1988). Strategy choice procedures and the development of multiplication skill. *Journal of Experimental Psychology: General, 117,* 258–275.

Siegler, R. S. (1989). Hazards of mental chronometry: An example from children's subtraction. *Journal of Educational Psychology, 81,* 497–506.

Siegler, R. S., & Booth, J. L. (2005). Development of numerical estimation: A review. In J. I. D. Campbell (Ed.), *Handbook of mathematical cognition* (pp. 197–212). New York: Psychology Press.

Siegler, R. S., & Jenkins, E. (1989). *How children discover new strategies.* Hillsdale, NJ: Erlbaum.

Siegler, R. S., & Mu, Y. (2008). Chinese children excel on novel mathematics problems even before elementary school. *Psychological Science, 19,* 759–763.

Siegler, R. S., & Robinson, M. (1982). The development of numerical understandings. In H. Reese & L. P. Lipsitt (Eds.), *Advances in child development and behavior* (Vol. 16, pp. 241–312). New York: Academic Press.

Siegler, R. S., & Shrager, J. (1984). Strategy choice in addition and subtraction: How do children know what to do? In C. Sophian (Ed.), *Origins of cognitive skills* (pp. 229–293). Hillsdale, NJ: Erlbaum.

Silver, E. A. (1985a). Research on teaching mathematical problem solving: Some underrepresented themes and needed directions. In E. A. Silver (Ed.), *Teaching and learning mathematical problem solving: Multiple research perspectives* (pp. 247–266). Hillsdale, NJ: Erlbaum.

Silver, E. A. (Ed.). (1985b). *Teaching and learning mathematical problem solving: Multiple research perspectives.* Hillsdale, NJ: Erlbaum.

Silver, E. A. (1986). Using conceptual and procedural knowledge: A focus on relationships. In J. Hiebert (Ed.), *Conceptual and procedural knowledge: The case of mathematics* (pp. 181–198). Hillsdale, NJ: Erlbaum.

Silver, E. A. (1987). Foundations of cognitive theory and research for mathematics problem-solving instruction. In A. H. Schoenfeld (Ed.), *Cognitive science and mathematics education* (pp. 33–60). Hillsdale, NJ: Erlbaum.

Silverman, I. W., & Rose, A. P. (1980). Subitizing and counting skills in 3-year-olds. *Developmental Psychology, 16,* 539–540.

Simon, H. A. (1995). Foreword. In J. Glasgow, N. H. Narayanan, & B. Chandrasekaran (Eds.), *Diagrammatic reasoning: Cognitive and computational perspectives* (pp. xi–xiii). Cambridge, MA: MIT Press.

Simon, M., & Blume, G. (1992). Understanding multiplicative structures: A study of prospective elementary teachers. In W. Geeslin & K. Graham (Eds.), *Proceedings for the Sixteenth Psychology of Mathematics Education Conference* (Vol. 3, pp. 11–18), Durham, NC.

Simon, T. J. (1997). Reconceptualizing the origins of number knowledge: A "nonnumerical account." *Cognitive Development, 12,* 349–372.

Simon, T. J. (1999). Numerical thinking in a brain without numbers? *Trends in Cognitive Science, 3,* 363–364.

Simon, T. J., Hespos, S. J., & Rochat, P. (1995). Do infants understand simple arithmetic? A replication of Wynn (1992). *Cognitive Development, 10,* 253–269.

Simonton, D. K. (1994). *Greatness: Who makes history and why?* New York: Guilford Press.

Sinclair, A. (1991). Children's production and comprehension of written numerical representations. In K. Durkin & B. Shire (Eds.), *Language in mathematical education: Research and practice.* (pp. 59–68). Buckingham, UK: Open University.

Sinclair, A., Siegrist, F., & Sinclair, H. (1982). Young children's ideas about the written number system. In D. Rogers & J. A. Sloboda (Eds.), *The acquisition of symbolic skills* (pp. 535–542). New York: Plenum Press.

Sinclair, A., & Sinclair, H. (1984). Preschool children's interpretation of written numbers. *Human Learning, 3,* 173–184.

Singer, H. D., & Low, A. A. (1933). Acalculia (Henschen): A clinical study. *Archives of Neurology and Psychiatry, 29,* 476–498.

Singh, S. (1997). *Fermat's enigma: The epic quest to solve the world's greatest mathematical problem.* New York: Anchor Books.

Singh, S., & Ribet, K. A. (1997). Fermat's last stand. *Scientific American, 277,* 68–73.

Skypek, D. H. (1984). Special characteristics of rational numbers. *Arithmetic Teacher, 25,* 22–25.

Slavin, R. E. (1980). *Using student team learning* (rev. ed.). Baltimore: Johns Hopkins University.

Slavin, R. E. (1987). Ability grouping and student achievement in elementary schools: A best evidence synthesis. *Review of Educational Research, 57,* 293–336.

Slavin, R. E. (1993). Ability grouping in the middle grades: Achievement effects and alternatives. *Elementary School Journal, 93,* 535–552.

Sleeman, D., & Brown, J. S. (Eds.). (1982). *Intelligent tutoring systems*. New York: Academic Press.

Slote, M. (1990). Rational dilemmas and rational supererogation. In P. K. Moser (Ed.), *Rationality in action: Contemporary approaches* (pp. 465–480). New York: Cambridge University Press. (Original work published 1986)

Smith, B. R., Piel, A. K., & Candland, D. K. (2003). Numerity of a socially housed hamadryas baboon (*Papio mamadryas*) and a socially housed squirrel monkey (*Saimiri sciureus*). *Journal of Comparative Psychology, 117*, 217–225.

Smith, C. L., Solomon, G. E. A., & Carey, S. (2005). Never getting to zero: Elementary school students' understanding of the infinite divisibility of number and matter. *Cognitive Psychology, 51*, 101–140.

Smith, D. E., & Ginsburg, J. (1937). *Numbers and numerals*. Washington, DC: National Education Association.

Smith, E., & Confrey, J. (1994). Multiplicative structures and the development of logarithms: What was lost by the invention of functions. In G. Harel & J. Confrey (Eds.), *The development of multiplicative reasoning in the learning of mathematics* (pp. 331–360). Albany: State University of New York Press.

Smith, E. E., & Medin, D. L. (1981). *Categories and concepts*. Cambridge, MA: Harvard University Press.

Smith, I. (1964). *Spatial ability: Its educational and social significance*. San Diego: Robert Knapp.

Smith, J. M. (2003). The mathematics of evolution. In G. Farmelo (Ed.), *It must be beautiful: Great equations of modern science* (pp. 161–179). New York: Granta Books.

Smith, M. U., & Good, R. (1984). Problem solving and classical genetics: Successful versus unsuccessful performance. *Journal of Research in Science Teaching, 21*, 895–912.

Smith, S. B. (1983). *The great mental calculators: The psychology, methods, and lives of calculating prodigies, past and present*. New York: Columbia University Press.

Smolin, L. (2001). *Three roads to quantum gravity*. New York: Basic Books.

Smolin, L. (2004). Atoms of space and time. *Scientific American, 290*, 66–75.

Snyder, T. (1982). *Search series*. New York: McGraw-Hill.

Sokol, S. M., Macaruso, P., & Gollan, T. H. (1991). *Patterns of impairment in developmental dyscalculia*. Presentation at the Society for Neuroscience Meeting, New Orleans.

Soloway, E., Lochhead, J., & Clement, J. (1982). Does computer programming enhance problem solving ability? Some positive evidence on algebra word problems. In R. J. Sediel, R. E. Anderson, & B. Hunter (Eds.), *Computer literacy* (pp. 171–215). New York: Academic Press.

Song, M. J., & Ginsburg, H. P. (1987). The development of informal and formal mathematical thinking in Korean and U.S. children. *Child Development, 58*, 1286–1296.

Sophian, C. (1988). Limitations on children's knowledge about counting: Using counting to compare two sets. *Developmental Psychology, 24*, 643–640.

Sophian, C. (1992). Learning about numbers: Lessons for mathematics education from preschool number development. In J. Bideaud, C. Meljac, & J.-P. Fischer (Eds.), *Pathways to number: Children's developing numerical abilities* (pp. 19–40). Hillsdale, NJ: Erlbaum.

Sophian, C. (1995). Representation and reasoning in early numerical development: Counting, conservation, and comparisons between sets. *Child Development, 66,* 559–577.

Sophian, C. (1998). A developing perspective on children's counting. In C. Donlan (Ed.), *The development of mathematical skills* (pp. 27–46). Hove, UK: Psychology Press.

Sophian, C., & Adams, N. (1987). Infants' understanding of numerical transformations. *British Journal of Developmental Psychology, 5,* 257–264.

Sorabji, R. (1982). Atoms and time atoms. In N. Kretzmann (Ed.), *Infinity and continuity in ancient and medieval thought* (pp. 37–86). Ithaca, NY: Cornell University Press.

Sowder, J. T. (1989). Research into practice: Developing understanding of computational estimation. *Arithmetic Teacher, 36,* 25–27.

Sowder, L. (1988). Children's solutions of story problems. *Journal of Mathematical Behavior, 7,* 227–238.

Spade, P. V. (1982). Quasi-Aristotelianism. In N. Kretzmann (Ed.), *Infinity and continuity in ancient and medieval thought* (pp. 297–307). Ithaca, NY: Cornell University Press.

Spearman, C. (1904). General intelligence objectively determined and measured. *American Journal of Psychology, 15,* 201–293.

Spearman, C. (1923). *The nature of intelligence and the principles of cognition.* London: Macmillan.

Spelke, E. S. (2005). Sex differences in intrinsic aptitude for mathematics and science? *American Psychologist, 60,* 950–958.

Spencer, S. J., Steele, C. M., & Quinn, D. M. (1999). Stereotype threat and women's math performance. *Journal of Experimental Social Psychology, 35,* 4–28.

Spiker, J., & Kurtz, R. (1987). Teaching primary-grade mathematics skills with calculators. *Arithmetic Teacher, 34,* 24–27.

Stallings, J. A. (1979). *Comparison of men's and women's behaviors in high school math classes.* Menlo Park, CA: SRI International.

Stanley, J. C. (1974). Intellectual precocity. In J. C. Stanley, D. P. Keating, & L. H. Fox (Eds.), *Mathematical talent: Discovery, description, and development* (pp. 1–22). Baltimore: Johns Hopkins University Press.

Starkey, P. (1992). The early development of numerical reasoning. *Cognition, 43,* 93–126.

Starkey, P., & Cooper, R. G., Jr. (1980). Perception of numbers by human infants. *Science, 210,* 1033–1035.

Starkey, P., & Gelman, R. (1982). The development of addition and subtraction abilities prior to formal schooling in arithmetic. In T. P. Carpenter, J. M. Moser, & T. A. Romberg (Eds.), *Addition and subtraction: A cognitive perspective* (pp. 99–116). Hillsdale, NJ: Erlbaum.

Starkey, P., & Klein, A. (2000). Fostering parental support for children's mathematical development: An intervention with Head Start families. *Early Education and Development, 11,* 659–680.

Starkey, P., Spelke, E. S., & Gelman, R. (1983). Detection of intermodal numerical correspondences by human infants. *Science, 222,* 179–181.

Starkey, P., Spelke, E. S., & Gelman, R. (1990). Numerical abstraction by human infants. *Cognition, 36,* 97–127.

Stavy, R., & Tirosh, D. (1993a). Subdivision processes in mathematics and science. *Journal of Research in Science Teaching, 30,* 579–586.

Stavy, R., & Tirosh, D. (1993b). When analogy is perceived as such. *Journal of Research in Science Teaching, 30,* 1229–1239.

Stazyk, E. H., Ashcraft, M. H., & Hamann, M. S. (1982). A network approach to mental multiplication. *Journal of Experimental Psychology: Learning, Memory and Cognition, 8,* 320–335.

Steele, C. M. (1992). Race and the schooling of black Americans. *The Atlantic Monthly, 69,* 68–78.

Steele, C. M., & Aronson, J. (1995). Stereotype threat and the intellectual test performance of African Americans. *Journal of Personality and Social Psychology, 69,* 797–811.

Steen, L. A. (1978). Mathematics today. In L. A. Steen (Ed.), *Mathematics today: Twelve informal essays* (pp. 1–12). New York: Springer-Verlag.

Steen, L. A. (1988). The science of patterns. *Science, 240,* 611–616.

Steen, L. A. (1990). Pattern. In L. A. Steen (Ed.), *On the shoulders of giants: New approaches to numeracy* (pp. 1–10). Washington, DC: National Academy Press.

Steffe, L. P. (1970). Differential performance of first-grade children when solving arithmetic addition problems. *Journal for Research in Mathematics Education, 1,* 144–161.

Steffe, L. P. (1991). The constructivist teaching experiment. In E. von Glasersfeld (Ed.), *Radical constructivism in mathematics education* (pp. 177–194). Boston: Kluwer Academic.

Steffe, L. P. (1994). Children's multiplying schemes. In G. Harel & J. Confrey (Eds.), *The development of multiplicative reasoning in the learning of mathematics* (pp. 3–39). Albany: State University of New York Press.

Steffe, L. P., & Cobb, P. (1988). *Construction of arithmetic meanings and strategies.* New York: Springer-Verlag.

Steffe, L. P., & Johnson, D. C. (1971). Problem-solving performances of first-grade children. *Journal for Research in Mathematics Education, 2,* 50–64.

Steffe, L. P., Spikes, W. C., & Hirstein, J. J. (1976). *Quantitative comparisons and class inclusion as readiness variables for learning first-grade arithmetical content.* Athens: The Georgia Center for the Study of Learning and Teaching Mathematics, University of Georgia.

Steffe, L. P., Thompson, D. W., & Richards, J. (1982). Children's counting in arithmetical problem solving. In T. P. Carpenter, J. M. Moser, & T. A. Romberg (Eds.), *Addition and subtraction: A cognitive perspective* (pp. 83–97). Hillsdale, NJ: Erlbaum.

Steffe, L. P., von Glasersfeld, E., Richards, J., & Cobb, P. (Eds.). (1984). *Children's counting types: Philosophy, theory, and applications.* New York: Praeger.

Steffe, L. P., & Wood, T. (Eds.). (1990). *Transforming children's mathematics education: International perspectives.* Hillsdale, NJ: Erlbaum.

Steinberg, R. M. (1985). Instruction on derived fact strategies in addition and subtraction. *Journal for Research in Mathematics Education, 16,* 337–355.

Stengel, A., LeBlanc, J. F., Jacobson, M., & Lester, F. K. (1977). *Learning to solve problems by solving problems: A report of a preliminary investigation* (Technical Report II. D. of the Mathematical Problem Solving Project). Bloomington, IN: Mathematics Education Development Center.

Stern, N. (1978). Age and achievement in mathematics. *Social Studies in Science, 8,* 127–140.

Sternberg, R. J. (1983). Criteria for intellectual skills training. *Educational Researcher, 12,* 6–12, 26.

Sternberg, R. J. (1996). What is mathematical thinking? In R. J. Sternberg & T. Ben-Zeev (Eds.), *The nature of mathematical thinking* (pp. 303–318). Mahwah, NJ: Erlbaum.

Sternberg, R. J., & Zhang, L-F. (Eds.) (2001). *Perspectives on thinking, learning, and cognitive styles.* Mahwah, NJ: Erlbaum.

Stevens, P. K. (1974). *Patterns in nature.* Boston: Little, Brown and Company.

Stevens, S. S. (1951). Mathematics, measurement, and psychophysics. In S. S. Stevens (Ed.), *Handbook of experimental psychology* (pp. 1–49). New York: John Wiley and Sons.

Stevenson, H. W., Chen, C., & Lee, S.-Y. (1993). Mathematics achievement of Chinese, Japanese, and American children: Ten years later. *Science, 259,* 53–58.

Stevenson, H. W., & Lee, S. (1990). Contexts of achievement. *Monographs of the Society for Research in Child Development, 55*(1–2, Serial 221).

Stevenson, H. W., & Lee, S. (1998). An examination of American student achievement from an international perspective. In D. Ravitch (Ed.), *The state of student performance in American schools* (pp. 7–52). Washington, DC: Brookings.

Stevenson, H. W., Lee, S. Y., Chen, C., Stigler, J. W., Hsu, C. C., & Kitamura, S. (1990). Contexts of achievement: A study of American, Chinese, and Japanese children. *Monographs of the Society for Research in Child Development, 55,* Serial No. 221.

Stevenson, H. W., Lee, S. Y., & Stigler, J. W. (1986). Mathematics achievement of Chinese, Japanese, and American children. *Science, 231,* 693–699.

Stevenson, H. W., & Stigler, J. W. (1992). *The learning gap.* New York: Simon & Schuster.

Stewart, I. (1987). *The problems of mathematics.* New York: Oxford University Press.

Stewart, I. (1990). *Does God play dice: The new mathematics of chaos.* New York: Penguin.

Stewart, I. (1995a). Fibonacci forgeries. *Scientific American, 272,* 102–105.

Stewart, I. (1995b). *Nature's numbers: The unreal reality of mathematics.* New York: Basic Books.

Stewart, I. (2003). The second law of gravitics and the fourth law of thermodynamics. In N. H. Gregsen (Ed.), *From complexity to life: Explaining the emergence of life and meaning: Proceedings of Templeton Symposium on Complexity, Information, and Design,* Santa Fe 1999 (pp. 114–150). Oxford: Oxford University Press.

Stewart, I., & Golubitsky, M. (1992). *Fearful symmetry.* Oxford: Blackwell.

Stigler, J. W. (1984). "Mental abacus": The effect of abacus training on Chinese children's mental calculation. *Cognitive Psychology, 16,* 145–176.

Stigler, J. W. (1988). Research into practice: The use of verbal explanation in Japanese and American classrooms. *Arithmetic Teacher, 36,* 27–29.

Stigler, J. W., Chalip, L., & Miller, K. F. (1986). Consequences of skill: The case of abacus training in Taiwan. *American Journal of Education, 94,* 447–479.

Stigler, J. W., Fuson, K. C., Ham, M., & Kim, S. M. (1986). An analysis of addition and subtraction word problems in American and Soviet elementary mathematics textbooks. *Cognition and Instruction, 3,* 153–171.

Stigler, J. W., & Hiebert, J. (1999). *The teaching gap: Best ideas from the world's teachers for improving education in the classroom.* New York: Free Press.

Stodolsky, S. S. (1985). Telling math: Origins of math aversion and anxiety. *Educational Psychologist, 20,* 33–39.

Støllum, H.-H. (1996). River meandering as a self-organization process. *Science, 271,* 1710–1713.

Stone, J., Lynch, C. I., Sjomeling, M., & Darley, J. M. (1999). Stereotype threat effects on Black and White athletic performance. *Journal of Personality and Social Psychology, 77,* 1213–1227.

Strauss, M. S., & Curtis, L. E. (1981). Infant perception of numerosity. *Child Development, 52,* 1146–1152.

Strauss, M. S., & Curtis, L. E. (1984). Development of numerical concepts in infancy. In C. Sophian (Ed.). *The origins of cognitive skill in infancy* (pp. 131–155). Hillsdale, NJ: Erlbaum.

Streefland, L. (1993). Fractions: A realistic approach. In T. P. Carpenter, E. Fennema, & T. A. Romberg (Eds.), *Rational numbers: An integration of research* (pp. 289–326). Hillsdale, NJ: Erlbaum.

Struik, D. J. (Ed.). (1969), *A source book in mathematics, 1200–1800.* Cambridge, MA: Harvard University Press.

Stump, E. (1982). Theology and physics in *De sacramento altaris*: Ockham's theory of indivisibles. In N. Kretzmann (Ed.), *Infinity and continuity in ancient and medieval thought* (pp. 207–230). Ithaca, NY: Cornell University Press.

Sullivan, J. W. N. (1956). Mathematics as an art. In J. R. Newman (Ed.), *The world of mathematics* (Vol. 3, pp. 2015–2021). New York: Simon & Schuster. (Original work published 1925)

Suppes, P., & Groen, G. J. (1967). Some counting models for first-grade performance data on simple addition facts. In J. M. Scandura (Ed.), *Research in mathematics education* (pp. 35–43). Washington, DC: National Council of Teachers of Mathematics.

Suppes, P., & Morningstar, M. (1972). *Computer-assisted instruction at Stanford, 1966–1968: Data, models, and evaluation of the arithmetic programs.* New York: Academic Press.

Sutherland, R. (1993). Connecting theory and practice: Results from the teaching of Logo. *Educational Studies in Mathematics, 24,* 1–19.

Sutherland, R., & Rojano, T. (1993). A spreadsheet approach to solving algebra problems. *Journal of Mathematical Behavior, 12,* 353–382.

Svenson, O., & Broquist, S. (1975). Strategies for solving simple addition problems: A comparison of normal and subnormal children. *Scandinavian Journal of Psychology, 16,* 143–151.

Svenson, O., & Hedenborg, M. L. (1979). Strategies used by children when solving simple subtractions. *Acta Psychologica, 43,* 477–489.

Sweller, J., Mawer, R. F., & Ward, M. R. (1983). Development of expertise in mathematical problem solving. *Journal of Experimental Psychology: General, 112,* 639–661.

Swets, J. A. (1964). *Signal detection and recognition by human observers.* New York: Wiley.

Sylla, E. (1975). Autonomous and handmaiden science: St. Thomas Aquinas and William of Ockham on the physics of the Eucharist. In J. E. Murdoch & E. D. Sylla (Eds.), *The cultural context of medieval learning* (pp. 349–396). Dordrecht: Reidel.

Sylvester, J. J. (1956). The study that knows nothing of observation. In J. R. Newman (Ed.), *The world of mathematics* (Vol. 3, pp. 1758–1766). New York: Simon & Schuster. (Excerpt of address to British Association, 1869)

Taal, M. (1994). How do mathematical experiences contribute to the choice of mathematics. *Sex Roles, 31,* 752–769.

Taleb, N. N. (2004). *Fooled by randomness: The hidden role of chance in life and in the markets* (2nd ed.). New York: Thomson-Texere.

Tamburino, J. L. (1982). *The effects of knowledge-based instruction on the abilities of primary grade children in arithmetic word problem solving.* Unpublished doctoral dissertation, University of Pittsburgh.

Tarski, A. (1956). Symbolic logic. In J. R. Newman (Ed.), *The world of mathematics* (Vol. 3, pp. 1901–1931). New York: Simon & Schuster. (Original work published 1941)

Tartre, L. A. (1990). Spatial skills, gender, and mathematics. In E. Fennema & G. H. Leder (Eds.), *Mathematics and gender* (pp. 27–59). New York: Teachers College Press.

Taves, E. H. (1941). Two mechanisms for the perception of visual numerousness. *Archives of Psychology, 37,* No. 265.

Taylor, R., & Wiles, A. (1995). Ring-theoretic properties of certain Hecke algebras. *Annals of Mathematics, 141,* 553–572.

Taylor, R. P. (2002). Order in Pollock's chaos. *Scientific American, 287,* 116–121.

Temple, C. M. (1989). Digit dyslexia: A category-specific disorder in developmental dyscalculia. *Cognitive Neuropsychology, 6,* 93–116.

Temple, C. M. (1991). Procedural dyscalculia and number fact dyscalculia: Double dissociation in developmental dyscalculia. *Cognitive Neuropsychology, 8,* 155–176.

Temple, C. M., & Marriott, A. J. (1998). Arithmetic ability and disability in Turner's syndrome: A cognitive neuropsychological analysis. *Developmental Neuropsychology, 14,* 47–67.

Temple, C. M., & Sherwood, S. (2002). Representation and retrieval of arithmetic facts: Developmental difficulties. *Quarterly Review of Experimental Psychology, 55A,* 733–752.

Temple, E., & Posner, M. (1998). Brain mechanisms of quantity are similar in 5-year olds and adults. *Proceedings of the National Academy of Sciences, 168,* 271–273.

Terrell, D. F., & Thomas, R. K. (1990). Number-related discrimination and summation by squirrel monkeys (*Saimiri sciureus sciureus* and *Saimiri boliviensus boliviensus*) on the basis of the number of sides of polygons. *Journal of Comparative Psychology, 104,* 238–247.

Tharp, R. G., & Gallimore, R. (1988). *Rousing minds to life: Teaching, learning, and Schooling in social context.* Cambridge, UK: Cambridge University Press.

Thomas, R. K., & Chase, L. (1980). Relative numerousness judgments by squirrel monkeys. *Bulletin of the Psychonomic Society, 16,* 79–82.

Thomas, R. K., Fowlkes, D., & Vikery, J. D. (1980). Conceptual numerousness judgments by squirrel monkeys. *American Journal of Psychology*, *93*, 247–257.

Thomas, R. K., & Lorden, R. B. (1993). Numerical competence in animals: A conservative view. In S. T. Boysen & E. J. Capaldi (Eds.), *The development of numerical competence: Animal and human models* (pp. 127–147). Hillsdale, NJ: Erlbaum.

Thompson, A. G. (1985). Teachers' conceptions of mathematics and the teaching of problem solving. In E. A. Silver (Ed.), *Teaching and learning mathematical problem solving: Multiple research perspectives* (pp. 281–294). Hillsdale, NJ: Erlbaum.

Thompson, C., & Van de Walle, J. (1980). Transition boards: Moving from materials to symbols in addition. *Arithmetic Teacher*, *28*, 4–8.

Thompson, P. W. (1982). Were lions to speak, we wouldn't understand. *Journal of Mathematical Behavior*, *3*, 147–165.

Thompson, P. W. (1985). Experience, problem solving, and learning mathematics: Considerations in developing mathematics curricula. In E. A. Silver (Ed.), *Teaching and learning mathematical problem solving: Multiple research perspectives* (pp. 189–236). Hillsdale, NJ: Erlbaum.

Thompson, P. W. (1994a, April 8–9). *Bridges between mathematics and science education*. Paper presented at the Project 2061 Conference on Developing a Research Blueprint, Baton Rouge, LA.

Thompson, P. W. (1994b). The development of the concept of speed and its relationship to the concept of rate. In G. Harel & J. Confrey (Eds.), *The development of multiplicative reasoning in the learning of mathematics* (pp. 179–234). Albany, NY: State University of New York Press.

Thompson, P. W. (1994c). Images of rate and operational understanding of the fundamental theorem of calculus. *Educational Studies in Mathematics*, *26*, 229–274.

Thomson, J. (2001a). Comments on Professor Benacerraf's paper. In W. C. Salmon (Ed.), *Zeno's paradoxes* (pp. 130–138). Indianapolis: Hackett Publishing.

Thomson, J. (2001b). Tasks and super-tasks. In W. C. Salmon (Ed.), *Zeno's paradoxes* (pp. 89–102). Indianapolis: Hackett Publishing. (Original work published 1954)

Thorndike, E. L. (1913). *Educational psychology: The psychology of learning* (Vol. 2). New York: Teachers College, Columbia University.

Thorndike, E. L. (1922). *The psychology of arithmetic*. New York: Macmillan.

Thornton, C. A. (1990). Strategies for learning the basic facts. In J. Payne (Ed.), *Teaching and learning mathematics for the young child*. Reston, VA: National Council of Teachers of Mathematics.

Thornton, C. A., & Wilson, S. J. (1993). Classroom organization and models of instruction. In R. J. Jensen (Ed.), *Research ideas for the classroom: Early childhood mathematics* (pp. 269–293). New York: Macmillan.

Threadgill-Sowder, J. (1985). Individual differences and mathematical problem solving. In E. A. Silver (Ed.), *Teaching and learning mathematical problem solving: Multiple research perspectives* (pp. 331–343). Hillsdale, NJ: Erlbaum.

Thurstone, L. L. (1938). Primary mental abilitities. *Psychometric Monographs*, No. 1.

Thurstone, L. L., & Thurstone, T. G. (1941). Factorial studies of intelligence. *Psychometric Monographs*, No. 2.

Tirosh, D., & Almog, N. (1989). Conceptual adjustments in progressing from real to complex numbers. In G. Vergnaud, J. Rogalski, & M. Artigue (Eds.), *Proceedings of the Thirteenth Annual Conference of the International Group for the Psychology of Mathematics Education* (Vol. 3, pp. 221–227). Paris: G. R. Didactique.

Tirosh, D., & Stavy, R. (1996). Intuitive rules in science and mathematics: The case of 'Everything can be divided by two.' *International Journal of Science Education, 18,* 669–683.

Tobias, S. (1978). *Overcoming math anxiety.* New York: Norton.

Tobias, S., & Donady, B. (1977). Counseling the math anxious. *Journal of National Association of Women Deans and Counselors,* Fall, 13–16.

Tobias, S., & Weissbrod, C. (1980). Anxiety and mathematics: An update. *Harvard Educational Review, 50,* 63–70.

Tocquet, R. (1961). *The magic of numbers* (D. Weaver, Trans.). New York: A. S. Barnes.

Todhunter, I. (2001). *A history of the mathematical theory of probability from the time of Pascal to that of Laplace.* Briston, UK: Thoemmes Press. (Original work published 1865)

Tomonaga, M., & Matsuzawa, T. (2002). Enumeration of briefly presented items by the chimpanzee (*Pan troglodytes*) and humans (*Homo sapiens*). *Animal Learning and Behavior, 30,* 143–157.

Tøndering, C. (2005). *Surreal numbers—An introduction.* Retrieved August 3, 2009 from www.tondering.dk/claus/sur15.pdf

Tourniaire, F. (1986). Proportions in elementary school. *Educational Studies in Mathematics, 17,* 401–412.

Towse, J., & Saxton, M. (1998). Mathematics across national boundaries: Cultural and linguistic perspectives on numerical competence. In C. Donlan (Ed.), *The development of mathematical skills* (pp. 129–150). Hove, UK: Psychology Press.

Trabasso, T., Isen, A. M., Dolecki, P., McLanahan, A. G., Riley, C. A., & Tucker, T. (1978). How do children solve class-inclusion problems? In R. S. Siegler (Ed.), *Children's thinking: What develops?* (pp. 151–180). Hillsdale, NJ: Erlbaum.

Travers, K. J., & Westbury, I. (1989). *The IEA study of mathematics. I. Analysis of mathematics curricula.* Oxford: Pergamon Press.

Treffert, D. A. (1988). The idiot savant: A review of the syndrome. *American Journal of Psychiatry, 145,* 563–572.

Treffert, D. A. (1989). *Extraordinary people: Redefining the idiot savant.* New York: Harper & Row.

Tsang, S. L. (1984). The mathematics education of Asian Americans. *Journal for Research in Mathematics Education, 15,* 114–122.

Tsang, S. L. (1988). The mathematics achievement characteristics of Asian-American students. In R. R. Cocking & J. P. Mestre (Eds.), *Linguistic and cultural influences on learning mathematics* (pp. 123–136). Hillsdale, NJ: Erlbaum.

Turnbull, H. W. (1929). *The great mathematicians.* London: Methuen.

Tyson, H., & Woodward, A. (1989). Why students aren't learning very much from textbooks. *Educational Leadership, 47,* 14–17.

Tzeng, O. J. L., & Wang, W. (1983). The first two R's. *American Scientist, 71,* 238–243.

Ulam, S. M. (1976). *Adventures of a mathematician.* New York: Charles Scribner Sons.

U.S. Department of Education (1991). *America 2000: An education strategy.* Washington, DC: Author.

Usiskin, Z., & Bell, M. (1983). *Applying arithmetic: A handbook of applications of arithmetic.* Chicago: University of Chicago Press.

Uspensky, J. V. (1937). *Introduction to mathematical probability.* New York: McGraw-Hill.

Vandenberg, S. G. (1966). Contributions of twin research to psychology. *Psychological Bulletin, 66,* 327–352.

Vandenbos, G. R. (2007). *APA dictionary of psychology.* Washington, DC: American Psychological Association.

Van den Brink, J., & Streefland, L. (1979). Young children (6–9): Ratio and proportion. *Educational Studies in Mathematics, 10,* 403–420.

Van de Walle, J. A. (1990). Concepts of number. In J. N. Payne (Ed.), *Mathematics for the young child* (pp. 63–87). Reston, VA: National Council of Teachers of Mathematics.

Van de Walle, J. A., & Watkins, K. B. (1993). Early development of number sense. In R. J. Jensen (Ed.), *Research ideas for the classroom: Early childhood mathematics* (pp. 127–150). New York: Macmillan.

VanLehn, K. (1983). On the representation of procedures in repair theory. In H. P. Ginsburg (Ed.), *The development of mathematical thinking* (pp. 197–252). New York: Academic Press.

VanLehn, K. (1986). Arithmetic procedures are induced from examples. In J. Hiebert (Ed.), *Conceptual and procedural knowledge: The case of mathematics* (pp. 133–179). Hillsdale, NJ: Erlbaum.

VanLehn, K. (1989). Problem solving and cognitive skill. In M. I. Posner (Ed.), *Foundations of cognitive science.* Cambridge, MA: MIT Press.

VanLehn, K. (1990). *Mind bugs: The origins of procedural misconceptions.* Cambridge, MA: MIT Press.

Van Loosbroek, E., & Smitsman, A. W. (1990). Visual perception of numerosity in infancy. *Developmental Psychology, 26,* 916–922.

Veblen, O., & Young, J. W. (1956). A mathematical science. In J. R. Newman (Ed.), *The world of mathematics* (Vol. 3, pp. 1696–1707). (Original work published 1910)

Vergnaud, G. (1983). Multiplicative structures. In R. Lesh & M. Landau (Eds.), *Acquisition of mathematics concepts and processes* (pp. 127–174). New York: Academic Press.

Very, P. S. (1967). Differential factor structures in mathematical ability. *Genetic Psychology Monographs, 75,* 169–207.

Von Glasersfeld, E. (1982). Subitizing: The role of figural patterns in the development of numerical concepts. *Archives de Psychologie, 50,* 191–218.

Von Glasersfeld, E. (1993). Reflections on number and counting. In S. T. Boysen & E. J. Capaldi (Eds.), *The development of numerical competence: Animal and human models* (pp. 225–243). Hillsdale, NJ: Erlbaum.

Von Helmholtz, H. (1956). On the origin and significance of geometrical axioms. In J. R. Newman (Ed.), *The world of mathematics* (Vol. 1, pp. 647–668). New York: Simon & Schuster. (Original lecture delivered 1870)

Von Mises, R. (1956). Mathematical postulates and human understanding. In J. R. Newman (Ed.), *The world of mathematics* (Vol. 3, pp. 1773–1783). New York: Simon & Schuster. (Original work published 1951)

Von Neumann, J. (1956). The mathematician. In J. R. Newman (Ed.), *The world of mathematics* (Vol. 4, pp. 2053–2063). New York: Simon & Schuster. (Original work published 1947)

Von Winterfeldt, D., & Edwards, W. (1986). *Decision analysis and behavioral research.* Cambridge, UK: Cambridge University Press.

Voss, J. F., Greene, T. R., Post, T. A., & Penner, B. C. (1983). Problem solving skill in the social sciences. In G. H. Bower (Ed.), *The psychology of learning and motivation: Advances in research and theory* (pp. 165–213). New York: Academic Press.

Vos Savant, M. (1993). *Is it solved? The world's most famous math problem: The proof of Fermat's last theorem and other mathematical mysteries.* New York: St. Martin's Press.

Vygotsky, L. S. (1962). *Thought and language.* Cambridge, MA: MIT Press.

Vygotsky, L. S. (1978). *Mind in society: The development of higher psychological processes* (M. Cole, V. John-Steiner, S. Scribner, & E. Souberman, Trans. & Eds.). Cambridge, MA: Harvard University Press.

Vygotsky, L. S. (1986). *Thought and language* (A. Kozulin, Trans.). Cambridge, MA: MIT Press.

Wainer, H. (1992). Understanding graphs and tables. *Educational Researcher, 21,* 14–23.

Wallace, D. F. (2003). *Everything and more: A compact history of* ∞. New York: W.W. Norton.

Wallas, G. (1945). *The art of thought.* London: Watts. (Original work published in 1926)

Walters, A. M., & Brown, L. M. (2005). The role of ethnicity on the gender gap in mathematics. In A. M. Gallagher & J. C. Kaufman (Eds.), *Gender differences in mathematics* (pp. 207–219). New York: Cambridge University Press.

Walton, G. M., & Cohen, G. L. (2003). Stereotype lift. *Journal of Experimental Social Psychology, 39,* 456–467.

Wang, M. C. (1973). Psychometric studies in the validation of an early learning curriculum. *Child Development, 44,* 54–60.

Wang, M. C., Resnick, L. G., & Boozer, R. F. (1971). The sequence of development of some early mathematics behaviors. *Child Development, 42,* 1767–1778.

Ward, K. (1996). *God, chance and necessity.* Oxford: Oneworld.

Warrington, E. K. (1982). The fractionation of arithmetical skills: A single case study. *Quarterly Journal of Experimental Psychology, 34A,* 31–51.

Washburn, D. A., & Rumbaugh, D. M. (1991). Ordinal judgments of numerical symbols by macaques (*Macaca mulatta*). *Psychological Science, 2,* 190–193.

Wassmann, J., & Dasen, P. R. (1994). Yupno number system and counting. *Journal of Cross-Cultural Psychology, 25,* 78–94.

Watanabe, S. (1998). Discrimination of "four" and "two" by pigeons. *Psychology Record, 48,* 383–391.

Watson, R. (2002). *Cogito, ergo sum: The life of René Descartes.* Jaffrey, NH: David R. Godine.

Watts, A. (1964). *Beyond theology*. New York: Pantheon.

Wearne, D., & Kouba, V. L. (2000). Rational numbers. In E. A. Silver & P. A. Kenney (Eds.), *Results from the Seventh Mathematics Assessment of the National Assessment of Educational Progress* (pp. 163–191). Reston, VA: National Council of Teachers of Mathematics.

Weaver, J. F. (1971). Some factors associated with pupil's performance levels on simple open addition and subtraction sentences. *Arithmetic Teacher, 18,* 513–519.

Weiner, B. (1979). A theory of motivation for some classroom experiences. *Journal of Educational Psychology, 71,* 3–25.

Wellman, H. M., & Miller, K. F. (1986). Thinking about nothing: Development of the concept of zero. *British Journal of Developmental Psychology, 4,* 31–42.

Wenger, R. G. (1994). Comments on Ed Dubinsky's chapter. In A. H. Schoenfeld (Ed.), *Mathematical thinking and problem solving* (pp. 244–247). Hillsdale, NJ: Erlbaum.

Wenger, R. H. (1987). Cognitive science and algebra learning. In A. H. Schoenfeld (Ed.), *Cognitive science and mathematics education* (pp. 217–251). Hillsdale, NJ: Erlbaum.

Werdelin, I. (1961). *The geometrical ability and space factor analysis in boys and girls.* Lund, Sweden: University of Lund.

Wertheimer, M. (1959). *Productive thinking*. New York: Harper and Brothers. (Original work published 1945)

Wesley, F. (1961). The number concept: A phylogenetic review. *Psychological Bulletin, 58,* 420–428.

West, D. J., & Goldberger, A. L. (1987). Physiology in fractal dimensions. *American Scientist, 75,* 354–365.

Weyl, H. (1949). *Philosophy of mathematics and natural science*. Princeton, NJ: Princeton University Press.

Weyl, H. (1956). The mathematical way of thinking. In J. R. Newman (Ed.), *The world of mathematics* (Vol. 3, pp. 1832–1849). New York: Simon & Schuster. (Original address given 1940)

Wheeler, J. A. (1968). Superspace and the nature of quantum geometrodynamics. In C. M. De Witt & J. A. Wheeler (Eds.), *Battelle Rencontres: 1967 lectures in mathematics and physics* (pp. 242–307). New York: W. A. Benjamin.

Whimbey, A., & Lochhead, J. (1982). *Problem solving and comprehension* (3rd ed.). Philadelphia: Franklin Institute Press.

White, L. A. (1956). The locus of mathematical reality: An anthropological footnote. In J. R. Newman (Ed.), *The world of mathematics* (Vol. 4, pp. 2348–2364). New York: Simon & Schuster. (Original work published 1947)

White, S., & Siegel, A. (1999). Cognitive development in time and space. In B. Rogoff & J. Lave (Eds.), *Everyday cognition: Its development in social context* (pp. 238–277). Cambridge, MA: Harvard University Press.

Whitehead, A. N. (1911). *An introduction to mathematics*. London: Thornton Butterworth.

Whitehead, A. N. (1929). *The aims of education and other essays*. New York: The Free Press.

Whitehead, A. N. (1956). Mathematics as an element in the history of thought. In J. R. Newman (Ed.), *The world of mathematics* (Vol. 1, pp. 402–416). New York: Simon & Schuster. (Original work published 1925)

Whitehead, A. N. (1963a). On mathematical method. In R. M. Hutchins, M. J. Adler, & C. Fadiman (Eds.), *Gateway to the great books: Mathematics* (Vol. 9, pp. 51–67). Chicago: William Benton. (Original work published 1911)

Whitehead, A. N. (1963b). On the nature of a calculus. In R. M. Hutchins, M. J. Adler, & C. Fadiman (Eds.), *Gateway to the great books: Mathematics* (Vol. 9, pp. 68–78). Chicago: William Benton. (Original work published 1898)

Whitehead, A. N., & Russell, B. (1910–1913). *Principia mathematica* (3 vols.). Cambridge, UK: Cambridge University Press.

Wickelgren, W. A. (1974). *How to solve problems.* San Francisco: W.H. Freeman.

Wiener, N. (1953). *Ex-prodigy: My childhood and youth.* New York: Simon & Schuster.

Wiener, N. (1956). *I am a mathematician: The later life of an ex-prodigy.* Garden City, NY: Doubleday.

Wiener, N. (1964). *God and Golem, Inc.: A comment on certain points where cybernetics impinges on religion.* Cambridge, MA: MIT Press.

Wigfield, A., Eccles, J. S., Yoon, K. S., Harold, R. D., Arbreton, A. J. A., Freedman-Doan, C., et al. (1997). Change in children's competence beliefs and subjective task values across the elementary school years: A 3-year study. *Developmental Psychology, 89,* 451–468.

Wigner, E. P. (1980). The unreasonable effectiveness of mathematics in the natural sciences. *Communications on Pure and Applied Mathematics, 13,* 1–14. (Original work published 1960)

Wilczek, F. (2003). A piece of magic: The Dirac equation. In G. Farmelo (Ed.), *It must be beautiful: Great equations of modern science* (pp. 132–160). New York: Granta Books.

Wilder, R. L. (1956). The axiomatic method. In J. R. Newman (Ed.), *The world of mathematics* (Vol. 3, pp. 1647–1667). New York: Simon & Schuster. (Original work published 1952)

Wiles, A. (1995). Modular elliptic curves and Fermat's last theorem. *Annals of Mathematics, 141,* 443–551.

Wilkinson, L. C., & Marrett, C. B. (Eds.). (1985). *Gender influences in classroom interaction.* New York: Academic Press.

Williams, J. E. (1994). Gender differences in high-shool students efficacy-expectation performance discrepancies across four subject-matter domains. *Psychology in the Schools, 31,* 232–237.

Willingham, W. W., Cole, N. S., Lewis, C., & Leung, S. W. (1997). Test performance. In W. W. Willingham & N. S. Cole (Eds.), *Gender and fair assessment* (pp. 55–126). Mahwah, NJ: Erlbaum.

Wills, K. E. (1993). Neuropsychological functioning in children with spina bifida and/or hydrocephalus. *Journal of Clinical Child Psychology, 22,* 247–265.

Wills, K. E., Holmbeck, G. N., Dillon, K., & McLone, D. G. (1990). Intelligence and achievement with myelomeningocele. *Journal of Pediatric Psychology, 15,* 161–176.

Wilson, K. M., & Swanson, H. L. (2001). Are mathematical disabilities due to a domain-general or a domain-specific working memory deficit? *Journal of Learning Disabilities, 34,* 237–248.

Wisdom, J. O. (2001). Achilles on a physical racecourse. In W. C. Salmon (Ed.), *Zeno's paradoxes* (pp. 82–88). Indianapolis: Hackett Publishing. (Original work published 1951)

Wise, L. L. (1985). Project TALENT: Mathematics course participation in the 1960s and its career consequences. In S. F. Chipman, L. R. Brush, & D. M. Wilson (Eds.), *Women and mathematics: Balancing the equation* (pp. 25–58). Hillsdale, NJ: Erlbaum.

Wittgenstein, L. (1972). *On certainty* (G. E. M. Anscombe & G. H. von Wright, Eds.; D. Paul & G. E. M. Anscombe, Trans.). New York: Harper Torchbooks. (Original work published 1950)

Woit, P. (2006). *Not even wrong: The failure of string theory and the search for unity in physical law*. New York: Basic Books.

Wollman, W. (1983). Determining the sources of error in a translation from sentence to equation. *Journal for Research in Mathematics Education, 14,* 169–181.

Woodruff, G., & Premack, D. (1981). Primative mathematical concepts in chimpanzee: Proportionality and numerosity. *Nature, 293,* 568–570.

Woods, S. S., Resnick, L. B., & Groen, G. J. (1975). An experimental test of five process models for subtraction. *Journal of Educational Psychology, 67,* 17–21.

Woodward, A., & Elliot, D. L. (1990). Textbook use and teacher professionalism. In Textbooks and schooling in the United States (Eighty-Ninth Yearbook of the National Society for the Study of Education, Part 1, pp. 178–193). Chicago: University of Chicago Press.

Woodworth, R. S., & Schlosberg, H. (1954). *Experimental psychology* (rev. ed.). New York: Henry Holt and Company.

Wynn, K. (1990). Children's understanding of counting. *Cognition, 36,* 155–193.

Wynn, K. (1992a). Addition and subtraction by human infants. *Nature, 358,* 749–750.

Wynn, K. (1992b). Children's acquisition of the number words and the counting system. *Cognitive Psychology, 24,* 220–251.

Wynn, K. (1995). Origins of numerical knowledge. *Mathematical Cognition, 1,* 35–60.

Wynn, K. (1996). Infants' individuation and enumeration of sequential actions. *Psychological Science, 7,* 164–169.

Wynn, K. (1998). Numerical competence in infants. In C. Donlan (Ed.), *The development of mathematical skills* (pp. 3–25). Hove, UK: Psychology Press.

Xia, L., Emmerton, J., Siemann, M., & Delius, J. D. (2001). Pigeons (*Columba livia*) learn to link numerosities with symbols. *Journal of Comparative Psychology, 115,* 83–91.

Xia, L., Siemann, M., & Delius, J. D. (2000). Matching of numerical symbols with number of responses by pigeons. *Animal Cognition, 3,* 35–43.

Xie, Y., & Shauman, K. (2003). *Women in science: Career processes and outcomes.* Cambridge, MA: Harvard University.

Xu, F. (2003). Numerosity discrimination in infants: Evidence for two systems of representation. *Cognition, 89,* B15–B25.

Xu, F., & Spelke, E. S. (2000). Large number discrmination in 6-month-old infants. *Cogniton, 74,* B1–B11.

Yancey, B. F., & Calderhead, J. A. (1886). New and old proofs of the Pythagorean theorem. *American Mathematical Monthly, 3,* 65–67, 110–113, 169–171, 299–300.

Yancey, B. F., & Calderhead, J. A. (1887). New and old proofs of the Pythagorean theorem. *American Mathematical Monthly, 4*, 11–12, 79–81, 168–170, 250–251, 267–269.

Yancey, B. F., & Calderhead, J. A. (1888). New and old proofs of the Pythagorean theorem. *American Mathematical Monthly, 5*, 73–74.

Yancey, B. F., & Calderhead, J. A. (1889). New and old proofs of the Pythagorean theorem. *American Mathematical Monthly, 6*, 33–34, 69–71.

Yandell, B. H. (2002). *The honors class: Hilbert's problems and their solvers.* Natick, MA: A. K. Peters.

Yates, F. A. (1964). *Giordano Bruno and the Hermetic tradition.* Chicago: University of Chicago Press.

Yee, D. K., & Eccles, J. S. (1988). Parent perceptions and attributions for children's math achievement. *Sex Roles, 19*, 317–333.

Young, R. M., & O'Shea, T. (1981). Errors in children's subtraction. *Cognitive Science, 5*, 153–177.

Zaslavsky, C. (1973). *Africa counts.* Boston: Prindle, Weber and Schmidt.

Zech, L., Vye, N. J., Bransford, J. D., Swink, J., Mayfield-Stewart, C., Goldman, S. R., & the Cognition and Technology Group at Vanderbilt. (1994). Bringing the world of geometry into the classroom with videodisc technology. *Mathematics Teaching in the Middle School, 1*, 228–233.

Zeldin, A. L., & Pajares, F. (2000). Against the odds: Self-efficacy beliefs of women in mathematical, scientific, and technological careers. *American Educational Research Journal, 37*, 215–246.

Zhang, J., & Norman, D. A. (1995). A representational analysis of numeration systems. *Cognition, 57*, 271–295.

Ziman, J. (1978). *Reliable knowledge.* Cambridge, UK: Cambridge University Press.

Zimmermann, W., & Cunningham, S. (Eds.). (1991). *Visualization in teaching and learning mathematics.* Washington, DC: The Mathematical Association of America.

Zorzi, M., & Butterworth, B. (1999). A computational model of number comparison. In M. Hahn & S. C. Stoness (Eds.), *Proceedings of the Twenty-First Annual Conference of the Cognitive Science Society* (pp. 778–783). Mahwah, NJ: Erlbaum.

Zorzi, M., Stoianov, I., & Umilta, C. (2005). Computational modeling of numerical cognition. In J. I. D. Campbell (Ed.), *Handbook of mathematical cognition* (pp. 67–83). New York: Psychology Press.

Appendix: Notable (Deceased) Mathematicians, Logicians, Philosophers, and Scientists Mentioned in the Text

Abel, Niels (1802–1829)

Agnesi, Maria Gaetana (1718–1799)

Alighieri, Dante (1265–1321)

Ampére, André-Marie (1775–1836)

Antiphon (480–411 BC)

Apollonius of Perga (c. 262–c. 190 BC)

Aquinas, Thomas (1225–1274)

Archimedes of Syracuse (287–212 BC)

Argand, Jean-Robert (1768–1822)

Aristotle (384 322 BC)

Aryabhatta (476–550)

Augustine of Hippo (354–430)

Bachet, Claude Gaspar (1581–1638)

Barrow, Isaac (1630–1677)

Bell, Eric Temple (1883–1960)

Bentham, Jeremy (1748–1832)

Bergson, Henri (1859–1941)

Berkeley, (Bishop) George (1685–1753)

Bernoulli, Jacob (aka James and Jacques) (1654–1705)

Bernoulli, Johann (1667–1748)

Bertrand, Joseph Louis Francois (1822–1900)

Bhäskara II (1114–1185)

Birkhoff, George David (1884–1944)

Bishop, Errett Albert (1928–1983)

Bolyai, János (1802–1860)

Bolzano, Bernhard Placidus Johann Nepomuk (1781–1848)

Bombelli, Raphael (1526–1572)

Bonaventure, St. (1221–1274)

Boole, George (1815–1864)

Bradwardine, Thomas (c. 1290–1349)

Brahmagupta (c. 598–c. 670)

Brouwer, Luitzen Egbertus Jan (1881–1966)

Brownell, William Arthur (1895–1977)

Bürgi, Joost (1552–1632)

Burley (sometimes Burleigh), Walter (c. 1275–c. 1344)

Cajori, Florian (1859–1930)

Cantor, Georg Ferdinand Ludwig Philipp (1845–1918)

Cardano (sometimes Cardan), Girolamo (sometimes Gerolamo) (1501–1576)

Carnap, Rudolph (1891–1970)

Cauchy, Augustin Louis (1789–1857)

Cayley, Arthur (1821–1895)

Chandrasekhar, Padma Vibhushan Subrahmanyan (1910–1995)

Chebyshev, Pafnuty (1821–1894)

Ch'ung-chih, Tsu (sometimes Chongzhi, Zu) (429–500)

Chuquet, Nicolas (1445–1488)

Church, Alonzo (1903–1995)

Clifford, William Kingdom (1845–1879)

Court, Nathan Altshiller (1881–1968)

Couturat, Louis (1868–1914)

Cox, Gertrude Mary (1900–1978)

Cronus, Diodorus (4th century BC)

d'Alembert, Jean le Rond (1717–1783)

Dantzig, George (1914–2005)

Darboux, Jean-Gaston (1842–1917)

David, F. N. (Florence Nightingale) (1909–1993)

Dedekind, Julius Wilhelm Richard (1831–1916)

de Fermat, Pierre (1601–1665)

Delaunay, Charles Eugene (1816–1872)

de la Vallée-Poussin, Charles Jean (1866–1962)

Democritus of Abdera (c. 460–c. 370 BC)

de Montmort, Pierre Rémond (1678–1719)

De Morgan, Augustus (1806–1871)

de Roberval, Gilles (1602–1675)

Desargues, Gérard (1591–1661)

Descartes, René (1596–1650)

Diophantus of Alexandria (c. 200–c. 284)

Dirichlet, Johann Peter Gustav Lejeune (1805–1859)

Einstein, Albert (1879–1955)

Epicurus of Samos (c. 341–c. 270 BC)

Erdös, Paul (1913–1996)

Euclid of Alexandria (c. 325–c. 265 BC)

Eudoxus of Cnidus (408–355 BC)

Euler, Leonard (1707–1783)

Fechner, Gustav Theodor (1801–1887)

Feynman, Richard Phillips (1918–1988)

Figulus, Publius Nigidius (c. 98–45 BC)

Fisher, Ronald (1890–1962)

Fontana, Nicola (sometimes Niccolo) of Brescia (aka Tartaglia) (1499–1557)

Fourier, Jean Baptiste Joseph (1768–1830)

Fraenkel, Abraham (1891–1965)

Frege, Gottlob (1848–1925)

Galileo, Galilei (1564–1642)

Galois, Évariste (1811–1832)

Gauss, Carl Friedrich (1777–1855)

Gerbert of Aurillac (Pope Sylvester II) (946–1003)

Germain, Marie-Sophie (1776–1831)

Girard, Albert (1595–1692)

Gödel, Kurt (1906–1978)

Goldbach, Christian (1690–1764)

Grandi, Luigi Guido (1671–1742)

Hadamard, Jacques (1865–1963)

Halayudha (10th century or before)

Halmos, Paul Richard (1916–2006)

Hamilton, William Rowan (1805–1865)

Hardy, Godfrey Harold (1877–1947)

Harriot, Thomas (1560–1621)

Heawood, Percy (1861–1955)

Heisenberg, Werner (1901–1976)

Hempel, Carl (1905–1997)

Heron of Alexandria (c. 10–c. 75)

Hilbert, David (1862–1943)

Hipparchus (190–120 BC)

Hippocrates (c. 470–c. 410 BC)

Hui, Yang (1238–1298)

Hypatia of Alexandria (c. 370–c. 415)

Jeans, James Hopwood (1887–1946)

Jourdain, Philip Edward Bertrand (1879–1919)

Kac, Mark (1914–1984)

Kant, Immanuel (1724–1804)

Kasner, Edward (1878–1955)

Kempe, Arthur (1849–1922)

Kline, Morris (1908–1992)

Klügel, George (1739–1812)

Kolmogorov, Andrey Nikolaevich (1903–1987)

Kovalevskaya, Sofia (aka Sonya) (1850–1891)

Kronecker, Leopold (1823–1891)

Lagrange, Joseph Louis (1736–1813)

Lakatos, Imre (1922–1974)

Lambert, Johann Heinrich (1728–1777)

Lamé, Gabriel (1795–1870)

Landau, Edmund (1877–1938)

Laplace, Pierre-Simon (1749–1827)

Legendre, Adrien-Marie (1752–1833)

Leibniz, Gottfried Wilhelm (1646–1716)

Leonardo of Pisa (Fibonacci) (c. 1170–1250)

Leucippus of Miletus (5th century BC)

Littlewood, John Edensor (1885–1977)

Lobachevsky, Nikolai (1792–1856)

Lorenzen, Paul (1915–1994)

Lucas, Francois Édouard Anatole (1842–1891)

Mariotte, Edme (c. 1620–1684)

Menger, Karl (1902–1985)

Mersenne, Marin (1588–1648)

Möbius, August Ferdinand (1790–1868)

Mohr, Georg (1640–1697)

Napier, John (1550–1617)

Newton, Isaac (1643–1727)

Neyman, Jerzy (1894–1981)

Nicomachus of Gerasa (c. 60–c. 120)

Noether, Emmy Amalie (1882–1935)

Oresme, Nicole (1323–1382)

Oughtred, William (1575–1660)

Pacioli, Luca (1445–c. 1515)

Pascal, Blaise (1623–1662)

Peano, Guiseppe (1858–1932)

Pearson, Egon (1895–1980)

Pearson, Karl (1857–1926)

Peirce, Charles Sanders (1839–1914)

Philoponus, John (c. 490–c. 570)

Piaget, Jean (1986–1980)

Plato (c. 427–c. 348 BC)

Playfair, John (1748–1819)

Plotinus (204–270)

Poincaré, Jules Henri (1854–1912)

Poisson, Siméon-Denis (1781–1849)

Polya, George (1887–1985)

Priestly, Joseph (1733–1804)

Ptolemy (c. 85–c. 165)

Pythagoras (of Samos) (c. 575–c. 495 BC)

Ramanujan, Srinivasa (1887–1920)

Recorde, Robert (1510–1558)

Ricmerstop, Albert (Albert of Saxony) (c. 1316–1390)

Riemann, Georg Friedrich Bernhard (1826–1866)

Rota, Gian-Carlo (1932–1999)

Rouse Ball, Walter William (1850–1925)

Rudolff, Christoff (1499–1545)

Russell, Bertrand (1872–1970)

Saccheri, Giovanni Girolamo (1667–1733)

Schröder, Friedrich Wilhem Karl Ernst (1841–1902)

Schweikart, Ferdinand Karl (1780–1857)

Scotus, Duns (c. 1266–1308)

Seki, Kowa (or Kowa, Seki) (c. 1640–1708)

Selberg, Atle (1917–2007)

Sheffer, Henry (1882–1964)

Sierpinski, Waclaw (1882–1969)

Skewes, Stanley (1899–1988)

Socrates (c. 469–399 BC)

Stevin, Simon (1548–1620)

Stifel, Michael (1487–1567)

Sundman, Karl Frithiof (1873–1949)

Sylvester, James Joseph (1814–1897)

Taniyama, Yutaka (1927–1958)

Tarski, Alfred (1901–1983)

Thales of Miletus (c. 624–c. 546 BC)

Theano (c. 546 BC–?)

Thom, René (1923–2002)

Thorndike, Edward Lee (1874–1949)

Torrecelli, Evangelista (1608–1647)

Turing, Alan (1912–1954)

Ulam, Stansilav (1909–1984)

Van Ceulen, Ludolph (1540–1610)

Veblen, Oswald (1880–1960)

Viète (sometimes Vieta), Francois (1540–1603)

Von Helmholtz, Hermann (1821–1894)

Von Koch, Niels Helge (1870–1924)

Von Lindemann, Carl Louis Ferdinand (1852–1939)

Von Mises, Ludwig (1881–1973)

Von Neumann, John (1903–1957)

Vygotsky, Lev Semenovich (1896–1934)

Wallis, John (1616–1703)

Weierstrass, Karl Theodor Wilhelm (1815–1897)

Wessel, Caspar (1745–1818)

Whitehead, Alfred North (1861–1947)

Widman, Johann (1462–1498)

Wiener, Norbert (1894–1964)

Wingate, Edmund (1593–1656)

Wittgenstein, Ludwig (1889–1951)

Zeno of Elea (c. 490–425 BC)

Zermelo, Ernst Friedrich Ferdinand (1871–1953)

AUTHOR INDEX

SUBJECT INDEX